Advances in Info-Metrics: Information and Information Processing across Disciplines

Advances in Info-Metrics: Information and Information Processing across Disciplines

Editors

MIN CHEN, J. MICHAEL DUNN,
AMOS GOLAN, AND AMAN ULLAH

OXFORD
UNIVERSITY PRESS

OXFORD
UNIVERSITY PRESS

Oxford University Press is a department of the University of Oxford.
It furthers the University's objective of excellence in research, scholarship,
and education by publishing worldwide. Oxford is a registered trade mark of
Oxford University Press in the UK and certain other countries.

Published in the United States of America by Oxford University Press
198 Madison Avenue, New York, NY 10016, United States of America.

Library of Congress Cataloging-in-Publication Data

Names: Chen, Min, 1960 May 25- editor. | Dunn, J. Michael, 1941- editor. | Golan, Amos, editor. |
Ullah, Aman, editor.
Title: Advances in info-metrics : a cross-disciplinary perspective of information and information
processing / editors, Min Chen, J. Michael Dunn, Amos Golan, Aman Ullah.
Description: New york : Oxford University Press, 2020. | Includes bibliographical references and index. |
Identifiers: LCCN 2020021172 (print) | LCCN 2020021173 (ebook) | ISBN 9780190636685 (hardback) |
ISBN 9780190636715 (epub)
Subjects: LCSH: Information theory—Statistical methods. | Uncertainty (Information theory)
Classification: LCC Q386 .A38 2020 (print) | LCC Q386 (ebook) | DDC 003/.54—dc23
LC record available at https://lccn.loc.gov/2020021172
LC ebook record available at https://lccn.loc.gov/2020021173

1 3 5 7 9 8 6 4 2

Printed by Integrated Books International, United States of America

Contents

PART IV. INFO-METRICS IN ACTION I: PREDICTION AND FORECASTS

PART V. INFO-METRICS IN ACTION II: STATISTICAL AND ECONOMETRICS INFERENCE

PART VI. INFO-METRICS, DATA INTELLIGENCE, AND VISUAL COMPUTING

Preface

Info-metrics is a framework for rational inference. It is the science of modeling, reasoning, and drawing inferences under conditions of noisy and insufficient information. Info-metrics has its roots in information theory (Shannon, 1948), Bernoulli's and Laplace's principle of insufficient reason (Bernoulli, 1713), and its offspring, the principle of maximum entropy (Jaynes, 1957). It is an interdisciplinary framework situated at the intersection of information theory, statistical inference, and decision making under uncertainty. Within a constrained optimization setup, info-metrics provides a simple way to model and understand all types of systems and problems. It is a framework for processing the available information with minimal reliance on assumptions and information that cannot be validated. Quite often, a model cannot be validated with finite data. Examples include biological, social, and behavioral models, as well as models of cognition and knowledge. The info-metrics framework naturally tackles these common problems.

In somewhat more technical words, the fundamental idea behind info-metrics is that the available information for modeling, inference, and decision making is insufficient to provide a unique answer (or solution) for most decisions or inferences of problems across all disciplines. This implies that there is a continuum of solutions (models, inferences, decisions, "truths") consistent with the information (or evidence/data) we have. Therefore, the final solution of choosing one of the infinitely many possible solutions must be within a constrained optimization setting. All information enters as constraints, and the decision function that chooses the "optimal" solution must satisfy certain properties. Within info-metrics, that decision function is an information-theoretic one. In the more commonly used terminology, it is the joint choice of the information used (within the optimization setting) and the decision function that determines the likelihood (known as the "likelihood function").

The info-metrics framework provides a consistent way to model all problems and arrive at honest and unbiased solutions or theories. It is an approach that is based on the minimally needed information for fully describing the system we model, or solving the problem at hand, at the needed level, or resolution, of interest.

Info-metrics encompasses the class of information-theoretic methods of inference, which has a long history. The recent interest in info-metrics has arisen not just within data and information sciences; rather it is in all areas of science (physics, chemistry, biology, ecology, computer science, and artificial intelligence), the social and behavioral sciences (economics, econometrics, statistics, political science, and psychology), and even applied mathematics and the philosophy of science.

In a recent book on the foundations of info-metrics, Golan (2018) develops, examines, and extends the theoretical underpinnings of info-metrics and provides extensive interdisciplinary applications. The present volume complements Golan's book by developing ideas that are not discussed in Golan's book and by providing many case studies and empirical examples. This volume also expands on the series of studies (mostly within the natural sciences) on the classical maximum entropy and Bayesian methods published in various proceedings starting with the seminal collection of Levine and Tribus (1979) and continuing annually. Our volume includes nineteen cross-disciplinary chapters touching on many topics whose unifying theme is info-metrics. It is a collection of interconnected chapters resulting from a decade of interdisciplinary work in info-metrics.

The objective of this volume is to expand the study of info-metrics and information processing across the sciences and to further explore the basis of information-theoretic inference and its mathematical and philosophical foundations. This volume is inherently interdisciplinary. It comprises new and previously unpublished research, and it contains some of the most recent developments in the field. Furthermore, it provides a balanced viewpoint on previous research works in various areas of the information sciences and info-metrics. Our emphasis here is on the interrelationship between *information* and *inference* where we view the word "inference" in its most general meaning—capturing all types of problem solving. That includes model building, theory creation, estimation, prediction, and decision making. The volume contains nineteen chapters in seven parts.

We start in Part I with the more philosophical notion of information: its meaning and value. The meaning of information is also discussed within the context of computational theory. Part II touches on information theory and behavior, exploring the interconnection between information, information-theoretic inference, and individual and collective behavior. Part III demonstrates the use of info-metrics in basic theory construction. The following two parts show info-metrics in action. Part IV deals with info-metric inference, with an emphasis on prediction and forecasts for problems across the sciences, whereas Part V discusses statistical and econometrics inference, connecting recent developments in info-metrics with some of the cutting-edge classical inferential methods. Part VI extends the framework to data intelligence and visualization. Finally, Part VII uses the tools of info-metrics to develop new approaches for nonparametric inference—inference that is based solely on the observed data.

Although the chapters in each part are related, each chapter is self-contained and includes a list of references. Each chapter provides the necessary tools for using the info-metrics framework for solving the problem that is the focus of that chapter. The chapters complement each other, but they do not need to be read in the order in which they appear. To get the full picture, however, we recommend reading all the chapters within each part.

This volume is of interest to researchers seeking to solve problems and do inference with incomplete and imperfect information. It is also of interest for those who are interested in the close connection between information, information-theoretic quantities, and inference. As such, it perfectly complements the existing body of knowledge on the info-metrics framework (e.g., Golan, 2018). The present volume is designed to be accessible to researchers, graduate students, and practitioners across the disciplines, requiring only some basic quantitative skills. As suggested earlier, this work is interdisciplinary and applications oriented, thereby providing a hands-on experience for the reader.

References

Bernoulli, J. (1713). *Art of Conjecturing.* Basel: Thurneysen Brothers.

Golan, A. (2018). *Foundations of Info-Metrics: Modeling, Inference, and Imperfect Information.* New York: Oxford University Press.

Jaynes, E. T. (1957). "Information Theory and Statistical Mechanics." *Physics Review,* 106: 620–630.

Levine, R. D., and M. Tribus (eds.). (1979). *The Maximum Entropy Formalism.* Cambridge, MA: MIT Press.

Shannon, C. E. (1948). "A Mathematical Theory of Communication." *Bell System Technical Journal* 27: 379–423.

—Min Chen, J. Michael Dunn, Amos Golan, and Aman Ullah
(Editors are ordered alphabetically by surname)

The Info-Metrics Institute

The Info-Metrics Institute (established in 2009 at American University in Washington, DC) is an interdisciplinary, multi-institutional entity that focuses on all areas of research within info-metrics. The overall objective of establishing the institute was to fill a gap in the research related to all types of information and data processing. That gap was not only in the methods themselves, but also in the lack of interactions among the different disciplines and the need to learn from one another. Since information processing is one of the connecting threads among the disciplines, establishing such an institute seemed both logical and timely.

Though young, the Institute has established itself as a leader in the info-metrics field. In addition to research, its core activities include workshops, conferences, seminars, hands-on tutorials, and fellowships (visiting scholars and graduate students). It is an exceptional data science (virtual) center. Its primary source of support has come from the Office of the Comptroller of the Currency (U.S. Department of Treasury). For more information on the Institute and its past and future activities, see https://www.american.edu/info-metrics.

This volume emerged from some of the Institute's more recent research activities. Most of the contributors of this volume are affiliated with the Institute or have participated in at least one of the Institute's activities.

Acknowledgments

This volume is the result of collective learning, discussions, workshops, and conferences, many of which took place at the Info-Metrics Institute and other locations around the globe during the last decade. We owe thanks and gratitude to all of those who helped us with this project. First, we owe a great debt to the authors of the chapters for their marvelous support and cooperation in preparation of this volume and most of all for the new insights into the info-metrics tools and framework they have provided. Second, we thank the reviewers who went through the chapters carefully, ensuring the desired level of quality.

We owe a special thanks to our former editor from Oxford University Press, Scott Parris. Two of us (Golan and Ullah) have worked with Scott for quite a while (in three previous books), and we are grateful to him for his wisdom, thoughtful suggestions, recommendations, and patience. We also thank James Cook, our new editor, for his patience, suggestions, and effort. In addition, we are most thankful to Damaris Carlos, University of California, Riverside, for the efficient assistance she provided. Finally, we are grateful for the Info-Metrics Institute and the other institutions that hosted and supported us.

Min Chen, J. Michael Dunn, Amos Golan, and Aman Ullah

Contributor List

Adriaans, Pieter University of Amsterdam, The Netherlands

Andrews, Martyn University of Manchester, UK

Caticha, Ariel University at Albany, USA

Chen, Min University of Oxford, UK

Choi, Hwan-sik Binghamton University, USA

Daniels, Bryan C. Arizona State University, USA

Dunn, J. Michael Indiana University Bloomington, USA

Durham, Garland California Polytechnic State University, USA

Feixas, Miquel University of Girona, Spain

Fernandez-Vazquez, Esteban University of Oviedo, Spain

Foster, F. Douglas The University of Sydney, Australia

Geweke, John University of Washington, USA and University of Technology Sydney, Australia

Golan, Amos American University and Santa Fe Institute, USA

Hall, Alastair R. University of Manchester, UK

Harte, John University of California, Berkeley, USA

Judge, George University of California, Berkeley, USA

Khatoon, Rabeya University of Bristol, UK

Kravchenko-Balasha, Nataly The Hebrew University of Jerusalem, Israel

Lahiri, Kajal University at Albany, USA

Lincoln, James University of Manchester, UK

Mao, Millie Yi Azusa Pacific University, USA

Nelson, Kenric P. Photrek, LLC. and Boston University, USA

Papalia, Rosa Bernardini University of Bologna, Italy

Sbert, Mateu University of Girona, Spain

Stutzer, Michael University of Colorado, Boulder, USA

Tu, Yundong Peking University, China

Ullah, Aman University of California, Riverside, USA

Wang, Wuwei Southwestern University of Finance and Economics, China

Wen, Kuangyu Huazhong University of Science and Technology, China

Wu, Ximing Texas A&M University, USA

PART I

INFORMATION, MEANING, AND VALUE

Preface

This part deals with two basic properties of information: its meaning and its value. In Chapter 1, Dunn and Golan focus on information and its value within the context of information used for decision making. The discussion is a philosophical one, yet it connects some of the more commonly used notions (utility, risk, and prices) with information and its value. In Chapter 2, Adriaans takes the notion of meaning in a new direction: he develops and discusses the notion of a computational theory of meaning. The interest here is in understanding "meaning" in terms of computation. In that way, Chapter 2 connects nicely to the more familiar concept of Kolmogorov complexity and its relationship to information.

Overall, these two chapters, though very different, touch on some basic issues in information and information processing. One central issue is that the quantity of information can be quantified (in, say bits), but it is semantic free. Yet, computationally, one can argue that some of the meaning may be (partially) captured via its complexity. For example, the shortest code (or program) needed to generate the input information is used. This can extend to the idea that the set of all programs produces a data set (our input information) on a computer as the set of meanings of that information. In that way, these two chapters are directly connected to the other chapters in this volume and provide a way to think about the information used for inference.

1
Information and Its Value

J. Michael Dunn and Amos Golan

1. Background and Basic Questions

In this chapter, we are interested in the information we use to make decisions. By making a decision, we mean selecting a course of action, or inaction, given all known possible choices of action. This also includes the more involved actions such as inference, scientific model building, and all types of predictions. Decision-making information is at the heart of scientific inquiries and is employed in different ways to solve mysteries and problems across disciplines. Every scientific research, policy analysis and decision, including the authoring of every chapter in this book, uses such information in one way or another. Understanding this information, its meaning and its value, within an interdisciplinary perspective, can prove useful for all modeling, information processing, and data analyses.

Broadly speaking, *information* is that "thing" that contributes to our intelligence and that is used by an agent, be it a human or any other species (maybe even robots) to interact with the world around it. But it is also the "thing" that is encoded in our genes and DNA. Information shapes both our intelligence and its evolution, where we can view intelligence as an organism's ability to sense, understand, process, and react to all types of information in order to improve its abilities and survival probabilities in a certain environment. The mechanisms by which intelligence evolves, such as the well-known "natural selection," have been studied for generations and across many fields.

In this chapter, we are interested in the first "thing"—the information around us that we, as observers, agents, and creators (of that information), use to make improved decisions, including better inference and predictions. That type of information is required for understanding and coping with the world around us, gaining knowledge, making informed decisions and predictions, and solving problems. It is information that can be transformed to knowledge. Employing a mostly nonquantitative approach, we discuss this information, focusing on what part of it is quantifiable and its value. But regardless of the way *information* is defined, it does help us make improved decisions. That by itself gives it a value.

J. Michael Dunn and Amos Golan, *Information and Its Value* In: *Advances in Info-Metrics: Information and Information Processing across Disciplines.* Edited by: Min Chen, J. Michael Dunn, Amos Golan, and Aman Ullah, Oxford University Press (2021). © Oxford University Press.
DOI: 10.1093/oso/9780190636685.003.0001

For information to do that, it of course must "inform" us; it has to become part of our mentality; and it has to be in a form that we can process, understand, and convert to useful knowledge. Yet it seems natural to expect that this value is relative and subjective. We will argue that, indeed, that value is not objective, and it is not unique.

To provide us with some initial context for discussing information, its meaning and its value, we need to think about information and its meaning in a somewhat more philosophical, yet practical, way. Consider a variation of the famous question: "If a tree falls in a forest and no one is around to hear it, does it make a sound?"[1]

We view sound as a special kind of information, and we build on this question to raise three fundamental questions about the ontological character of information:

1. If a tree falls in a forest and you are there to observe it, would that give you information?
2. If a tree falls in a forest with no one around, would there be information?
3. If no tree falls in a forest and you are there to observe it, or no one is around, would there be information?

Information is often discussed in the context of communication (Shannon, 1948; Dretske, 1981; Barwise and Seligman, 1997). There is a sender who communicates information to a receiver. The communicated information is quantifiable. The point of (1) is that there is no "known" sender, unless one counts the forest, or perhaps the universe or God. The point of (2) is that there is no receiver (or observer) either. The answer to both questions is "yes." Information does not exist only in the mind of the receiver, the sender, or any observer. For us, information is an abstraction that can exist without either a sender or a receiver. It can lie like nuggets of gold under the soil waiting to be gleaned. The point of (3) is that, like (1), there is no sender, but nothing new (whether observed or not) "happened." The answer to this question is somewhat more involved, yet the answer can be "yes" as well under certain scenarios.

As we discuss these questions and our interpretation of information within the context of decision making, we keep connecting it to our main topic of interest: the value of that information. Our interest here is to connect our answers to the following set of axiological questions about the value of information.

[1] This question seems to trace back to the philosopher George Berkeley (1685–1753). Berkeley actually argued something even stronger, that not even the tree would exist without a perceiver. "The objects of sense exist only when they are perceived: the trees therefore are in the garden, or the chairs in the parlour, no longer than while there is some body by to perceive them" (*A Treatise Concerning the Principles of Human Knowledge*, sec. 45). Berkeley famously created a theory of subjective idealism, in which nothing exists but perceptions. The parts of the world not perceived by humans exist only because of perceptions in the mind of God.

1. If there is information, must it always have a value, and if so, (*i*) to whom? And (*ii*) is that value absolute (unique)?
2. If there is information and it is used for decision making, is its value related to the utility and risk of the decision maker?
3. If there is information and it has a value, is that value related to market prices?

In this chapter, we discuss the three fundamental ontological questions about information with an interdisciplinary perspective. Our overall goal is not just to provide our views about this initial set of questions, but also to try to understand the answers in a more general way—a way that is similar across disciplines. Within that goal, we also study the next set of questions relating to the value of information under the three different "tree-falling" scenarios presented. It is a chapter about information used for decision making, including decisions about modeling and inference and the value of that information. (The fascinating research on how to process such information is outside the scope of this chapter. Different approaches are discussed in the other chapters of this volume and elsewhere.)

In the next section, we discuss the notion of information and its meaning within the context of the previous fundamental three ontological questions and decision making.[2] Then, our discussion moves toward understanding the notion of the value of information. To achieve that objective, we must first touch on the concepts of utility and risk. We do so in section 3. In section 4, we connect our earlier discussions, and the concept of information, to decisions, prices, risk, and value. In section 5, we provide a concise discussion of our understanding of disinformation, or misinformation, and its value and utility to the individual and to society as a whole. We conclude (Section 6) with some thoughts on open questions.

2. Information

2.1 Definition and Types

"*Information*" can be defined as anything that informs us. However, this is a kind of circular definition. What does it mean to be "informed?" From a decision-making and inference point of view, it means a certain input (call it "input

[2] It is, however, important to emphasize the view that practically any piece of information may have some value outside of decision making, especially for pure learning, joy, curiosity, or amusement. Perhaps an extreme example of this is Voorhees's (1993) *The Book of Totally Useless Information*. But here we are interested primarily in information used for making decisions.

information") enters our decision process and affects our inference and decision. That input information may be (*i*) objective, such as established physical laws, an observed action, or an undisputed fact, or it may be (*ii*) subjective, such as our intuitions, assumptions, interpretation of the state of nature or of an observed event, value judgments and other soft information. We also take the view that, overall, information is true and is not intended to be false; it is not deceitful information that is intentionally communicated to mislead the receiver or the decision maker. However, in most cases, the information is noisy and imperfect, and its meaning may be subject to interpretational and processing errors. It may also contain errors and mistakes. Although our discussion here is on true information, due to the increasing interest in deceitful information, we will confront it in section 5 where we also discuss the way it manifests itself when considering the value of that information. To start the discussion, we briefly examine the more practical interdisciplinary aspect of information.

The word *information* is packed with seemingly many meanings and interpretations, making it somewhat vague in its more ordinary usage. One practical interpretation is that it is this "thing" that allows us to reduce the bounds of possibility or concentrates probability of possible outcomes, when we do not know a fact or a potential outcome with certainty. It is this "thing" that informs us; it puts us in the condition of "having information." But "having information" is a weaker notion than having knowledge or even beliefs.[3] We can have information because we observed something, we were told something, or even because it is in a book we own. This does not mean that we have it in our heads or that we understand it, especially in this age of the "extended mind" (see Clark and Chalmers, 1998).

Even though having information is not the same thing as having knowledge, having information may end up contributing to one's stock of knowledge—however measured and of whatever quality. For the applied researcher who is interested in modeling, inference, and learning, this means that *information* is anything that may affect one's estimates, the uncertainties about these estimates, or decisions. It is "meaningful content."

These definitions allow *information* to be quantified, but this quantification is only in terms of the amount of the information—not its meaning. For example, the amount of *informational content* in the random variable A about (the random variable) B is the extent to which A changes the uncertainty about B. When A is another random prospect, a particular outcome of it may increase or

[3] As Dunn (2008) put it, "think of information, at least as a first approximation, as what is left from knowledge when you subtract, justification, truth, belief, and any other ingredients such as reliability that relate to justification. Information is, as it were, a mere 'idle thought.' Oh, one other thing, . . . subtract the thinker."

decrease the uncertainty about *B*. On average, however, if *A* and *B* are related, knowledge of outcomes of *A* should decrease uncertainty about the prediction of outcomes of *B*. More technically, the amount of *informational* content of an outcome (of a random variable) is an inverse function of its probability. Consequently, anything "surprising" (an outcome with a low probability, such as a symphony written by the authors of this chapter) must contain a great deal of *information*. The idea of measuring the amount of information as an inverse of its probability is the intuitive basis of Shannon's (1948) quantitative theory of information transmission, originally called communication theory and now known as information theory.

We can also think of another, though similar, interpretation of information. It is some kind of a force that induces a change in motion; the force is the "thing" and the motion is the "decision." Stating this notion from a human or behavioral perspective, we can view *information* as whatever induces an agent, supposedly rational, to update her or his beliefs. Note that this includes the notion that new information may strengthen our beliefs and actions in our already existing decision, but it will reduce the bounds of uncertainty about that decision. It also magnified our certainty about that decision. (See also the discussion in Caticha, 2012.)

This discussion sheds some light on the meaning of information and the way it is often understood in science and among different decision makers. Naturally, it is not a comprehensive discussion; rather, it is just a first step in a longer voyage into the meaning of the word *information*. For an illuminating discussion of information and the philosophy behind it, see Floridi's (2011) seminal text. For further general discussions on the different facets of information, see, for example, Adriaans (2013), van Benthem and Adriaans (2008), Dunn (2001, 2008), the review by Dunn (2013), and Golan (2018).

Although all disciplines, scientists, and decision makers may have their own interpretation of the meaning of information, in practice, we all talk about the practical information used as input for modeling, inference, and decision making. For the rest of our discussion, we concentrate on that practical definition. For whatever purpose we use our information, we can think of two types of information: *hard* and *soft*. A third type, *priors*, is a mix of the first two and can emerge from theoretical arguments or certain observed information (Golan, 2018, Chapter 8).

Hard information is our observed information, commonly called "data." It most often appears in terms of numbers, symbols, and sentences. That type of information can be quantified as was done by Hartley (1928), Shannon (1948), and Wiener (1948). However, the quantified information, measured often in bits of information, is free of its semantics/meaning.

Soft information, however, is more challenging. It is composed of the assumptions, conjectures, axioms, and other theoretical arguments imposed by the observer, modeler, or decision maker. It also captures nonquantifiable beliefs, interpretations, values, and intuition. A few examples will convey the problems with soft information. Consider the interpretation of evidence in a court of law. All judges and jurors are presented with the same evidence (information). But in addition to the hard, observed (most often, circumstantial) evidence, they incorporate in their decision process other soft information such as their own values, their own subjective information and personal impression, and their own interpretation of the evidence presented. They end up disagreeing even though they face the same hard evidence. Similarly, think of scientists and decision makers trying to understand certain behaviors or the state of a social or biological system based on the hard data. They impose some structure that is based on their understanding of the way the system works. In this respect, consider an example from biology. In this case, the imposed structure is the gene expression networks in healthy and disease tissues, a structure that is based on the authors' understanding of the way the system works (Vasudevan et al., 2018). Whether or not that structure is correct, this information is part of the input information. The resulting solution, inference, and decisions (output information) depend on the complete set of input information that includes the soft information (which is quantified as noise in the experimental gene expression of Vasudevan et al., 2018). Although it is at times hard or even impossible to quantify, that softer information affects our inference and decision.

Prior information, on the other hand, can be soft, or hard, or a combination of both. It is everything we know about the system but without the observed sample. It is the information that is available in advance of estimation or inference. Prior information can come from anything that potentially influences the specification of a problem but arises outside of the determining system. It can be constructed from pure theoretical arguments (say, Shannon's grouping property), or it can be composed from empirical observations. These empirical observations may come from previous experiments or from other similar systems. Priors are important in any inferential, modeling and decision analyses. Although we defined that concept very briefly, we do not discuss it further as it falls outside the objectives of this chapter. For a comprehensive discussion of prior information and ways to construct these priors see Golan (2018, Chapter 8), or Golan and Lumsdaine (2016).

2.2 Processing, Manipulating, and Converting to Knowledge

Since we are interested here in the information we use to make decisions, we must also discuss that information within the context of problem solving and inference. In most cases, and regardless of the size of our data (or sample), we do not have enough hard information or evidence (often, in terms of circumstantial evidence) about the problem we are trying to solve or the decision we are trying to make. The problem is logically underdetermined: there is more than a single inference that can be logically consistent with that information. Therefore, we must take soft (and uncertain) information into account in our problem solving and inference. Thus, our inference is complicated by the high level of uncertainty surrounding the information we use; the uncertainty in the data itself (the data are random) and the uncertainty about the model itself. This means that, practically, all problems are underdetermined. The solution to such problems is within a constrained optimization framework (e.g., Jaynes, 1957a, b; Golan, 2018), but it is outside the scope of this chapter.

Utilizing our decision-making information in the most efficient way, be it for inference, decision making, or just for joy means, however, that we must manage that information. Managing the information is of course an instrumental issue of its own. This is true even with "Big Data." The goal of the information processor is to convert, in the most efficient way, the input information into knowledge. This means that at times we must discard unnecessary or redundant information, or most often, we need to aggregate our detailed—micro-level—information with minimal loss of informational content. The objective of finding the real information in the data—be it small, complex, or big—demands much effort. The tools of information theory are often used for identifying that information but most often when solving ill-behaved data or ill-posed problems. The same tools are useful for modeling and for information processing. They are also useful for finding the redundant or least informative parts of the data, as well as for efficiently aggregating data.

The motivation for aggregating information is not just for reasons of efficiency, however. The aggregation level depends on the objective of the decision maker or modeler. It is a direct function of the questions we are trying to solve, the systems we try to model and understand, or the prediction we are trying to make. Norretranders (1998, pp. 30–31) illustrates this with a simple example. Consider a supermarket, where in the cash register all prices are registered and then added by goods. According to information theory, there is more information in the long list of prices than in the final "equilibrium" prices and quantities, or in the means (and other aggregated measures—moments) of the prices. (Similarly, there is always more information in the complete sample than in moments of that sample.) Therefore, it may seem more efficient not to aggregate these values and thus prevent loss of information. But in this case the decision maker is interested in the overall summary statistics, say averages or totals, and not in the single prices. The decision maker is not interested in the detailed micro-level information but rather in the macro-detailed of these data.

We aggregate information because the necessary information is determined by the decision maker's use, for example, the customer in the supermarket. It is easier and more efficient as all decisions must be made in finite (usually fast) time. Sorting through all the micro-level details may become inefficient, confusing, and often will not allow the decision maker to understand the basic information, but rather she/he will be bogged down by the confusing, less important details. Stating this differently, often the small micro details are not essential for understanding the full macro structure of a system. More technically, these micro details may wash out on average so the global quantities (say moments or expected values) capture the full story. Nonetheless, the aggregation process should be done with care. It must preserve the real story (meaning) of the information, and it must be done with minimal loss of information. (See also the related, and relatively new and innovative, literature within economics on rational inattention, introduced by Sims (2003).

Unlike information aggregation, discarding information should be done only if the information is redundant or in cases where we know that the discarded part is independent of the information needed for our decision or inference. At times, however, when the human mind, or the machine, cannot handle such a large information set, discarding part of it needs to be done thoughtfully.

Consequently, when thinking of information, we are not merely interested in collecting as much information as possible, but rather we need to process it efficiently in order to make informed decisions, better models, and improved inferences. See other chapters in this volume dealing with the principle of maximum entropy, Bayesian methods, and other efficient information processing rules. You might also consult Golan (2008, 2018).

2.3 Observer and Receiver

Folding our discussion so far into a few sentences, we view information as anything that informs, or may inform, us; anything that may affect our preferences and decisions, or the uncertainties about these. It may cause us to change (or update) our previous decision, or it may direct us to stick with our earlier decision and therefore reduce our uncertainty about that choice. From the processing point of view, the hard information can be quantified in terms of bits of information, but that quantification is in terms of the quantity of information and is completely free of the meaning (semantic) of that information.

Going back to our fundamental first two questions (if a tree falls in a forest and you are there to observe it, would that give you information? and if a tree falls in a forest with no one around, would there be information?), we conclude the following.

The information is independent of the observer and/or the receiver. This statement holds under our above definitions and for all types of information. For example, we have argued that information can be that "thing" that induces an agent to change her beliefs or actions or the uncertainty about these actions. Does it satisfy the first question? Yes. The tree is the "thing," and the observer is an agent monitoring the forest. Does it satisfy the second question? Yes. The tree is, again, the "thing" but there is no observer—no agent with preferences or beliefs.

We provide two arguments to support our view. The first argument is that, if the answer to the first question is "yes," the answer to the second one must be "yes" as well. It is independent of the observer; the "thing" did happen whether we know it or not. The second argument is less philosophical. Consider the other trees in the forest to be the observer. We do not argue that they have beliefs, but they do have more sun now. The information conveyed to them via the fallen tree is that now they can grow stronger and taller.

Having answered the first two questions, it is time to tackle the third fundamental question. No "thing" happened, and maybe you were the observer or maybe there was no one. First, consider the case that you as an observer were there. What is your conclusion? From a strict "Shannon information" point of view, there would not ordinarily be much information, if any, that you would receive. Ordinarily, when walking through a forest, it is not surprising in any way that the various trees we see are still standing. However, if you were a ranger checking on the forest just after a hurricane had hit, you would be very surprised and gain much information. Or even if no hurricane occurred, every time we checked the same forest and saw that nothing had changed, we would have new information about the stability of that forest or about the reduction in our uncertainty about the life expectancy of the trees. Now, what if no one is there? Here we get into a more problematic territory, but our view is that there is still information there, just as in the only surviving copy of a dusty, unread book. Call this an abstract form of information, "potential information" if you will, but for us it is still information. Or, rather than think about "potential information," we can logically argue that the answer we provided for the second question holds here as well: the information is independent of the observer. That is, the answer must be that, "yes," even in that case there is information. From the inference, modeling, or decision-making point of view, it means that if we started with the view that the forest is stable, we ended up with that same view, but now our level of uncertainty about that view (or prediction), has decreased.

Now, we want to analyze the third question from a machine-learning point of view. If the machine learns from the past and current behavior of the entities of interest, what does it learn when the behavior did not change? It learns that there is a higher probability that the system is in equilibrium or in another stable state for a well-defined period of time. What will the machine do? It does not matter for our arguments. What matters is that, even for the machine, just as for any other observer, even if supposedly nothing "new" happened, there is new information. And if there is information, it has a value (at least potentially, for someone). We discuss this issue in the next section.

3. The Value of Information

Oscar Wilde in *Lady Windermere's Fan* wrote that a cynic is "a man who knows the price of everything, and the value of nothing." He also wrote, "a sentimentalist is a man who sees an absurd value in everything and doesn't know the market price of any single thing." It is tempting to equate the value of something, at least to an individual, to the price that this individual would be willing to pay for it. In fact, this is basically the way "value" is perceived in economic thinking. It is, therefore, natural to start our discussion of the value of information by looking into the concept of utility.

3.1 Utility and Value

We start by defining "value" and "utility" within the context of our discussion. There are different types, or classes, of values. A top-level classification might include ethical (moral), personal, and aesthetic values. Some might also want to add practical value. Others might want to add religious, political, sports, and other values. Nevertheless, one might argue that some, or all, of these are subclasses of the top three (ethical/moral, personal, and aesthetic). But even when we explicitly acknowledge the category of aesthetic values, the issues of aesthetic pleasure, collectability, utility, and price often hang in the background. The personal values, on the other hand, may at times be reduced to usefulness or even happiness. For example, valuing my family over my career or my career over my family is a common paradox. In terms of usefulness, it can be translated to what benefits me the most at a given period, my family or my career? But it is possible to pursue family values to the exclusion of one's own happiness, and the same holds with pursuing one's career.

Stated in common social sciences terminology, the benefit question can be framed in terms of utility. But in what way are value and utility related? Both emerge from the individual's preferences and beliefs. This means that value and utility exist only if there is at least one of these: observer, receiver, or sender (of the information). Both—value and utility—are positively correlated. However, utility by itself does not help the decision maker. Rather, it is the maximization of this utility subject to some constraints that helps; say twenty-four hours in a day when considering time spent at home versus work, or the family's resources when deciding on schooling versus purchasing a new car.

Jeremy Bentham, in his *Introduction to the Principles of Morals and Legislation* (1789), defined "utility" as follows. "By utility is meant that property in any object, whereby it tends to produce benefit, advantage, pleasure, good, or happiness, ... or ... to prevent the happening of mischief, pain, evil, or unhappiness to the party whose interest is considered."

The value of something, on the other hand, can be intrinsic or extrinsic. It is the "thing" that can be valued for itself or for what it is in relation to other things that we value. (We discuss this matter in the section on absolute vs. relative—section 3.5.) This basic idea is sometimes called "instrumental value" because, typically, the relation is its usefulness. Jeremy Bentham is famous for creating Utilitarianism: the philosophy that the value of any action can be measured by its utility to the individual or society. This is because utility can be viewed as the value (say, in units of "utils") that a certain action, decision, or behavior gives us or is likely to give us. In that sense, as previously noted, utility is a subjective quantity. The utility that Tom (a cat) gets from drinking a glass of milk is possibly much higher than the utility that Jerry (a mouse) receives, while Jerry's utility from being able to escape Tom is much larger than Tom's loss of the chase. However, unlike utility, in many cases, value can be expressed in monetary units. And as we argued earlier, since it is a relative measure, the monetary value an agent puts on a certain "thing" may be different from that of her brother's. The notion of a monetary value means that we must introduce prices into our discussion of values. We do so next.

3.2 Prices and Value

The American investor and philanthropist Warren Buffett is supposed to have said: "Price is what you pay; value is what you get." This is a nice saying, but we need to remember that prices are always non-negative even though values may be negative. If the value is negative, ideally one might pay to avoid it. Or to put it another way, the positive value would be in the avoidance.

Prices emerge from transactions. Every transaction involves at least two (buyer and seller), and each gets a positive (or non-negative) value from that transaction. Although transactions can be in nonmonetary units (say, two chickens are valued the same as four gallons of milk), in most cases the monetary price of the transaction (say, purchasing two chickens) reflects the value of these chickens to the buyer. These prices are determined via the forces of demand (through utility maximization subject to resources) and supply (through input costs and technological efficiency). But in what way is this related to information and its value?

If value is associated with price, it is helpful to consider some of the factors influencing the price of information. But first we must remember that information is a unique commodity. Unlike all other commodities that, once sold, are physically moved from the seller to the buyer, information is different. If *A* sells *B* a shirt, then *B* has the shirt and *A* does not have it anymore. If, on the other hand, *A* sells *B* information, then regardless of the price *B* paid, both *A* and *B* can be in possession of that information.

Having discussed utility, value, and prices in a more general way, we now turn our attention to the factors that affect the value of information. But value cannot always be directly related to prices. Even if one argues against this claim, then the question becomes: price for whom—the individual (say, the decision maker), or the observer, or society as a whole?

3.3 Hedonic Values: The Factors that Determine the Value

Many factors may affect the value and price of information when a price exists. Here we provide a partial and simplified list. Each reader can add to that list. Importance (to the buyer) is obviously the place to start. This could mean importance for survival, ethical importance, personal importance, or aesthetic importance. It could also be instrumental importance (the value that information has in helping us to obtain, or achieve, some other thing). Other factors might include quantity, scarcity, attractiveness, provenance, social impact, and environmental impact. For example, consider the provenance impact on the value of a piece of information. Suppose that you find a piece of information regarding climate change: "The Arctic Icecap is melting four times faster than expected." Knowing that this information was arrived at through the work of scientists using careful measurements and statistical analysis makes it much more valuable than simply finding it in a fortune cookie. Or consider a Bitcoin that depends on its provenance for value since its value lies in its "blockchain," which is essentially a public ledger recording its transactions from one owner to the next.

But information is a special good in that, in addition to these basic factors, other information-specific ingredients contribute to its value (positively or negatively). These ingredients include truth, degree of certainty, reliability, evidence, reproducibility in certain cases, analysis, presentation, accessibility, or even the power (to the user) embedded in that information. Each of these can get complicated in ways familiar to information scientists, decision makers, philosophers, scientists, and lawyers.

As a simple example, let us deal with just the lawyers to point out that information can be copyrighted, patented, or otherwise protected from free accessibility. Each of these actions its value. Consider a cable company or an Internet provider. That provider may not care about the content of the information and its value, but rather only about the megabits (and speed) it can transmit. This is true when the provider is supplying broadband to your home for your computer, but not when it is supplying the latest movie to your TV. The reason is obvious. In the latter case, it is supplying copyrighted material that it pays for and that has some value to the user. Therefore, the cable company sets a price based on these two factors: broadband (or speed of the Internet) and the copyrighted material (i.e., its value).

Finally, we must consider another special factor: the abundance of the information. As we discussed previously, the same information can be sold to many individuals. How will this affect the price or value? Will it always reduce the value of that information to the buyer? Furthermore, and regardless of the number of buyers, the seller can still, in principle, continue to possess that information. Or consider a related problem (when thinking of value) that has to do with the reproducibility of information. At least in its abstract sense, information can be reproduced perfectly (cloned). Of course, its analog representation can be reproduced only with some approximation of accuracy. Think of a painting that can be copied with perfect accuracy.

Having touched on one of the special features of digital information, we now connect it to the concept of value.

3.4 Digital Information and Value

We stated earlier that digital information can be perfectly reproduced or cloned. This is true at least in more abstract terms. This "cloning" problem is one reason copyright protection for software, digital music, movies, and so on is so difficult to control and enforce. It also explains why digital money, such as Bitcoin, is so complex. Aaronson (2009), building on work presented by Wiesner (1993), has proposed using quantum states as a way of encoding information in a way that does not allow for its reproduction and cloning. He writes that "quantum states, on the other hand, cannot in general be copied, since measurement is an irreversible process that destroys coherence" (Wiesner, 1993, p. 1). Similarly, information can be encrypted in a digital representation, so even though that representation might be cloned, no one without the encryption key would know what it means. Nevertheless, the identically same information would still be hiding inside each clone. But how does this relate to value?

There is a bit of a paradox in the fact that digital information can be perfectly reproduced. Naturally, a less than perfect reproduction (such as a fake Picasso or counterfeit money) will have lower value than the original. But when the copy is literally the same as the original, as it can be with digital information, one might argue that the copy would have the same value as the original. This argument can be misleading, however, since the very scarcity of a piece of information contributes to its value. But does the reproduction of a piece of information always decrease its value? It may well decrease its price, and possibly its value, but the two are different. Assuming for now that the value of information is subjective, the impact of the reproduction of information may be different to different users of that information.

But regardless of the individual user, we must always distinguish between the value to an individual and to society. This is because the individual's utility from the information is most often different from that of society. Although policymakers are interested in the long-run welfare of society and will value the information accordingly, the individuals will value it based on their own utility, preferences, and personal benefit from that information. It satisfies the same principle as the discount factor (how much we value the present relative to the future), where the social planner (or the policymaker) has a much lower discount factor (value the future as much—or almost as much—as the present), than the individual who cares more about the short run.

3.5 Absolute versus Relative Value

Having discussed some of the major factors behind the value of information, we now refine the discussion and investigate additional distinctions that should be considered when assessing the value of information. These distinctions are closely related to the fundamental definition of the general concept of "value" we discussed earlier. They are (*i*) objective versus subjective; (*ii*) intrinsic versus extrinsic; and (*iii*) absolute versus relative.

In a way, the first two distinctions are just special cases of the last. *Subjective* value is relative to the person making the value judgment, while *extrinsic* value is relative to something outside of the thing being valued. These distinctions may be applied to both the user of the information and the piece of information itself. Rather than develop and discuss all these eight possible distinctions of the value of information and the interconnection among them, we concentrate here on the concept of *relativity* (and subjectivity), which subsumes, to a high degree, the first two distinctions. A more comprehensive exploration of this complex topic is beyond the scope of this chapter.

We begin our discussion by going back to our earlier listing of some of the main factors affecting the value of information. These factors included the different facets of importance: instrumental importance, attractiveness, provenance, social impact, and even environmental impact. Each of these factors is relative to the decision maker (user: sender or receiver, or even observer or eavesdropper). For example, a piece of information may help Bill save his child's life but will have no value for Jill. Naturally, the relative value of that information is much higher for Bill. Or think of social impact, for example. Although an exact social impact may be measured, the value of that impact may be quite different for different individuals. This is why election results are upsetting for some and celebrated by others. Even a factor such as abundance is relative. Think of water in the Sahara Desert versus Washington State. The same arguments hold for the other factors. To prove that the value of information is relative or subjective, it is sufficient to provide the following argument. If at least one factor that determines the value of information is relative (subjective), the value of information must be relative (subjective) as well. Stating this differently, we can say there is no unique way of defining a relatively (subjectively) based absolute value of information.

We now provide a different argument by an analogy to value of life. This is a well-debated topic in the scientific and policy literature. (See, for example, the critical review of Viscusi and Aldy, 2003.) The value that we put on our lives as individuals is much different from that of the insurer, the actuarial, or the government. It is relative to our own closeness to the individual, age, background, location, and so forth. Now, if there exists an information set that will allow a doctor to cure someone's disease, theoretically she (and her family) will be willing to pay any amount for it. Others who do not have that disease will be willing to pay much less or nothing.

As mentioned earlier, the value of information is subjective and relative. It is not objective; it is not unique; it is relative to other information; and it is subjective to the decision maker, modeler, or the one who simply enjoys that information. We now provide a more quantified version of that statement. Consider trying to solve a decision or inference problem based on circumstantial evidence or, similarly, based on insufficient information. These are common problems for all decision makers, all scientists, and practically for everyone. Such problems are inherently underdetermined; there is a continuum of solutions that satisfy the information we have. A common way to solve such problems is to transform it to a well-behaved constrained optimization problem. The information we have is stated in terms of constraints, say the structural equations that we believe capture the underlying mechanism of the modeled system, or, in more statistical terms, the constraints may be some moments (say the mean and variance) of the distribution we wish to infer. But since many (actually, infinitely many) possible solutions satisfy these constraints, we have to choose a decision function (often called an objective function) that will pick up one of these many solutions. This is, in fact, the way all inferential methods work. But the contribution of each one of these constraints—pieces of information—to the final solution is different. Some have more impact, while others have less impact. But these are all relative. It is the marginal contribution of each piece of information relative to the *rest* of the information used. For example, if we now receive another independent piece of information and add it to the previous set, the marginal contribution of each piece of information may change. And, again, it will remain relative to the total information used. These values cannot be absolute, and they are not unique.

A complete quantitative derivation of this argument is provided in Golan (2018). That argument follows on the classical (Jaynes) maximum entropy method and its extension and generalization for all types of uncertainties—the info-metrics framework. In these modeling and inference problems, the information is specified as constraints within a constrained optimization problem. The objective function of choice is an information (entropy) one. In that case, the inferred parameters (the "solution") are just the well-known Lagrange multipliers, associated with each piece of information (constraint) in the model. These multipliers are in units of information, say bits. Nevertheless, the inferred informational values of these parameters are relative to the rest of the information used in that specific optimization problem. Moreover, they are not absolute for all similar problems and information sets. For a complete discussion, formulation, and examples, see Golan (2018, Chapters 2–4).

3.6 Prices, Quantities, Probabilities, and Value

Before concluding this section, we connect some of the ideas we discussed so far and the assumptions that govern that connection. Not all kinds of information have values that depend solely on the quantity. However, let us assume some context where the quantity of certain types of information can be priced, at least within some reasonable range. If we think of the pricing of materials, certainly knowing just the number of units, say mere kilograms or liters, is not enough. The pricing also depends on the type of materials—say gold, charcoal, water, or crude oil. Nevertheless, once we fix the type of information, its pricing might be based on certain characteristics such as the general usefulness of the source and the scarcity of that information. To illustrate this point, let us suppose that the source is the CIA. Such information might be very useful for certain kinds of decisions, such as foreign policy and other national security matters. We also assume–again restricting ourselves to certain types of information and a certain range of quantity—that a price, p, can be established for each unit of information ι. Thus, the value of a piece of information ι, $VoI(\iota)$, is

$$VoI(\iota) = p \times QoI(\iota),$$

where $QoI(\iota)$ is the quantity of information. Introducing Shannon's (1948) observation that the quantity of information can be expressed in terms of its improbability (or surprise value, called "surprisal"), we can rewrite the above value function as

$$VoI(\iota) = p \times QoI(\iota) = p \times \log_2 [1/Prob(\iota)]$$

where $QoI(\iota) = \log_2 [1/Prob(\iota)]$ and '$Prob(\iota)$' stands for the probability of information ι. Note that the $1/Prob(\iota)$ part has the effect of reversing "probability" to "improbability," or scarcity, and the $\log_2$ part provides the unit of measurement: it converts the improbability to binary digits (*bits*). With these definitions and under the unrealistic and simplified assumptions imposed, we have shown one way to construct the value of information, assuming we know a price per bit. We also ignored the issue of whether all types of information can be quantified. But because the meaning of information cannot be quantified, these arguments are not convincing, as it is easily seen that two different sets of information may have the same value, even if we knew the price. But the price itself is unknown; there is no objective price. We, therefore, must resort back to our earlier conclusion that the value of information is not unique: it is subjective and relative.

In terms of our three basic questions (the tree and an observer), we can say the following. Regardless of your answer to each question, whenever there is information, it has a value. Wherever there is information, it has a utility, or at least some potential utility. But that value and that utility are not absolute. Or stating this differently, they are subjective and intrinsic.

3.7 A Comment on Ethics versus Value

As we conclude this section, it is beneficial to comment on the recent literature related to the ethics of information. Although the topic of this chapter is the value of information, it is important to prevent confusion between the two: value and ethics. The ethics of information touches on ethical issues regarding information. Simply stated, ethics has primarily to do with rules of conduct and their value, but not the value of other things. It has no direct relevance to our discussion of the value of information. A nice discussion on the topic appears in Floridi's book, *The Ethics of Information* (2013), and his article (2002).

Although both of us, as well as Floridi, discuss information-related concepts, the only common element in our discussions is "*information*." However, in Chapter 6 of his book ("The Intrinsic Value of the Infosphere"), Floridi touches on a specific aspect of the value of information. As the abstract to that chapter states:

> The question I shall address is this: what is the least, most general common set of attributes that characterizes something as intrinsically valuable, and hence worthy of some moral respect, and that without which something would rightly be considered intrinsically worthless or even positively unworthy of respect? The answer that I shall develop and support here is that the minimal condition of possibility of an entity's least intrinsic value is to be identified with its ontological status as an informational entity. All entities, when interpreted as clusters of information—when our ontology is developed at an informational level of abstraction—have a minimal moral worth *qua* informational entities and so may deserve to be respected (p. 102).

That discussion gets very interesting and very complicated but goes beyond the objectives and scope of our chapter. Floridi's fundamental conclusion is that all information has intrinsic value, meaning that it is "worthy of respect." This conclusion might be true in some way (though Floridi himself seems to allow exceptions), but it does not distinguish the values of different pieces of information according to their usefulness for making decisions and inferences. We maintain that the value of information is extrinsic, subjective, and relative for this purpose. We also argue that the value is always relative, and it is *not* unique.

Let us give an analogy. It is a common moral assumption that all humans have intrinsic value, but this assumption does not mean that every human has equal instrumental value. Not everyone can perform the same jobs with equal ability, and not all abilities or jobs are valued the same by all. Similarly, not all information is of equal usefulness, especially in making decisions. Having discussed some of the main ingredients that determine the value of information, we turn to another component that may affect the value: risk.

4. Risk, Decisions, Prices, and Value

Decision making and strategic behavior are among the most frequently mentioned uses of information and its value. Early writings on the value of information by McCarthy (1956) and Stratonovich (1965) focused on the value of information for decision making. These works discussed the amount a decision maker would be willing to pay for an additional piece of information with the potential of improving the benefits following from the decision. These benefits capture both notions: an improved decision (from the decision maker's perspective) and less uncertainty about the information used for the decision. Similarly, one can think about improving the decision by lessening the risk involved. The reduction of risk can result from acquiring more information. See, for example, the classic work of Kahneman and Tversky (1979, 2013), Kahneman (1991), and Tversky and Kahneman (1985) on decisions under uncertainty, as well as the body of work by Machina and co-authors (e.g., Machina, 1987; Machina and Rothschild, 2008). There is also an extensive amount of work within game theory on decision strategies based on the players' different information. But this issue, though illuminating and innovative, falls outside the scope of our discussion, and so we do not discuss that literature here; we are interested in the information for decision and its value rather than in the decision process itself.

The literature's focus on decision making and prices certainly seems logical. Imagine being the CEO of a large corporation and trying to decide how much to spend on research and development or on marketing research. Or, similarly, consider a candidate running for office having to decide how much to spend on a survey. As we noted earlier, however, information can also be important even if it is not used for decision making. In such cases, although the information has a value, it is difficult to put a price, or an exact "market" value, on the information, especially in cases where there is no fully developed "market" for that information.

Within the context of decision making, at least one value of information is to inform a decision maker of the risk that is under consideration. As we have shown earlier, the quantity of information is standardly quantified (Shannon, 1948) via a formula that encodes the inverse of probability. It is defined as the (base 2) logarithm of the inverse of the probability of an event. It is a direct function of our surprise or the scarcity of that event. But that quantity is semantics free. For example, two different events or propositions, both of which have the same probability, possess the same quantity of information, even though they have very different meanings or possible values. Ignoring the semantical problem, in order to evaluate value and risk in a simplified way, we find it helpful to introduce prices.

Assuming the total price of something is affected positively by its quantity, then the total price (or "value") of a quantity of information is the product of that quantity (say, bits) and the price per unit (bit). Recalling that a major determinant of the price of any good is the scarcity of that good, we can use this fact in determining the price of information. But since information is proportional to the inverse of probability, a very rare, or scarce, event has a high value, just as rare diamonds have. We have shown this in the previous section. Risk, in contrast, is defined as the probability of an event times its unit price (or expected value per unit). Often risk is expressed in terms of loss or expected loss, say probability times expected loss. This classical definition of risk (Knight, 1921) means that the value of a piece of information is actually the inverse of its risk. We explore this idea next and highlight some subtleties in the definitions that emerge during our exploration.

Various proposals have been made about how to mathematize the concept of risk. In our discussion, we follow Kaplan and Garrick (1981), who introduced one of the best, most intuitive, and broadest ways of understanding risk. They argue that a risk analysis consists of an answer to the following three questions:

1. What can happen? (What can go wrong? Will the tree fall?)
2. How likely is it that that will happen? (What is the probability that the tree will fall?)
3. If it does happen, what are the consequences? (Will the observer see it? Will it fall on the observer?)

To answer these questions, we make an exhaustive list of possible outcomes or scenarios. Following Kaplan and Garrick, we define the triplet $\langle s_i, Prob_i, x_i \rangle$ where s_i is a scenario identification or description; $Prob_i$ is the probability of that scenario; and x_i is the consequence, or evaluation measure, that can be either negative (loss) or positive (benefit) for scenario s_i. In that setting and loosely speaking, risk is the set of triplets $R = \{\langle s_i, Prob_i, x_i \rangle\}$, $i = 1, \ldots, N$. In this way, risk is not just the "probability *multiplied* by the consequence," but rather the "probability *and* its consequence" (or the family of possible consequences and their probabilities). In their work, Kaplan and Garrick generalize this definition to a whole family of risk curves. Nevertheless, in the present discussion, it is more fruitful to concentrate on the simpler case, which is an extension of Keynes's earlier seminal work.

We want to make "risk" a single (normalized) number so that we can compare various degrees of risk. A standard way, discussed earlier, is to define risk as the probability of something happening multiplied by the resulting cost or gain.[4] Risk originally emphasized the cost (negative benefit), but clearly there are two sides to placing a bet, or for making decisions, on the future value of s_i. We can therefore use the value of the mathematical expectation to compare levels of risk, conditional on the available information. It is useful to go back to Keynes's classic definition (1921, p. 315). He argued that "risk" may be defined as follows. If A is the amount of good that may result, with a probability $Prob$, q is the complement

[4] This is of course a simplification. Let us call the definition above "singular risk." If one is tracking various scenarios, then one can add the total value obtained by adding together the values of the singular risk of the various scenarios and dividing by the number of scenarios (or multiply by the probability of each scenario). This will yield an average risk. It would complicate the math a bit (and the interpretation) but not by much. We will stick with singular risk so as to not unnecessarily complicate things. The more general notion is related to, though different from, what in economics is known as "expected utility." See the classical work of Machina on decisions, risk, and uncertainty.

of *Prob*, so $Prob + q = 1$, and E is the value of the mathematical expectation, so that $E = Prob \times A$, then the "risk" is $Risk \equiv Prob \times (A - E) = Prob \times (1 - Prob) \times A = Prob \times q \times A = q \times E$, where "$\equiv$" stands for "definition." This is just a simple case of our previous, more general definition. It is an interesting perspective on risk. Unlike the more traditional way of having risk defined on the basis of the "bad" or negative outcome, here the "bad" outcome is simply that the good outcome will not happen—something in the spirit of an opportunity cost. For example, if the value resulting from decision A is \$10 and we will get it with a probability of 0.9, then the risk is $0.9 = (1 - 0.9) \times (0.9 \times 10) = 0.1 \times 9$. Keynes also argued that, in this case, we can view E as measuring the net immediate sacrifice that should be made in the hope of gaining A, q is the probability of that sacrifice, and therefore $q \times E$ is the risk.

We want to connect this basic definition with information, utility, and value. As we previously saw, C. E. Shannon, in developing communication channels, was dealing with the quantitative notion of *information;* thus, a comparison between risk and information (in their quantitative forms) is entirely appropriate. As defined earlier, within the context of decision making, risk typically deals with events or outcomes. Therefore, we shall not distinguish between events, outcomes, pieces of information, and propositions. Following on Keynes's definition, we can thus write

$$Risk(\iota) = Prob(\iota) \times Util(\iota),$$

where $Util(\iota)$ *stands* for the utility of the information ι for the decision maker. Rather than $Util(\iota)$, we can use the expectation (expected utility), as Keynes did, but this will not change the logic of our derivations here. Using our previous definitions and derivations, we can rewrite risk in terms of the quantity of information and the value of information:

$$Risk(\iota) = Prob(\iota) \times Util(\iota) = 2^{-QoI(\iota)} \times Util(\iota) = 2^{-VoI(\iota)/p} \times Util(\iota),$$

where p denotes the price of a unit of information. This equation also demonstrates the inverse relationship between value of information and risk. In addition, it nicely demonstrates Marshall's (1920, p. 78) description of utility and its connection to prices and values. Marshall wrote: "Utility is taken to be correlative to Desire or Want. . . . Desires cannot be measured directly, but only indirectly, by the outward phenomena to which they give rise: and . . . in those cases with which economics is chiefly concerned the measure is found in the price which a person is willing to pay for the fulfillment or satisfaction of his desire."

In this discussion, we assumed that the information is quantifiable, the probabilities of events are known, and some price per unit of information is observable or known. In the real world, however, these assumptions do not hold in decision making, modeling, and inference. Not only are we always working with insufficient information, but that information is often noisy, and much of the necessary information cannot be observed. We only observe pieces of the information, and so we must infer the rest. For example, we do not observe preferences, but we do observe actions taken. We then need to infer the preferences from the actions. This is an underdetermined problem. However, there is another complication: for each act of decision making, the meaning of the information may be different and will affect its price. With this in mind, in this section we highlighted, under simplified assumptions, some of the basic issues related to information, its value, utility, and risk. For a complementary discussion and derivations on decision making and inference with complicated information, see the literature on info-metrics, such as Golan (2018), the other chapters in this volume, and other related literature.

In terms of our three basic ontological questions (the tree and an observer), we can make the following statement. Regardless of your answer to each of the questions, whenever there is information, it has a value. In terms of our second set of axiological questions, we can say that value is relative. It may be different by sender, receiver and observer, as well as by situation (space and time). For all values that are different from zero, a certain utility must be associated with it. If this information is used for decision making, its value is inversely related to the risk involved (from the decision). If a market exists, for that information prices will exist as well, and the value is positively correlated with these prices.

5. The Value of Disinformation

Must information be true? If we follow directly on our discussion so far, the answer is simply "no." But the notion of information and truth demands more than just a fast reply, for it is a fundamental issue for philosophers, scientists, and all decision makers.

Floridi (2011) has put forward the Veridicality Thesis, that is, the principle that information must be true. His thesis, in part, avoids what he calls the Bar-Hillel–Carnap Semantic Paradox (BCP). The BCP arises naturally in the context of classical logic, which requires that an arbitrary contradiction implies every sentence whatsoever, and thus the contradiction can be thought of as containing an infinite (or at least maximal) amount of information (see Carnap and Bar-Hillel, 1952 and Bar-Hillel, 1964). According to Floridi, because a contradiction is not true, it really contains no information whatsoever.

Dunn (2008, 2013) and Golan (2014), as well as Fetzer (2004), have argued against this view. We, however, take that point of view here: information is not required to be true. After all, if it is this "thing" that affects our decisions or preferences, why should it necessarily be true? Is it not just a matter of semantics? In fact, we think that the tendency to regard information as true is part of the pragmatics of the word, not its ordinary semantics in the sense of meaning. If we give you something and call it "information," we are suggesting that it is true (and informs you) unless we go out of our way to tell you otherwise.[5] To give an analogy from Dunn (2008), it is much as if we give you something and call it "food." This perspective certainly suggests that it is edible unless we go on to say that it is spoiled or poisoned. Simply stated, terms should be defined in a way that they are most useful and that communicate our thoughts. We choose to take the minimalist, yet practical (especially for decision making, modeling, and inference), approach here and not require that information be true. Still the fundamental purpose of information might be for communicating knowledge or, more precisely, for communicating the raw material (input) for knowledge. So, must information at least be true for it to have value? But value to whom? The sender, the receiver, or possibly an occasional observer? We have already discussed these issues under the assumption that information is true. But what if it is deliberately false information—disinformation? It clearly can have value to the sender, even though it is false or even because it is false, if it increases the sender's utility. A common motive for sending false information is to mislead or confuse. But false information may also be of value to the receiver. White lies, for example, are often told for the very purpose of avoiding hurting someone's feelings, whether or not that someone knows it. Presumably, this is ordinarily of value to the sender as well. Though of much less importance, value can also come accidentally. For example, say we unintentionally tell you something false, and you use it as evidence to bet on a horse. Perhaps you even win big time.

According to the Oxford English Dictionary (OED), *disinformation* is "false information which is intended to mislead, especially propaganda issued by a government organization to a rival power or the media." There is a subtle distinction between "*dis*information" and "*mis*information." The OED says that misinformation is "false or inaccurate information, especially that which is deliberately intended to deceive." So, both words suggest the intention to deceive, but only *dis*information actually requires it. For the sake of making a distinction, let us use "misinformation" as simply false information without suggesting any intention to deceive, but also without rejecting the possibility.

[5] The distinction between syntax, semantics, and pragmatics comes from Morris (1938), although in its essentials it goes back to C. S. Peirce. Syntax has roughly to do with grammar, semantics with meaning, and pragmatics with use. That information is true or that food is edible are what Grice (1989) calls "conversational implicatures," which he grounds in pragmatics by virtue of a "Cooperative Principle" that leads to certain "Maxims of Conversation" (Quantity, Quality, Relation, Manner).

As we have already suggested, we might think that misinformation, whether intended or not, would have no value. Maybe this would be so in an ideal world—just as in that ideal world counterfeit money would have no value. But clearly disinformation, like counterfeit money, can have value to its user, which is the very reason that motivates its deceptive use. That is why disinformation is so valuable in times of war.

If misinformation is false and disinformation is purposely false, there is still one more important category, and that has to do with the "informant" simply not caring whether what he or she is communicating is true or false. This frequently happens in works of fiction (though there are exceptions, e.g., historical novels). But we are talking about something that happens often in the lives of politicians and other salespeople. The person wants to persuade the listeners at all cost and does not care whether the statements are true or false. The philosopher Harry Frankfurt has written an article (1986) as well as a book (2005) on this topic and has entitled both "On Bullshit."

A stronger version of "bullshit" is the relatively new concept called "fake news," which has been very much in the nonfake news recently and is a form of disinformation. Normally, an agent (sender) intentionally gives *false* information only when that information has a certain value (positive utility) to her/him. For the receiver, however, it often may have negative utility—and that may be the value to the sender, who for some reason wants to disadvantage and confuse the receiver. But sometimes it may have no utility, or even positive utility, to the receiver, whether or not the sender intended it. Therefore, it is useful to think of disinformation in terms of its overall social benefits and costs. This also allows us to compare false information with true information. But the analysis is far from trivial, and the only way to arrive at a universal "solution" that false information always does some harm to society as a whole seems to be to assume that "truth" has a value in itself. We can relate this discussion to our earlier one on social versus individual values, or, rate of return, via the following conjecture. Our conjecture involves the relationship between the overall values (on average) of information and disinformation.

Conjecture: *The social value of true information is greater than that of false information, where social value means the societal "rate of return" from the information.*

We are not fully convinced that the above conjecture is always true. We would like to think that evolution, either of human animals or human cultures, would select for truth-telling. Nevertheless, as Bond and Robinson (1988) show, on the one hand, many examples can be selected for deception in the evolution of both plants and animals. On the other hand, we also believe that this conjecture is in line with Pinker's (2018) *Enlightenment*.

Finally, we note another form of disinformation, or fake news, known sometimes as "cooked data." This type of disinformation is quite rare, yet it happens in both science and politics. In this form of disinformation, someone provides you with some information (or data—the hard information) that was manipulated in such a way that a certain prediction or inference (say, policy) is obtained in the processing stage of the data. That certain prediction is the one the sender wants, but it is the incorrect one—just like disinformation. Luckily, it is often easier to identify that type of disinformation by using the correct statistical tools, or even empirical laws such as Benford's law (Benford, 1938; Newcomb, 1881). This law describes the frequency distribution of numerical data. That distribution satisfies a power law where the first digit in any data is observed in inverse frequency to its size—one is the most frequent, and nine is the least frequent.

In terms of our three basic questions (the tree and an observer), we can say the following. Regardless of your answer to each question, whenever there is information, and regardless of whether it is true, false, fake, or due to some disinformation campaign (someone told you the tree fell, but it did not), it has a value. But in this case, it could have a positive value to one person and a negative value to another, and it does not have to sum to zero.

In terms of our second set of questions, the axiological questions, we can say the following. All types of information have values. The values are relative. They may be different by sender, receiver, and observer, as well as by situation (space and time). All values that are different than zero have a certain utility that is associated with it. If a market exists for that information, or disinformation, prices will also exist, and the value is positively correlated with these prices. If the information is used for decision making, its value is inversely related to the risk involved from the decision.

6. Concluding Thoughts

In this chapter, we highlighted in a simple and qualitative way our thoughts on the meaning of information, our uses for it, and its value. We discussed all types of information, though we have concentrated on information used for decision making and on the related activities of modeling and inference. Since decision making involves risk and is dependent on the individual's utility and resources, we also touched on these concepts.

Overall, we argued that all information has a value, be it true, false, deliberately false, objective, subjective, hard, or soft. That value is usually positive, but overall it can be positive or negative, though we have conjectured that at the aggregated social level, it is positive. The value is related to our utility (or disutility) gained from that information, but it is not absolute.

We expressed some of our ideas by returning to our basic three questions about the falling tree. Although these questions dealt directly with the concept of information, at times it was helpful to apply it to the other concepts we discussed.

Are we convinced now that we know with certainty what information is? Probably not. But with probability one, we are convinced that we know what information is when we talk about information needed for decision making and related activities such as modeling and inference. It is this thing (true or false) that affects our preferences, actions, models, and inferences. It has a value, and that value is relative and not unique. It is in the domain of our utility function.

Did we solve the value of information question? No. But we have provided arguments that it is not unique or objective. We have also argued that *information* is a unique good: it is different from all other goods. Therefore, it is sometimes impossible to price it and therefore to calculate its absolute (or "market") value.

We have also shown that when it comes to inference and modeling, we can rank the different pieces of information in terms of their relative importance or values. But we can never construct their absolute values. Further, even the relative ranking is problem and information specific.

Finally, although we touched (in a simple and qualitative way) on different aspects of information and its value, many open questions are left unanswered. The answer to some questions may be found in other chapters of this volume, but we suspect that the answers to the most fundamental ones are still to be worked out. These include (*i*) the most fundamental question of what information is, (*ii*) how to combine the more mathematical definition of information (and entropy) with the meaning of that information (unify semantic and quantity), and (*iii*) whether there is a certain type of information that has an absolute (extrinsic) value? Maybe the answer to question (*iii*) can be found once question (*ii*) is solved.

Although there are many more open questions, we only highlighted those that are directly related to the subjects discussed in this chapter. For a nice discussion of a more inclusive list of open questions related to the philosophy of information, see Floridi (2011).

Acknowledgments

We thank the three referees for their helpful remarks on a previous version of this chapter.

References

Aaronson, S. (2009). "Quantum Copy-Protection and Quantum Money." *CCC '09 Proceedings of the 2009 24th Annual IEEE Conference on Computational Complexity*, (pp. 229–242). Washington, DC, USA: Computer Society.

Adriaans, P. (2013). "Information." In Edward N. Zalta (ed.), *The Stanford Encyclopedia of Philosophy*. https://plato.stanford.edu/archives/fall2013/entries/information.

Bar-Hillel, Y. (1964). *Language and Information: Selective Essays on Their Theory and Application*. Reading, MA: Addison-Wesley.

Barwise, J., and Seligman, J. (1997). *Information Flow. The Logic of Distributed Systems. Cambridge Tracts in Theoretical Computer Science*, vol. 44. Cambridge: Cambridge University Press, 1997.

Benford, F. (1938). "The Law of Anomalous Numbers." *Proceedings of the American Philosophical Society*, 78: 551–572.

Bentham, J. (1789). *An Introduction to the Principles of Morals and Legislation*. (First printed in 1780). London: T. Payne, and Son.

Berkeley, G. (1710). *A Treatise Concerning the Principles of Human Knowledge*. Dublin: Aaron Rhames.

Bond, C. F., and Robinson, M. (1988). "The Evolution of Deception." *Journal of Nonverbal Behavior*, 12: 295–307.

Carnap, R., and Bar-Hillel, Y. (1952). *An Outline of a Theory of Semantic Information*. Technical Report No. 247. Cambridge, MA: MIT Press. Reprinted in Bar-Hillel (1964).

Caticha, A. (2012). *Entropic Inference and the Foundations of Physics*. Monograph commissioned by the 11th Brazilian Meeting on Bayesian Statistics. EBEB-2012 (USP Press, São Paulo, Brazil, 2012); online at http://www.albany.edu/physics/ACaticha-EIFP-book.pdf.

Clark, A., and Chalmers, D. J. (1998). "The Extended Mind." *Analysis*, 58: 7–19.

Dretske, F. (1981). *Knowledge and the Flow of Information*. Cambridge, MA: MIT Press.

Dunn, J. M. (2001). "The Concept of Information and the Development of Modern Logic." In W. Stelzner and Manfred Stöckler (eds.), *Zwischen traditioneller und moderner Logik: Nichtklassische Ansatze (Non-classical Approaches in the Transition from Traditional to Modern Logic)* (pp. 423–447). Paderborn: Mentis-Verlag.

Dunn, J. M. (2008). "Information in Computer Science." In J. van Benthem and P. Adriaans (eds.), Philosophy of Information (pp. 581–608). Amsterdam: North Holland.

Dunn, J. M. (2013). "Guide to the Floridi Keys." Essay Review of Luciano Floridi's book *The Philosophy of Information*, *MetaScience*, 22: 93–98.

Fetzer, J. (2004). "Information: Does It Have to Be True?" *Minds and Machines*, 14: 223–229.

Floridi, L. (2002). "On the Intrinsic Value of Information Objects and the Infosphere." *Ethics and Information Technology*, 4: 287–304.

Floridi, L. (2011). *The Philosophy of Information*. Oxford: Oxford University Press.

Floridi, L. (2013). *The Ethics of Information*. Oxford: Oxford University Press.

Frankfurt, H. (1986). "On Bullshit." *Raritan Quarterly Review*, 6: 81–100.

Frankfurt, H. (2005). *On Bullshit*. Princeton, NJ: Princeton University Press.

Golan, A. (2008). "Information and Entropy Econometrics—A Review and Synthesis." *Foundations and Trends® in Econometric*, 2: 1–145. (Also appeared as a book, Boston: Now Publishers, 2008.)

Golan, A. (2014). "Information Dynamics." *Minds and Machines*, 24: 19–36.

Golan, A. (2018). *Foundations of Info-metrics: Modeling and Inference with Imperfect Information*. Oxford: Oxford University Press.
Golan, A., and Lumsdaine, R. (2016). "On the Construction of Prior Information—An Info-Metrics Approach." *Advances in Econometrics,* 36: 277–314.
Grice, P. (1989). *Studies in the Way of Words*. Cambridge, MA: Harvard University Press.
Hartley, R. V. L. (1928). "Transmission of Information." *Bell System Technical Journal*, 7, no. 3: 535–563.
Howard, R .A. (1966). "Information Value Theory." *IEEE Transactions on Systems Science and Cybernetics*, (SSC-2): 22–26.
Jaynes, E. T. (1957a). "Information Theory and Statistical Mechanics." *Physics Review*, 106: 620–630.
Jaynes, E. T. (1957b). "Information Theory and Statistical Mechanics II." *Physics Review*, 108: 171–190.
Kahneman, D. (1991). "Commentary: Judgment and Decision Making: A Personal View." *Psychological Science*, 2: 142–145.
Kahneman, D., and Tversky, A. (1979). "Prospect Theory: An Analysis of Decision under Risk." *Econometrica*, 4: 263–291.
Kahneman, D., and Tversky, A. (2013). Prospect Theory: An Analysis of Decision under Risk. *Handbook of the Fundamentals of Financial Decision Making*, (pp. 99–127). Hackensack, New Jersey: World Scientific Publication.
Kaplan, S., and Garrick, B. (1981). "On the Quantitative Definition of Risk." *Risk Analysis*, 1: 11–27.
Keynes, J. M. (1921). *A Treatise on Probability*. New York: Macmillan.
Knight, F. H. (1921). *Risk, Uncertainty, and Profit*. Boston: Houghton Mifflin.
Machina, M. J. (1987)."Decision Making in the Presence of Risk." *Science*, 236: 537–543.
Machina, M. J., and Rothschild, M. (2008). "Risk." In S. N. Durlauf and L. E. Blume (eds.), *The New Palgrave Dictionary of Economics*, 2nd ed. Basingstoke, Hampshire, UK; New York, NY: Palgrave Macmillan.
Marshall, A. (1920). *Principles of Economics. An Introductory Volume*. 8th ed. London: Macmillan.
McCarthy, J. (1956). "Measures of the Value of Information." *Proceedings of the National Academy of Sciences of the United States of America*, 42: 654.
Morris, C. W. (1938). "Foundations of the Theory of Signs." In O. Neurath (ed.), *International Encyclopedia of Unified Science*, vols. 1, no. 2. Chicago: University of Chicago Press.
Newcomb, S. (1881). "Note on the Frequency of Use of the Different Digits in Natural Numbers." *American Journal of Mathematics*, 4(1/4): 39–40.
Nørretranders, T. (1998). *The User Illusion: Cutting Consciousness Down to Size*. New York: Viking Penguin. Translated by Jonathan Sydenham from the original published in Danish (1991) as *Maerk Verden* (*Mark the World*). Gyldendalske Boghandel.
Pinker, S. (2018). *Enlightenment Now: The Case of Reason, Science, Humanism, and Progress*. New York: Viking.
Shannon, C. E. (1948). "A Mathematical Theory of Communication." *Bell System Technical Journal*, 27: 379–423.
Sims, C. A. (2003). "Implications of Rational Inattention." *Journal of Monetary Economics*, 50: 665–690.
Stratonovich, R. L. (1965). "On Value of Information." *Reports of USSR Academy of Sciences, Technical Cybernetics*, 5: 3–12. [In Russian.]

Tversky A., and Kahneman, D. (1985). "The Framing of Decisions and the Psychology of Choice." In V. T. Covello, J. L. Mumpower, P. J. M. Stallen, and V. R. R. Uppuluri (eds.), *Environmental Impact Assessment, Technology Assessment, and Risk Analysis*. NATO ASI Series (Series G: Ecological Sciences), vol. 4. Berlin: Springer.

van Benthem, J., and Adriaans, P. (2008). *Philosophy of Information*. Amsterdam: North-Holland.

Vasudevan, S., Efrat Flashner-Abramson, F. Remacle, R. D. Levine, and Nataly Kravchenko-Balasha. (2018). "Personalized Disease Signatures through Information-Theoretic Compaction of Big Cancer Data." PNAS, 115(30): 7694–7699.

Viscusi, W. K., and Aldy, J. E. (2003). "The Value of a Statistical Life: A Critical Review of Market Estimates Throughout the World." *Journal of Risk and Uncertainty*, 27: 5–76.

Voorhees, D. (1993). *The Book of Totally Useless Information*. New York: MJF Books.

Weiner, N. (1948). *Cybernetics or Control and Communication in the Animal and the Machine*. New York, NY: John Wiley.

Wiesner, S. (1983). "Conjugate Coding." *SIGACT News*, ACM, 15: 78–88. (Original manuscript written circa 1970.)

2
A Computational Theory of Meaning

Pieter Adriaans

1. Introduction

This chapter describes some results in the context of a long-term research project that aims to understand learning as a form of data compression (Adriaans, 2008). We will use the theory of Kolmogorov complexity as our basic tool (Li and Vitányi, 2019). This theory measures the complexity of data sets in terms of the length of the programs that generate them; such an approach induces a theory of information that in a sense is orthogonal to Shannon's classical theory. This theory defines information in terms of probability, which consequently leads to a theory of optimal codes and data compression. In Kolmogorov complexity, we start at the other end of the spectrum with the notion of optimal compression in terms of a short program, and we derive a theory of probability and model selection from it. From the perspective of info-metrics, we will show that many familiar concepts such as entropy, randomness, sufficient statistics, signal, and noise have interesting algorithmic counterparts and that a meticulous mapping of central concepts of the two disciplines can lead to fruitful new insights.

1.1 Chapter Overview

In the first part of the chapter, we study a conglomerate of proposals for optimal model selection using compression techniques: one-part-, two-part code optimization, minimum description length, and Kolmogorov's structure function and facticity. We show that none of these proposals adequately solves the task of optimal model selection. There is no compression-based technique for optimal model selection because:

1. Such techniques cannot be invariant over the choice of reference Turing machines, which block the development of an adequate measurement theory.

Pieter Adriaans, *A Computational Theory of Meaning* In: *Advances in Info-Metrics: Information and Information Processing across Disciplines.* Edited by: Min Chen, J. Michael Dunn, Amos Golan, and Aman Ullah, Oxford University Press (2021).
DOI: 10.1093/oso/9780190636685.003.0002

2. These techniques are subject to a phenomenon that we call polysemy: two or more models might generate optimal compression with no or little mutual information.
3. Total recursive functions exist that allow us to define vacuous two-part code models that have no statistical significance.

Consequently, there is no pure mathematical justification for two-part code optimization, although in the right empirical context the methodology will work. In the second part of the chapter, we develop a contextual theory of computational semantics based on an interpretation of Turing machines as possible worlds. We define the notion of the *variance of a Turing frame* as the largest distance between the complexities assigned to the same string in a set of possible worlds. We show that the concepts of randomness and compressibility lose their meaning in contexts with infinite worlds. The theory allows us to define the *degree of informativeness* of semantic predicates for individual agents. We show that this notion is fundamentally unstable with regard to the choice of a reference world/machine. We sketch some application paradigms of this theory for various types and groups of agents.

1.2 Learning, Data Compression, and Model Selection

Many learning algorithms can be understood as lossy compression algorithms; that is, they generate a compressed model of the data set from which the original data set cannot be reconstructed, but that can be used as a classifier to predict unknown values in the data set. The C4.5 decision tree induction algorithm,[1] for example, uses the notion of *information gain* to separate a high-dimensional search space into hyper cubes of compressible data. Here information gain of an attribute is defined as the number of bits saved when encoding an arbitrary member of the data set by knowing the value of this attribute. The resulting decision tree can be seen as a *model* of the database generated by a process of *lossy compression*, but with the expectation that the data set as a whole can be compressed lossless with this model. When the data is collected in an appropriate general setting, the tree can be used to predict values in new data sets generated under comparable conditions.

[1] A well-known learning routine that, using the data stored in a data base table, constructs a classifier in the form of a decision tree for a target column, based on the values of the other columns. Technically, the algorithm constructs a subspace of the multidimensional hyperspace defined by the table, where the probability of a specific value of the target column is high, provided that the table was constructed using an appropriate probability distribution.

An abundance of practical applications of learning by compression are available. Well-known machine-learning algorithms such as decision tree induction, support vector machines, neural networks, and many others can be interpreted as processes that incrementally exploit regularities in the data set to construct a model (Adriaans, 2010). Classification and clustering algorithms based on standard industrial compression software perform surprisingly well in many domains (Cilibrasi and Vitanyi, 2005; Adriaans, 2009), and data compression seems to play an essential role in human cognition (Chater and Vitányi, 2003).

A computational theory of meaning tries to explain aspects of the notion of a data set in terms of computation. With the development of the relatively new discipline of the philosophy of information in the past twenty years, the contours of such theory have emerged. The philosophical aspects of the theory developed in this chapter are treated extensively in the entry on information in the *Stanford Encyclopedia of Philosophy* (Adriaans, 2018), so in this chapter we will concentrate on the issues related to the development of a *theory of measurement* for various forms of information.

Here we will analyze learning as data compression from the perspective of the theory of Kolmogorov complexity (Li and Vitányi, 2019). Understanding the phenomenon of learning was Ray Solomonoff's original research motivation in the 1950s. He was the first to discover what we now know as algorithmic information theory or, with some historical injustice, Kolmogorov complexity:

> As originally envisioned, the new system was very similar to my old "inductive inference machine" of 1956. The machine starts out, like a human infant, with a set of primitive concepts. We then give it a simple problem, which it solves using these primitive concepts. In solving the first problem, it acquires new concepts that help it solve the more difficult second problem, and so on. A suitable training sequence of problems of increasing difficulty brings the machine to a high level of problem solving skill. (Solomonoff, 1997, p. 81)

The Kolmogorov complexity of a data set is the length of the smallest program that produces this data set on a universal Turing machine. Kolmogorov complexity is a *quantitative* measure of information that is associated with the concept of the theoretical maximal compression of a data set by a program. In this sense, we call it *one-part code optimization*. In this chapter, we investigate the possibilities of generalizing this quantitative theory into a version that also covers a *qualitative* notion of information: meaningful information. This amounts to a compression in two parts or *two-part code optimization*: a program and its input.

1.3 Some Examples

In some cases and for some data sets, such a meaningful separation is trivial.

Example 1. Take the concept of a Fibonacci number. The Fibonacci numbers grow fast with their index, so in the limit these numbers are extremely sparse. We can compress the description of large Fibonacci numbers by specifying a program f that generates the Fibonacci sequence and an index i. The expression $f(i)$ describes the ith Fibonacci number, and since the code for the program f is finite above some threshold value of i, all descriptions of the form $f(i)$ will be shorter than the numbers they describe. From this point on, it is *useful* to know that a number is Fibonacci since it gives us a more efficient description of the number. For these numbers we can say that the predicate or program f is (part of) the *meaning* of the number x given by $f(i) = x$, where i is the *input* for the *program* f.

We are interested in the question of whether such an approach can be generalized into a computational theory of meaning. We do not intend to provide full coverage of the notion of meaning or even of the concept of meaningful information in this chapter. The underlying research program is also not reductionist. We do not defend the view that all meaning should be reduced to computation. In fact, we believe that analysis of the notion of meaning largely belongs to the humanities and cannot be studied empirically. However, a substantial number of semantic issues are of interest to the empirical sciences and can be studied in terms of computation. The idea that a program can be the meaning of a data set is fairly abstract; therefore, let us give an example.

Example 2. The astronomical observations of Tycho Brahe (1546–1601) enabled Johannes Kepler (1571–1630) to formulate his laws. The planets move around the sun according to a certain mathematical model, and this model generates regularities in the data set of observations. If Brahe had lived in another planetary system or at another time, he could have made different observations that still would have enabled Kepler to formulate the same laws. Consequently, this type of empirical data set has a structural part (the general underlying mathematical model) and an ad hoc part (the specific time and place of the individual observations). The task of a learning algorithm or an empirical scientist is to separate the two types of data or extract the mathematical model from the data set. The underlying model can be seen as providing meaningful information that is "hidden" in the data set. One would expect that, in principle, this process could be automated, and it has indeed been shown that with the genetic programming technique Kepler's laws can be learned from data sets with observations (Koza, 1990).

These observations lead to the following basic intuition: Researchers and learning algorithms extract meaningful information from data sets by observing patterns and regularities. These regularities are associated with invariant models that underlie the observations, but they are blurred with the noise associated with the conditions under which the data sets were created. If the data sets have maximum entropy, we can observe no regularities, and we can extract no meaningful information. In this concept, the notion of meaningful information is associated with the compressibility of the data set. Being compressible is a necessary condition for extracting meaningful information.

2. From Computation to Models

Turing machines (Hopcroft, Motwani, and Ullman, 2001) are abstract models of (mostly female) human computers such as those who worked in offices in the first half of the twentieth century. They had a desk with an in and an out tray. Problems were submitted in the in tray, and the computer did her work: she solved the problem by writing symbols on a paper according to fixed mathematical rules. After that, the solution was put in the out tray. Motivated by this process, Alan Turing (1912–1954) defined the concept of an *abstract computational agent*:

- He reduced the symbol set to only three symbols: 1, 0, and *b* (blank).
- The sheets of paper were replaced by a two-way infinite tape.
- The computation rules could be stored in a transition table.
- The computer could only be in a finite number of different states, and for each state the transition table would specify exactly what the computer had to do.

The concept is illustrated in Figure 2.1.

We give some notational conventions: the set $\{0,1\}^*$ contains all finite binary strings. $\mathbb{N}$ denotes the natural numbers, and we identify $\mathbb{N}$ and $\{0,1\}^*$ according to the correspondence:

$$(0,\varepsilon),(1,0),(2,1),(3,00),(4,01),\ldots$$

Here ε denotes the *empty word*. This equivalence allows us to apply number theoretical predicates like "… is prime" to strings and complexity theoretic

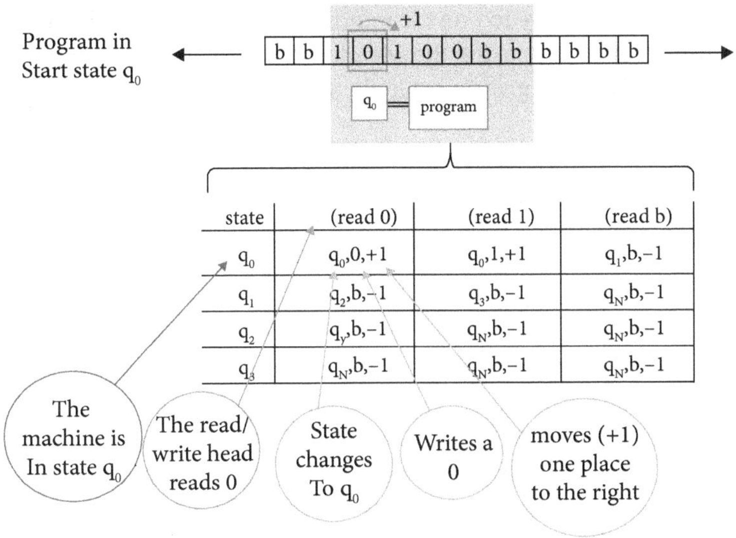

state	(read 0)	(read 1)	(read b)
q_0	q_0,0,+1	q_0,1,+1	q_1,b,−1
q_1	q_2,b,−1	q_3,b,−1	q_N,b,−1
q_2	q_y,b,−1	q_N,b,−1	q_N,b,−1
q_3	q_N,b,−1	q_N,b,−1	q_N,b,−1

Figure 2.1 Schematic overview of a Turing machine. There is an infinite tape with the symbols 1, 0 and *b*. There is a read-write head that scans the tape. At any moment the machine is in a definite state. The actual program is stored in the transition table that is illustrated in the exploded view. The machine makes discrete computation steps: it reads a cell, selects the row in the transition table according to the cell and the state it is in, then writes a symbol in the cell and moves the tape one cell. The machine in the image is in the start state q_0. It ends in an accepting computation (state q_Y) when the string of zero's and one's on the tape ends with 00.

predicates like "... is compressible" to numbers. Any data set is a number and any number is a data set. The *length* $l(x)$ of x is the number of bits in the binary string x. We make some important observations concerning Turing machines:

- The set of all Turing machines is countable. This means that we can index the machines by the set of natural numbers: T_i is the ith Turing machine, where $i \in \mathbb{N}$. We write $T_i(x) = y$ to indicate the computation of machine T_i on input x that halts with output y. In this sense, T_i is a computable function.
- One can show that the index of the machine can be computed from the definition of the machine and vice versa. Turing showed that there are so-called universal machines, U_j, that can emulate any other Turing machine T_i.
- A universal Turing machine takes its input in two parts: (1) the index i of the machine T_i and (2) x the input for the computation. Note that we cannot simply concatenate the code for i and x to form ix, since concatenation is

associative. We have to provide information about the split between i and x. This is done by making the code for i *self-delimiting*. There are various methods. Next we give an example of a simple routine to make a string self-delimiting.

Example 3. Suppose we have a string of 21 bits:

$$i = 101100001100100011100$$

The binary number for 21 is coded by the string 10011, which is 5 bits long. We program our universal Turing machine in such a way that it first reads the length of the number that indicates the length of the string i that follows in terms of a prefix of zeros followed by a demarcation bit 1. In this case, 5 zeros are followed by 1, which gives: 000001. After reading the prefix, the machine knows the length of the code for the number that follows and reads the string 10011 coding the number 21 itself. With this information, it can demarcate the end of the string i. The self-delimiting code for the string i, written as $\bar{i}$, then becomes:

$$\bar{i} = 00000110011101100001100100011100$$

The length of the code for the self-delimiting version of i, written as $l(\bar{i})$, is, in this way, limited by $n + 2\log n + 1$, where $n = |\ i\ |$. In our example case, we need $\lceil \log_2 21 \rceil = 5$ bits for the length of the number 10011 and the five zeros 00000 plus the demarcation bit 1, which gives 11 extra bits. We can now concatenate any other string, say $x = 01010100$ to $\bar{i}$ and still parse the string:

$$\bar{i} = 0000011001110110000110010001110001010100$$

into its constituents 101100001100100011100 and 01010100.

There are many other (and more efficient) ways to make strings self-delimiting, but the takeaway message is that we do not have to create more overhead than $2\log n + 1$, where n is the length of the original string. Making data sets self-delimiting is a relatively harmless operation in terms of descriptive complexity. We can now write:

$$U_j(\bar{i}x) = T_i(x) = y$$

That is, running the universal machine U_j on the input $\bar{i}x$ is equivalent to running machine T_i on input x, which computes y. U_j first reads the self-delimiting code for the index i and then emulates T_i on input x.

- The existence of universal Turing machines indicates that the set of Turing machines defines a *universal concept of computation*: any finite method of manipulating discrete symbols in sequential order can be emulated on Turing machines. For a discussion of the consequences, see Adriaans (2018).
- It is easy to prove that there are *uncomputable numbers*:
  - $T_i(x)$ is defined if i stops on input x in an accepting state after a finite number of steps.
  - We specify a recursive function g as: $g(x,y) = 1$ if $Tx(y)$ is defined, and 0 otherwise.
  - Since g is recursive, there will be a Turing machine r such that $T_r(y) = 1$ if $g(y,y) = 0$ and $T_r(y) = 0$ if $g(y,y) = 1$.
  - But then we have $T_r(r) = 1$ if $g(r,r) = 0$ and since $T_r(r)$ is defined $g(r,r) = 1$.
  - We have a paradox: ergo g(x,y) cannot be recursive.
- No computer program can decide for all other programs whether or not they will eventually stop in an accepting state. This implies that there is an uncomputable mathematical object, the so-called Halting Set, that codes for every combination of natural numbers i and x whether or not $T_i(x)$ halts in an accepting state.

We are now ready to define Kolmogorov complexity. The standard reference (Li and Vitányi, 2019) for the definitions concerning Kolmogorov complexity is followed. K is the prefix-free Kolmogorov complexity of a binary string. It is defined as follows:

Definition 1. $K(x|y) = \min_i \{l(\bar{i}) : U(\bar{i}y) = x\}$
that is, the shortest self-delimiting index of a Turing machine T_i that produces x on input y, where $i \in \{1,2,...\}$ and $y \in \{0,1\}^*$.

The actual Kolmogorov complexity of a string is defined as the one-part code that is the result of what one could call the forcing operation:

Definition 2 (Forcing operation). $K(x) = K(x \mid \varepsilon)$

Here all the compressed information, both model and noise, is forced onto the model part. Note that by defining complexity in this way, two-part optimization seems more natural than the one-part version. The forcing to one-part optimization is an essential step in the definition of Kolmogorov complexity as an asymptotic measure, since it allows the switch to another universal

Turing machine to be "hidden" in the code, for example, a transformation from $U(\bar{i}y) = x$ to $U(\overline{\bar{i}y\varepsilon}) = x$. In general, two-part code is more expressive, and thus more efficient, than one-part code. Observe, for example, that as a result of the forcing operation we have to make the whole input prefix-free.

Definition 3. *x_U^* is the shortest program that generates x on U. We have $U(x^*) = x$ and $K(x) = l(x^*)$.*

Observe that Kolmogorov complexity is not computable. Because of the Halting Set we cannot systematically run all programs with length $< l(x)$ to find x^*. We know, from a simple counting argument, that in the limit a dense class of these strings is incompressible. Incompressible strings have maximal entropy. In this sense, random strings are *typical.* A string is typical if it has no definable qualities that make it stand out from the rest of the crowd. That is, no program compresses it. A useful concept in this context is the randomness deficiency of a string (in bits):

Definition 4. *The randomness deficiency of a string x is*

$$\delta(x) = l(x) - K(x)$$

The randomness deficiency of a string is the amount of bits by which it can be compressed; that is, it is nonrandom. The concept can be extended to elements of sets:

Definition 5. *The randomness deficiency of an element of a finite set S is:*

$$\delta(x|S) = \log |S| - K(x|S)$$

The random deficiency $\delta(x|S)$ of an element x of a set S is a measure for the *typicality* of x in S. The expression $\log |S|$ specifies the length of an index we need to give to specify a typical element of S:

Example 4. Suppose that Amsterdam has 2^{20} inhabitants and 2^{10} lawyers. Let A be the set of Amsterdam's inhabitants, LA be the set of lawyers, and j the name of John Doe. Then the amount of bits I need to identify John knowing that John lives in Amsterdam is 20, which, for the sake of argument, is longer than the coding of the name "John Doe." The amount of bits we need to give to identify John, knowing that he is a lawyer who lives in Amsterdam, is 10, which again, for the sake of argument, is shorter than coding the name "John Doe." In short, John is not a typical inhabitant of Amsterdam, but he could be a typical lawyer. In that case, the set LA "explains" everything we need to know about him to identify him, even more than his name. *Mutatis mutandis*: $\delta\left(j|A\right) = 10$ and $\delta\left(j|LA\right) = 0$.

The basis for Kolmogorov complexity as a theory of asymptotic measurement is the *Invariance Theorem*. Since the result is vital for this chapter, we give the theorem with its proof:

Theorem 1. *The complexities assigned to a string x on the basis of two universal Turing machines U_i and U_j will never differ more than a constant:*

$$| K_{U_i}(x) - K_{U_j}(x) | \leq c_{U_i U_j} \tag{2.1}$$

Proof: Suppose we have two different universal Turing machines U_j and U_k. Since they are universal, they can both emulate the computation $T_i(x)$ of Turing machine T_i on input x:

$$U_j\left(\overline{T}_i^j x\right)$$
$$U_k\left(\overline{T}_i^k x\right)$$

Here $l(T_i)$ is the length of the code for T_i on U_j and $l(T_i)$ is the length of the code for T_i on U_k. Suppose $l\left(\overline{T}_i^j x\right) \ll l\left(\overline{T}_i^k x\right)$, that is, the code for T_i on U_k is much less efficient that on U_j. Observe that the code for U_j has constant length, that is, $l\left(\overline{U}_j^k\right) = c$. Since U_k is universal, we can compute:

$$U_k\left(\overline{U}_j^k \overline{T}_i^j x\right)$$

The length of the input for this computation is:

$$l\left(\overline{U}_j^k \overline{T}_i^j x\right) = c + l\left(\overline{T}_i^j x\right)$$

Consequently, the specification of the input for the computation $T_i(x)$ on the universal machine U_k never needs to be longer than a constant. □

The invariance theorem also holds for the randomness deficiency of a string:

Lemma 1. $| \delta_{U_i}(x) - \delta_{U_j}(x) | \leq c_{U_i U_j}$

Proof: $| \delta_{U_i}(x) - \delta_{U_j}(x) | = | (l(x) - K_{U_i}(x)) - (l(x) - K_{U_j}(x)) | = | K_{U_i}(x)) - K_{U_j}(x))$
$| \leq c_{U_i U_j}$ □

Note that the fact that the theory of computing is universal is of central importance to the proof of the invariance theorem: the proof deploys the fact that any machine can emulate any other machine. This proof works because emulations of universal Turing machines like $\overline{U_j}^k$ can be hidden in the one-part code that measures the Kolmogorov complexity. This "trick" is not open to us when we turn to two-part code optimization, as we will see in paragraph 2.3. A string is Kolmogorov random with reference to a universal Turing machine if there is no program that compresses it. Using randomness deficiency, we have the following definition:

Definition 6 (Kolmogorov randomness).

$$Random_U(x) \text{ iff } \delta_U(x) \leq c_r$$

Usually, the constant c_r is taken to be 0 in this definition, but $\log l(x)$ is better. Note that *Random* (...) is a meta-predicate.

Example 5. To illustrate the power and limitations of these proposals, suppose a demon runs a Kolmogorov oracle service. For a consult, you have to bring two binary objects: (1) your reference Turing machine and (2) your data set:

- Suppose you have done some experiment and you have a data set S of 10 Mb. As Turing machine of choice, you bring a C compiler of size 10 Mb. The demon is not interested in the circumstances under which you collected the data or the structure of it. He simply runs all possible C programs smaller than 10 Mb for an infinite amount of time and gives you the length $K_C(S)$ of the smallest program f that, when fed to the C compiler, produces your data set S. Note that the demon does not give you the program. You are only interested in a qualitative measure of complexity.
- Suppose you would have brought a Cobol compiler of 100 Mb. Cobol programs are more verbose. This will affect the demon's answer. In general, the smallest Cobol program that generates your data set S will be bigger than the comparable C program. $K_{Cobol}(S) > K_C(S)$. You complain about this to

the demon. His answer is: "I cannot help that. The only guarantee I can give you is that, when you come with bigger and bigger data sets, the difference in the limit between the C and the Cobol program that compresses the data will be constant." This is because, when the data set becomes so big that the difference between the C code and the Cobol code is bigger than the Cobol code needed to implement a C compiler, it becomes viable to do so. In that case, the smallest Cobol program that compresses your data set would be an implementation of a C compiler in Cobol followed by the C program that generates your data set. This trick is the essence of the invariance proof.

- This approach works well in the limit but creates problems for small data sets. Suppose you go to the demon with your C compiler and data set S of 5 Mb. The demon tells you the data can be compressed. The next day you bring the same data set S with a Cobol compiler as reference machine. Now the demon tells you the data set S is random. You complain: the same data set cannot be random one day and compressible the next. "Oh that is perfectly possible," the demon explains: "your C compiler is 10 Mb, your data set is 5 Mb. Your data set is not big enough to do the compiler emulation trick of the invariance proof, and Cobol is too inefficient to compress it.
- So the next day you do the test again with the same 5 Mb data set but now with a very small-reference universal Turing machine U that can be coded in 100 bits. Your guess is that, by selecting a very lean machine, the bias introduced by the selection of the reference machine will be minimized. The demon runs the test and tells you that your data set is random. You are surprised: how is it possible that a data set is compressible when we code in C and not compressible when we use a small machine? Again the demon expains: "the C compiler has a lot of predifined mathematical functions. It can call these functions by name, but your machine U has to provide them explicitly in the code for S. There is not enough room for that, since your data set is only 5 Mb."
- Finally you ask the demon: "since you know all the possible Turing machines, could you make the choice for an optimal reference machine for me?" "No problem," says the demon. "I would choose a universal Turing machine U that already contains your data set S and produces it when the first bit of the input is 1." In this way, your data set is compressed to 1 bit by U: $K_U(X) = 1$. This possibility is called the *Nickname Problem* (see Gell-Mann and Lloyd, 2003; Adriaans, 2012).

By stressing the fact that Kolmogorov complexity is an asymptotic theory of measurement, the research in computer science has largely ignored the problems described earlier in this chapter (Li and Vitányi, 2019). Unfortunately, for the application of the theory to real-life finite problems, the issues are problematic and relevant. In the second half of this chapter, we will present a formal theory that allows us to study these phenomena in great detail.

2.1 Two-Part Code Optimization

Let $\mathfrak{H}$ be a set of hypotheses and let x be a data set. By using Bayes' law, the optimal computational model under this distribution would be:

$$M_{map}(x) = argmax_{M \in \mathfrak{H}} \frac{P(M)P(x|M)}{P(x)} \tag{2.2}$$

This is equivalent to:

$$argmin_{M \in \mathfrak{H}} -\log P(M) - \log P(x|M) \tag{2.3}$$

Here $-\log P(M)$ can be interpreted as the length of the optimal *model code* in Shannon's sense and $-\log P(x|M)$ as the length of the optimal *data-to-model code*—that is, the data interpreted with help of the model. This insight is canonized in the so-called Minimum Description Length Principle.

Definition 7. *Minimum Description Length (MDL) Principle: The best theory to explain a data set is the one that minimizes the sum in bits of a description of the theory (model code) and of the data set encoded with the theory (the data-to-model code).*

MDL is a powerful tool for learning on the basis of positive examples.

Example 6. We give an example from a routine to learn Deterministic Finite Automata (DFA). A DFA is a simple grammar model that is represented as a directed graph. It has a set of states, including one start state and possibly a number of accepting states. The arrows between the states are associated with symbols from the language of the DFA. If one moves from one state to another, one writes the associated symbol. The class of DFAs is equivalent to the class of regular languages. Figure 2.2 gives two examples. Here we find $S+$ a finite set of example sentences of a language. We have two possible model DFAs, $L1$ and $L2$. $L1$ generates all finite sequences of the letters a, b and c, whereas $L2$ only generates the sentences in $S+$ and longer ones. Observe that the $L1$ has one state in which we have to choose between four options. $L2$ has three states, but we can only make a choice between two options in state 2.

S+ = {c, cab, cabab, cababab, cababababab}

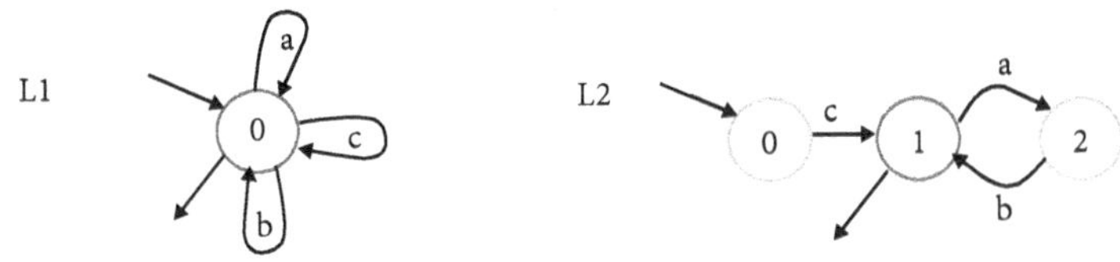

Coding in bits:
$|L1| \approx 5 \log_2 (3+1)\ 2 \log_2 (1+1) = 20$
$|L2| \approx 5 \log_2 (3+1)\ 2 \log_2 (1+3) = 40$

arrows · # letters · Empty letter · Outside world · # states

Figure 2.2 Selecting the best model for a finite language example using MDL. *S+* is a set of positive examples. Intuitively it is the language that consists of *c* followed by a repetition of the sequence *ab*. We have two possible DFA's (see text for a definition) that both generate the language: *L*1 and *L*2. *L*1 has one state (which is the start-state as well as the accepting state) and 5 arrows: it can be coded with 20 bits. *L*2 has 3 states (0 is the start-state and 1 the accepting state) and 5 arrows: it can be coded with 40 bits. Intuitively *L*2 is a better model, since it only generates the sentences from the sample *S+*. This is also the model MDL would choose. See text.

We can estimate the conditional probability of *S+* in terms of the amount of choices we have to make to generate it:

$$| L1(S+) | = 26\log_2 4 = 52$$
$$| L2(S+) | = 16\log_2 = 16$$

Comparing the MDL code for both models, we get:

$$| L2 | + | (L2(S+) | = 40 + 16 < | L1 | + |(L1(S+) | = 20 + 52$$

Here $| L1 |$ is the length of the *model code*, and $| L1(S+) |$ is the length of the *data-to-model code*: the number of bits we need to code *S+* given *L*1. So, following the MDL approach, we select *L*2 as the better model for *S+*.

Note that *L*1 generates a set of sentences that is much richer than the set generated by *L*2. One can say that *L*2 does a better job of explaining the structure of *S+* exactly *because* it is a smaller set. The intuition that arises is the following: the best theory that explains a set of examples is the simplest/smallest set *S* that still contains all of the examples. Exactly this intuition is captured by the

Kolmogorov structure function. Note that the class of DFAs is much smaller than the general class of recursive functions. If we extend MDL to this class, we enter the realm of Kolmogorov complexity. In this context, it is natural to interpret formula 2.3 as (see, e.g., Vitányi, 2006):

$$argmin_{M \in \mathfrak{H}} K(M) + K(x|M) \tag{2.4}$$

In this way, one obtains, within the limits of the asymptotic accuracy of Kolmogorov complexity, a theory of optimal model selection. A similar proposal that literally combines computational and stochastic elements is the structure function proposed by Kolmogorov in 1973 (Vereshchagin and Vitányi, 2004). The structure function specifies a two-part description of a data set: (1) the regular part—a predicate that describes a set of strings, measured in terms of its Kolmogorov complexity $K(S)$ and (2) an ad hoc part—the index of the data set in this set measured as the corresponding Hartley function log | S |.

$$h_x(\alpha) = \min_S \{\log|S| : x \in S, K(S) \leq \alpha\} \tag{2.5}$$

The structure function is a remarkable object, and it has taken a number of decennia to understand its workings. Observe that it facilitates a parameter sweep operation over the value of α. This value limits the Kolmogorov complexity of a set S that contains the data set x that we want to model. The objective is to minimize the cardinality and descriptive complexity of S. The smaller the set is, the smaller the length of index of x in S given by log | S | will be. The smaller the descriptive complexity of the model S, the more general the set S will be. The structure function allows us to look for *the simplest set with the lowest descriptive complexity for which x is typical.*

From the perspective of info-metrics, one can view the structure function as a tool for separating noise from a signal. The set S gives an optimal description of the data set S if it is typical with respect to the set, that is, x is just a random element of the set with no outstanding qualities. The implied methodology for model selection then would be to find a set of possible messages S that make the message x look random conditional to the set. In terms of info-metrics, one would say that we try to maximize the entropy of the data set, conditional to the description of the set of messages, while keeping this description as short as possible. In this way, we balance descriptive power with predictive power. Note that the model class we use here is the set of recursive functions. We therefore define a tool for model selection in the context of the most general concept of computation we know. To understand the power of this proposal, consider the following line of reasoning.

Example 7. Consider a Kolmogorov demon that operates a structure function oracle. Again we bring a data set, say D, and a reference universal Turing machine, U. We ask the demon: "what is the optimal model according to the structure function." The demon gives us set S. "How do you know that this is the best model," we ask. "I'll explain," the demon responds: "the set S is the smallest set with the smallest descriptive complexity that still contains x and minimizes the randomness deficiency $\delta(x|S)$." "So," I ask, "how can you say it is a good model? You know nothing of the conditions under which the data set D was sampled." "I don't need to," the demon answers; "the data set you gave me is compressible. That is a mathematical fact. The structure function simply makes the underlying structure that explains the compression visible in terms of the set S. This is also a mathematical fact. I cannot guarantee you that your model will generalize beyond your data set in the experiments that you are performing, but I can prove to you that, given your choice of reference Turing machine U, the set S is a best model: it gives an uncomputable but mathematically optimal generalization of x given U."

Along these lines, Vereshchagin and Vitányi (2004) prove a very strong result. The model selected by the structure function is optimal: "We show that the structure function determines all stochastic properties of the data: for every constrained model class it determines the individual best fitting model in the class irrespective of whether the true model is in the model class considered or not." (p. 3265) The basis for the optimality claim is the fact that the set S defines an algorithmic sufficient statistic:

Definition 8. *The set S defines an algorithmic sufficient statistic for an object x if:*

- *S is a* model; *that is, we have* $K(x|S) = \log|S| + O(1)$, *which gives a constant* randomness deficiency $\log|S| - K(x|S) = O(1)$. *This implies that x is typical in S; that is, S defines all statistical relevant aspects of x.*
- *We have* information symmetry $K(x,S) = K(x) + O(1)$, that is, *S should not be too rich in information but should specifiy just enough to identify x.*

This makes the structure function prima facie a sort of holy grail of learning theory. It suggests that compressible data sets have an inherent "meaning," that is, the optimal predicate that describes their essential structure. It also gives us an interpretation of the notion of randomness: a data set x is random if and only if it has nothing but the singleton set $\{x\}$ as a model. Based on such observations, some authors have claimed Kolmogorov complexity to be a "universal solution to the problem of induction" (Solomonoff, 1997; Rathmanner and Hütter, 2011). Unfortunately, attempts to develop a theory of model selection and meaningful information based on these results have not been very successful (Gell-Mann and Lloyd, 2003; McAllister, 2003; Vitányi, 2006; Wolpert and Macready 2007; Adriaans, 2012; Bloem, de Rooij, and Adriaans, 2015). With hindsight, this does not come as a surprise because in the following paragraphs we will prove that these conclusions are wrong:

- Many different models could be selected as optimal by two-part code optimization.
- Two-part code optimization selects models that have no useful stochastic interpretation, Consequently:
- There is no pure mathematical foundation for two-part code optimization, althoughfrom an empirical point of view it is a very useful methodology.

2.2 Reasons to Doubt the Validity of Two-Part Code Optimization

The MDL principle is often referred to as a modern version of Occam's Razor (Spade and Panaccio, 2016), although, in its original form, Occam's Razor is an ontological principle and has little to do with data compression. In many cases, MDL is a valid heuristic tool, and the mathematical properties of the theory have been studied extensively (Grünwald, 2007, p. 31). Still, MDL, Occam's Razor, and two-part code optimization have been the subject of considerable debate in the past decade (e.g., McAllister, 2003). MDL defines the concept of "the best theory," but it does not explain why the definition should work.

Domingos has argued against the validity of Occam's Razor in machine learning: he maintains that the smallest model is not necessarily the best (predictive, comprehensible; (Domingos, 1998). Domingos explicitly targets MDL in his analysis: "Information theory, whose goal is the efficient use of a transmission channel, has no direct bearing on KDD (Knowledge Discovery in Databases), whose goal is to infer predictive and comprehensible models from data. Having assigned a prior probability to each model in the space under consideration, we can always recode all the models such that the most probable ones are represented by the shortest bit strings. However, this does not make them more predictive, and is unlikely to make them more comprehensible." Domingos makes a valid point about comprehensibility, which is not an ambition of two-part code optimization.

Domingos's objections to the selection of short models, however, are misguided for various reasons: (1) two-part code does not select the shortest model but a model that optimizes the balance between a compressed model of the theory and the data coded, given the theory, that is, it selects a theory that is short but not too short; and (2) the probability assigned to the model under

Kolmogorov theory is related to the length of its optimal code and as such it is an objective asymptotic measure, not some ad hoc choice. Thus, the claim for a general theory of induction that some authors have made (Solomonoff, 1997; Rathmanner and Hütter, 2011) is stronger than the objections suggest.

At the same time, the idea behind two-part code optimization as a method for selecting the best model also seems to be at variance with the no free lunch theorems (Wolpert and Macready, 2005): "[*A*]*ny two optimization algorithms are equivalent when their performance is averaged across all possible problems.*"

A number of related issues that have been resolved elsewhere are as follows:

- What is the risk that we dismiss a set as random while in fact it is compressible? The answer is analyzed in depth in Bloem, Mota, de Rooij, Antunes, and Adriaans (2014): the probability that we mistakenly classify a data set as random, while it is in fact compressible, decays exponentially with the amount of bits the data set can be compressed. So under most distributions this risk is negligible. We will not treat this issue further in this chapter, but note that it illustrates the fact that the much feared noncomputability of the Kolmogorov complexity, in the context of practical application of learning algorithms, is not a big issue. Most data sets that are compressible are trivially compressible.
- Can we use an incremental compression process to approximate an optimal model? The answer in the context of information theory is no, although in practice many successful learning algorithms work this way. In Adriaans and Vitányi, 2009), it is proved that smaller models do not necessarily lead to smaller randomness deficiency. Many data sets can be compressed to local minima that have little or no predictive value.
- Adriaans (2012) has attempted to formulate a definitive theory of meaningful information based on data compression. Here we used, with hindsight, the erroneous idea that separating a data set into a structural and ad hoc part forces the introduction of additional code with a length in the order of the logarithm of the length of the whole data set. This can be compared to the additional code needed for making a string self-delimiting, as discussed earlier. This extra code would block the introduction of "vacuous" splits without any useful meaning, which would facilitate finding the right balance between predictive and descriptive power. A related attempt is presented in Antunes and Fortnow (2003). A flaw of this approach is that, for very large data sets (with a size that is exponential in the length of small Turing machines), universal Turing machines would be automatically selected as optimal models, which is counterintuitive. In the meantime, other doubts about the validity of this approach have been raised (Bloem, de Rooij, and Adriaans, 2015). In this chapter, we show that there is no hope

of eliminating vacuous splits without any valid stochastic interpretation for a large class of functions (see par. 2.4).

2.3 There Are Many Different Models That Could Be Selected as Optimal by Two-Part Code Optimization

One of the basic issues causing unclarity is the relative ad hoc character of the Kolmogorov structure function. It is based on a parameter sweep function, and it introduces the concepts of finite sets as models. These decisions make it hard to prove more general results about the function. In order to overcome these problems, we analyze a version of a balanced two-part code, not using Kolmogorov complexity, but expressed directly in terms of a two-part version of it. This idea was already introduced in Adriaans (2012). We get this version of Kolmogorov complexity when we skip the forcing operation. It allows us to analyze the concept of an optimal model in terms of a balanced two part code in more detail.

According to definition 1, the conditional complexity of x given y is: $K(x|y) = \min_i \{l(\bar{i}) : U(\bar{i}y) = x\}$. The final definition of prefix-free Kolmogorov complexity is reached by forcing the two-part code representation into a one-part version by defining: $K(x) = K(x \mid \varepsilon)$. An alternative route is to define a slight variant where the forcing operation is not required. We get a two-part variant Kolmogorov complexity:

Definition 9. $K_2(x) = \min_{i,y} \{l(\bar{i}y) : U(\bar{i}y) = x\}$

The expression $\bar{i}$ in $U(\bar{i}y) = x$ is the sef-delimiting code for the index i. It allows us to separate the concatenation $\bar{i}y$ into its constituent parts, i and y. Here i is the index of a Turing machine which can be seen as capturing the *regular* part of the string x, and y describes the input string for the machine, that is, the *irregular* part, such as errors in the model, noise, and other missing information. Based on this insight, one can define a balanced two-part code complexity. Figure 2.3 helps us understand the situation. It shows the various entities involved in two-part code optimization. We compare the "performance" of two different reference universal Turing machines U_1 and U_2:

- On the y-axis, we measure the structural or program parts of the information.
- On the x-axis, we map the ad hoc or irregular measurements.
- The counter diagonals mark points of equal compressibility.

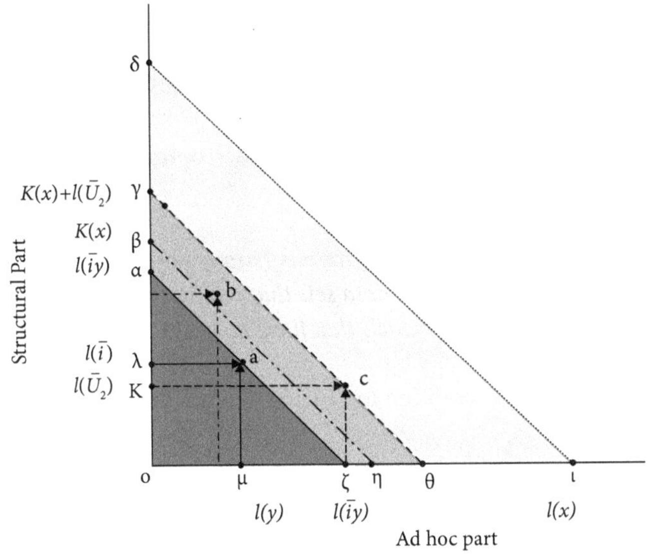

Figure 2.3 The balance between structural and ad hoc information in two part code optimization for a specific string x, given that the optimal conditional compression of x is defined by $U_1(\bar{i}y) = x$.

- The line $o\iota$ gives the length $l(x)$ of the original string x. No points correspond to two-part compression beyond the counter diagonal $\delta\iota$.
- The line $\alpha\zeta$ gives the length of the smallest two-part compression $\bar{i}y$ with $l(\bar{i}y)$. The actual optimal compression is defined by point a. Here the distance $o\lambda$ is the length of the structural part $l(\bar{i})$, and the distance $o\mu$ is the length of the nonstructural ad hoc input y. There are no points that correspond to two-part compression below the counter diagonal $\alpha\zeta$,but there might be many optimal solutions on this line.
- In general, two-part code will be more expressive and consequently more efficient than one-part code, so the counter diagonal $\beta\eta$, corresponding to $K(x)$, lies somewhat further from the origin.
- The distance $o\gamma$ specifies the length of the original Kolmogorov complexity of x + the extra code for a potentially more efficient Turing machine U_2, as specified by the construction used in the invariance proof. It is the constant $l(U_2)$ that in the limit makes Kolmogorov complexity invariant, between the Turing machines U_1 and U_2. This defines the area $\alpha\gamma\theta\zeta$ as a domain of (asymptotic) uncertainty.
- Point a identifies the optimal two-part code compression of x. Point b is a competing model with a somewhat less efficient compression that is still

in the asymptotic uncertainty domain between models U_1 and U_2. There may be many such competing models for which we cannot decide on the goodness of fit within the accuracy of our measurement system.

This overview shows that two-part code optimization is essentially polysemantic:

Definition 10. *A computational system is (strongly) polysemantic under two-part code optimization if it has data sets that within the variance of a constant c have at least two optimal models that have no mutual information.*

Polysemy is a natural condition for Turing equivalent systems:

Theorem 2. *The class of Turing machines is polysemantic under two-part code optimization.*

Proof: This is a direct consequence of the invariance theorem itself. Suppose $U1$ and $U2$ are two universal Turing machines and x is a data set. We select $U1$ and define the standard complexity: $K_{U1}(x|y) = \min_i \{l(i) : U1(\bar{i}y) = x\}$ with $K_{U1}(x) = K_{U1}(x \mid \varepsilon)$. Suppose we interpret $\bar{i}y$ as a two-part code with i the index of a machine and y its input. We can also select $U2$ as our reference machine. We get: $K_{U2}(x|y) = \min_i \{l(\bar{j}) : U2(\bar{j}z) = x\}$ and $K_{U2}(x) = K_{U2}(x \mid \varepsilon)$. We know that $K_{U1}(x) = K_{U2}(x) + O(1)$ because we can always emulate $U1$ on $U2$ and vice versa. This also holds for our two-part code theory: $U1(\overline{U2}\,\bar{j}z) = x$ and $U2(\overline{U1}\,\bar{i}y) = x$. Suppose that $U2$ is much more efficient for generating x than $U1$. Then the conditions for the invariance theorem are there, and we'll have $\overline{U2}\,\bar{j}z = \bar{i}y$, that is, $\overline{U2} = \bar{i}$ and $\bar{j}z = y$. Under two-part code optimization, we choose $\overline{U2}$ as the model for x. If we now change our choice of reference machine to $U2$, we will choose by $U2(\bar{j}z) = x$ the program $\bar{j}$ as our model. Since $\overline{U2}\,\bar{j}z$ is optimal and thus incompressible, $\overline{U2}$ and $\bar{j}$ have no mutual information. □

Lemma 2. *In the class of Turing equivalent systems, the size of the optimal models selected by means of two-part code optimization is not invariant.*

Proof: immediate consequence of theorem 2. Selecting another reference Turing machine implies the emergence of a new optimal model that has no mutual information with the initial optimal model. This includes information about its size. Theorem 2 shows that such mutually exclusive models indeed exist. □

2.4 Two-Part Code Optimization Selects Models That Have no Useful Stochastic Interpretation

One might be tempted to argue that polysemy is caused by the extreme power of Turing equivalent systems and that one could control the effects by selecting weaker models of computation. It seems, however, that polysemy is a fairly natural condition. A restriction to total functions does not change the situation. Because two-part code is more expressive than one-part code, there are many trivial total functions that split up a data set in two parts that cannot be interpreted in any way as specifying some structural or ad hoc model. A good example is the so-called Cantor packing function:

$$\pi(x,y) = \frac{1}{2}(x+y+1)(x+y)+y \tag{2.6}$$

The formula defines a two-way polynomial time computable bijection between $\mathbb{N}$ and $\mathbb{N}^2$. A useful concept in this context is information efficiency. We use the shorthand $f(\overline{x})$ for $f(x_1, x_2, \dots, x_k)$. We consider functions on the natural numbers. If we measure the amount of information in a number n as:

$$I(n) = \log n$$

then we can measure the information effect of applying function f to n as:

$$I(f(n)) = \log f(n)$$

This allows us to estimate the *information efficiency* as:

$$\Delta(f(n)) = I(f(n)) - I(n))$$

More formally:

Definition 11 (Information Efficiency of a Function). *Let $f : \mathbb{N}^k \to \mathbb{N}$ be a function of k variables. We have:*

- *the* input information $I(\overline{x})$ *and*
- *the* output information $I(f(\overline{x}))$.
- *The information efficiency of the expression $f(\overline{x})$ is*

$$\Delta(f(\overline{x})) = I(f(\overline{x})) - I(\overline{x})$$

- *A function f is* information conserving *if* $\Delta(f(\overline{x})) = 0$. That is, *it contains exactly the amount of information in its input parameters.*
- *It is* information discarding *if* $\Delta(f(\overline{x})) < 0$ *and*
- *it has* constant information *if* $\Delta(f(\overline{x})) = 0$.
- *It is* information expanding *if* $\Delta(f(\overline{x})) > 0$.

We can measure the information efficiency of such functions in terms of the balance between the information in the input and the output:

$$\Delta(\pi(x,y)) = \log_2 \pi(x,y) - \log_2 x - \log_y \tag{2.7}$$

One can prove that splitting up numbers (and consequently data sets, since they can be expressed as numbers) for some functions *always* gives a data gain of at least 1 bit. For the majority of the points in the space $\mathbb{N}^2$, the function π has an information efficiency close to 1 bit. On every line through the origin $y = hx(h > 0)$, the information efficiency in the limit is constant positive:

$$\lim_{x\to\infty} \Delta(\pi(x,hx)) = \tag{2.8}$$

$$\lim_{x\to\infty} \log\left(\frac{1}{2}(x+hx+1)(x+hx)+hx\right) - \log x - \log hx =$$

$$\log\left(1/2(h+1)^2\right) - \log h$$

This observation is illustrated in Figure 2.4, which shows the smooth behavior of the information efficiency of the Cantor function. For an in-depth discussion, see Adriaans (2019).

As an example, take any random number, say 1491212. Since the Cantor function is a bijection, this number can be expressed as a combination of two numbers, in this case (811,915):

$$\pi(811,915) = 1/2((915+811+1)(915+811)) + 811 = 1491212$$

When we compute the information efficiency, we get:

$$\Delta(\pi(811,915)) = \log_2 1491212 - \log_2 811 - \log_2 915 \approx 1$$

In binary numbers: 1491212 becomes 101101100000100001100 (with length 21), 811 becomes 1100101011 (with length 10), and 915 is written as 1110010011 (with length 10). So splitting the numbers 1491212 into 811 and 915 gives us an information gain of one bit. Of course, this bit is not really lost, but it reflects the fact that π is not a commutative operation: $\pi(x,y) \neq \pi(y,x)$. The two ordered

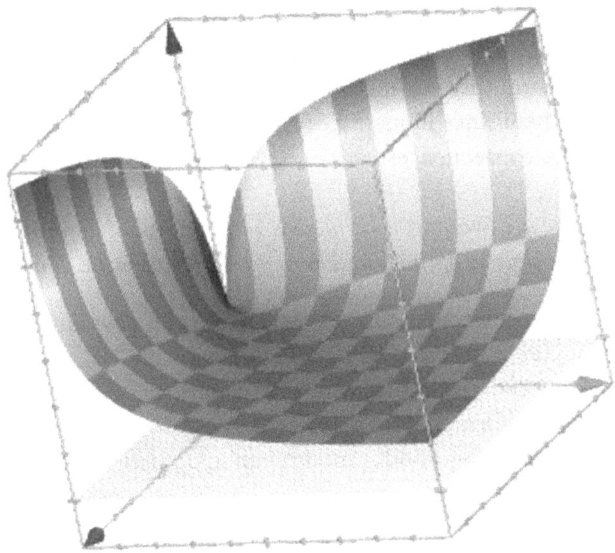

Figure 2.4 The information efficiency of the Cantor packing function $\Delta(\pi(x,y))$, $0 < x < 10^9, 0 < y < 10^9, -1 < z < 7$. The shaded area is the $z = 0$ surface.

pairs of arguments, (x,y) and (y,x), are transformed into one set, x,y. This generates a loss of information of 1 bit. When we select an order of the arguments to form either $\pi(x,y)$ or $\pi(y,x)$ we create this bit of information again.

One consequence is that we can define universal Turing machines from which the whole machinery of prefix-free code with its $1 + 2\log n$ penalty for making strings self-delimiting can be removed. We simply fit our machine with the Cantor function. Let U_C be such a machine. We get:

$$U_C(101101100000100001100) = T_{811}(1110010011)$$

That is, when read in the binary code for 1491212 the machine will split this in the binary codes for 811 and 915. Compare this with our standard U_1 machine that takes the prefix 000011010 to make the separation, which leads to an input that is 8 bits longer:

$$U_1(00001101011001010111110010011) = T_{811}(1110010011)$$

The takeaway message is that there exist information efficient mathematical routines that split data sets into two parts. Consequently, the idea that some natural penalty of a log factor is involved with splitting data sets and that this penalty could be used to separate noise from data in an MDL context is wrong.

It is an illusion that is partly caused by the tunnel vision of concentrating on concatenation and the splitting of strings. Mathematics has much more efficient functions to perform these operations, and when necessary Kolmogorov complexity will use them since it quantifies all possible programs. For our research program, this has consequences:

Lemma 3. *There are two-part code models without mutual information that are asymptotically close to optimal and have no valid stochastic interpretation.*

Consider the Cantor function $\pi(a,b) = c$. Define two new functions $f_{\pi,a}(x) = \pi(a,x)$ and $f_{\pi,b}(x) = \pi(x,b)$. Since π is defined on $\mathbb{N}^2$, we have a dense set of pairs (a,b) for which $K(a|b) = K(a)$ and $K(b|a) = K(b)$; that is, they have no mutual information. We have $K(f_{\pi,b}) = K(b) + O(1) K(f_{\pi,a}) = K(a) + O(1)$. Yet, they are both "models" of c: $f_{\pi,b}(a) = c$ and $f_{\pi,a}(b) = c$. We have:

$$U\left(\overline{f_{\pi,b}}a\right) = c$$

where $\overline{f_{\pi,b}}$ is the structural part and a the ad hoc part, and

$$U\left(\overline{f_{\pi,a}}b\right) = c$$

where $\overline{f_{\pi,a}}$ is the structural part and b the ad hoc part. Since we can perform this operation for any set of pairs of natural numbers (x,y) with $\pi(x,y) = z$ the separation of z into x and y has no useful stochastic interpretation.

Now consider a data set c. Suppose c^* is the smallest program that generates c on our reference machine U. We have $K(c) = l(c^*)$. Compute $c^* = \pi(a,b)$. We have $\overline{f_{\pi,a}}$ and $\overline{f_{\pi,b}}$ as vacuous models for c and we have $l(a) + l(b) < l\left(\overline{f_{\pi,b}}a\right) < l(c*) + O(1)$. □

We can always find two-part "models" close to the Kolmogorov complexity of any data set. Observe that many other recursive functions compute such efficient data splits. A restriction to total functions is not sufficient to rule out polysemy and vacuous models. In many cases, two-part code is more expressive or as expressive as its one-part variant. This explains how the results of (Vereshchagin and Vitányi, 2004) concerning the "optimality" of the Kolmogorov structure function can be reconciled with the no free lunch-like theorems in for example, Wolpert and Macready (2005). The proofs in Vereshchagin and Vitányi (2004) are not wrong from a formal point of view, but their interpretation is incorrect. The models selected by the Kolmogorov structure function have minimal randomness deficiency, but

1. there might be many such models,
2. they may have little or no mutual information, and
3. although formally their randomness deficiency is minimal, in many cases they have no valid statistical interpretation.

Polysemy is already implied by the notion of invariance, which is central to Kolmogorov complexity: Several different algorithms may give an optimal compression, but the length of these programs always varies within a constant.

The concept of polysemy is important in many domains besides machine learning. It plays a role in empirical sciences, it is the basis for art as imitation, and it is an important feature of human cognition. In psychology, the concept of polysemy is known as a Gestalt switch: a shift in interpretation of a data set where structural data and ad hoc data switch roles.

Example 8. Consider the image Wittgenstein used in his Philosophical Investigations of the rabbit that, when turned 90 degrees, appears to be a duck (see Figure 2.5). One could say that, from a cognitive point of view, the predicate "rabbit" is an optimal description of the image, as well as "duck," but that the description as an "image of a rabbit that appears to be a duck when turned 90 degrees" is too rich and does not help to compress the image as well as the shorter descriptions. In this case, a description that gives all the structural aspects of the image is not the best description to compress it. The image really has two mutually exclusive meanings, although the meanings have mutual elements, for example, the eye. The ad hoc arrangement of the rabbit's ears is essential for the structural description as a duck's beak, while the ad hoc irregularities at the back of the duck's head are vital for its interpretation as a rabbit's mouth. Obviously, since the models have restricted

Figure 2.5 Wittgenstein's Duck-Rabbit. The image can be interpreted in two mutually exclusive ways.

mutual information, their capacity for generalization (duck vs. rabbit) and prediction (flying vs. jumping) varies. Moreover, if a data set has at least two optimal models with no or little mutual information, it will also have a spectrum of related, nearly optimal models that share information with both of them. All of these models will have different capacities for generalization. This issue is true everywhere (linguistics, info-metrics, etc.) whenever one has to provide an interpretation of information.

2.5 Empirical Justification for Two-Part Code Optimization

Data compression is an important heuristic principle: if a data set is not compressible, there are no regularities to be observed and thus no predictive models can be constructed. It is a necessary but not a sufficient condition for selecting good models. This does not imply that every compression leads to useful models. Two-part code optimization does tell us that when we are confronted with two different theories that explain a data set, we should choose the one that compresses the data set better. Still, even in this weak reading, it is prima facie not clear why this should be a good directive.

A justification for two-part code optimization emerges if we study it in the context of empirical science. The empirical method forces us to generalize the notion of meaning in two dimensions: (1) invariance over observers and (2) invariance over time. There is a connection between these notions of invariance and data compression. If one collects various observations over time of the same phenomenon by various observers in a data set, general invariant patterns will emerge that can be separated from all kinds of ad hoc aspects of the observations. The invariant patterns will allow us to compress the data set. In this conception of empirical science, there is a real relation between compression of data and the construction of useful models. Note that the fact that generalization over observers in time gives us reliable models is itself empirical. There is no a priori mathematical motivation for it:

Observation 1. *The justification for two-part code optimization is ultimately empirical and not mathematical.*

With these observations in mind, we can adapt definition 2.1 of MDL to one that better fits the empirical aspects.

Definition 12. *Empirical Minimum Description Length (EMDL) Principle.*
If we select a database on a sufficient number of descriptions of observations of the same phenomenon under different circumstances, then, with high probability, the theory that minimizes the sum in bits of a description of the

theory (model code) and of the data set encoded with the theory (the data-to-model code) provides the best description of the invariant aspects of the phenomenon and thus gives the best generalized model with the best predictive capacities.

MDL works because the empirical method world works in the world. The empirical method itself is designed to separate ad hoc phenomena from structural ones. Applied to the Wittgenstein gestalt switch example, the polysemy would disappear if we interpreted the picture in the context of a series of observations of ducks or rabbits.

In the rest of this chapter, we investigate the consequences of these insights. In a world in which compression does not necessarily give a better model, the context of such a compression operation becomes important. In order to cope with this situation, we develop a computational theory of semantics in which agents have access to various sets of Turing machines that can be seen as possible worlds in terms of a modal logic.

3. Meaning as Computation

Modern research into the notion of meaningful information starts with Frege (1879). He saw that, even in an abstract discipline like mathematics, the meaning of a description of an object should be distinguished from the object itself. As an example consider a concept such as "[t]he first number that violates the Goldbach conjecture," where Goldbach's conjecture states that every even integer greater than 2 can be expressed as the sum of two primes (see Adriaans, 2018, for a discussion). If this number exists, we have an effective (but very inefficient) procedure to verify its existence: enumerate all the even numbers and all primes and check whether there is a number that cannot be written as the sum of two primes. If this number does not exist, clearly the phrase still has a definite meaning for us. This meaning cannot be a number because we are not sure it exists, so the meaning must actually be specified as the specific computation that would allow us to find the number. Yet if this number exists, it has a very special meaning for us.

Frege uses the terms *extension* and *intension* in this context. The computations "1 + 4" and "2 + 3" have the same extension "5," but they describe different intensions. There are contexts in which such a distinction is necessary. Consider the sentence "John knows how to prove that $\log_2 2^2 = 2$." Clearly, the fact that $\log_2 2^2$ represents a specific computation is relevant here. The sentence "John knows how to prove that 2 = 2" has a different meaning. We call the idea that different computations form different meanings of numbers a *Fregean theory of meaning*. This intuition is related to the extraction of computational models from data sets described in our introduction. In this sense, one mathematical object can have an infinity of different meanings, and by necessity only a finite fraction of these possibly compress the object.

A Fregean theory of meaningful information allows us to unify philosophical approaches to the analysis of meaningful information (Bar-Hillel; 1964; Floridi, 2011) with the computational approach to complexity theory. In the context of such a theory of computational semantics, a number of philosophical issues have a clear interpretation. We refer to Adriaans's paper (2018) for an extensive discussion. We shortly discuss the following two instances:

1. Floridi's criterion of truthfulness states that untrue sentences do not contain any information (2011). The association with computation allows us to interpret this issue in a specific way. The phrase $f(x)$ specifies a meaning, but until we know that the computation actually stops, we cannot say that it gives us information. Given the halting problem, there is no way for us to decide this question without running the program $f(x)$. We can give an upper bound for the amount of semantic information we would get if the program were to halt; this is $K(f(x))$. So we can say the expression $f(x) = y$ has an objective meaning $f(x)$ but that it only contains semantic information when it is true. It is easy to write a simple program that will find *the first number that violates the Goldbach conjecture*, but when the conjecture is true, such a program will never stop. In this case, the expression has a meaning but no denotation; that is, we have a program that explains what we are looking for, even if there is nothing to find.
2. An infinite number of different computations may generate the same data set. Consequently, a data set has an inifinite amount of different meanings. Kolmogorov complexity asymptotically measures the amount of meaningful information in a data set by measuring the length of the shortest program (i.e., meaning) that generates the set. Since various programs may give an optimal compression of the data set, there is not one unique intrinsic meaning for a data set. Still we can measure the amount of information in a meaning $f(x)$ as $K(f(x))$, even if the program has no accepting computation, that is no denotation. Dunn and Golan also discuss the notion of false and untrue information in Chapter 1, this volume (see also Golan, 2018).

Observation 2. *In a Fregean theory of meaning, we can objectively quantify the amount of meaning in a data set, even if the meaningful information is not truthful, but we cannot specify one intrinsic meaning for a data set.*

In essence, the conception of meaning is polyvalent, complex, and subjective. A poem or a painting, a fragment of a song or even a blob of paint on the wall, may be meaningful for one person and not for another. In its crude form, the concept of meaning seems to be purely subjective: any object can have any meaning for any subject. A Fregean theory of meaningful information is at variance with the monolithic platonic interpretation of meaning. Meanings are complex. Take Frege's famous example of the unique descriptions of the planet Venus as "the morning star" and "the evening star." At least three dimensions seem to be in play:

1. The object: the planet Venus itself.
2. The name "Venus" that refers to this object.
3. Unique descriptions of this object such as "The morning star."

Intuitively, the planet Venus is a more meaningful object than the average pebble on the beach. One way to measure the notion of meaningfulness would be to analyze the complexity of the set of all possible unique descriptions of Venus, but in principle this set is infinite (e.g., consider the unique description "The star that was visible in London but not noticed by John on Friday the 13th at 8 o'clock"). The meaning of an object is not intrinsic to that object.

In this context, a simple object might have more meanings than the actual information it contains, and compression does not help us to reconstruct it. As we have seen, this changes as soon as we collect observations in a systematic way. From a series of observations, we can in some cases construct a real general model. This aspect of model selection is not yet addressed by the theory as we presented it. We get a version of the two-part code when we interpret the program f in the expression $f(x) = y$ as a predicate or model for y. The rationale behind this proposal is that in many cases, when $f(x) = y$ is true, the program f contains information about the data set y.

Note that, in many cases, f cannot be said to be a model of y; for example, consider $y = x$ and f is the print program. After execution of the program, we know that it is true that y has been printed, which is semantic information, but does not tell us anything about the content of y. At the other extreme, suppose that f is a program that generates Fibonacci numbers and x is an index. Then we can say that f really is a model of y as the xth Fibonacci number.

4. From a Single Computing Agent to Communities of Interactive Learning Agents

In the beginning of this chapter, we presented Turing's idea of a computing machine as an abstract model of a real-life human computer. We showed how Solomonoff used this concept and others to define the notion of a learning agent and develop a general theory of induction. In the previous paragraphs, we analyzed many shortcomings of these proposals. We believe many of these problems can be removed by a more general notion of learning agents. In this part of the chapter, we invite the reader to study not one isolated computing agent, but communities of agents that can cooperate and learn from each other. Our single human computer is working in a whole office of workers that together try to solve a set of problems. Extensions we will consider include the following:

- The agents have a naming function. They can call a specific Turing machine by its name. This is in fact the upgrade of the nickname issue from a bug to a useful feature.
- The agents have memory. They can access past results in a database under easy accessible keys that act as names.
- They have, under various levels of transparency, access to each other's database and computational routines.

The access relation is modeled in terms of a theory of Turing machines as possible worlds and Turing frames as a tool for measuring the amount of semantic information in a data set. As a result, we can measure the amount of informativeness of a certain predicate for an individual agent, but we can not specify an intrinsic universal meaning. Consequently, we defend the view that the class of admissible reference machines should always be specified a priori since it plays an important role in the boundary estimates of model complexities.

For learning agents that have access to the general class of universal Turing machines, model selection by means of compression is inherently unstable. In such a rich world, data sets have no intrinsic models. As soon as a different reference Turing machine is selected, a completely different model that might not even share mutual information with the original model might be selected as "optimal." On the other hand, the optimal models really do give an exhaustive description of the stochastic qualities of a data set from a certain perspective. Data that under one model is seen as ad hoc may be interpreted as structural in another context, and vice versa. This result blocks the development of any general theory of measurement of semantic information, such as Sophistication (Koppel, 1987, 1995), Meaningful Information (Vitányi, 2006), and Facticity (Adriaans, 2012). Such a measurement theory can be developed but only in the context of some restricted "reasonable" set of accessible computational models.

In order to get some grip on these intuitions, we introduce the abstract notion of a computational agent—an entity with some general computational facilities. It could be an abstract universal Turing machine, a human being, a robot, or an alien. We allow such a computational agent to draw upon past experiences in a very general way: it has access to a database with subroutines that it can call by a short index or a name. Such a database may contain mathematical functions (i.e., a routine to compute Fibonacci numbers) as well as cultural objects (i.e., a function that produces the theme of Bach's Musical Offering). On the basis of this simple setup, one can model a rich set of agent types (creative agents, learning agents, student–teacher relationships).

4.1 Semantics for Turing Machines

Here are some notational conventions:

- Fraktur capitals indicate sets:
  - $\mathfrak{T}$ is the set of Turing machines.
  - $\mathfrak{U}$ is the set of universal Turing machines.
  - $\mathfrak{M}$ is a meaning and a set of codes for Turing machines.
  - $\mathfrak{W}$ is a set of accessible universal machines.
  - $\mathfrak{P}$ is the set of informative predicates.
  - $\mathfrak{O}$ is the set of optimally informative predicates.
- Lowercase letters like x, y, and z will be used as variable names indicating strings; a, b, i, and j are indexes of variables, sets, machines, and so on. Other lowercase letters (or combinations) will be used in denote functions f, fp, v, s, or when no confusion is possible as names for propositions p, q, r.
- Uppercase is reserved for special objects: T is a Turing machine, U is a universal Turing machine, and P is a logical predicate.
- Computational meta-predicates are predicates that might be true of an object but are not computable, such as $Random(x)$, $\delta(x)$, the randomness deficiency of x and $K(x)$, and its Kolmogorov complexity.

Computational predicates allow us to reason about true statements associated with computations:

Definition 13 (Computational Predicates).

- *$P(x)$ is a proposition that can be true or false, it denotes the predicate P applied to variable x: x is P.*
- *Any Turing machine T_i has an associated computational predicate P_i with the same index with the following semantics:*

$$v(P_i(x)) \iff \exists(k)T_i(k) = x$$

Here v is the valuation function, $P_i(x)$ is a proposition.
- *The extension of P_i is $\{x|\exists(k)T_i(k) = x\}$, where k is an index of x in this set.*

In the following, we will refer to Turing machines as *Turing predicates* if we are predominantly interested in the qualities of an object they compute. This construction allows us to formulate true propositions that are associated with computations. For example, if T_i is a machine that generates Fibonacci numbers, we can say that $P_i(x)$ is true; that is, x is a Fibonacci number if there is some index k such that $T_i(k) = x$.

Note that (1), in general, the extension of such a predicate P_j is uncomputable; (2) this undecidability by itself has no consequences for proof theoretical results of the modal systems as long as we assume that the valuation function v is defined (as is the case in the theory of Kolmogorov complexity introduced below); (3) this definition generalizes over the set of all possible universal Turing machines; and (4) the restriction to one-place predicates operating on strings does not reduce the expressiveness. Any n-place predicate can be implemented via a Turing machine that separates its input-string into n parts via whatever technique we can think of (self-delimiting code, enumeration of sets, etc.).

4.2 Descriptions of Turing Machines

If x is a string, then $\overline{x}^U$ is the self-delimiting code for this string in the format of the universal Turing machine U. The type of self-delimiting code used is not relevant here, apart from the fact that the length of $\overline{x}^U$ is limited for practical purposes by $n + 2\log n + 1$, where $n = l(x)$. When no confusion is possible, we will write $\overline{x}$ for $\overline{x}^U$

T_i is an *abstract mathematical entity*; that is, the Turing machine with index i. $\overline{T_i}^U$ is a *description* of T_i for U, which means that a prefix-free program that emulates T_i on a universal Turing machine U. $\overline{T_i}^U$ can be processed directly on U. Suppose $T_i(y) = x$; then we have:

$$U\left(\overline{T_i}^U y\right) = T_i(y) = x \tag{2.9}$$

The emulation of machine T_i on input y by universal machine U performs the same computation as when T_i operates on y directly. It computes x. Equation 2.9 can be interpreted as a piece of semantic information in the context of Floridi's concept of semantic information (Floridi, 2011). We follow Adriaans (2018), where this issue is discussed in depth:

- The universal Turing machine U is a *context* in which the computation takes place. It can be interpreted as a *possible computational world* in a modal interpretation of computational semantics.

- The sequences of symbols T_i, x, and y are *well-formed data.*
- The sequence T_i is a self-delimiting description of a program, and as such it can be interpreted as a piece of *well-formed instructional data.*
- The sequence $T_i(x)$ is an *intension.*
- The sequence y is the corresponding *extension.* The expression $U\left(T_i{}^U y\right) = T_i(x) = y$ states that the result of the program $T_i(x)$ in world U is y. It is a true sentence and as such a piece of well-formed, meaningful, truthful data, that is, semantic information.

4.3 Names for Turing Machines

- $\hat{T}_i^U$ is a name for the Turing machine T_i, accessible from the universal Turing machine U. That is, a prefix-free code that is associated with the program $\overline{T}_i^U$ for the universal machine U. $\hat{T}_i^U$ can only be processed on U if it has access to a naming function.
- A naming function, n : *names* $\rightarrow$ *descriptions*, for a universal Turing machine U^n assigns a description to a name: $n\left(\hat{T}_i^U\right) = \overline{T}_i^U$. Naming functions are part of the definition of the universal machine and thus are always finite.
- U_j^n is the universal Turing machine with index j and naming function n. Turing functions with a naming function are a proper subset of the class of universal Turing machines. At the start of the computation, the prefix-free code on the tape will be checked against the naming function. If it is defined, the computation will run on the associated description; if not, the prefix-free code will be interpreted as a proper program.
- $U_j^{[T_i, T_j, \dots T_k]}$ is the universal Turing machine with index j and a specification of the naming function as an explicit list of Turing machines $\left[T_i, T_j, \dots T_k\right]$. The sequence is meaningful: If $U_j^{[T_i, T_j, \dots T_k]}$ then $\hat{T}_i^U < \hat{T}_j^U < \cdots < \hat{T}_k^U$ according to the association between strings and numbers defined above.
- $T(y) = x$ indicates an accepting computation: the Turing machine T produces output x given input y. The same holds for $U(y) = x$. The computation $T(x)$ does not necessarily have to end in an accepting state; it may be undecidable.
- $U\left(\hat{T}^U y\right) = U\left(\overline{T}^U y\right) = z$ denotes the universal Turing machine U emulating a Turing machine T computing z on input y. In $U\left(\hat{T}^U y\right)$, the machine is called by its name; in $U\left(\overline{T}^U y\right)$ it is called by its description. We'll write

$U(\hat{T}y) = U(\overline{T}y) = T(y) = z$, leaving the superscript U out if no confusion is possible.

- Universal Turing machines can emulate each other; that is, the emulation relation is symmetrical:

$$U_i\left(\overline{U_j}^{U_i}\overline{T}^{U_j}y\right) = U_j\left(\overline{U_i}^{U_j}\overline{T}^{U_i}y\right) = T(y) = z$$

The subscripts and superscripts are important in this notation. If U_j is a Java interpreter, then $\overline{T}^{U_j}$ is Java code for T; if U_i is a Prolog interpreter, then $\overline{T}^{U_i}$ is Prolog code for T. The computation $U_j\left(\overline{U_i}^{U_j}\overline{T}^{U_j}y\right)$ erroneously tries to run Java code on a Prolog interpreter. Emulations can be stacked indefinitly:

$$U_1\left(\overline{U_2}^{U_1}\overline{U_3}^{U_2}\dots\overline{U_n}^{U_{n-1}}\overline{T}^{U_n}y\right) = T(y) = z$$

Here U_i reads the self-delimiting code for $U_{i+1}^{U_i}$ and starts to emulate it.

The exact details of the implementation of a naming function are unimportant, but one could think of a finite *naming table* stored in the form of tuples $\langle name, description \rangle$ on a separate tape. One description may have various names. Note that (1) naming functions for universal machines cannot be reflexive or symmetrical, since that would lead to infinite regresses in the naming function: a name *a* for a description *b* containing a naming function *f* associating a name *a* to a description *b* containing a naming function *f*, etc.; and (2) once the naming function is defined, names can in principle be used by any other program on *U* during its execution. In the context of Kolmogorov complexity, this will happen automatically, when needed, because it searches over all possible programs. As soon as a short name for a binary object is defined, it will have consequences for the Kolmogorov complexity of a number of binary objects. The downside is that the naming function increases the complexity of our universal machine. Defining naming functions is finding a balance between computational power and descriptive complexity.

Note that we can reduce the Kolmogorov complexity of any string x to one bit by means of the naming function: just define $T_i(\varepsilon) = x$ and $T_i^{\,U}$ = "1," then $U(1\varepsilon) = x$. This shows that the naming function influences the Kolmogorov complexity of objects and the corresponding universal distribution. In general, we will have $l(\hat{T}_i) \ll l(\overline{T}_i)$. A learning agent, on the basis of his experience, can update his probability distribution over the world by specifying short names for long strings that occur frequently.

4.4 Turing Frames

Universal Turing machines can emulate all other Turing machines, but whether they always have access to the code of other machines is not clear. In this section, we define an abstract accessibility relation that regulates this availability. This relation can be reflexive and symmetric and thus is more abstract than the naming function described earlier.

- A Turing frame $\mathfrak{F} = \langle \mathfrak{U}_i, R\rangle$ is a tuple where $\mathfrak{U}_i$ is a set of universal Turing machines and $R : \mathfrak{U}_i \times \mathfrak{U}_i$ is an accessibility relation. The restrictions on R are called the frame conditions.
- $R : \mathfrak{U} \times \mathfrak{U}$ is an accessibility relation between universal Turing machines. Some possible constraints on the accessibility relation are:
  - reflexive iff $R(U_w, U_w)$, for every U_w in $\mathfrak{U}$
  - symmetric iff $R(U_w, U_u)$ implies $R(U_u, U_w)$, for all U_w and U_u in $\mathfrak{U}$
  - transitive iff $R(U_w, U_u)$ and $R\left(U_u, U_q\right)$ together imply $R\left(U_w, U_q\right)$, for all U_w, U_u, U_q in $\mathfrak{U}$.
  - serial iff, for each U_w in $\mathfrak{U}$ there is some U_u in $\mathfrak{U}$ such that $R(U_w, U_u)$.
  - Euclidean iff, for every U_u, U_t, and U_w, $R(U_w, U_u)$ and $R(U_w, U_t)$ implies $R(U_u, U_t)$ (Note that it also implies: $R(U_t, U_u)$.

This definition allows us to formulate relations between universal machines in terms of various systems of modal logic (Hughes and Cresswell, 1996). If there are no constraints on R, each machine just has its own database with a haphazard collection of accessible systems. The associated modal logic is K. If each machine has access to a database with the same set of machines, then R is an equivalence relation. The associated modal system is S5.

- A Transparent- or S5 Turing frame is associated with S5 modal logic; that is, the accessibility relation is reflexive, symmetric and transitive.
- $\mathfrak{W}_{U,\mathfrak{F}} = \{U_i | R(U, U_i)\}$ is the set of universal machines accessible from U according to R specified in $\mathfrak{F}$.
- A model is a 3-tuple $M = \langle \mathfrak{U}_i, R, v\rangle$.
- Here v is a valuation function. We recursively define the truth of a formula relative to a set of universal Turing machines $\mathfrak{U}_i$ and an accessibility relation R:
  - if $v\left(U_w, p\right)$ then $U_w \vDash p$
  - $U_w \vDash \neg p$ if and only if $U_w \nvDash p$
  - $U_w \vDash (p \wedge q)$ if and only if $U_w \vDash p$ and $U_w \vDash q$
  - $U_w \vDash \Box p$ if and only if for every element U_u of $\mathfrak{U}$, if $R(U_w, U_u)$ then $U_u \vDash p$

- $U_w \vDash \Diamond p$ if and only if for some element U_u of $\mathfrak{U}$, it holds that $R(U_w, U_u)$ and $U_u \vDash p$
- $\vDash p$ if and only if $U \vDash p$, where U is the reference universal machine.

Since we only discuss mathematical statements that are true in all worlds, the valuation function v will be the same for all models. We can abstract from individual models $M = \langle \mathfrak{U}_i, R, v \rangle$ and concentrate on the associated Turing frame $\mathfrak{F} = \langle \mathfrak{U}_i, R \rangle$.

4.5 Variance of Turing Frames

In the following paragraphs, we will only consider S5 Turing frames. The variance of a Turing frame is the largest distance between the different Kolmogorov complexities assigned to strings by the associated universal Turing machines in the frame:

Definition 14 (Variance of a frame). *The variance of a frame* $\mathfrak{F} = \langle \mathfrak{U}_i, R \rangle$ *is*

$$Var(\mathfrak{F}) = \min c_{U_x U_y} \left(\forall \left(U_x, U_y \in \mathfrak{U}_i \right) \forall \left(z \in \{0,1\}^* \right) |K_{U_x}(z) - K_{U_y}(z)| \leq c_{U_x U_y} \right)$$

There are frames with unbounded variance. Any infinite S5 Turing frame has unbounded variance:

Theorem 3. *If* $\mathfrak{U}$ *is a countable infinite set of Turing machines and* $\mathfrak{F} = \langle \mathfrak{U}, R \rangle$ *is an S5 Turing frame, that is, a frame defined on R that is symmetric, reflexive, and transitive, then* $Var(\mathfrak{F})$ *is unbounded.*

Proof: Suppose that $Var(\mathfrak{F}) = c$. Select a fixed universal machine U_i. For any other universal machine U_j we have:

$$U_i\left(\overline{U_j}^{U_i} \overline{T}^{U_j} \varnothing\right) = U_j\left(\overline{U_i}^{U_j} \overline{T}^{U_i} \varnothing\right) = T(\varnothing) = x$$

Since the length of the indexes for an infinite class of objects cannot be bounded and R is S5, there will be an infinite number of universal machines U_j with index $\overline{U_j}^{U_i}$ such that:

$$l\left(\overline{U_j}^{U_i} \overline{T}^{U_j}\right) > l\left(\overline{U_j}^{U_i}\right) > l\left(\overline{U_i}^{U_j} \overline{T}^{U_i}\right) + c$$

which gives:

$$| \, l\left(\overline{U_j}^{U_i}\overline{T}^{U_j}\right) - l\left(\overline{U_i}^{U_j}\overline{T}^{U_i}\right) \, | > c$$

and

$$| \, K_{U_i}(x|\varnothing) - K_{U_j}(x|\varnothing) \, | = | \, K_{U_i}(x) - K_{U_j}(x) \, | > c$$

So $Var(\mathfrak{F}) > c$.

In such a frame, the notions of Kolmogorov complexity and randomness deficiency are de facto undefined. We can always select a machine that compresses a string or that makes a random string compressible. The smaller the variation, the more exact our measures will be. By Solomonoff's invariance theorem, the variation is limited by the size of the Turing machines in the frames, so this is a more formal variant of the class of "natural" machines proposed by Hütter (2005) and Rathmanner and Hütter, 2011). To show how the choice of a universal Turing machine affects the notions of randomness, compressibility and variance, we give an elaborate example.

Example 9. Suppose we have a simple universal Turing machine U^n with a naming function n. $U^{[]}$, with an empty list, is the machine U with an empty naming table; and U^{max} is U with a large naming table that contains all known mathematical functions. $U^{[T_{fp}]}$ is the version of U that only has a name for the machine T_{fp} that generates Fibonacci primes. Consider the set of Fibonacci primes: Fibonacci numbers that are also prime. $FB(i)$ denotes the ith Fibonacci prime. It is not known whether or not this set is finite.[2] When we want to study this set with algorithmic complexity theory, the choice of a universal machine is not neutral:

- In the world $U^{[]}$, a possibly infinite number of Fibonacci primes will be compressible, but there will be quite a large initial segment of smaller Fibonacci primes that is still regarded as random (e.g., for sure the elements of the set: $\{2, 3, 5, 13, 89, 233, 1597\}$). Since $U^{[]}$ is minimal, the code for Fibonacci primes has to be stored explicitly in the program T_{fp} that recognizes Fibonacci primes. Thus, an example like $FP(11) = 2971215073$ will be regarded as incompressible since for $U\left(\overline{T_{fp}}11\right) = 2971215073$ we will have $l\left(\overline{T_{fp}}11\right) > l(2971215073)$.

[2] In September 2015, the largest known certain Fibonacci prime was F_{81839}, with 17103 digits. See https://en.wikipedia.org/wiki/Fibonacci_prime, retrieved October 9, 2015.

- When we shift to $U^{[T_{fp}]}$, the concept of Fibonacci primes can be called as a name $\widehat{T}_{fp}$ and thus smaller examples up to an initial segment will be compressible, for example, $U^{[T_{fp}]}\left(\hat{T}_{fp}11\right) = 2971215073$ and $l\left(\hat{T}_{fp}11\right) < l(2971215073)$.
- When we study invariance between $U^{[]}$ and $U^{[T_{fp}]}$, this advantage will vanish because of the length of the extra codes in the expression $U^{[T_{fp}]}\left(\overline{U^{[]}}\ \overline{T_{fp}}11\right) = U^{[]}\left(\overline{U^{[T_{fp}]}}\ \hat{T}_{fp}11\right) = 2971215073$. The Kolmogorov complexity measured with $U^{[T_{fp}]}$ in this frame gets a penalty of at least $l\left(U^{[]}\right)$ and 2971215073 again is a random number.
- If we select U^{max} as our reference machine, not much will change with regard to the set of Fibonacci primes that is compressible. However, the variance between $U^{[T_{fp}]}$ and U^{max} will be big because of the length of the code for U^{max} in the expression $U^{max}\left(\overline{U^{[T_{fp}]}}\ \hat{T}_{fp}11\right) = U^{[T_{fp}]}\left(\overline{U^{max}}\ \hat{T}_{fp}11\right) = 2971215073$. The Kolmogorov complexity measured with $U^{[T_{fp}]}$ in this frame gets a penalty of $l\left(\overline{U^{max}}\right)$. If the set of Fibonacci primes is indeed finite and $l\left(\overline{U^{max}}\right)$ is big, it might be the case that the whole set is not "visible" any more from the perspective of $U^{[T_{fp}]}$.

Since we do not know whether the set of Fibonacci primes is infinite, there might be choices of universal Turing machines for which the set of numbers that can be adequately modeled as Fibonacci primes is empty. A rich world with more information is certainly not always better than a poor world. In a world in wich we are only interested in Fibonacci primes, $U^{[T_{fp}]}$ is a better choice than U^{max} or $U^{[]}$. In the frame $\mathfrak{F} = \left\langle\left\{U^{[T_{fp}]}, U^{max}\right\}, R\right\rangle$, there might be no strings that are compressed by T_{fb} due to the large variance.

4.6 Meaningful Information

Here we investigate various sets of predicates relevant to the notion of meaning. The set of informative predicates relative to a world U_p and a threshold function $f: \mathfrak{T} \times \mathfrak{U} \rightarrow \mathbb{N}$ is:

Definition 15 (Turing predicates informative relative to a world).

$$\mathfrak{P}_{U_p}(x) = \left\{T_i | \exists(k)\left(U_p(T_i k) = x \wedge l\left(\overline{T_i} k\right) < l(x)\right)\right\}$$

The amount of information the associated predicate P_i of a Turing machine T_i carries about an object x relative to U_p is:

Definition 16 (amount of information in a predicate).

$$I_{U_p}(T_i,x) = \begin{cases} \left(l\left(T_i^{U_p}\right)\right. & \text{if } \exists(k)\left(U_p\left(\overline{T}_i k\right) = x \wedge l\left(\overline{T}_i k\right) < l(x)\right) \\ 0 & \text{otherwise.} \end{cases}$$

that is, the length of the index for T_i on U_p if it compresses x and 0 otherwise.

This notion of informativeness is not stable. In some cases, we can make a string x compressible by a predicate T_i simply by defining a short name T_i for it:

Theorem 4 (instability of informativeness). *There are combinations of strings x, Turing machines T_i, and universal Turing machines U_m and U_n such that $I_{U_m}(T_i,x) = 0$ and $I_{U_n}(T_i,x) > 0$.*

Proof: Choose T_i and x so that $\exists(k)\left(U_m\left(\overline{T}_i k\right) = x\right) \wedge (k \ll l(x))$ and $I_{U_m}(T_i,x) = 0$, that is, x is not compressible for U_m due to the length of T_i. Define $U_n = U_m^s$, a machine that is equal to U_m except for the naming function $s\left(\hat{T}_i\right) = \overline{T}_i$, where $\hat{T}_i \ll \overline{T}_i$ so that $U_m^s\left(\hat{T}_i k\right) = x$ and $l(k) + l\left(\hat{T}_i\right) < l(x)$. □

The optimally informative predicates are a subset of the informative sets.

Definition 17 (optimally informative Turing predicates relative to a world).

$$\mathfrak{O}_{U_p}(x) = \left\{T_i \mid \exists(k)\left(U_p\left(\overline{T}_i k\right) = x \wedge l\left(\overline{T}_i k\right) < l(x)\right) \wedge \right.$$
$$\left. \neg\left(\exists\left(T_j, m\right)\right\}\left(U_p\left(\overline{T}_j m\right) = x \wedge \left(\overline{T}_i k\right) > \left(\overline{T}_j m\right)\right)\right\}$$

The first condition $\exists(k)\left(U_p\left(\overline{T}_i k\right) = x \wedge l\left(\overline{T}_i k\right) < l(x)\right)$ ensures that $T_i \in \mathfrak{P}_{U_p}(x)$. The optimally informative predicates are a subset of the informative sets. The second condition $\neg\left(\exists\left(T_j, m\right)\left(U_p\left(\overline{T}_j m\right) = x \wedge \left(\overline{T}_i k\right) > \left(\overline{T}_j m\right)\right)\right)$ ensures that T_i gives maximal compression of x. This makes $\mathfrak{O}_{U_p}(x)$ super unstable, on top of the instability of $\mathfrak{P}_{U_p}(x)$:

Lemma 4 (super instability of optimal informativeness). *There are combinations of strings x, Turing machines T_i, and universal Turing machines U_m and U_n such that $\mathfrak{O}_{U_n}(x) = \{T_i\}$, whatever the content of $\mathfrak{O}_{U_m}(x)$.*

Sketch of Proof: Along the lines of the proof of theorem 4. As soon as we shift from U_m to a new world $U_n = U_m^s$ that is similar except for the fact that we have a naming function $s(T_i) = T_i$, that compresses the string x one bit better than the ones in the current set $\mathfrak{D}_{U_m}$ of optimally informative predicates, the whole set is replaced by $\mathfrak{D}_{U_n} = \{T_i\}$.

Super instability in this case implies that a small change in the definition of our universal machine makes the whole set of optimally informative predicates collapse onto one new predicate. For the set of informative predicates, only individual predicates jump out of the set, so they are more resilient.

The instability that is generated by selecting short names for objects is also known as the *Nickname Problem* (see Gell-Mann and Lloyd, 2003; Adriaans, 2012). From the perspective of theory of learning and methodology of science, such instabilities are not so much a bug of the theory but a feature. They explain the phenomenon of a *paradigm shift* in science: older models are replaced by more powerful ones, and by one stroke the previous models lose their explanatory power.

4.7 Types of Agents

Our definition of the meaningfulness of a data set is related to a specific agent. A data set A is meaningful for a computational agent B if B has access to a description C that compresses A. The *amount of informativeness of description C for agent B* is $l(A) - l(C)$, that is, the difference in length of the descriptions. In this context, a data set can be meaningful for an agent in several ways:

- *On the basis of density.* Suppose a data set is coded as a string and there are regions with significantly higher density of certain symbols than others. In such a case, the data set can be compressed by means of basic Shannon information theory. This notion of compressibility is essentially spatial and is related to comparable phenomena in thermodynamics. The appropriate measurement theory is Shannon information theory.

We illustrate this with an example. The basic notion of Shannon information is that of a sequence of messages. Observe the binary representations of the numbers 811 and 915, which are 1100101011 (with 6 one bits) and 1110010011 (with 6 one bits), respectively. Concatenation is a spatial operation, so when we concatenate the two strings, neither the local organization of the bits nor the total bit count changes. We get the string:

11001010111110010011

with 12 one bits. Consequently, the estimated Shannon entropy of the messages does not change. One could say that a bit string from the perspective of Shannon information is a one-dimensional discrete space in which bits can move freely without affecting the overall entropy when measured over the whole space. For an in-depth analysis and a discussion of some related issues in thermodynamics, see Adriaans, 2009, 2018).

- *On the basis of computation.* Suppose a data set does not show any obvious spatial regularities, but it can be described as a representation of a special mathematical object, say an expansion of the number π. In such a case, we can use this insight to compress the data set. Such a compression is inherently computational. Such data sets are in the context of thermodynamics and random processes so rare that we can safely ignore the possibility that they occur. In any practical situation, such data sets only are constructed by agents with a specific goal in mind (say, encryption). The appropriate measurement theory is Kolmogorov complexity.

To continue the example of the previous paragraph. Another way to compute a string that contains the information of the original strings 1100101011 and 1110010011 (and add one bit to code the order of the concatenation) is to compute $\pi(811, 915) = 1491212$, which is in binary code:

101101100000100001100

This string has only 8 one bits, and the local organization of the bits has changed completely. Shannon information cannot explain this phenomenon, but Kolmogorov complexity can.

- *On the basis of experience.* Suppose the agent has a data set stored in a database accessible by a short index or key. The agent is then able to produce the data set easily, and what is more he can compress any other data set that is similar. This is the way agents deal with cultural objects like pieces of music, novels, and paintings. In a world of computational agents, having easy access to large random data sets has value. For example, two agents with access to the same random string can exchange a message in an absolutely undecipherable code. The appropriate measurement theory is Turing frames.

This theory has the advantage that it allows an agent to discover the meaning of a string incrementally by identifying predicates that compress the set partially. This allows for incremental discovery of meaning and solves many of the problems described in Adriaans and Vitányi (2009). Meaning is essentially composite for example, the number 2971215073 is meaningful because it is prime *and* it is a Fibonacci number. The two different predicates are in most contexts more informative than a single one. This is especially because we can discover the fact that it is prime *independent* of the fact that it is a Fibonacci number. Another example is the Mona Lisa, which is a rich work of art because we can see many different meanings in it, not because it represents some monolithic Platonic idea of "Mona-Lisa-ness."

We specify various *application paradigms* of the theory.

- A *learning agent* that tries to detect the meaning of a string by searching for predicates that compress the string. As soon as it discovers such a predicate, it has to do two things: (1) check whether the predicate is intrinsic (i.e., holds in all relevant worlds) and (2) check whether it is not subsumed by predicates it already knows. If the agent is confronted with a sequence of data sets, it might hone its expectation by formulating names for strings that occur frequently.
- A *creative agent* tries to maximize the meaning in a data set. Such a *factic process* is by nature unstable. A creative agent manipulates data sets in such a way that the meaningfulness increases. That is, it will be compressed by an increasingly rich set of predicates that are mutually independent. Such a process might be subject to sudden phase transitions where the set collapses into randomness.
- *Mixed groups of these agents* can be defined so as to model various social, economical, and other processes (pupil–teacher, artist–public). A mixed group of creative and learning agents can develop what one could call a "culture": a set of expectations about (i.e., names for) the artifacts that agents produce. For agents that do not share this culture, a lot of artifacts will seem random because they do not have access to the naming functions that compress them.
- In this context, notions such as ethics, esthetics, trust, and theory of other minds can be studied. For a group of agents living in a certain environment, the accessibility relation will not be reflexive, symmetrical, or transitive. That is, they can observe their own and others' behavior, but they do not have access to the code to emulate themselves or others.

5. Discussion

The discovery of the algorithmic information theory or Kolmogorov complexity was partly motivated by Carnap's ambition (1945) to assign a proper a priori probability to an individual statement, given an infinite logical description of the world (Solomonoff, 1964, 1997). Solomonoff formulated the idea that the set of prefix-free input strings for a universal Turing machine would provide such a probability distribution (Solomonoff, 1964, 1997; Li and Vitányi, 2019). Carnap's concept was further developed by Kripke (1963) into what we now know as possible world semantics. Solomonoff's proposals, though hugely influential in computer science, never had a big impact on philosophical research. This chapter bridges part of the gap between modal logic and complexity theory. It describes a computational theory of meaning in which universal Turing machines are interpreted as descriptions of possible worlds and their associated universal distributions. This approach clarifies some of the inherent difficulties in the interpretation of Kolmogorov complexity, specifically the issue of the selection of a universal reference machine.

Many attempts have been made to define the meaning of a string in terms of the *most informative predicate*, the one that generates the highest compression, but according to lemma 4 it is impossible to prove any form of invariance for this notion. In fact, the very construction behind the invariance proof blocks an interpretation of the notion of a program as an invariant optimal model, since it shows that Turing machines can emulate each other at a constant cost. This implies that, in the context of the invariance proof, universal Turing machines would function as optimal models for arbitrary data sets; that is, the most informative predicate would be a universal machine, which destroys the intuition that these models have any meaningful stochastic interpretation.

This analysis also shows that meta-predicates such as "... is random," "... is informative" and "... is compressible" are contingent. For any string given a set of universal machines, there might be predicates that are necessarily informative in all worlds, possibly informative only in some worlds or necessarily uninformative in all worlds. Moreover, theorem 3 tells us that we have to select an a prioiri finite set of reference machines in a frame, even if we want to define classical Kolmogorov complexity. It is misleading to concentrate on small machines when defining Kolmogorov complexity. In principle, on one hand, a frame with small machines can still have a variance of the size of the biggest machine; on the other hand, we could define a frame rich with information about our domain with low variance. For most induction problems, a rich frame with low variance seems a better option.

The less you specify about your discourse, the more vague the notion of meaning becomes. If we specify nothing about the relation R, the corresponding frame is associated with modal system *K*. In such a frame, the notion of meaning is very unstable because any shift to a new world can generate a new set of accessible worlds. In principle, one could investigate the notion of meaning in the whole landscape of modal logics in this way, but that is not the goal of the present study.

Acknowledgments

I thank Amos Golan and the anonymous referees for many helpful comments on earlier versions. This research was partly supported by the Info-Metrics Institute of the American University in Washington, the Commit project of the Dutch science foundation NWO, the Netherlands eScience center, the IvI of the University of Amsterdam, and a Templeton Foundations Science and Significance of Complexity grant supporting the Atlas of Complexity Project.

References

Adriaans, P. (2001). "Learning Shallow Context-Free Languages under Simple Distributions." In Ann Copestake and Kees Vermeulen (eds.), *Algebras, Diagrams and Decisions in Language, Logic and Computation, CSLI/CUP*, 1–17. London: Imperial College Press.

Adriaans, P. W. (2008). "The Philosophy of Learning, the Cooperative Computational Universe." In P. W. Adriaans and J. F. A. K. van Benthem (eds.), *Handbook of Philosophy of Information*. Amsterdam: Elsevier Science.

Adriaans, P. W. (2009). "Between Order and Chaos: The Quest for Meaningful Information." *Theory of Computing Systems*, 45(4): 650–674. Special Issue: Computation and Logic in the Real World; Guest Editors: S. Barry Cooper, Elvira Mayordomo, and Andrea Sorbi. Basel: Springer-Verlag

Adriaans, P. W. (2010). "A Critical Analysis of Floridi's Theory of Semantic Information." In Hilmi Demir (ed.), *Knowledge, Technology and Policy, Luciano Floridi's Philosophy of Technology: Critical Reflections*. Basel: Springer-Verlag.

Adriaans, P. W. (2012). "Facticity as the Amount of Self-Descriptive Information in a Data Set." "arXiv:1203.2245 [cs.IT], 2012.

Adriaans, P. W. (2018). *Information, Stanford Encyclopedia of Philosophy*. In E. N. Zalta (ed.). Stanford: Metaphysics Research Lab, Stanford University. https://plato.stanford.edu/entries/information.

Adriaans, P. W. (2019). *Semi-Countable Sets and Their Application to Search Problems*. https://Arxiv.Org/Abs/1904.03636

Adriaans, P. W., and Vitányi, P. M. B. (2009). "Approximation of the Two-Part MDL Code. *IEEE Transactions on Information Theory*, 55(1): 444–457.

Antunes, L., and Fortnow, L. (2003). "Sophistication Revisited." In Proceedings of the 30th International Colloquium on Automata, Languages and Programming, vol. 2719 of *Lecture Notes in Computer Science*, pp. 267–277. New York: Springer.

Bar-Hillel, Y. (1964). *Language and Information: Selected Essays on Their Theory and Application*. Reading, MA: Addison-Wesley.
Bloem, P., de Rooij, S., and Adriaans, P. (2015). Two Problems for Sophistication. In K. Chaudhuri, C. Gentile, and S. Zilles (eds.), *Algorithmic Learning Theory*. Lecture Notes in Computer Science, vol. 9355. New York: Springer.
Bloem, P., Mota, F., de Rooij, S., Antunes, L., and Adriaans, P. (2014). *A Safe Approximation for Kolmogorov Complexity*. ALT 2014: 336–350.
Carnap, R. (1945). "The Two Concepts of Probability: The Problem of Probability." *Philosophy and Phenomenological Research*, 5(4): 513–532.
Chater, N., and Vitányi, P. (2003). Simplicity: A Unifying Principle in Cognitive Science? *Trends in Cognitive Sciences*, 7(1): 19–22.
Cilibrasi, R., and Vitanyi, P. M. B. (2005). "Clustering by Compression." *IEEE Transactions in Information Theory*, 51(4): 1523–1545.
Domingos, P. (1998). "Occam's Two Razors: The Sharp and the Blunt." Proceedings of the Fourth International Conference on Knowledge Discovery and Data Mining (pp. 37–43). New York: AAAI Press.
Floridi, L. (2011). *The Philosophy of Information*. Oxford: Oxford University Press.
Frege, G. (1879). *Begriffsschrift: eine der arithmetischen nachgebildete Formelsprache des reinen Denkens*. Halle: Verlag von Louis Nebert.
Gell-Mann, M., and Lloyd, S. (2003). "Effective Complexity." In M. Murray Gell-Mann and Constantino Tsallis (eds.), *Nonextensive Entropy–Interdisciplinary Applications*. Oxford, UK: Oxford University Press, pp. 387–398.
Golan, A. (2018). *Foundations of Info-Metrics*. New York: Oxford University Press.
Grünwald, P. (2007). *The Minimum Description Length Principle*. Cambridge, MA: MIT Press.
Hopcroft, J. E., Motwani, R., and Ullman, J. D. (2001). *Introduction to Automata Theory, Languages, and Computation*. 2nd ed. Reading, MA: Addison-Wesley, 2001.
Hughes, G. E., and Cresswell, M. J. (1996). *A New Introduction to Modal Logic*. New York: Routledge.
Hütter, M. (2005). *Universal Artificial Intelligence: Sequential Decisions Based on Algorithmic Probability*. Berlin: Springer, 2005.
Koppel, M. (1987). "Complexity, Depth, and Sophistication." *Complex Systems*: 1087–1091.
Koppel, M. (1995). *Structure, the Universal Turing Machine: A Half-Century Survey* (2nd ed., pp. 403–419). Berlin: Springer Verlag.
Koza, J. R. (1990). "Genetic Programming: A Paradigm for Genetically Breeding Populations of Computer Programs to Solve Problems." Stanford University Computer Science Department Technical Report STAN- CS-90-1314.
Kripke, S. (1963). "Semantical Considerations on Modal Logi." *Acta Philosophica Fennica*, 16: 83–94.
Li, M., and Vitányi, P. M. B. (1991). "Learning Simple Concepts under Simple Distributions." *SIAM Journal on Computing*, 20(5): 911–935.
Li, M., and Vitányi, P. M. B. (2019). *An Introduction to Kolmogorov Complexity and Its Applications*, 4th ed. New York: Springer-Verlag.
McAllister, J. W. (2003). "Effective Complexity as a Measure of Information Content." *Philosophy of Science*, 70(2): 302–307.
Rathmanner, S., and Hütter, M. (2011). "A Philosophical Treatise of Universal Induction." *Entropy*, 13(6): 1076–1136.

Solomonoff, R. (1997). "The Discovery of Algorithmic Probability." *Journal of Computer and System Sciences*, 55(1): 73–88.

Solomonoff, R. J. (1964). "A Formal Theory of Inductive Inference: Parts 1 and 2." *Information and Control*, 7: 1–22 and 224–254.

Spade, P. V., and Panaccio, C. (2016). "William of Ockham." In Edward N. Zalta (ed.), *Stanford Encyclopedia* of *Philosophy*. Stanford: Metaphysics Research Lab, Stanford University.

Vereshchagin, N. K., and Vitányi, P. M. B. (2004). "Kolmogorov's Structure Functions and Model Selection." *IEEE Transactions in Information Theory*, 50(12): 3265–3290.

Vitányi, P. M. B. (2006). "Meaningful Information." *IEEE Transactions on Information* Theory, 52(10): 4617–4626.

Wolpert, D. H., and Macready, W. (2007). "Using Self-Dissimilarity to Quantify Complexity: Research Articles. *Complexity*, 12(3):77–85.

Wolpert, D. H., and Macready, W. G. (2005) "Coevolutionary Free Lunches." *IEEE Transactions on Evolutionary Computation*, 9(6): 721–735.

PART II
INFORMATION THEORY AND BEHAVIOR

This part deals with information theory and behavior. It touches on the interconnection between information, information-theoretic inference, and individual and collective behavior. In Chapter 3, Daniels discusses ways to infer the logic of collective behavior. It focuses on common challenges in inferring models of complicated distributed systems and examines how the perspective of information theory and statistical physics is useful for understanding collective behavior. The approach taken in that chapter combines machine learning with dimensionality reduction techniques.

In Chapter 4, Choi develops a direct connection between human ability and information theory. The chapter proposes an information-theoretic perspective on the role of human ability in decision making. Personality is defined as a response system of an agent that maps a situation to a behavior. The agent-specific response system is characterized by various constraints. Based on the principle of maximum entropy, these constraints are treated as information that reduces the entropy of the system.

In Chapter 5 Judge discusses information recovery and inference as they relate to adaptive economic behavior and choice. The emphasis here is on the connection between adaptive economic behavior and causal entropy maximization. That connection allows linking the data and the unobservable system behavioral parameters.

Overall, these three chapters provide new insights into individual and collective behavior and choice. The insight is the information observed and the inference of that information as they relate to the behavior.

3

Inferring the Logic of Collective Information Processors

Bryan C. Daniels

1. Introduction

Biology is full of examples of complicated, carefully regulated, and often slightly mysterious examples of information processing: A housefly senses a gust of air and moves its wings, evading my flyswatter; a cascade of signaling proteins translates a hormone detected at a cell's surface into the production of specific genes in the nucleus; a colony of ants finds a new suitable rock crevice to use as a nest after its old one is destroyed. In each case, large sets of individual components must coordinate to carry out different actions depending on an environmental input.

A major challenge for modern science is to connect the small-scale dynamics of these individual components to the information processing consequences at the larger scale of the aggregate whole. If we think of biological systems as performing computations, transforming sensory input into coordinated and adaptive output behavior (Figure 3.1), the goal is to comprehend the logic of these distributed computers.

In this chapter, I summarize a new approach for understanding collective information processing that is emerging at the interface of machine learning, statistical physics, information theory, and more traditional biological and social science. This approach can be viewed as expanding on existing notions of collective computation and distributed computing, which in the past focused mainly on theoretical results in cognitive science and neuroscience (Rumelhart and McClelland, 1996), in order to make them more data-rich and broaden them to include other systems such as collections of fish or ants or people or proteins (Figure 3.1; Couzin, 2009; Solé et al., 2016). Using extensive data sets, we are able to focus on how *specific* collective systems operate, going beyond generalized theory.

The challenge is formidable for at least three reasons: (1) the large number of interacting parts in each system means there are many potential contributors to

Bryan C. Daniels, *Inferring the Logic of Collective Information Processors* In: *Advances in Info-Metrics: Information and Information Processing across Disciplines.* Edited by: Min Chen, J. Michael Dunn, Amos Golan, and Aman Ullah, Oxford University Press (2021).
DOI: 10.1093/oso/9780190636685.003.0003

	Individual scale input		Aggregate output
Ants:	Nest quality	⟶	Nest choice
Neurons:	Sensory signals	⟶	Decision, action
Proteins:	Chemical signal	⟶	Gene expression
Fish:	Predator location	⟶	Group direction
Politicians:	Individual opinions	⟶	Law
Genes:	Developmental cues	⟶	Cell type

Figure 3.1 Interpreting adaptive collective behavior as computation. Across diverse biological and social systems, individuals detect information (input), such as the quality of a potential nest site visited by individual ants, that is combined to determine a single behavior of the aggregate (output), such as the nest choice of an entire ant colony. In each case, the aggregate output is the outcome of complicated interactions among a large number of individuals. Our goal is to understand how the aggregate behavior is produced by the behavior of individual components.

the final output, leading to large, unwieldy models; (2) the number and nonlinearity of interactions means model outputs are dependent on their component-level parameters in highly nontrivial ways, making parameter estimation difficult even if the structure of interactions is known; and (3) different examples of the same adaptive system often vary in their particular structure, making it necessary to build a new model for every new case.

In this chapter, I argue for a two-step process to best extract the logic of collective information processors (Figure 3.2). First, a suite of machine-learning approaches are used to infer a detailed model from data. This step gathers data into a predictive framework that encompasses the full complexity of the system at the level at which data is available. Second, and just as challenging, dimensionality reduction techniques are applied to the full model in order to produce simplified explanations. This abstraction step prunes away detail to more intuitively explain low-dimensional behavior at the aggregate scale.

Information measures are key to this perspective. At the most fundamental level, information processing provides a basic conceptual framework for what living systems are doing. In the case of collective systems, measures such as the Fisher information can be used to connect individual components to aggregate scales. Bayesian inference techniques benefit from concepts in information geometry, and Bayesian model selection can be interpreted as matching the amount of information encoded in the data to the models that describe the data. Finally, information compression is a natural interpretation of dimension reduction methods that lead to simplified understanding.

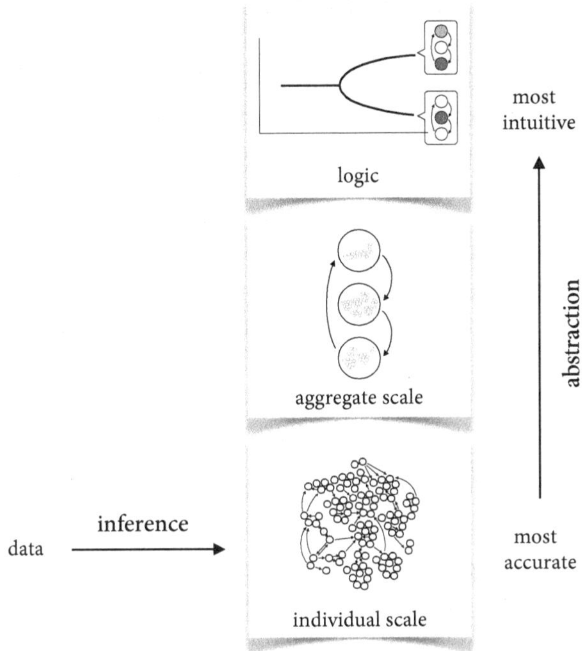

Figure 3.2 A two-step strategy for extracting the logic of biological collective information processors. First, a detailed model is inferred from data taken at the level of individual components. Next, successively more abstract representations allow a simplified understanding of the most important drivers of aggregate output.

The goals of machine learning, artificial intelligence, big data, and particularly data science partially overlap with what we seek in the science of collective behavior: predictive and simplified models of complex data sets. Yet, we are primarily motivated here by fundamental questions about living systems: How do collective systems remain adaptive? How do individual components manage to successfully regulate behavior that involves many other components? What is carefully tuned by evolutionary selection, and what is compensated for through active adjustment? What general strategies do distributed biological systems use to create adaptive logic?

Following the two-step framework of Figure 3.2, let us consider in more detail how to accomplish the tasks of inference and abstraction. We will look at each task in turn, reviewing the challenges that arise and the corresponding state-of-the-art methodologies being developed to address them.

2. First Task: Infer Individual-to-Aggregate Mapping

In biological collectives, the mapping from individual components to aggregate behavior is often unknown, difficult, intricate, degenerate, and nonlinear. The first challenge, then, is to transform data into a predictive model that connects the individual to the aggregate scale.

Performing inference, we search for the most accurate model we can find, evaluating our success by how well we can predict everything we can measure about the system's behavior. With the goal of encompassing all measured variables, often large in number in collective systems, this inference step is likely to produce a model that is particularly complex. This is represented at the bottom of Figure 3.2: a complicated model that acts as an accurate summary of our knowledge, a gathering of all relevant information.

Three major difficulties arise in the inference step:

- Heterogeneity and a large amount of potentially relevant detail at the individual scale (approached using big computers, big data, and a nuanced perspective on the relationship between modeling and theory)
- Unknown interaction structure and limited data (approached using maximum entropy and other effective modeling techniques, making use of model selection and regularization)
- Parameter compensation and emergence (approached using the concept of sloppiness and learning to live with parameter uncertainty)

Each of these issues has been the subject of extensive effort in the past few decades, which we review in this section.

2.1 Inference Challenge 1: An Abundance of Potentially Relevant Detail—Solved by Large-Scale Reverse Engineering

The fundamental challenge in understanding collective systems is clearly that they involve lots of parts. It is often impossible to measure and daunting to reason about the large number of individual-level properties that could be important.

In some cases, this complexity can be cleverly circumvented. The field of statistical physics produces simple mathematical explanations of material properties in terms of the collective behavior of atoms and molecules, using techniques like the renormalization group (Wilson, 1979; Goldenfeld, 1992). But the cleanest explanations rely on multiple assumptions that often do not hold in biological and social systems. First, explanations in statistical physics often assume that we know the interactions between individual components and that these interactions are uniform across the system.[1] Second, in materials science, we are interested in scales that are extremely large compared to individual components—humans typically care about collective properties involving trillions of trillions of molecules. In short, we know how to parsimoniously describe systems with many contributing parts when interactions are simple and when we zoom out to include a huge number of them.

But the individual neurons in my brain are wired up differently than yours; the likely votes by Congress depend on changing individuals and changing opinions; a large number of distinct and overlapping gene-regulatory networks exist in a given cell. A typical number of individuals in these systems is hundreds to millions or at most billions, always vastly less than a trillion trillion. So what happens when individual components are diverse and changing, when there are a large but not huge number of them, and when each specific system involves myriad contingencies and historical contexts?

Such challenges are not new to science. We approach them as usual with experimentation and hypotheses, building models using carefully reasoned intuition and then testing them and gradually refining them. What is uniquely challenging about these systems is the volume of potentially important detail, the richness of information. It is difficult to hold all the important variables in one's head at once, and given the number of diverse systems we might want to understand, it takes too long to model each specific system anew.

Much effort has been aimed at the problem of systematically producing predictive models of such complicated collective behavior. The dominant conceptual framework is that of the network (Newman, 2010; Natale, Hoffmann, Hernández, and Nemenman, 2018). The proliferation of network explanations for collective biological behavior has ranged in scale from gene-regulatory systems (e.g., Bonneau, 2006; Peter and Davidson, 2017) and neurons (e.g., Bassett et al., 2011) to groups of organisms (e.g., Rosenthal et al., 2015). The implicit and reasonable assumption has been that the best way to proceed is simply to enumerate all the complicated details, pinning down the behavior of every individual and how it interacts with others. When this is tractable, the attitude is to be agnostic as to which details are most important to the aggregate behavior.

Network science is a large and successful scientific enterprise, and a huge number of methods have been developed to "reverse-engineer a network" from data (for a few representative cases and reviews, see Natale et al., 2018, and Bonneau et al., 2006). The zoo of existing methods embody a range of

[1] Or, if we do not know specific individual interactions, we assume any non-uniformities average out in such a way that we can understand the system in terms of a typical average individual.

typical modeling assumptions. For instance, we may assume that individuals are characterized by binary, discrete, or continuous states; that individuals either have dynamic states that update in discrete or continuous time, or that their joint states are described by an equilibrium function; and that interactions among individuals are described by linear or nonlinear functions. These modeling choices define the space of models over which an inference routine must search, which typically proceeds using a minimization algorithm that matches the model to statistics of the data. The results are often interpreted statistically in a Bayesian framework.

If one starts with abundant individual scale data, there is not much to decide in setting up network inference besides these initial modeling assumptions, and the strength of predictions will be determined by the veracity of the assumptions. Fast computers with large memories allow for inferring models of collective behavior with unprecedented detail.

In contrast to inferring detailed networks for specific systems, much of the initial theory of adaptive distributed systems relied on simplifying assumptions that did not require knowing all individual-level details. For instance, we may assume that gene-regulatory networks or neural networks are randomly connected, and then we may ask about the properties of these random networks (e.g., Kauffman, 1969; Amit, Gutfreund, and Sompolinsky, 1985). Classic results in parallel distributed computing ("neural networks") have shown that *arbitrary* computations can be carried out by abstracted neural units, if we are allowed to impose specific interactions.

Automated network inference is reaching an exciting point at which we can begin to theorize about collective behavior without having to rely on these assumptions. We instead can infer and use as our starting point a model of the full, messy, heterogeneous system, using data from, say, simultaneously measured neurons or simultaneously measured genes, as is now becoming routine. We may even study an ensemble representing typical cases of these networks, as are presently being accumulated in online repositories (Daniels et al., 2018).

Data and knowledge of the space of possible interactions are often limited, however, requiring new concepts, which we discuss in the following inference sections. Machine learning has faced this same problem and, in a limited sense, has already solved the problem of getting predictive power in complicated heterogeneous systems. The trade-off for gathering such extensive knowledge is that, in the extreme case, we must resign ourselves to being unable to have a human check all the details. Machine learning has already surpassed the speed at which humans can construct predictive models: automated language translation or image recognition involves incomprehensibly large sets of data, variables, and interactions, and the resulting learned models are represented in a way that would not be easily understood or checked by a human. The potential usefulness to the science of collectives is that machine-learning techniques are already forming predictive models that capture important aggregate properties. These inscrutable but predictive descriptions provide a useful starting point for constructing understandable scientific theories of collective behavior.

Thus, the increasing speeds of machine learning and statistical inference give us the opportunity to model a large number of distinct systems. But beyond predictions and parsimonious descriptions of the aggregate scale, we also want to understand adaptive computations in terms of the actual individual-level interactions in each system. We want to know how a brain classifies images *using neurons*. How do we incorporate important but limited knowledge about mechanistic interactions at the scale of individual components?

2.2 Inference Challenge 2: Structural Uncertainty Due to Limited Data—Solved by Hierarchical Model Selection and Regularization

In our network inference problem, we want to know how individuals influence each other, yet often, even with detailed measurements, we do not have enough data to pick out the correct interactions from the space of all possible interactions. Even after selecting a particular class of models, the space of all possible models grows combinatorially with the number of individuals N—that is, it becomes huge. This means that even with a large amount of data, when N is moderately large, the space of models can easily overrun the space of all possible data. In this underdetermined regime, in which a model can perfectly fit any possible data, the model becomes useless. The challenge is then to efficiently make use of the detail present in the data while avoiding overfitting.

A particularly productive approach to this problem is to use models that match their level of complexity to the data and questions at hand. Complicated models can be produced if necessary, but when data is limited the model stays simple, in a way that produces better predictions. Intuitively, this works by ignoring smaller signals in the data that are likely caused by nonsystematic noise.

In this way, we match the level of detail, or dimensionality, that is supported by the data. Machine learning may be interpreted in this way more generally, where the process of restricting a model to lie in or close to a lower dimensional subspace is called regularization. This is dramatically realized, for instance, in reservoir computing, where input is nonlinearly transformed into a high-dimensional space and only the most predictive low-dimensional linear combination is retained (Lukoševicius and Jaeger, 2009). Note that such regularization can also be interpreted as compression or model simplification (explored in section 3 of this chapter). Regularization can be used to find the appropriate level of complexity that produces the maximally accurate model (bottom of Figure 3.2), and similar techniques can then be used to throw away more detail and further reduce the dimensionality (moving toward the top of this figure).

Here, I highlight two approaches that use this conceptual structure. First, in a stochastic equilibrium modeling framework, we can construct a model with output that is as random as possible, adding interactions until statistics of the data are sufficiently well fit. This leads to maximum entropy approaches. Second, in a setting of deterministic dynamics, complexity in the form of nonlinear interactions and "hidden" unmeasured dynamical variables can be added until the system produces dynamics that fit time series data sufficiently well.

As a side note, one might object that putting a lot of effort into dealing with limited data is silly in that we should instead emphasize simply taking more or better data. Though a new experiment is often a good option, having "limited data" can sometimes be interpreted not as a problem with the experiment but a fact of life in the system.[2] Some systems, like a macaque society, have a stable structure only over a limited timescale. Taking more data is not an option, and asking for the "true" structure is not a well-defined question. Yet we may still want to characterize the interactions that are strong enough to have predictable effects.

2.2.1 Maximum Entropy Modeling

One powerful modeling approach in collective behavior is to treat observed states of the system as independent snapshots and then infer the probabilities with which all possible states of the system arise. This makes the most sense when dynamics are fast compared to the phenomena we are interested in and therefore ignorable. This type of model is common in equilibrium statistical physics.

A typical case starts with a system with N individuals, each of which can be either active or inactive—for instance, neurons that are firing or silent, or fish that are startled or calm. A static model produces the probability $p(\vec{x})$ of any given N-dimensional binary state $\vec{x}$ of active and inactive individuals. With enough data, we could estimate these probabilities by simply using the frequency with which every possible aggregate state occurs:

$$p(\vec{x}) \approx \frac{\text{number of observations of state } \vec{x}}{\text{total number of observations}}. \qquad (3.1)$$

[2] This is also related to the problem of sloppiness discussed in section 2.3.

The glaring problem for large N is that there are 2^N possible states, so that getting an accurate estimate of these probabilities requires more than 2^N observations. The idea of the maximum entropy approach is to instead force the model to reproduce only statistics that we *can* measure accurately. Specifically, the entropy of $p(\vec{x})$ is maximized given the constraint of fitting some given statistics.

If we can make many measurements of the simultaneous states of individuals, then natural, easily observed statistics are the frequencies with which individuals are active, the frequencies with which pairs of individuals are jointly active, and so on. For instance, the probability that individuals i and j are jointly active puts a specific constraint on a marginal of $p(\vec{x})$:

$$p(i \text{ and } j \text{ active}) = \sum_{\vec{x} \text{ with } x_i=1, x_j=1} p(\vec{x}) \approx \frac{\text{number of observations of } i \text{ and } j \text{ jointly active}}{\text{total number of observations}}. \quad (3.2)$$

Typically, pairwise correlations will be most accurately captured by data, and higher-order correlations will require progressively more data. This motivates the typical form for a maximum entropy expansion, the form of which turns out to be straightforward to derive (Schneidman, Berry, Segev, and Bialek, 2006; Mora and Bialek, 2011; Daniels, Krakauer, and Flack, 2012, 2017):

$$p(\vec{x}) \propto \exp\left(-\sum_i h_i x_i - \sum_{ij} J_{ij} x_i x_j - \sum_{ijk} K_{ijk} x_i x_j x_k + \ldots\right). \quad (3.3)$$

The parameters J_{ij}, K_{ijk}, ... represent effective interactions between individuals that make specific subgroups more or less likely to be simultaneously active. While the form of the maximum entropy distribution (Eq. [3.3]) can be written down analytically, finding the parameters that match the statistics for a particular data set is a difficult inverse problem. Many approaches have been proposed for solving these inverse problems efficiently and in various approximations (Lee and Daniels, 2019).

The expansion after adding each term in Eq. (3.3) is the distribution with maximum entropy that fits those correlations. This is a form of hierarchical model selection: we add degrees of freedom (interactions) to the model until we fit the data well enough, but we do not go back to remove weak or unimportant lower-order interactions. Then each successive model in the list includes strictly more structure than the last, implying that the entropy monotonically decreases as we include more terms and thereby incorporate more information from the data. In cases for which we can estimate the entropy of the full distribution, we can track how much of the information the model captures as we add more terms (Schneidman et al., 2006). The information captured by pairwise models is in many cases large, with not much left to be fit by higher-order interactions (Schneidman et al., 2006; Merchan and Nemenman, 2016).

Including all possible pairwise interactions can go too far and lead to overfitting if, for instance, some individuals are rarely active, meaning that joint activations are prohibitively rare. More sophisticated versions of the maximum entropy approach instead include only the most important interactions (Ganmor, Segev, and Schneidman, 2011) or use a cluster expansion that incorporates pairwise statistics only among clusters that contribute most to the joint entropy (Cocco and Monasson, 2012).

The maximum entropy approach is also useful in collective behavior in more general contexts than binary states and correlations. In principle, any state space and set of constrained statistics can be incorporated, such as the mean, variance, and correlations of the velocities of flocking birds (Bialek et al., 2014).

Maximum entropy models have been successfully applied to the collective behavior of multiple biological systems, including neurons (Schneidman et al., 2006), flocking birds (Bialek et al., 2014), and animal conflict (Daniels et al., 2012) (though see also warnings about extrapolating pairwise maximum entropy results to larger systems [Roudi, Nirenberg, and Latham, 2009] and the dangers of inferring interaction structure in the case of common input [Schwab, Nemenman, and Mehta, 2014]).

2.2.2 Dynamical Inference

In inferring dynamical systems, too, model selection can be used to adapt to the amount of information in the data. Imagine starting with data from a system that responds to an input with reproducible dynamics. This could be a cellular signal transduction cascade (Daniels et al., 2008), metabolic oscillations (Daniels and Nemenman, 2015), or a worm responding to sensory input (Daniels, Ryu, and Nemenman, 2019). The goal will be to represent the observed dynamics using a set of differential equations. In a very general form,

$$\frac{d}{dt}\vec{x}(t) = \vec{f}\left(\vec{x}(t), \vec{y}(t), \vec{\theta}_x\right)$$
$$\frac{d}{dt}\vec{y}(t) = \vec{g}\left(\vec{x}(t), \vec{y}(t), \vec{\theta}_y\right), \tag{3.4}$$

where $\vec{x}$ is the vector of observed dynamical variables, $\vec{y}$ represents unobserved "hidden" variables, and $\vec{\theta}$ contains parameters controlling the dynamics of individual variables and their interactions. Of course, the important modeling decisions arise in defining the forms of f and g, setting the space of possible models over which the inference scheme should search.

The general dynamical inference problem can quickly become unwieldy in that the space of possible models is enormous and impossible to search comprehensively. One possibility is to use no hidden variables, adding descriptive power (information) by increasing only the complexity of the function $\vec{f}$. The search space of possible functions can be constrained to a set of preset functions that combine to form $\vec{f}$, either using mathematically convenient functions (x, x^2, y, xy, $\cos(x)$, ...) (Schmidt and Lipson, 2009; Brunton, Proctor, and Kutz, 2016) or functions with biologically inspired nonlinearities ($\tanh(x)$, $(1+\exp(-x))^{-1}$, ...) (Daniels and Nemenman, 2015).

Another possibility is to add complexity both in the form of the right-hand sides of Eq. (3.4) and the addition of hidden dynamical variables $\vec{y}$, providing the opportunity to predict the existence of important unmeasured system components (Daniels and Nemenman, 2015). This approach is particularly useful when data are only available at an aggregate scale (as in Figure 3.4) or some individual-level details are missing. Unfortunately, this only adds to the enormity of the potential search space. Hierarchical model selection can be used as a counteraction: analogously to the maximum entropy expansion in Eq. (3.3), we make the model selection process more efficient by predefining a set of models of increasing complexity that can eventually fit any data, stopping the search when the model fits the data within experimental uncertainty (Daniels and Nemenman, 2015).

These dynamical inference approaches have been successful in producing predictive models from time series in physical systems (Schmidt and Lipson, 2009), simulated data from glycolysis oscillations (Daniels and Nemenman, 2015), and animal locomotion (Daniels et al., 2018).

2.3 Inference Challenge 3: Parameter Uncertainty Due to Scale Separation and Sloppiness—Solved by Bayes and not Focusing on Individual Parameters

Besides the difficulties encountered in the last section in constraining the interaction structure due to having a limited number of datapoints, there is another fundamental challenge to the inference of predictive models in collective systems: a problem that arises with data at the aggregate scale even if we have a huge amount of it. We think of models here as mappings from individual-level mechanisms to aggregate-scale consequences (Figure 3.3). The problem lies in

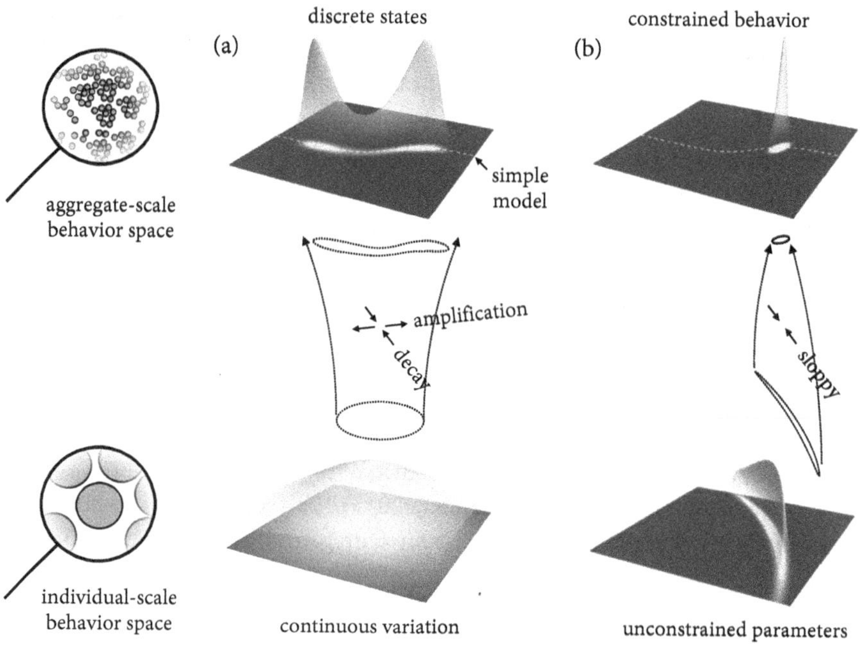

Figure 3.3 Parameter space compression, sloppiness, and simplified emergent models. A schematic representing the idea that some information is lost and some gets amplified in coarse-graining. The 3D plots represent probability distributions over behavior, both at the scale of individuals (bottom) and the scale of the aggregate (top). a) The large space of possible behaviors at the individual scale (bottom) typically maps onto a lower-dimensional output space at the aggregate scale (top). This can be understood as parameter space compression, the idea that moving to larger scales (up in this diagram) causes some effective parameter directions to decay. The dashed line on the top plot represents a simpler, lower-dimensional model that ignores the decaying parameter direction, but still well represents the collective behavior. This case also depicts a collective instability leading to amplification away from the center point. This creates two distinct possible aggregate behaviors, and relates to the idea of phases in statistical physics. See Section 3.2.1. b) For the same system, we now focus on a case in which we make a precise measurement at the aggregate scale. The same parameter space compression implies that even highly constrained aggregate behavior (top) typically corresponds to large regions of possible individual-scale parameters (bottom), leading to the phenomenon of sloppiness. See Section 2.3.

the fact that the typical properties of these mappings make it difficult to constrain parameters at the individual scale.

Not surprisingly, changes at the individual scale can have nonobvious effects—this is what makes complex systems interesting. Perturbing an individual fish may do nothing, while perturbing three fish simultaneously has huge effects (synergy). Removing a protein from a system entirely may modify an existing signaling pathway to take over and give similar output (compensation and robustness). These effects clearly make parameter fitting more challenging by virtue of being highly nonlinear.

What may be surprising, however, is that (1) there is some statistical regularity to the types of nonlinearities we find in these mappings in real systems (the phenomenon of sloppiness); and (2) the ubiquity of these phenomena point us toward an approach that deemphasizes finding the "correct" individual-scale parameters and instead uses Bayesian ensembles over parameters.

The idea is represented schematically in Figure 3.3. Parameters at the individual scale may be very unconstrained (wide probability distribution representing a huge space of possibilities at the bottom of Figure 3.3a), while they get mapped onto only a small space of possible aggregate behaviors (effectively lower-dimensional, thin-wedge probability distribution at the top of Figure 3.3a). This means that some directions in the individual-level parameter space are unimportant or "sloppy," while only a few are important or "stiff." The effect can be dramatic in that even taking lots of measurements to highly constrain the aggregate behavior (Figure 3.3b top) still leaves the possibility of huge swaths of parameter space at the individual level (Figure 3.3b bottom).

Properties of this mapping can be interpreted information-theoretically in terms of the Fisher information matrix. The Fisher information is a generalized measure of sensitivity of model outputs (what we call aggregate level properties) on model parameters. The Fisher information can be interpreted as the curvature of the Kullback-Leibler divergence of the distribution of model behavior as a model parameter is varied. With units of bits divided by the parameter's units squared, it answers the question of how quickly the output behavior becomes distinguishable when that parameter is varied. The Fisher information for parameter μ and aggregate state x is the average squared derivative of the log-likelihood function:

$$\mathcal{I}_x(\mu) = \int \left(\frac{\partial \log p(x)}{\partial \mu} \right)^2 p(x) dx. \tag{3.5}$$

The more general Fisher information matrix describes sensitivity to individual parameters and to simultaneous changes to pairs of parameters:

$$\mathcal{I}_x(\mu, \nu) = \int \frac{\partial \log p(x)}{\partial \mu} \frac{\partial \log p(x)}{\partial \nu} p(x) dx. \tag{3.6}$$

The Fisher information is also useful (see section 3.2.1) in quantifying functional sensitivity. Within model inference, it is a convenient means of describing parameter uncertainty.

The phenomenon of sloppiness describes properties of the Fisher information matrix that are common across a large number of complex dynamical models from systems biology and elsewhere (particularly those with many interacting components, the exact sort we encounter in modeling collective behavior). In most large nonlinear models, eigenvalues of the Fisher information matrix span many orders of magnitude and are roughly evenly spaced in log space. There are typically a few large "stiff" eigenvalues, with corresponding eigenvectors that describe directions in parameter space that are tightly constrained by experimental data. Remaining are a plentiful number of extremely small eigenvalues deemed "sloppy," corresponding to directions in parameter space that can be varied a large amount without changing the aggregate behavior (see Figure 3.3b) (Transtrum et al., 2015).

Roughly speaking, sloppiness is caused by nonlinear compensation: in large systems, it is usually the case that the aggregate scale effects of varying one parameter can be approximately canceled by varying some combination of other parameters. Typically, this sloppy compensation is highly nonlinear, so it is not easy to redefine parameters in such a way to remove sloppiness (as it would be if the mapping were linear).

This is important to inferring models of collective behavior because, even if we know the individual-scale interaction topology, typically many related parameters (rate constants, etc.) are unmeasured. We are then forced to fit them to the data, often using measurements at the aggregate scale. Sloppiness then implies that we will be unable to tightly constrain many directions in parameter space. The practical implication is that the aggregate-level data will be compatible with large swaths of parameter space (Figure 3.3b), leaving certain details of the individual scale unknown. Because sloppiness usually becomes extreme in large systems, this parameter uncertainty persists *even as we take lots of data* at the aggregate scale, becoming a fundamental problem for fitting models of collective behavior.

The solution: stop worrying about fitting a precise set of parameters. (See Daniels, Dobrzynski, and Fey, 2018) for a more detailed discussion of parameter estimation in systems biology.) One tactic is simply to choose a specific set of parameters that sufficiently well fits the data. Surprisingly, this often produces predictive models (Transtrum et al., 2015) but can be dangerous: it is possible that predictions other than those used to constrain the model depend strongly on the position within the sloppy subspace of possible parameters. Safer is a Bayesian approach, which characterizes the entire sloppy subspace of parameters that fit the data within statistical uncertainty. The most straightforward way to do this is to use Monte Carlo methods to sample from the posterior over parameters; measuring a model output of interest using each member of the parameter ensemble then estimates the posterior distribution of the output.

More broadly, sloppiness suggests that lower-dimensional descriptions should be possible that succinctly capture the behavior controlled by stiff parameter directions. That brings us to our second task: finding simplified descriptions of collective behavior.

3. Second Task: Find Abstract System Logic

We don't model the trillion trillion molecules in a glass of water to predict what it does; instead we use effective models at a different scale. In the past one hundred years, statistical physics and high-energy physics have made great strides in understanding how this works in simple systems. Much of the excitement in the field of biosocial collective behavior is in learning how to analogously build simple but relevant effective models in more complicated systems.

What we get from inference procedures described in section 2 are explanatory models that may be arbitrarily complicated and not easily interpreted. This is the "black box" problem. Generalized methods from machine learning, such as neural networks and reservoir computing, are most prone to this problem, as they tend to intentionally overcompensate with the amount of included detail (Denker et al., 1987). But even approaches that explicitly favor simplicity (as in section 2.2) end up looking like a spaghetti tangle[3] when there is enough data to support it. That is, even a well-characterized system can be hard to understand. The recent excitement in neuroscience about ever-larger and more detailed data sets provides a good motivating example: Suppose we will someday be able to simultaneously measure and faithfully model every neuron in the human brain. Then what next?

Abstraction and simplification are crucial elements in the story of collective behavior. Across all the examples presented in Figure 3.1, the magic of collective behavior lies in large collections of individuals producing coherent

[3] Also known as a hairball (Lander, 2010) or ridiculogram (attributed to Marc Vidal).

low-dimensional dynamics. Our aim is to discover what drives this low-dimensional aggregate behavior and connect it to what we know about the complicated behavior of individuals. The challenge, then, is to find ways to compress existing detailed explanations into simplified understanding.

3.0.1 Why Do We Want to Do This? Advantages of Coarse-Grained, Dimensionality-Reduced Description

We put effort into simplifying models because systems become easier to understand when one finds the right coarse-grained representation. In some cases, we can put this in explicit information-theoretic terms: how many bits do I need to remember to predict system behavior (Daniels et al., 2012)? In some cases, it is a loose qualitative statement that, for instance, categorizing a system as implementing a Hopf bifurcation is easier to conceptualize than a detailed network representation. At a different level, simplified models are also important to understand in that compression is often happening in the system itself, with individual components cognitively adapting to their collective environment (Daniels et al., 2012; Flack, 2017).

This is part of a broader epistemological stance that recognizes "effective" models as legitimate and at times preferable to models defined at a detailed scale (Shalizi and Moore, 2003; Wolpert, Groschow, Libby, and DeDeo, 2015). It is also part of an ongoing debate about how to understand evolutionary forces acting at aggregate scales different than the familiar individual genome scale.[4] In the context of other complex, multiscale systems—for example, modeling whole cells (Babtie and Strumpf, 2017), animal behavior (Stephens, Osborne, and Bialek, 2011), or even abstract systems like cellular automata (Crutchfield and Mitchell, 1995)—this point is well appreciated: the goal is not only to encapsulate all of our knowledge in the most detailed model possible, but also to create approximations that are easier to work with, analytically and intuitively. There is a tension between our "best current understanding" or most accurate model and the model that gives the best intuition.

[4] This can turn into a lofty philosophical argument about epistemology and ontology—is our effective understanding ontologically "real" or just "an accurate description of our pathetic thinking about nature" (Gunawardens, 2014)? Is there an objectively "true" level at which aggregate objects and phenomena exist (Shalizi and Moore, 2003; Hoel, Albantakis, and Tononi, 2013)? Here we will instead focus pragmatically on predictive modeling—the best description is the one that makes the best predictions (which can depend on the question being asked). As was famously quipped by George Box: "All models are wrong but some are useful" (Box, 1979).

3.0.2 Do We Expect to Be Able to Compress? What Does "Logic" Look Like?

A modern understanding of why effective models work (Transtrum et al., 2015; Machta, Chachra, Transtrum, and Sethna, 2013) stems from renormalization group ideas used in statistical physics to characterize phases of matter.[5] The basic idea is to track how different types of interactions become more or less important as we "zoom out" from a system. Analogously to the Central Limit Theorem, the effects of some interactions are washed out, whereas other "relevant" interactions become more important. The relevant interactions are kept in the effective model, and we forget about the irrelevant ones, making for a simpler model. In this way, we have explicitly constructed the model so that it predicts the aspects that we deem most important. In physics, the most important aspects are typically those that occur at large spatial scales or low-energy scales.

In contrast to defining aggregate states in terms of space or energy, in adaptive collective behavior, the most important aspects are those that define informational properties. The key first question is then: What are the important aggregate states that we think of the system as computing? In the theory of computation, these aggregate logic states that define the computation are known as "information-bearing degrees of freedom" (Landauer, 1961). We refer to them here as informational states.

In aiming for simpler representations of adaptive systems, we can therefore be more focused in that we need not care about the simplest explanation of the system in general, but the simplest one that captures the informational states. In this sense, we want a parsimonious description of the "logic" or "algorithm" being implemented by the system. In neuroscience, cognitive science, and systems biology, it is common to talk about computations being performed by a system at the aggregate scale (Flack, 2017; Marr and Poggio, 1976; Dennett, 2014). The "logic" or "algorithm" that we want is precisely a simplified, compressed model of the mapping from the information contained in individuals to the information contained in the aggregate state (Marr and Poggio, 1976; Flack and Krakauer, 2011), one that might be used for control (Tomlin and Axelrod, 2005).

The focus here is on information and computation because, by definition, adaptive systems use relevant information about the state of the world to behave appropriately. Many aspects of adaptive systems are best understood in terms of maximizing relevant information (Nemenman, 2012; Sharpee, 2017), and biology is commonly conceptualized as being fundamentally informational

[5] In high-energy physics, renormalization explains how the laws of physics appear different when average energies are much different, as in, for instance, cosmological epochs just after the Big Bang.

(Krakauer et al., 2011; Davies and Walker, 2016). As one specific example, the visual system has been shown in multiple ways to be informationally optimized: the retina adapts to the statistics of incoming stimuli in a way that maximizes information transfer (Smirnakis et al., 1997), and the properties of retinal cell types produce optimal information transfer for images with statistics found in natural scenes (Kastner, Baccus, and Sharpee, 2015).

Essentially, our goals are the same as those of rate-distortion theory. We want to throw away information that is least important for the computation (lowering the "rate") and then to measure how well the reduced model performs (measuring the "distortion"). In the ideal case, the compressed representation exactly preserves all the aggregate properties that we care about predicting. Doing this efficiently—getting the most power for predicting aggregate properties given a limited amount of retained model information—is precisely the aim of rate-distortion theory and the closely related information bottleneck framework (Still, 2014).

In general, it is not guaranteed that model compression will work. We can imagine situations in which we are unlucky and predictions are impossible without knowing the precise state of every element of the system.[6] Happily for scientists, a (somewhat mysterious) property of our universe is that many details are often unimportant to what we care about (Transtrum, 2015). Compression techniques rely on systems being low-dimensional in some representation, and the trick is to find the right representation. As an example, the JPEG compression format preserves information that is most salient to human observers, and it does a good job for the typical sorts of images we encounter. Similarly, compression techniques for models need to know what aggregate-level features are important and cannot be "one size fits all."

Three broad approaches in particular have been useful in the case of collective behavior in living systems:

- Grouping into modules
- Focusing on aggregate-scale transitions: bifurcations, instability, and criticality
- Explicit model reduction

Note, however, that in general this endeavor of approximation and simplification is rather hopelessly broad and all-encompassing. A huge number of other related approaches exist, more or less specific to particular models and

[6] For instance, in computer engineering, a hash function produces output that changes dramatically with any small change to the input.

systems (e.g., searching through a set of possible structural forms in cognitive science (Kemp and Tenenbaum, 2008) or groupings of species (Feret et al., 2009) in biochemical signaling networks). There is no single correct way to do compression—this is the art of science.

3.1 Logic Approach 1: Emergent Grouped Logic Processors: Clustering, Modularity, Sparse Coding, and Motifs

One simple way to reduce dimensionality in models of collective behavior is to group components into distinct modules. This grouping can be accomplished using a variety of closely related concepts, including clustering, modularity, sparse coding, and community detection. The notion that groups can have distinct collective properties is well understood in, for instance, solid-state physics, where a phonon has a definite identity and physical effects but cannot be understood in terms of any individual molecule. Similarly, in many collective systems explanations formulated in terms of individual components are not well posed. For example: Which gene causes disease X? Which neuron causes decision Y? Which senator was responsible for passage of bill Z?

The phenomenon of modularity—whereby biological systems often consist of relatively independent subgroups—suggests that discovering these groups will produce more parsimonious descriptions. This can also be viewed as finding the important or natural scales of a system (Daniels, Ellison, Krakauer, and Flack, 2016). Particularly in neuroscience (e.g., Bassett et al., 2013) and genomics (e.g., Segal et al., 2003), clustering or searching for modules is used to interpret high-dimensional networks. Often, network inference procedures explicitly start with clustering before doing inference, effectively forcing the inference step to happen at a higher-order scale (Bonneau et al., 2006). Clustering can also be useful when performed at the higher level of dynamics and transitions among aggregate states. For instance, an inferred model of fly motion partitions behaviors into a hierarchical set of stereotypical movements (Berman, Bialek, and Shaevitz, 2016).

Most basic is clustering based on some intuitive notion of similarity—for instance, finding groups of individuals whose behavior is most correlated. This is also called community detection in networks. Similarly to network inference, a huge number of methods have been developed, and the best performing method will depend on the question being asked. Broadly, "hard clustering" methods separate components into nonoverlapping sets, and "soft clustering" allows components to be part of multiple groups. Some common general-purpose methods include k-means, hierarchical clustering, and multivariate Gaussians.

These basic clustering methods find intuitive groupings, but this does not yet necessarily give insight into the logic that produces a system's output. For instance, a simple clustering method will ask for the number of desired clusters k, but without further specifying the problem, we can only choose k arbitrarily. Instead, we want to choose k based on the aspects of the system we care most about, which in this case is its informational output. This leads us toward techniques written in the language of information theory, such as sparse coding (Daniels et al., 2012) or rate-distortion clustering methods (Slonim, Atwal, Tkacik, and Bialek, 2005). For instance, sparse coding can describe a system parsimoniously in terms of commonly appearing active subgroups, and then information theory can measure how much this reduces the information we need to remember in order to best predict future co-occurring individuals. The general idea is to specify a "distortion function," defining what information we want to retain,[7] and then to vary a single parameter λ that determines the complexity of the representation. For small λ, we favor more accuracy and more clusters, in the extreme case putting each component into its own cluster (at the bottom of Figure 3.2). As λ increases, we group components into larger clusters in a way that retains the most information about the system's output (moving toward the top of Figure 3.2).

Another approach, currently less automated, looks for patterns in network connectivity and dynamics that have known informational or logical functionality. These are known as functional motifs (Alon, 2007) or logical subcircuits (Peter and Davidson, 2017). The overabundance of some types of motifs is suggestive that they are more useful for information processing, and the computational properties of these motifs has been explored at length (e.g., Alon, 2007; Payne and Wagner, 2015). In this way, we can think of clusters and motifs as intuitive parts from which more complicated computations are built.

3.2 Logic Approach 2: Instability, Bifurcations, and Criticality

Grouping individual components is a useful step, but it may not tell us much about the system's logic. Another useful approach is to describe the system's behavior in terms of collective transitions or instabilities that control changes among aggregate informational states.

[7] Note that the optimal clusters depend on the distortion function. This captures the fact that the best representation of a system depends on which aspects of the system we consider to be relevant (Shalizi and Moore, 2003). In the case of systems for which we wish to understand the origins of a particular aggregate function, we can take this aggregate "output" to define the relevant informational states.

This focus on higher-level logic is also advantageous in that it sidesteps issues of the inability to constrain parameters (section 2.3), especially when there are unmeasured but important components (as is often the case in, for instance, neuroscience and cell biology). As one example, the identity of stable functional states in a gene-regulatory network has been shown to be robust to parameter changes (Jia, Jolly, and Levine, 2018).

Informational states often correspond to those that include many components or those seen at long timescales. These lead to two mathematical limits and two domains of theory: (1) the limit of many components leads to transitions studied in statistical physics using the language of phases and criticality, and (2) the limit of large time leads to transitions among attractors studied in dynamical systems using the language of bifurcations.

3.2.1 Fisher Information and Criticality

In an equilibrium system, we can make an analogy with statistical physics and talk about a system's phases: What are the coarse-grained aggregate states that characterize the system? As in Figure 3.3, information that washes out at the aggregate scale allows us to ignore some individual-scale details. Information that grows can produce well-separated, distinct aggregate states. These states are primed to become important at the aggregate scale as they carry specific information about the individual scale—they become informational states.

What causes these emergent phases? In physics and in collective behavior more generally, we think about this by considering how the system changes as we move away from the individual level—more atoms, more molecules, more people. Intuitively, whether information is amplified or decays depends on how perturbations spread through a system (Daniels, Krakauer, and Flack, in preparation). A perturbation may die out or be overwhelmed by noise, becoming smaller as it spreads to more individuals. This corresponds to an irrelevant, compressed direction in parameter space in Figure 3.3. Or the perturbation can be amplified, becoming larger in magnitude as it spreads, corresponding to a relevant, growing direction in Figure 3.3. It is this latter case that corresponds to a collective instability that can produce distinct aggregate states (Daniels et al., 2017; Daniels, in preparation). In renormalization group flows, this instability comes from an effective parameter value being amplified as the scale increases.[8] Importantly, these instabilities can be connected to computational functions such as consensus formation and decision making (Daniels, Flack, and Krakauer, 2017). This process through which instabilities create distinct

[8] More general cases of collective behavior are often trickier to represent formally in the renormalization group language because the aggregate states we are interested in are not always simple sums over individuals.

aggregate states, the transition between components behaving as if "anything goes" versus "we have all settled on this particular arrangement," is known as symmetry breaking in physics (Sharpee, 2017; Anderson, 1972; Sethna, 2006). The notion that symmetry breaking and the creation of distinct attractors are fundamental to defining meaning in biology has been explored by numerous authors (Solé et al., 2016; Anderson, 1972; Brender, 2012). Cell types are explained as attractors of gene-regulatory networks (Lang, Li, Collins, and Mehtr, 2014), and swarming and milling states of fish schools are interpreted as collective phases (Tunstrom et al., 2013).

As inputs or individual behavior change, how do they control changes in these aggregate states? This is key to describing information processing and can be measured using the same Fisher information that we used to measure sensitivity to parameters in section 2.3 (Eq. [3.5]). A crucial insight in making a connection to statistical physics is that phase transitions are defined by extreme system sensitivity (Daniels et al., 2017; Daniels et al., in preparation). This is intuitively clear in that small changes in control parameters lead to systemwide changes in behavior at a phase transition: changing your freezer's temperature from −1 degree C to +1 degree C creates very different behavior of the water molecules in the ice cube tray. This intuition is made sharp by the Fisher information, which has been shown to become infinite precisely at phase transitions (Prokopenko, Lizier, Obst, and Wang, 2011).[9]

We can think of the Fisher information as measuring amplification: the degree to which information at the small scale has large, aggregate-scale effects. This can measure the sensitivity of the structure in a social animal group to changes in individual bias toward conflict (Daniels et al., 2017) or the sensitivity of a group of fish to an individual who detects a predator (Sosna et al., 2019). In this way, the informational perspective is useful for framing the idea of phase transitions in biology. Even in finite systems (away from the limit of an infinite number of components that produces sharp "true" phase transitions), the Fisher information connects with biological function as a generalized measure of functional sensitivity.

When viewing biological collectives as computers whose output must be sensitive to changing input, it is perhaps unsurprising that many are found to lie near such instabilities. "Nearness to criticality" has been found across many collective systems (Mora and Bialek, 2011), ranging from neurons (Cocchi, Gollo, Zalesky, and Breakspear, 2017) to flocks (Bialek et al., 2014) to societies

[9] Technically speaking, this is true only at continuous-phase transitions because an energy barrier is associated with discontinuous-phase transitions (leading to hysteresis) that prevents a given (symmetry broken) state from being easily poked into a new aggregate state. Still, the long-time equilibrium state becomes infinitely sensitive to perturbations at discontinuous transitions.

(Daniels et al., 2017). Studies demonstrating criticality typically start with an inferred model and show that small changes to parameters can bring the system to a peak in sensitivity, or measure other indications of self-similarity arising from the system being at a marginal point between information amplification and decay (Daniels et al., in preparation).

It is not necessarily the case that maximal sensitivity is best, and it may be advantageous for adaptive systems to move closer or further from criticality. There is some evidence, for instance, that distance from criticality in the brain varies over the sleep–wake cycle (Priesemann, Valderrama, Wibral, and van Quyen, 2013). Also, fish schools change their density to respond to the perceived level of threat of a predator (Sosna et al., 2019). This can be interpreted as moving closer to criticality when fish feel threatened (letting changes spread more quickly through the school) and staying further from criticality otherwise (to avoid responding to random uninformative changes in individual behavior). In a social system, this distance has been measured in biologically meaningful units as the number of individuals whose behavior would have to change to get to a point of maximal sensitivity (Daniels et al., 2017).

3.2.2 Dynamical Systems and Bifurcations

If our model is deterministic and dynamic (as in dynamical inference, section 2.2.2), it often makes sense to think of the stable attractors as defining logic states of the system. This viewpoint, asking about the system's behavior in the limit as time $t \to \infty$, is analogous to the large number limit that defines phases in equilibrium models. And analogously to the mathematics of phase transitions, here the machinery of nonlinear dynamics (Strogatz, 1994) gives us the language of bifurcations for describing the changes of state that describe a system's logic. (This view of thinking of dynamics as computation has been explored and debated at length in cognitive science; e.g., Michell, 1994; Beer, 2014.)

As an example, consider a system that performs a reproducible dynamic response to a stimulus, such as a worm responding to the application of heat by changing its speed and direction of motion (Daniels, Ryu, and Nemenman, 2019). We are certain that this behavior is controlled by neurons and so we could attempt to infer a model at the level of individual neurons, but we will consider a case in which data at this level of detail is not available. Instead, as shown in Figure 3.4, we have data describing the speed of the worm as a function of time for trials with varying heat intensity, which we treat as input to the system. The adaptive model selection approaches described in section 2.2 can then be used to infer an effective model that uses the types of saturating interactions that are common to biological systems, capturing the dynamics at a coarse-grained level.

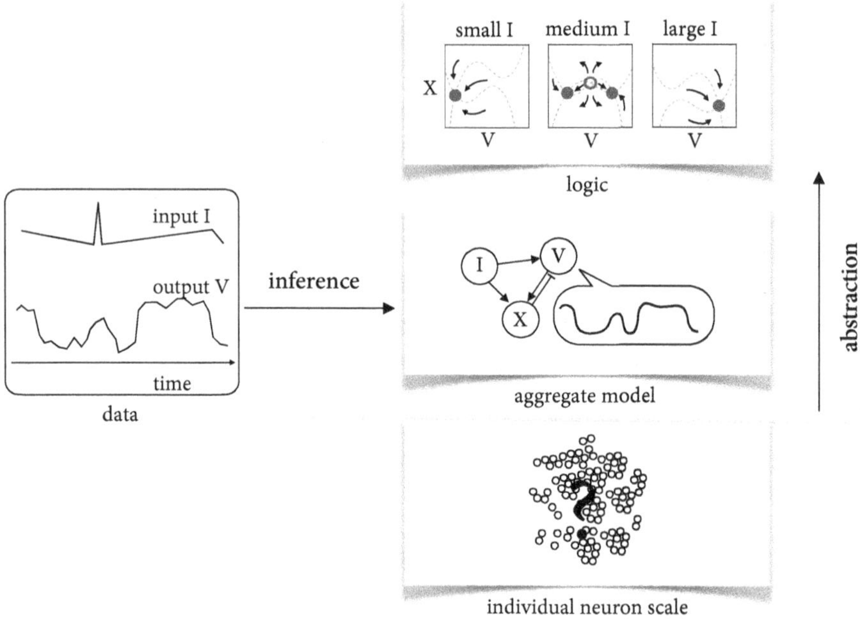

Figure 3.4 Phase space structure as logic. In this caricatured example, time series data leads to an effective dynamical model (represented as nodes and arrows in the middle row) that can predict the result of arbitrary dynamical input. The inference procedure makes use of effective dynamical variables, which can be interpreted as encapsulating the behavior of groups of individual neurons, and in this case includes an additional hidden variable *X*. Examining the phase space structure (top) produces a simple logic in terms of steady-state fixed points (filled dots, with an unstable fixed point as an unfilled dot, nullclines as dotted lines, and dynamical flow lines as solid arrows): the structure of the dynamics can be traced to a pair of saddle-node bifurcations.

Even this coarse-grained representation can be used as a starting point to describe the logic of the system. We abstract to the level of logic by examining the phase space structure of the inferred model, shown in the top row of Figure 3.4.[10] In this scenario, the heat input induces a bifurcation that switches the system between distinct patterns of motion. Even if the system does not ever

[10] In the case shown in Figure 3.4, the inferred model is two-dimensional. This case is particularly simple because, using Morse–Smale theory (Palis and de Melo, 1982), it is possible to uniquely classify almost all (compact) two-dimensional-phase portraits according to the number and types of attractors. It is not always possible to perform such a classification of dynamical systems in higher-dimensional cases. Particularly in cases of deterministic chaos, simpler abstract descriptions may not be possible.

saturate to fully reach one of the fixed-state attractors, examining the model in this way provides a succinct explanation for the switch-like behavior: it arises in this imagined example from a pair of saddle–node bifurcations. Furthermore, to the extent that the effective model is a sufficiently realistic representation of the behavior of coarse-grained groups of neurons, any future explanation in terms of individual neurons will be consistent with the inferred logic.

3.3 Logic Approach 3: Explicit Model Reduction

In the previous sections, we looked for specific reduced representations: grouping individual components, characterizing the system using sensitivity with respect to certain directions in parameter space, or deriving the structure of attractors. In a more general context, it has long been a dream to produce automated approximations and model simplifications that begin with a complicated model and produce a simpler model of the same type. The hope is to find explicit approximations of known detailed models.

Dynamical models written in the form of a Markov process can be analyzed using the ε machine formalism (Shalizi and Crutchfield, 2001). Starting with a known Markov process, this formalism defines the minimally complex Markov model that exactly reproduces the behavior of the original process. Recent developments have generalized this reasoning to the more typical case in which we cannot produce a smaller model that produces the exact same predictions, but instead we look to maximize predictive power while restricting the model size (Marzen and Crutchfield, 2017).

As we saw in section 2.3, sloppiness in models suggests that lower-dimensional representations should be good approximations. Another particularly elegant method uses the same information geometry that defines sloppiness to find such approximate models. Treating the Fisher information matrix as a metric tensor, following geodesics corresponds to mapping out the model manifold, the space of possible outputs of the model. Following the sloppiest direction corresponds to changing parameters in a way that minimally affects the measurable outputs and often approaches boundaries on the model manifold, places where taking combinations of parameters to infinity does not change the model output (Transtrum and Qiu, 2016). In this way, mathematical limits corresponding to simplifying approximations can be found in a semiautomated way. For example, starting with a more complicated mechanistic model for enzyme kinetics, the method can automatically discover the approximations that lead to the widely used Michaelis–Menten model, which assumes that the substrate is in instantaneous chemical equilibrium (Transtrum and Qiu, 2016).

4. The Future of the Science of Living Systems

We observe that proteins, neurons, ants, fish, politicians, and scientists each create structures that process information and perform impressive collective feats. Obtaining a deep understanding of how parts come together to act as an adaptive whole is a worthy challenge for modern science.

The relationship between simplified theories, statistical physics, and machine learning has been emphasized for many decades and continues to be fruitful (Denker et al., 1987; Seung, Sompolinsky, and Tishby, 1992; Mehta and Schwab, 2014). The intent here is to use these ideas to understand many specific cases of collective behavior observed in nature. Since each case is different and involves so many details, it will likely be necessary to construct a new model for every system. Automated methods will be key to expedite this process.

We might dream of a time in the not-too-distant future when all tasks in inferring abstracted models can be accomplished by an automated machine. Such a machine might use fish school data to locate a critical instability and demonstrate how information about predators is maximized when the school is closer to the instability. With spiking data from millions of neurons, it might group them into functional clusters and show how they create an aggregate-level gated sensory classifier. Using insulin expression data from millions of patients, it might produce both patient-specific predictions and an abstract dynamical framework for how interventions control the phase and frequency of oscillations. In each case, the two-step process of Figure 3.2 means that we get both the accuracy of the full messy predictive model and the parsimonious abstracted theory describing the aggregate level logic.

This is not an outrageous goal. It will require conceptual ingenuity both on the side of efficient inference and model selection and on paring down and interpreting predictive models once they are built. The payoff to practitioners is highly predictive models, and the payoff to science is a much improved position to understand overarching principles in biology.

What are we expecting to find? There are already hints that broader principles or strategies for collective information processing are at work. For instance, a two-phase picture for robust collective decision making, moving from a distributed uncertain phase to a redundant consensus phase, arises naturally when using the criticality framework of section 3.2.1. This two-phase implementation of decision-making corresponds to a logical structure that could be tested across a variety of biological and social systems (Daniels et al., 2017; Arehart, Jin, and Daniels, 2018). We are beginning to see how a variety of systems regulate distance from criticality at the individual scale, and how to classify behavioral dynamics based on whether they include bifurcations or instabilities that induce qualitative changes to the phase space structure based on sensory input (e.g., Daniels et al., 2018).

What might this approach be missing? First, inferring a full generative, predictive model before using it to ask questions about logic and mechanism may be overkill. In at least one case of dynamical inference, some qualitative results about mechanisms that map from microscopic to macroscopic can be obtained without using a specific generative model (Barzel, Liu, and Barabási, 2015). For instance, aggregate states and distance from a symmetry-breaking transition can be identified and tracked using only observed variance in the states of individual players (Mojtahedi et al., 2016). A second related point is that this approach treats inference and abstraction as two separate steps. This creates a clean caricature, with the two parts having different (and competing) goals: inference favors predictability even if it means more complexity, and abstraction favors simplicity even if it means less accuracy. Advances on these two fronts can in some respects be made independently, with inference searching for details to add to make a model more predictive (a common perspective in biology and machine learning) and abstraction searching for ways to throw details away (a common perspective in information theory and statistical physics). Yet the two perspectives are not always clearly separable. We often have to find the right abstract level to do inference at all, as in regularized models, and we may be able to describe logic without having to infer lower-level mechanisms. Ultimately, it is in the tension of combining these two perspectives that science progresses.

Acknowledgments

The ideas presented here have been influenced by numerous fruitful discussions with the C4 Collective Computation Group, particularly Jessica Flack, David Krakauer, Chris Ellison, and Eddie Lee. Thanks to Ken Aiello for helpful comments on an earlier draft.

References

Alon, U. (2017). "Network Motifs: Theory and Experimental Approaches." *National Reviews Genetics*, 8(6): 450–461.

Amit, D. J., H. Gutfreund, and H. Sompolinsky. (1985). "Spin-Glass Models of Neural Networks." *Physical Review A*, 32(2): 1007–1018.

Anderson, P. (1972). "More Is Different." *Science*, 177.4047 (1972), 393–396.

Arehart, E., T. Jin, and B. C. Daniels. (2018). "Locating Decision-Making Circuits in a Heterogeneous Neural Network." *Frontiers in Applied Mathematics and Statistics*, 4: 11.

Babtie, A. C., and M. P. H. Stumpf. (2017). "How to Deal with Parameters for Whole-Cell Modelling." *Journal of the Royal Society Interface*, 14(133): 20170237.

Barzel, B., Y.-Y. Liu, and A.-L. Barabási. (2015). "Constructing Minimal Models for Complex System Dynamics." *Nature Communications*, 6: 7186.

Bassett, D. S., et al. (2011). "Dynamic Reconfiguration of Human Brain Networks during Learning." Proceedings of the National Academy of Sciences, 108(18): 7641–7646.

Bassett, D. S., et al. (2013). "Robust Detection of Dynamic Community Structure in Networks." *Chaos*, 23: 013142.

Beer, R. D. (2014). "Dynamical Systems and Embedded Cognition." In K. Frankish and W. Ramsey (eds.), *The Cambridge Handbook of Artificial Intelligence*, (pp. 128–148). Cambridge: Cambridge University Press, 2014.

Berman, G. J., W. Bialek, and J.W. Shaevitz. (2016). "Hierarchy and Predictability in *Drosophila* Behavior." *Proceedings of the National Academy of Sciences*, 113(42): 11943.

Bialek, W., et al. (2014). "Social Interactions Dominate Speed Control in Poising Natural Flocks near Criticality." *Proceedings of the National Academy of Sciences*, 111: 7212–7217.

Bonneau, R., et al. (2006). "The Inferelator: An Algorithm for Learning Parsimonious Regulatory Networks from Systems-Biology Data Sets De Novo." *Genome Biology*, 7(5): 1.

Box, G. E. P. (1979). "Robustness in the Strategy of Scientific Model Building." Army Research Office Workshop on Robustness in Statistics, pp. 201–236.

Brender, N. M. (2012). "Sense-Making and Symmetry-Breaking: Merleau-Ponty, Cognitive Science, and Dynamic Systems Theory." *Symposium: Canadian Journal of Continental Philosophy*, 17(2): 246–270.

Brunton, S. L., J. L. Proctor, and J. N. Kutz. (2016). "Discovering Governing Equations from Data by Sparse Identification of Nonlinear Dynamical Systems." *Proceedings of the National Academy of Sciences*, 113(15): 3932–3937.

Cocchi, L., L. L. Gollo, A. Zalesky, and M. Breakspear. (2017). "Criticality in the Brain." *Progress in Neurobiology*, 158: 132–152.

Cocco, S., and R. Monasson. (2012). "Adaptive Cluster Expansion for the Inverse Ising Problem: Convergence, Algorithm and Tests." *Journal of Statistical Physics*, 147: 252.

Couzin, I. D. (2009). "Collective Cognition in Animal Groups." *Trends in Cognitive Sciences*, 13(1): 36–43.

Crutchfield, J. P., and M. Mitchell. (1995). "The Evolution of Emergent Computation." *Proceedings of the National Academy of Science*, 92: 10742–10746.

Daniels, B. C., et al. (2018). "Criticality Distinguishes the Ensemble of Biological Regulatory Networks." *Physical Review Letters*, 121(13): 138102.

Daniels, B. C., and I. Nemenman. (2015). "Automated Adaptive Inference of Phenomenological Dynamical Models." *Nature Communications*, 6: 8133.

Daniels, B. C., C. J. Ellison, D. C. Krakauer, and J. C. Flack. (2016). "Quantifying Collectivity." *Current Opinion in Neurobiology*, 37: 106–113.

Daniels, B. C., D. C. Krakauer, and J. C. Flack. (2017). "Control of Finite Critical Behavior in a Small-Scale Social System." *Nature Communications*, 8: 14301.

Daniels, B. C., D. C. Krakauer, and J. C. Flack. (in preparation). "Distance from Criticality in Adaptive Collective Behavior."

Daniels, B. C., D. C. Krakauer, and J. C. Flack. (2012). "Sparse Code of Conflict in a Primate Society." *Proceedings of the National Academy of Sciences*, 109(35): 14259.

Daniels, B. C., et al. (2008). "Sloppiness, Robustness, and Evolvability in Systems Biology." *Current Opinion in Biotechnology*, 19(4): 389–395.

Daniels, B. C., J. C. Flack, and D. C. Krakauer. (2017). "Dual Coding Theory Explains Biphasic Collective Computation in Neural Decision-Making." *Frontiers in Neuroscience*, 11: 313.

Daniels, B. C., M. Dobrzynski, and D. Fey. (2018). "Parameter Estimation, Sloppiness, and Model Identifiability." In B. Munsky, L. Tsimring, and W. Hlavacek (eds.), *Quantitative Biology: Theory, Computational Methods, and Models*. Cambridge, MA: MIT Press, 271.

Daniels, B. C., W. S. Ryu, and I. Nemenman. (2019). "Automated, Predictive, and Interpretable Inference of *Caenorhabditis Elegans* Escape Dynamics." *Proceedings of the National Academy of Sciences*, 116(15), 7226–7231.

Davies, P. C. W., and S. I. Walker. (2016). "The Hidden Simplicity of Biology." *Reports on Progress in Physics*, 79: 102601.

Denker, J., et al. (1987). "Large Automatic Learning, Rule Extraction and Generalization." *Complex Systems*, 1: 877–922.

Dennett, D. (2014). "The Software/Wetware Distinction: Comment on 'Unifying Approaches From Cognitive Neuroscience And Comparative Cognition' by W. Tecumseh Fitch." *Physics of Life Reviews*, 11: 367–368.

Feret, J., et al. (2009). "Internal Coarse-Graining of Molecular Systems." *Proceedings of the National Academy of Sciences*, 106(16): 6453–6458.

Flack, J. (2017). "Life's Information Hierarchy." In S. I. Walker, P. C. W. Davies, and G. F. R. Ellis (eds.), *From Matter to Life: Information and Causality*. Cambridge: Cambridge University Press, 283.

Flack, J. C., and D. C. Krakauer. (2011). "Challenges for Complexity Measures: A Perspective from Social Dynamics and Collective Social Computation." *Chaos: An Interdisciplinary Journal of Nonlinear Science*, 21(3): 037108.

Ganmor, E., R. Segev, and E. Schneidman. (2011). "Sparse Low-Order Interaction Network Underlies a Highly Correlated and Learnable Neural Population Code." *Proceedings of the National Academy of Sciences*, 108(23): 9679–9684.

Goldenfeld, N. (1992). *Lectures on Phase Transitions and the Renormalization Group*. New York: Westview Press.

Gunawardena, J. (2014). "Models in Biology: 'Accurate Descriptions of Our Pathetic Thinking'. " *BMC Biology*, 12: 29.

Hoel, E. P., L. Albantakis, and G. Tononi. (2013). "Quantifying Causal Emergence Shows That Macro Can Beat Micro." *Proceedings of the National Academy of Sciences*, 110(49): 19790–19795.

Jia, D., M. K. Jolly, and H. Levine. (2018). Uses of Bifurcation Analysis in Understanding Cellular Decision-Making." In B. Munsky, L. Tsimring, and W. Hlavacek (eds.), *Quantitative Biology: Theory, Computational Methods, and Models*. Cambridge, MA: MIT Press, 357.

Kastner, D. B., S. A. Baccus, and T. O. Sharpee. (2015). "Critical and Maximally Informative Encoding between Neural Populations in the Retina." *Proceedings of the National Academy of Science*, 112: 2533–2538.

Kauffman, S. (1969). "Metabolic Stability and Epigenesis In Randomly Constructed Genetic Nets." *Journal of Theoretical Biology*, 22(3): 437–467.

Kemp, C., and J. B. Tenenbaum. (2008). "The Discovery of Structural Form." *Proceedings of the National Academy of Sciences*, 105(31): 10687–10692.

Krakauer, D. C., et al. (2011). "The Challenges and Scope of Theoretical Biology." *Journal of Theoretical Biology*, 276(1): 269–276.

Landauer, R. (1961, July). "Irreversibility and Heat Generation in the Computational Process." *IBM Journal of Research and Development*, 5: 183–191.

Lander, A. D. (2010). "The Edges of Understanding." *BMC Biology* 8: 40.

Lang, A. H., H. Li, J. J. Collins, and P. Mehta. (2014). "Epigenetic Landscapes Explain Partially Reprogrammed Cells and Identify Key Reprogramming Genes." *PLoS Computational Biology*, 10(8):

Lee, E. D., and B. C. Daniels. (2019). "Convenient Interface to Inverse Ising (ConIII): A Python 3 Package for Solving Ising-Type Maximum Entropy Models." *Journal of Open Research Software*, 7: 3.

Lukoševicius, M., and H. Jaeger. (2009). "Reservoir Computing Approaches to Recurrent Neural Network Training." *Computer Science Review*, 3(3): 127–149.

Machta, B. B., R. Chachra, M. K. Transtrum, and J. P. Sethna. (2013). "Parameter Space Compression Underlies Emergent Theories and Predictive Models." *Science*, 342 (6158): 604–607.

Marr, D. C., and T. Poggio. (1976). "From Understanding Computation to Understanding Neural Circuitry." Massachusetts Institute of Technology Artificial Intelligence Laboratory A.I. Memo 357.

Marzen, S. E., and J. P. Crutchfield. (2017). "Nearly Maximally Predictive Features and Their Dimensions." *Physical Review*, E 95 (2017), 051301(R).

Mehta, P., and D. J. Schwab. (2014). "An Exact Mapping between the Variational Renormalization Group and Deep Learning." arXiv preprint 1410.3831.

Merchan, L., and I. Nemenman. (2016). "On the Sufficiency of Pairwise Interactions in Maximum Entropy Models of Networks." *Journal of Statistical Physics*, 162: 1294–1308.

Mitchell, M. (1998). "A Complex-Systems Perspective on the 'Computation vs. Dynamics' Debate in Cognitive Science." *Proceedings of the Twentieth Annual Conference of the Cognitive Science Society*. Mahwah, NJ: Lawrence Erlbaum Associates, 710.

Mojtahedi, M., et al. (2016). "Cell Fate Decision as High-Dimensional Critical State Transition." *PLoS Biology*, 14(12): (2016), e2000640.

Mora, T., and W. Bialek. (2011). "Are Biological Systems Poised at Criticality?" *Journal of Statistical Physics*, 144 (2011), 268–302.

Natale, J. L., D. Hofmann, D. G. Hernández, and I. Nemenman. (2018). "Reverse-Engineering Biological Networks from Large Data Sets." In B. Munsky, W. Hlavacek, and L. Tsimring (eds,), *Quantitative Biology: Theory, Computational Methods, and Models*. Cambridge, MA: MIT Press, 213.

Nemenman, I. (2012). "Information Theory and Adaptation." In M. E. Wall (ed.), *Quantitative Biology: From Molecular to Cellular Systems*, (Chapter 5). Boca Raton, FL: Taylor and Francis.

Newman, M. (2010). *Networks: An Introduction*. New York: Oxford University Press, 2010.

Palis, J., Jr., and W. de Melo. (1982). *Geometric Theory of Dynamical Systems*. Berlin: Springer-Verlag.

Payne, J. L., and A. Wagner. (2015). "Function Does not Follow Form in Gene Regulatory Circuits." Scientific Reports 5, 13015.

Peter, I. S., and E. H. Davidson. (2017). "Assessing Regulatory Information in Developmental Gene Regulatory Networks." *Proceedings of the National Academy of Sciences*, 114(23): 5862–5869.

Priesemann, V., M. Valderrama, M. Wibral, and M. Le Van Quyen. (2013). "Neuronal Avalanches Differ from Wakefulness to Deep Sleep." *PLoS Computational Biology*, 9(3): e1002985.

Prokopenko, M., J. T. Lizier, O. Obst, and X. R. Wang. (2011). "Relating Fisher Information to Order Parameters." *Physical Review E*, 84(4): 41116.
Rosenthal, S. B., et al. (2015). "Revealing the Hidden Networks of Interaction in Mobile Animal Groups Allows Prediction of Complex Behavioral Contagion." *Proceedings of the National Academy of Sciences*, 112(15): 4690–4695.
Roudi, Y., S. Nirenberg, and P. E. Latham. (2009). "Pairwise Maximum Entropy Models for Studying Large Biological Systems: When They Can Work and When They Can't." *PLoS Computational Biology*, 5(5): e1000380.
Rumelhart, D. E., and J. L. McClelland. (1986): *Parallel Distributed Processing.* Vol. 1. Cambridge, MA: MIT Press.
Schmidt, M., and H. Lipson. (2009). "Distilling Free-Form Natural Laws from Experimental Data." *Science*, 324(5923): 81–85.
Schneidman, E., M. J. Berry II, R. Segev, and W. Bialek. (2006). "Weak Pairwise Correlations Imply Strongly Correlated Network States in a Neural Population." *Nature*, 440: 1007.
Schwab, D. J., I. Nemenman, and P. Mehta. (2014). "Zipf's Law and Criticality in Multivariate Data without Fine-Tuning." *Physical Review Letters*, 113(6): 068102.
Segal, E., et al. (2003). "Module Networks: Identifying Regulatory Modules and Their Condition-Specific Regulators from Gene Expression Data." *Nature Genetics*, 34(2): 166–176.
Sethna, J. (2006). *Entropy, Order Parameters, and Complexity.* New York: Oxford University Press.
Seung, H., H. Sompolinsky, and N. Tishby. (1992). "Statistical Mechanics of Learning from Examples." *Physical Review A*, 45(8): 6056.
Shalizi, C. R. and C. Moore. (2003). "What Is a Macrostate? Subjective Observations and Objective Dynamics." arXiv preprint cond-mat/0303625.
Shalizi, C. R., and J. P. Crutchfield. (2001). "Computational Mechanics: Pattern, Prediction Strucutre and Simplicity." *Journal of Statistical Physics*, 104: 817–879.
Sharpee, T. O. (2017). "Optimizing Neural Information Capacity through Discretization." *Neuron*, 94(5): 954–960.
Slonim, N., G. S. Atwal, G. Tkacik, and W. Bialek. (2005). "Information-Based Clustering." *Proceedings of the National Academy of Sciences*, 102(51): 18297–18302.
Smirnakis, S. M., et al. (1997). "Adaptation of Retinal Processing to Image Contrast and Spatial Scale." *Nature*, 386(6620): 69–73.
Solé, R., et al. (2016). "Synthetic Collective Intelligence." *BioSystems*, 148: 47–61.
Sosna, M. M. G., et al. (2019). "Individual and collective encoding of risk in animal groups." *Proceedings of the National Academy of Sciences*, 116(41): 20556–20561.
Stephens, G. J., L. C. Osborne, and W. Bialek. (2011). "Searching for Simplicity in the Analysis of Neurons and Behavior." *Proceedings of the National Academy of Sciences*, 108: 15565–15571.
Still, S. (2014). "Information Bottleneck Approach to Predictive Inference." *Entropy*, 16(2): 968–989.
Strogatz, S. H. (1994). *Nonlinear Dynamics and Chaos.* New York: Perseus.
Tomlin, J. C., and J. D. Axelrod. (2005), "Understanding Biology by Reverse Engineering the Control." *Proceedings of the National Academy of Sciences*, 102(8): 4219–4220.
Transtrum, M. K., and P. Qiu. (2016). "Bridging Mechanistic and Phenomenological Models of Complex Biological Systems." *PLoS Computational Biology*, 12(5): 1–34.

Transtrum, M. K., et al. (2015). "Sloppiness and Emergent Theories in Physics, Biology, and Beyond." *Journal of Chemical Physics*, 143(1): 010901.
Tunstrom, K., et al. (2013). "Collective States, Multistability, and Transitional Behavior In Schooling Fish." *PLoS Computational Biology*, 9: e1002915.
Wilson, K. G. (1979). "Problems in Physics with Many Scales of Length." *Scientific American*, 241(2): 158–179.
Wolpert, D. H., J. A. Groschow, E. Libby, and S. DeDeo. (2015). "Optimal High-Level Descriptions Of Dynamical Systems." arXiv preprint 1409.7403v2.

4
Information-Theoretic Perspective on Human Ability

Hwan-sik Choi

1. Introduction

Since Shannon's seminal work (1948) was published, information theory and the concept of entropy have been useful in various fields of science, including physics, computer science, statistics, and economics. In information theory, entropy is a measure of uncertainty or lack of information of a system. The principle of maximum entropy (Jaynes, 1957a,b) states that an appropriate model given a set of constraints is the one with the maximum entropy satisfying those constraints. Since constraints usually reduce the maximum entropy of a system, a set of entropy-reducing constraints may be thought of as information.

Humans are quite heterogeneous in their ability, and as a result they behave differently in a given situation. The unique set of abilities that each person possesses imposes constraints on behavior. Combining this idea with the maximum entropy principle, this chapter proposes an information-theoretic perspective on the role of human ability.

I model a person as an entity that interacts with a situation to produce a behavior. From this point of view, a person is a response system that takes a situation as an input to produce a behavior as an output. This response function represents a person-specific mapping from situations to behaviors. To the extent that different people have different response functions, a person is uniquely identifiable by this mapping. In the broadest sense, this mapping is defined as personality. The idea of treating personality as a response system is not new. For example, see Roberts (2006, 2009) and Almlund et al. (2011). The main objective of this chapter is to provide an information-theoretic framework for a human's response system as a common ground for psychology and economics.

Let X be a situation and Y be an output of a human response system. Based on the maximum entropy principle, we may choose a model that maximizes the joint entropy of X and Y under the constraints that the response function imposes. The total entropy is maximized if X and Y are independent, that is, if

Hwan-sik Choi, *Information-Theoretic Perspective on Human Ability* In: *Advances in Info-Metrics: Information and Information Processing across Disciplines*. Edited by: Min Chen, J. Michael Dunn, Amos Golan, and Aman Ullah, Oxford University Press (2021).
DOI: 10.1093/oso/9780190636685.003.0004

the response function ignores X completely. Any relation between X and Y that reduces the joint entropy is considered to be information.

A human response system is characterized by many factors. At a rudimentary level, physical laws govern a person's movements. Moreover, the human genome provides constraints in the response system. For example, humans are programmed by their genes to consume water and food for survival. Because of this genetic constraint, we would find people around an oasis in a desert rather than all over the desert uniformly.[1] These physical and genetic constraints reduce the entropy of a system and are considered to be information.

In addition to these constraints, at the advanced level, human beings have a complex decision-making structure that depends on infinitely many person-specific characteristics and abilities. Since a person's various abilities are an essential component of the human response function, human ability imposes constraints on how X influences Y. Therefore, I treat the entropy-reducing human ability as information.

In the neoclassical economic model, rationality is the central information-theoretic assumption. Under this assumption, an economic agent possesses unlimited information capabilities in the following sense: the agent is assumed to fully absorb and understand all available information, subjective beliefs on the state of nature are coherent, and behavioral outcomes are always optimal under a certain preference structure.

Rationality is thought to be an ideal version of personality that contains a large set of constraints, which reduces entropy. Therefore, rationality possesses rich information. For this reason, we can get a simple and robust behavioral model with rationality. However, just as Simon (1955, 1956) points out that rationality may be bounded by limited human capabilities, this chapter argues that it is important to incorporate humans' limited ability in dealing with information, which I call *information capacity*, into the neoclassical economic framework. More specifically, I consider that individuals differ greatly in the following three components of information capacity: the ability to acquire information, the ability to process information, and the ability to discern incorrect information. These three components are explained in detail in section 2.1.

Sims (2003, 2006) proposes an extension to the neoclassical model by adding information capability as a constraint for a rational agent. His idea is briefly explained in section 2.2. In section 2.3, I approximate information capacity from the perspective of social psychology. I provide a heuristic discussion that relates information capacity with broad domains of noncognitive ability modeled by the Big Five personality traits: Openness (to Experience), Conscientiousness,

[1] See Benjamin et al. (2012) for a review of recent developments in genoeconomics.

Extraversion, Agreeableness, and Neuroticism. This discussion focuses on the connection between information capacity, which is an unobservable quantity, and the Big Five personality model, for which there is a well-developed set of measures. See McCrae and Costa (1999) and John et al. (2008) for an introduction to the Big Five personality model.

Among the Big Five personality traits, I consider the following three traits to be related to information capacity: Openness, Conscientiousness, and Agreeableness. Openness is related to curiosity and intelligence; Conscientiousness is related to being careful, hardworking, and thorough; and Agreeableness is related to being trustful and softhearted. I hypothesize the following: Openness is positively related to information acquisition; Conscientiousness is positively related to information acquisition and processing; and Agreeableness is negatively related to discerning incorrect information. Therefore, an open, conscientious, and disagreeable person would have a high level of information capacity. I discuss further details in section 2.3.

To examine these hypotheses, an empirical study is presented in section 3. I find that those three traits have expected correlations with information-related behaviors. I also find empirical evidence that information capacity might be related to long-term positive consequences in wealth accumulation and economic well-being.

2. Information-Theoretic Perspective on Human Ability

2.1 The Maximum Entropy Principle and Information Capacity

Consider a response system of an entity such as an atom, a cell, a plant, or a human being, which takes an external situation or an environment as an input X and produces behaviors as an output Y. The state of nature is defined as the bundle (X, Y). As shown in Figure 4.1, for example, energies and masses surrounding an entity may characterize its physical environment.

In addition to the physical environment, the situation that a person faces includes incentives, memories, and past experiences. In economics, market prices and institutions may be important inputs to the response function of an individual. Given the situation, an entity produces behaviors as an outcome. The output of a human response system includes physical behaviors, choices, feelings, and thoughts as in the framework that Roberts (2006, 2009) proposes.

The response function is characterized by many factors. For example, natural laws govern all types of entities. Genes provide additional constraints for organisms such as plants and animals. For humans, personality characterizes further constraints. These constraints are shown in Figure 4.1 as the boxes under

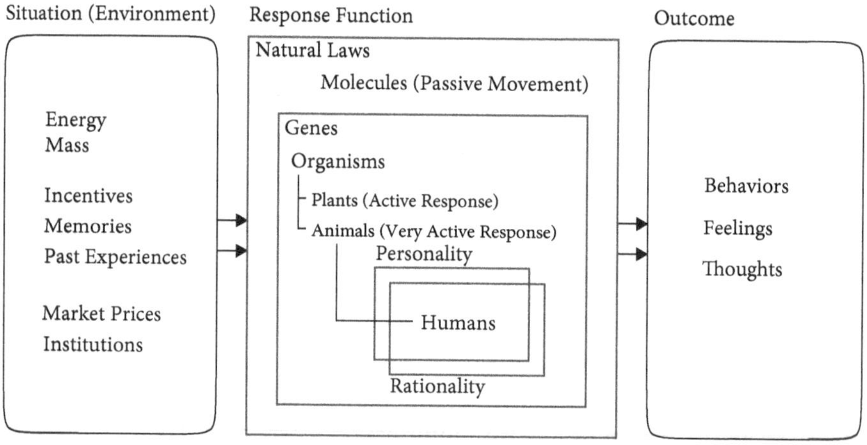

Figure 4.1 An entity is uniquely identified by its response function that maps a situation to an outcome. The sources of constraints imposed by a response function are shown as the boxes under "Response Function." Inner boxes represent the additional sources of constraints over those imposed by outer boxes. At a rudimentary level, the response function is characterized by natural laws that govern all types of entities and genes that provide constraints for organisms such as plants and animals. For humans, personality defines the response function and rationality is an ideal personality used in the neoclassical economic theory.

"Response Function." Rationality is shown in Figure 4.1 as constraints given by an ideal personality.

Let **P** be a probability measure of the state of nature (X, Y). Let $H(X, Y)$ be the joint entropy of (X, Y) defined as the Lebesque integral

$$H(X,Y) \equiv -\int \ln f_{X,Y}(X,Y)\; d\mathbf{P}$$

with respect to the probability measure **P**, where $f_{X,Y}(\cdot,\cdot)$ is the joint density or probability mass function of X and Y depending on whether (X, Y) is continuous or discrete. When (X, Y) is continuous, the joint entropy may also be written as the Riemann integral $H(X,Y) = -\int f_{X,Y}(x,y) \ln f_{X,Y}(x,y) dx dy$ using the joint density function $f_{X,Y}(x,y)$. For easy exposition, I assume that (X, Y) is continuous in the rest of this chapter.

The entropy of X and Y is defined as $H(X) \equiv -\int \ln f_X(X)\, d\mathbf{P}$ and $H(Y) \equiv -\int \ln f_Y(Y)\, d\mathbf{P}$, where $f_X(\cdot)$ and $f_Y(\cdot)$ are the marginal density functions of X and Y, respectively. The conditional entropy $H(Y|X)$ of Y given the information X is defined as $H(Y|X) \equiv -\int \ln f_{Y|X}(Y|X)\, d\mathbf{P}$, where $f_{Y|X}(\cdot|x)$ is the conditional density function of Y given $X = x$. We have the equality

$$\begin{aligned} H(X,Y) &= H(X) + H(Y|X) \\ &= H(Y) + H(X|Y). \end{aligned} \tag{4.1}$$

See Cover and Thomas (2006) for further properties of various entropy concepts.

The response function of an entity maps a situation X to an output Y. If the output of a response function is invariant with respect to X, it imposes no relation between X and Y. In this case, X and Y are independent, and the maximum joint entropy is

$$H(X,Y) = H(X) + H(Y). \tag{4.2}$$

Any relational constraints between X and Y from a response function reduce $H(X,Y)$. Subtracting (4.1) from (4.2), we measure the reduction of the joint entropy due to the constraints by

$$I(X,Y) \equiv H(Y) - H(Y|X),$$

which is called the mutual information of X and Y. The mutual information can also be calculated by

$$I(X,Y) = \int \ln \frac{f_{X,Y}(X,Y)}{f_X(X) f_Y(Y)} \, d\mathrm{P},$$

which is the Kullback–Leibler information criterion from the joint density $f_{X,Y}(\cdot,\cdot)$ to the product $f_X(\cdot)f_Y(\cdot)$ of the marginal densities of X and Y. The mutual information is a non-negative quantity, and equals zero if X and Y are independent. The mutual information is symmetric in the sense that $I(X,Y) = H(Y) - H(Y|X) = H(X) - H(X|Y)$.

Let Π_0 be the constraints given by the laws that govern all entities universally such as physical laws. The information in Π_0 is measured by the mutual information

$$I(X,Y;\Pi_0) = H(Y;\Pi_0) - H(Y|X;\Pi_0).$$

Since Π_0 is uniform across all entities, it does not identify an individual entity. Since nonorganic matter responds to external environments passively, the responses of nonorganic matter are governed by natural laws that describe physical constraints on the matter. These physical constraints reduce the joint entropy by $I(X,Y;\Pi_0)$ and therefore, are considered to be information.

Unlike nonorganic matter, organisms actively react to a situation. One of the important behavioral constraints in the response function is imposed by the genes that each organism possesses. Therefore, genes are information that reduces the entropy of the system. Simple organisms such as bacteria or plants have limited ability to react. They react to the situation less actively than more complex organisms. Advanced organisms have a complex response function in which various situational factors play a role.

The reaction ability of simple organisms is relatively homogeneous across entities within a species compared to more complex organisms, such as animals. For complex organisms, each entity has a unique character even within the same species. I define the character as an encompassing concept that includes the characteristics of an organism that may not interact with its environment as well as those that may shape its response function interacting with a situation. Following Roberts (2006) and Almlund et al. (2011), I define personality as a person's response function. Therefore, personality is a subset of a human character in this definition.

Let Π_i be a set of constraints given by an organism i with the property $\Pi_0 \subset \Pi_i$. Since we can identify each organism uniquely by its reaction function, an organic entity is considered to have a unique character or ability that provides a set of constraints Π_i. This ability reduces the entropy of the output further in addition to what natural laws and genes do. The information that Π_i provides is given by the mutual information

$$I(X, Y; \Pi_i) = H(Y; \Pi_i) - H(Y|X; \Pi_i).$$

Since each organism has a different set of abilities, $I(X, Y; \Pi_i)$ varies across i.

Among all organisms, humans seem to have the most complex and heterogeneous response system. Since human ability characterizes constraints on a person's behaviors, thoughts, and feelings, personality comprises human ability. Therefore, human ability is information that *reduces* the entropy. More concretely, cognitive and noncognitive abilities are specific attributes of personality. On the one hand, higher levels of cognitive and noncognitive abilities reduce entropy and provide larger mutual information $I(X, Y; \Pi_i)$ with a larger set of constraints Π_i. On the other hand, if an individual possesses little ability as a result, for example, of being in a coma, the person is not very responsive to X, and therefore, $I(X, Y; \Pi_i)$ would be close to $I(X, Y; \Pi_0)$. When the person dies, we have $I(X, Y; \Pi_i) = I(X, Y; \Pi_0)$.

Let Π^* be the maximal set of constraints from an ideal personality with unlimited information capabilities, which gives the maximum mutual information $I(X, Y; \Pi^*)$, and let $\Pi_i \subset \Pi^*$ be a set of constraints imposed by the personality of an individual i having the mutual information $I(X, Y; \Pi_i)$. Define

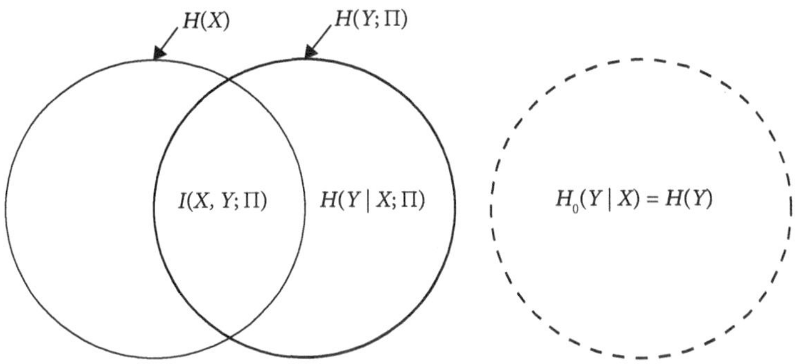

Figure 4.2 $H(X)$ is the entropy of a situation X. $H(Y;\Pi)$ is the entropy of an output Y given the constraints Π imposed by a response function. $H(Y|X;\Pi)$ is the conditional entropy of the output given X under Π. The mutual information $I(X,Y;\Pi) = H(Y;\Pi) - H(Y|X;\Pi)$ is the reduction of entropy of Y given X, which is equal to the information contained in X and Y for each other. The joint entropy of the state of nature is $H(X,Y) = H(X) + H(Y|X;\Pi)$. As an extreme case, $H_0(Y|X) = H(Y)$ represents the maximum entropy when outcomes are completely independent of X and free from restrictions of Π. Assuming $H(Y) = H(Y;\Pi)$, the information in Π is the reduction of the conditional entropy $H_0(Y|X) - H(Y|X;\Pi)$ due to Π.

$J(X,Y;\Pi_i) \equiv I(X,Y;\Pi_i) - I(X,Y;\Pi_0)$ to be the additional mutual information that Π_i gives over what Π_0 does. Thus, $J(X,Y;\Pi_i)$ represents the person-specific part of $I(X,Y;\Pi_i)$. I define information capacity $\mathbb{C}_i$ of an individual i as

$$\mathbb{C}_i \equiv -\ln\left(1 - \frac{J(X,Y;\Pi_i)}{J(X,Y;\Pi^*)}\right) = \ln\left(\frac{I(X,Y;\Pi^*) - I(X,Y;\Pi_0)}{I(X,Y;\Pi^*) - I(X,Y;\Pi_i)}\right), \tag{4.3}$$

where $J(X,Y;\Pi^*) = I(X,Y;\Pi^*) - I(X,Y;\Pi_0)$. We have $\mathbb{C}_i = 0$ if $\Pi_i = \Pi_0$ and $\mathbb{C}_i = \infty$ if $\Pi_i = \Pi^*$.

Figure 4.2 illustrates the mutual information $I(X,Y;\Pi)$ derived from a set of constraints Π. The case with no constraint is shown as the dashed circle with $H_0(Y|X) \equiv H(Y)$, which gives maximum entropy in the absence of the mutual information. As shown in the figure, Π decreases the joint entropy $H(X,Y)$ and increases the information capacity defined in Eq. (4.3) since the mutual information $I(X,Y;\Pi)$ has increased compared to the case with no constraint.

2.1.1 Neoclassical Models and Bounded Rationality

In neoclassical economic theory, the human response function is modeled as a process of optimal choice under constraints. An example is the optimal consumption model shown in Figure 4.3. An economic agent has a preference

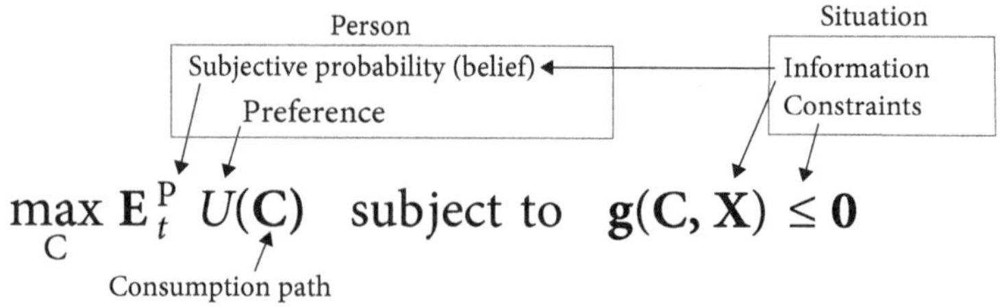

Figure 4.3 The neoclassical rational choice model. The agent with a utility function $U(\cdot)$ and a belief **P** maximizes the expected utility to choose the optimal consumption **C** under some constraints. The situation is given by the variable **X** and a set of constraints $\mathbf{g}(\mathbf{C},\mathbf{X}) \leq \mathbf{0}$ that depends on **X**.

structure represented by a utility function $U(\cdot)$ over a set of consumption paths C. The agent also forms a subjective belief, which is represented by the probability measure **P** over all consumption paths. The agent chooses the best consumption by maximizing the expected utility under a set of budget constraints, $\mathbf{g}(\mathbf{C},\mathbf{X}) \leq \mathbf{0}$, where **X** is situational factors, such as endowments and market prices. In this framework, the situation is given by the budget constraints and the information that influences formation of the belief. The response function is the subjective expected utility maximization under the constraints.

In the neoclassical economic framework, rationality is a key component of personality. A rational economic agent is assumed to possess unlimited ability to acquire and process information. Although the neoclassical model provides a solid and rigorous mathematical foundation of the human decision-making process, it puts less emphasis on the diversity of human ability and preference. Stigler and Becker (1977) argue that we can still capture the salient features of human behaviors without considering the diversity of preference across individuals importantly.

Nonetheless, humans have limited ability in acquiring and processing information. Simon (1955) argues that rationality is bounded because of limited access to information and finite computational capabilities. Kahneman (1973) considers attention a limited resource. Moreover, discerning incorrect information is a crucial factor in making optimal decisions when we have conflicting information such as opposite behavioral suggestions from other people or the media. Distinguishing relevant information from incorrect information requires significant effort and ability. However, the neoclassical economic model does not properly account for these informational burdens.

In the following sections, I focus on human information capacity as an important set of constraints. Specifically, I consider three components of information capacity: the ability to acquire information, the ability to process information, and the ability to discern incorrect information.

2.1.2 Information Acquisition

Information acquisition is the process of searching, asking for, and receiving information through our five sensory organs. The information acquisition process makes information available for processing. Information processing occurs only after the information becomes available through information acquisition.

In personality, curiosity and thoroughness might be the abilities that are most closely related to information acquisition. People who are neither curious nor thorough would have a low level of information acquisition ability, since they would not collect information thoroughly or would not be interested in new information. For example, when a person needs to make the best choice among some products, one has to choose how many products to compare. A bank may offer dozens of credit card products, an employer may offer numerous retirement plans, an investment company can provide dozens of mutual funds to choose from, or an insurance company can sell many medical insurance plans. Acquiring the information on all these products would be a difficult task because it requires thoroughness and curiosity.

We face a similar problem in collecting information to buy a car or a house, choose a mortgage product, or maximize tax returns. People with a low level of information acquisition ability would choose to stop receiving more information, even though they could get it for free or have no problem in understanding the information if it is received. In contrast, people with a high level of information acquisition ability might be curious enough to search for more comparable products or be willing to put more effort into asking for the details of the products to be thorough.

I distinguish between being thorough, which is part of personality or a response function, and completing a task thoroughly, which is a behavioral output. We are concerned primarily with personality rather than behaviors. A person who is not thorough may search for information thoroughly if his or her situation includes a strong incentive. A thorough person requires little incentive to perform the search thoroughly.

To be more specific, consider the following example. Let $X = (X_1, X_2, Z)$ be the situation where X_1 and X_2 are relevant information for completing a task Y, and Z is the level of situational incentive to search for information thoroughly. Suppose that the information X_1 is acquired by agents regardless of the incentive level Z, but the acquisition of X_2 requires Z to be larger than a person-specific threshold. An agent with a high level of thoroughness would have a low level of threshold to acquire X_2.

Let μ be the person-specific ability level that increases as thoroughness or more generally information acquisition ability increases. We can model that X_2 is acquired only if $Z + \mu > 0$. The acquired information vector of the agent is written as $(X_1, 1\{Z+\mu > 0\} \times X_2)$. The probability that X_2 is acquired by the agent is $P(Z+\mu > 0)$ which increases as the information-acquisition-ability level μ increases.

Figure 4.4 illustrates the information-theoretic interpretation of information acquisition ability. The circles in the figure represent the entropy of the situation $X = (X_1, X_2)$ and the behavior Y. We have two circles for personalities Π_H and Π_L with high and low levels of information acquisition ability, respectively. The agent with personality Π_H acquires both X_1 and X_2, whereas Π_L acquires only X_1. Since $H(X, Y; \Pi_H) \leq H(X, Y; \Pi_L)$ as shown in the figure, Π_H is thought to provide more constraints or information than Π_L. Equivalently, the information capacity $\mathbb{C}_H$ defined in Eq. (4.3) with Π_H is larger than $\mathbb{C}_L$ defined with Π_L since $I(X, Y; \Pi_H) \geq I(X, Y; \Pi_L)$.

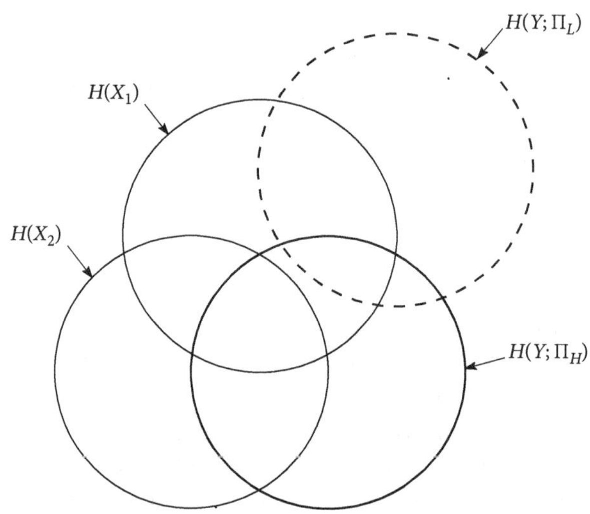

Figure 4.4 Personalities with high and low levels of information acquisition ability are shown as Π_H and Π_L, respectively. $H(X_1)$ and $H(X_2)$ are the entropy of the situation $X = (X_1, X_2)$. $H(Y; \Pi_H)$ and $H(Y; \Pi_L)$ represent the entropy of the output Y given Π_H and Π_L, respectively. Π_H acquires both X_1 and X_2 whereas Π_L acquires X_1 only. The mutual information $I(X_2, Y; \Pi_L)$ for Π_L is zero. Therefore, Π_H leads to a smaller joint entropy than Π_L, that is, $H(X, Y; \Pi_H) \leq H(X, Y; \Pi_L)$. The personality Π_H is thought to provide more entropy-reducing constraints or information than Π_L. Equivalently, Π_H gives a larger information capacity as defined in Eq. (4.3) than Π_L since $I(X, Y; \Pi_H) \geq I(X, Y; \Pi_L)$.

2.1.3 Information Processing

Once information is received, people with a high level of information processing ability have a better understanding of the acquired information. They digest the acquired information fully by examining and studying the content carefully. More importantly, they perceive what the information truly implies without misunderstanding. On the other hand, people with a low level of information processing ability may skip the details in the acquired information. When they do not understand the content of the information fully, they might fill in the gap with their own beliefs rather than rely on canvassing the information more meticulously.

In personality, being careful, organized, thorough, and meticulous would be closely related to information processing ability. Also, being patient and hardworking might be useful for digesting information.

Figure 4.5 illustrates the information processing ability. The figure shows the entropy $H(X)$ of a situation X and two entropy circles for the behavior Y with personalities Π_H and Π_L with high and low levels of information processing ability, respectively. The personality Π_H processes more information in X than Π_L does. The better information processing ability of Π_H is shown as the larger area of intersection between $H(X)$ and $H(Y;\Pi_H)$ than $H(Y;\Pi_L)$ in the figure. Since $H(X,Y;\Pi_H) \leq H(X,Y;\Pi_L)$, the personality Π_H is thought to provide more constraints or information than Π_L. Also, Π_H gives a larger information capacity $\mathbb{C}_H$ defined in Eq. (4.3) than Π_L since $I(X,Y;\Pi_H) \geq I(X,Y;\Pi_L)$.

Figure 4.5 Personalities with high and low levels of information processing ability are shown as Π_H and Π_L, respectively. $H(X)$ is the entropy of the situation X. $H(Y;\Pi_H)$ and $H(Y;\Pi_L)$ represent the entropy of the output Y given Π_H and Π_L, respectively. The intersection of $H(X)$ and $H(Y;\Pi_H)$ is the mutual information $I(X,Y;\Pi_H)$, which is larger than $I(X,Y;\Pi_L)$. Therefore, Π_H gives a smaller total entropy than Π_L, that is, $H(X,Y;\Pi_H) \leq H(X,Y;\Pi_L)$. The personality Π_H is thought to provide more entropy-reducing constraints or information than Π_L, and gives a larger information capacity defined in Eq. (4.3) than Π_L since $I(X,Y;\Pi_H) \geq I(X,Y;\Pi_L)$.

2.1.4 Discerning Incorrect Information

People are commonly exposed to superstitions, fortune telling, and political and marketing campaigns with fake information. People might also have friends or colleagues who have misinformation or incorrect beliefs. However, relying on incorrect information does not reduce the entropy of the system since incorrect information adds only noise. Therefore, the ability to ignore incorrect information is an important component of information capacity.

Discerning incorrect information is particularly important when people have contradictory information, including conflicting thoughts, messages, statements, or opinions. Since nature cannot produce contradictions, conflicting information is generated by humans' erroneous judgments, cognitive biases, incorrect interpretations, misunderstanding, overconfidence, spontaneous thoughts, ignorance, or deliberate lies. In this chapter, I am agnostic on the motives and mechanisms behind the creation of incorrect information.

When a decision maker receives two conflicting pieces of information, he or she must choose a single source of information since it is impossible for all information to be true. People encounter conflicting information in the following two forms.

First, conflicting information may arise from a conflict between person's internal beliefs or judgments and external observable information. For example, a person may believe that an insurance plan on a new television would not be necessary, but the salesperson might give the opposite advice; or one may judge that the future of cryptocurrency is not promising, but a friend might try to convince a person that it is unwise not to invest in cryptocurrency now.

Second, we may have multiple conflicting external information sources. For example, we may have conflicting earnings forecasts or economic outlooks from different sources; financial experts in the media and privately hired financial advisers might suggest opposite investment strategies; or when a girl decides to get married, both a fortuneteller and the girl's mother may have opposite opinions on her fiancé.

In our framework, the ability to choose the correct information source is part of the constraints that shape the response function. Although one can still choose the best action based on incorrect information that does not reflect the reality, the action would not be optimal conditional on correct information or the reality. We can interpret this in our information-theoretic framework in the following way. When the personality Π_i relies on the correct information X, the mutual information $I(X, Y; \Pi_i)$ is large. If Π_i ignores X and chooses to use the incorrect information ε which is irrelevant for Y, it would not decrease the joint entropy at all. Therefore, choosing the true information, X, is considered to provide more entropy-reducing constraints than choosing the false information source. Note that incorrect information is thought to be noise in this chapter. Therefore, it does not provide any constraint that reduces entropy.

In the following, I consider two cases in which the ability to discern incorrect information may play an important role: the first case is concerned with a too-small mix of correct and incorrect information, and the second case is about having a too-large mix of correct and incorrect information.

For the first case, suppose that there is one piece of information, which may be correct or incorrect, for a matter about which a person has little knowledge. Having little information on the matter, people who lack the ability to discern incorrect information might rely on this initial information with incorrect trust of the information source. Therefore, inability to discern incorrect information may lead to the anchoring effect explained in Tversky and Kahneman (1974). For example, as reported in Stewart (2009), the minimum payment amount shown on a credit card statement may be an anchor amount for their actual monthly payment amount. Those who have a low level of ability to discern incorrect information may be easily influenced by the minimum payment amount.

For another example, suppose there are two conflicting sources of information with no other tangible or objective evidence that might resolve or reduce conflicts. People who cannot discern incorrect information might choose the first source or even a random information source if the selection of true information is not evident. Also, those people might see only what they want to see from the information sources regardless of its validity. For example, consider how people form a belief on the value of life, which influences their political stances on safety regulations, abortion, and death penalty. Since there is no objective or agreed-upon measure for the value of life, there are many conflicting opinions. People with a low level of ability to discern incorrect information might be easily persuaded by a close friend or would flip a coin to accept someone's opinion rather than thinking critically and developing their own educated opinion, or even reading research papers on this issue.

The second case is concerned with the situation for which there is a large mix of correct and incorrect information. In the last several decades, we have experienced an unprecedented development of information technology. Creating, storing, and sharing information have become very inexpensive. Most people in the United States have access to the Internet, and many use social media services. Consequently, we often have too many sources of information to choose from. It has become a challenge to discern high-quality information or correct information in the Internet. For example, when people choose the best information source for controversial political issues, product reviews, or even Thanksgiving turkey recipes, search engines give them a dizzying array of options. People often have difficulty choosing where to start reading and get overwhelmed by a large number of conflicting opinions.

When there are many sources of information, conventional information search models assume that all information is noisy but correct. Typically, the key component of the conventional models is to incorporate the trade-off between the value of additional information and the marginal cost of time and effort in acquiring and processing the information. I distinguish the internal cost determined by person-specific ability from the external cost given by the situation surrounding the information sources such as the market price of labor, leisure, and information itself.

In the perspective of information capacity as proposed in this chapter, the internal cost of information will be low if one has a high level of information acquisition and processing ability. To provide a general framework, this chapter is agnostic on the exact mechanism of information acquisition and processing and does not develop an information search model.

A more challenging case arises when available information contains conflicts. With limited ability to discern incorrect information, people might be confused or frustrated by having to go through all sources of conflicting information and decide which sources to choose. For example, for purposes of discussing a new economic policy, one might experience more confusion and frustration when having numerous opposite and conflicting expert opinions than when having no information. If one has no ability to discern misinformation, the expected value of the conflicting information might be zero because having two opposite expert opinions would be the same as having no information at all. If we consider the opportunity cost of time and leisure, the net value of the conflicting information would be negative.

Because of fear of confusion or frustration, inability to discern incorrect information might create aversion to more information. I provide an insight into people's aversion to having too much information with a story in Thaler and Sunstein (2008) and Thaler (2015). They note that a bowl of cashews just before dinner would challenge people's self-control, and many people would appreciate eliminating the cashews from their choice set, even though they could simply ignore the cashews in front of them. This behavioral tendency suggests that self-control is a limited resource that people do not want to deplete. Likewise, a large amount of information mixed with incorrect information challenges people's ability to process the information, and more importantly, discern incorrect information. When people have limited information capacity, they prefer not having any more information to having to exercise their ability to process it and discern incorrect information. Ignorance is bliss when one cannot discern incorrect information, where more information sources might bring additional incorrect information. People with a high level of ability to discern incorrect information would work hard to figure out the reasons for conflicts and have a better chance of dismissing incorrect information. Obviously, the ability to discern incorrect information involves the ability for information acquisition and processing.

More importantly, the ability to discern incorrect information requires the presumption that any information might be incorrect and information sources cannot be trusted. It is also important to be able to reject or refuse others' opinions or suggestions without feeling too much guilt or empathy. It would be difficult to persuade or nudge those who are mistrustful or those who tend to reject and disagree with any given information. Therefore, with regard to personality, being critical, distrustful, cold-hearted, and not empathetic would be the abilities that improve discerning incorrect information. People with these abilities are less susceptible to fake news, marketing campaigns, or salespeople's persuasion, and they are also less superstitious.

Figure 4.6 shows the information-theoretic view of the ability to discern incorrect information. In the figure, the circle $H(X)$ represents the entropy of the correct information X. $H(\varepsilon)$ is the entropy of incorrect information ε, which is shown as irrelevant noise, which should have zero mutual information with Y if only correct information is used. There are two entropy circles for Y with the

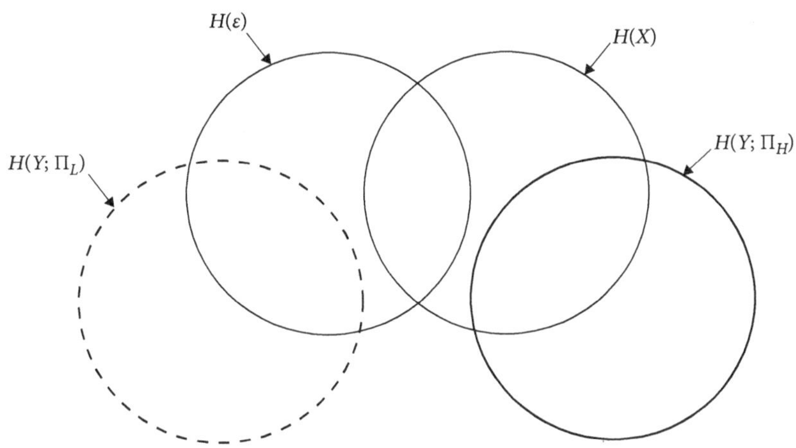

Figure 4.6 Personalities with high and low levels of the ability to discern incorrect information are shown as Π_H and Π_L, respectively. $H(X)$ is the entropy of the relevant situation X, and $H(\varepsilon)$ is the entropy of incorrect information ε. $H(Y;\Pi_H)$ and $H(Y;\Pi_L)$ represent the entropy of the output Y given Π_H and Π_L, respectively. The personality Π_H discerns incorrect information ε and the relevant information is X only, whereas the information for the agent with Π_L ignores X and uses incorrect information ε only. The mutual information $I(X,Y;\Pi_L)$ given Π_L is zero. Therefore, Π_H gives a smaller total entropy than Π_L, that is, $H(X,Y;\Pi_H) \leq H(X,Y;\Pi_L)$. The personality Π_H is thought to provide more entropy-reducing constraints or information than Π_L, and gives a larger information capacity defined in Eq. (4.3) than Π_L since $I(X,Y;\Pi_H) \geq I(X,Y;\Pi_L)$.

personalities Π_H and Π_L with high and low levels of ability to discern incorrect information, respectively. The personality Π_H discerns incorrect information ε and uses the relevant information X, but Π_L relies on incorrect information ε only. The personality Π_H reduces the joint entropy by dismissing ε in the figure, and Π_L ignores X to have zero mutual information $I(X, Y; \Pi_L)$ between Y and X. Since $H(X, Y; \Pi_H) \leq H(X, Y; \Pi_L)$, the personality Π_H is thought to provide more information than Π_L. In other words, the information capacity $\mathbb{C}_H$ defined in Eq. (4.3) with Π_H is larger than $\mathbb{C}_L$ defined with Π_L since $I(X, Y; \Pi_H) \geq I(X, Y; \Pi_L)$.

In the next section, I briefly introduce Sims's (2003, 2006) rational inattention theory. His theory explicitly incorporates the cost of information processing as a constraint. I also discuss potential directions for extending the neoclassical models.

2.2 Rational Inattention Theory and Extensions

Many researchers have emphasized modeling limited capacity for information processing. Simon (1955, 1978), for example, emphasizes that limits on human computational capacity can be an important constraint in the rational choice model. His seminal work has led to the literature on bounded rationality (Kahneman, 2003) and behavioral economics (Camerer et al., 2011).

Sims (2003, 2006) has proposed yet another extension of the neoclassical model by explicitly incorporating information processing cost as a constraint in the expected utility maximization problem. In Sims's (2003) rational inattention theory, an agent receives a signal s for the information about w only through a channel with finite capacity. The capacity limit of the information channel is thought to be inversely related to the information processing cost. The rational inattention model uses the mutual information between w and s as a measure of the channel capacity. It assumes that the mutual information $I(\mathrm{w}; \mathrm{s})$ between w and s is bounded, that is,

$$I(\mathrm{w}; \mathrm{s}) \leq C, \tag{4.4}$$

where C represents the maximum ability of the channel to convey the signal. When an inattentive decision maker faces a choice in an optimization problem with state-dependent payoffs, the agent has the ability to choose the signal structure between w and s under the constraint in (4.4). For example, one might select the signal structure with the most favorable conditional distribution of w given s satisfying (4.4). Therefore, the existence of the finite upper bound C in (4.4) implies that the agent has a limited information processing capacity.

The rational inattention theory has been further developed and applied in recent years. Some examples are Matějka and McKay (2014) and Caplin and Dean (2015). Applying our framework to the rational inattention theory, we may say that people have different levels of C depending on their levels of information capacity.

Our framework is concerned with mutual information as in Sims's model. However, there are several differences. The situation X in our framework includes both observed and unobserved situations surrounding an individual, whereas Sims's model considers the observed signal s only. Moreover, our framework considers general constraints on a response system, including what an individual cannot choose such as the physical laws or genes as well as those imposed by the limited ability of individuals. Therefore, the ability to be attentive in Sims's model is only part of the general constraints considered in our framework. More importantly, our framework distinguishes three different components through which the mutual information may be limited by the information capacity. These components are the information acquisition, information processing, and discernment of incorrect information.

Although our framework is more general than Sims's model, our theory does not provide a tractable model as Sims's model does. In fact, the goal of our theory is to provide an encompassing perspective for more specific models that incorporate information capacity such as Sims's model. Although it is challenging to develop a tractable model of information capacity in the neoclassical economic framework, it is a promising area of research. In the following, I discuss several general directions for potential extensions of neoclassical economic models.

First, limited information capacity could be in the preference structure. Curiosity may be modeled as a strong preference for new experiences. Curious people may prefer receiving more information because learning something new pleases them. For information acquisition, one may prefer reading books or watching documentaries to other activities.

The level of time discount factor in the neoclassical preference model represents the level of patience or the ability to delay gratification or sacrifice today's consumption for future consumption. Heuristically, the time discount factor may account for part of being thorough and industrious, which requires patience. Another example of preference is that a person may prefer a clean state or well-organized thoughts rather than a messy outcome or disorganized memories. Some people may dislike complexity and avoid complicated situations. Also, a fear or phobia of something may limit their information capacity. For example, a person who is afraid of insects may not want to think about anything related to insects; or a person who dislikes a specific type of people may not want to communicate with or get advice from them. Almlund et al. (2011, section 6) discuss the relation between personality and preference parameters.

Second, limited information capacity may influence the formation of beliefs, thus changing expectations. People with a high level of information capacity may form beliefs coherently based on observed evidence. They may understand evidence without distortion or prejudice. Therefore, information is accurately reflected in beliefs. As discussed earlier, people who cannot discern incorrect information might rely on superstitions to form a belief, or avoid additional information because they are concerned about having more incorrect information potentially. When information about a situation is incomplete, people who are neither curious nor thorough might form their belief by filling the information gap with guessing rather than collecting more information.

Third, it could be modeled as part of the constraints in the optimization process. The rational inattention model takes this approach by assuming that the capacity of information channels is constrained. In general, we may consider a cost function of exercising information capacity and impose an endowment or finite upper bound on the total cost. Traditionally, using a cost function has been a popular method in various economic models.

Lastly, we might limit the accuracy of expected utility maximization, which leads to a suboptimal choice. Calculating the exact future utility from a future consumption path may require contemplation. Figuring out the exact expected future payoffs is not trivial even if the true probability distribution of payoffs is known. Also, if taking different actions leads to similar payoffs, figuring out the exact optimal action requires accuracy in maximizing the expected utility, which demands a high level of information capacity.

People with a low level of information capacity might use heuristics as proposed by Tversky and Kahneman (1974). They might not even recognize the relative loss from the nonoptimal choice if their information capacity is limited. Or they might not care about the loss because people are indifferent about similar payoffs and achieving the exact maximum is never their goal. People might not regret their actions even if it were shown ex post that their actions were slightly suboptimal.

Figure 4.7 shows the decision process of a neoclassical model. In the figure, the information environment is shown as the input of the decision process and the behavior as the output. It shows the four aspects of the neoclassical model: belief, preference, constraints, and optimization. These four aspects determine a person's response function. At the center of the four aspects, we have the three components of information capacity: the ability to acquire information, the ability to process information, and the ability to discern incorrect information. The figure illustrates that these three components influence the four aspects of the neoclassical decision-making process.

In the next section, I consider limited information capacity from the perspective of social psychology. Specifically, this chapter examines the domains of noncognitive ability that are related to information capacity.

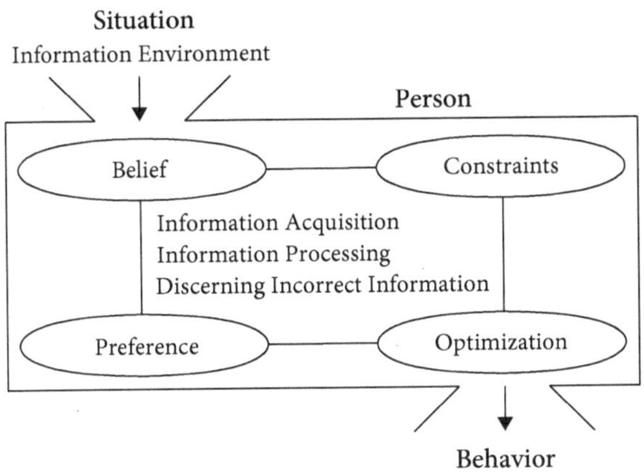

Figure 4.7 The neoclassical economic framework is shown as a response system with four aspects: belief, preference, constraints, and optimization. Three components of information capacity are information acquisition and processing capacity, and the ability to discern incorrect information. The figure shows that the three components of information capacity influence the four aspects of the neoclassical model.

2.3 Information Capacity and the Big Five Personality Traits

Since information capacity is a construct that is not directly observed, operationalizing information capacity through empirical observation of human behaviors is as difficult as measuring preference parameters in economics. Instead of providing a method of measurement of information capacity directly, this chapter provides a heuristic discussion to relate information capacity with existing measures of human ability. The heuristic approach given in this chapter is expected to be a starting point of a more accurate measurement of information capacity by taking advantage of established methods of operationalizing important domains of human ability.

Information capacity is related to both cognitive and noncognitive abilities. Therefore, it is reasonable to represent information capacity as an element in broad domains of cognitive and noncognitive ability. If cognitive and noncognitive abilities are a coordinate system that maps human ability, a level of information capacity is a coordinate in this system.

Although cognitive ability is closely related to information capacity, this chapter emphasizes that noncognitive ability is as important as cognitive ability in representing information capacity. From this point of view, I focus on approximating information capacity with a vector of broad domains of noncognitive ability. Although not discussed in this section, I include a measure of cognitive ability as well in the empirical analysis in the next section to account for the role of cognitive ability in determining information capacity.

In the economics literature, noncognitive ability is also referred to as soft skills or personality traits (Heckman and Kautz, 2012). In this chapter, I use the term *personality traits* for noncognitive ability since I adopt a model from social psychology. In addition, I consider that personality traits determine the shape of the human response function.

I begin by discussing the taxonomy of personality traits. Over the past several decades, the Big Five personality traits model (McCrae and Costa, 1986, 1987, 1999) has gained popularity and has been widely used in social psychology. The five traits are Openness, Conscientiousness, Extraversion, Agreeableness, and Neuroticism. John et al. (2008, p. 120) define them conceptually as follows: Openness is "the breadth, depth, originality, and complexity of an individual's mental and experiential life"; Conscientiousness is "socially prescribed impulse control that facilitates task and goal-directed behavior, such as ... planning, organizing and prioritizing tasks"; Extraversion is "an energetic approach toward the social and material world"; Agreeableness is "a prosocial and communal orientation toward others . . . such as altruism, tender-mindedness, trust, and modesty"; and Neuroticism "contrasts emotional stability and even-temperedness."

The Big Five personality traits are designed to cover a wide range of human personality. Personality as a response function has infinite degrees of freedom. The Big Five personality traits are regarded as five large domains in the functional space that personality lies in. The measured Big Five personality traits of a person are the five projections of the individual's personality onto the five subspaces. Although the five domains would not span the entire personality space, they are supposed to cover a broad part of the functional space and provide a comprehensive map for identifying human personality.

Personality as the whole response function can be distinguished from the Big Five personality traits because the Big Five personality model is only one of many models that approximate human personality. Another approximating personality model would span a different set of subdomains, and measured personality traits of this model would be a set of projections of personality onto this new set of subdomains.

The Big Five personality traits are identified by obtaining the projections. The most popular method to operationalize the projections is to use a self-reported survey. The survey questions are designed to approximately construct the five domains. Then, the survey responses are averaged to a numerical value, which gives the measured Big Five personality traits. For example, the survey items used in the Health and Retirement Study (HRS, 2012, https://hrsonline.isr.umich.edu) consist of twenty-six adjectives. The survey asks a respondent "Please indicate how well each of the following describes you" for each of the twenty-six adjectives. The adjectives used in the HRS for the Big Five personality traits are as follows:

- Openness (seven items): creative, imaginative, intelligent, curious, broad-minded, sophisticated, adventurous.
- Conscientiousness (five items): organized, responsible, hardworking, careless (-), thorough.
- Extraversion (five items): outgoing, friendly, lively, active, talkative.
- Agreeableness (five items): helpful, warm, caring, softhearted, sympathetic.
- Neuroticism (four items): moody, worrying, nervous, calm (-).

Each answer is coded from 1 ("not at all") to 4 ("a lot") and those with "(-)" are inversely coded. For each of the five traits, the average value of the answers represents the measured personality trait, which is the operationalized version of the projected point onto the trait's domain.

Among the Big Five personality traits, as suggested earlier, I consider Openness, Conscientiousness, and Agreeableness to be the most relevant traits to information capacity. Since each of the Big Five personality traits is a broad domain of noncognitive ability with many subfacets, some facets within one domain may be positively correlated with information capacity, while other facets are negatively correlated. To focus our attention on the Big Five personality model rather than its subfacets, I consider the total correlation between information capacity and each of the Big Five personality traits. In this chapter, I give a heuristic discussion on the aggregate correlation between information capacity and the following three traits: Openness, Conscientiousness, and Agreeableness.

As indicated by the definitions of the Big Five traits and the list of adjectives, the aggregate correlation between information acquisition ability and Openness and Conscientiousness would be positive. People who are curious and intelligent tend to seek more information and quickly receive information. Moreover, people who are careful, thorough, hardworking, and organized would collect more relevant information.

Information processing ability would also be positively related to Openness and Conscientiousness. People who are intelligent, broadminded, and sophisticated may understand information better because, even if the information is not complete, they could fill the gap coherently. People who are organized, hardworking, careful, and thorough would digest information better than those who are disorganized, lazy, careless, and not thorough.

For the ability to discern incorrect information, I conjecture that both Conscientiousness and Agreeableness would be important. Conscientiousness would have a positive relation with the ability to discern incorrect information, whereas Agreeableness is negatively related. Conscientious people would be able to identify the inconsistency between information sources and evaluate the reliability of the information carefully and thoroughly. The high level of information processing ability of conscientious people would help discerning incorrect information.

When we are concerned about distinguishing incorrect information given by another person in social interactions, Agreeableness will be negatively related to the ability to discern incorrect information. When softhearted and trusting people receive conflicting information or misinformation, it might be difficult to investigate and examine the information by asking questions directly to the information provider thoroughly and aggressively because of their emphasis on social harmony and prosocial orientation. Distinguishing misinformation may require interrogation of the information provider. However, agreeable people would feel it difficult to inquire thoroughly because of their trusting nature and their propensity to be nice. Moreover, to start the investigation to discern incorrect information, one must be mistrustful, antagonistic, and critical rather than sympathetic and softhearted. For agreeable people, it might be difficult to mistrust and reject someone's opinion because of the fear of creating conflicts with other people, being rude, or violating cultural norms. All else being equal, disagreeable people would not be easily persuaded by incorrect information.

Therefore, open, conscientious, and disagreeable people may have a higher information capacity. Heuristically, we may consider Openness, Conscientiousness, and Agreeableness to be correlated with the constant C in inequality (4.4) in rational inattention theory.

In the next section, I empirically study the role of Openness, Conscientiousness, and Agreeableness in various information-related behaviors.

3. An Empirical Study on Information Capacity

In section 2 of this chapter, I defined information capacity conceptually. Since we do not have direct measures of information capacity, the empirical analysis in this section attempts to operationalize information capacity by using existing measures of cognitive and noncognitive ability. If information capacity is related to conventional ability measures such as IQ and the Big Five personality traits, we may study the role of information capacity indirectly by examining empirically whether these ability measures have expected effects as a vector of proxy for information capacity.

In this section, I take a "reduced-form" approach. Unlike a structural model, a reduced-form model does not impose a specific mechanism in the neoclassical economic model. I first examine Openness, Conscientiousness, and Agreeableness as a proxy vector for information capacity. More specifically,

I study the predictive power of those three traits for some behavioral outcomes that are thought to be positively correlated with information capacity. Second, using Openness, Conscientiousness, and Agreeableness as a proxy vector for information capacity, I analyze the role of these traits in long-term financial outcomes that might require a large amount of information capacity over a long time horizon.

3.1 Data

I use data from the Health and Retirement Study (HRS), a biennial longitudinal survey conducted from 1992 and ongoing. The survey is designed to study the aging population in the United States, and its sample is nationally representative for people over the age of fifty. In each of the biennial survey waves, 50 percent of the respondents are selected for an enhanced face-to-face interview, and the rest are interviewed via a telephone. In the next wave, the other 50 percent are selected for a face-to-face interview. After the face-to-face interview, an in-depth psychosocial and lifestyle questionnaire is given as a "leave-behind" survey that the respondent is requested to complete and mail back to the HRS. Since 2006, the HRS's personality questionnaire that was explained in the previous section has been administered in the leave-behind survey.

Since personality traits have been measured for 50 percent of the respondents in each biennial wave, we have repeated personality measurements every four years. Therefore, we have 2006/2010/2014 or 2008/2012 measurements for most people in the survey. I assume that the personality traits are stable for our sample. Although the personality traits may change over the life cycle, their rank-order in a population becomes increasingly stable (Roberts and DelVecchio, 2000). Therefore, our assumption seems to be reasonable as our sample consists of an aging population. Based on this assumption, for all individuals with multiple personality measurements, I take the time-average of the measurements. Then, I normalize the time average by subtracting the sample mean and dividing by the standard deviation from the 2010 personality traits distribution. The resulting measurements are the standardized measurements (z-score) of the personality traits, for which a one-unit increase of a personality trait is interpreted as a one-standard-deviation increase of the trait in our sample.

The HRS contains detailed demographic and educational information as well as psychosocial measures for each respondent. Also, it has very detailed financial variables at the household level. In this section, the unit of analysis is an individual. Since financial variables are measured at the household level, I consider single households only to avoid the confounding effect of spouses for couples.

Our main sample consists of all single individuals in the 2012 and 2014 survey waves. Among the 11,051 individuals in the sample, 25 percent are male,

69 percent are white, and 18 percent have college education or higher. The average net wealth of the individuals is $251,937, and the average income is $34,698.

In the next section, I first examine the hypothesis that Openness, Conscientiousness, and Agreeableness are a proxy for information capacity. Then, I hypothesize that information capacity has accumulative financial consequences. To test this hypothesis, I study the role of Openness, Conscientiousness, and Agreeableness in wealth accumulation and a subjective financial hardship measure.

3.2 Information Acquisition and Processing

To examine whether Openness, Conscientiousness, and Agreeableness are a proxy vector for information capacity, I study the predictive power of the three personality traits for the four variables listed below. Note that these four variables are behavioral outputs of a human response function that would depend on both situation and personality traits, and they are not measures of ability. On the other hand, a good set of personality measures should be independent of situational specificity. The measured Big Five personality traits in the HRS are designed to be invariant with situations. I assume that the measures of the Big Five personality traits developed for the HRS are valid and stable. See Mroczek and Kolarz (1998) and Prenda and Lachman (2001) for the construct validity of these measures. Here are the four variables.

1. Whether respondents read books, magazines, or newspapers daily: People with a high level of information acquisition ability would be more likely to seek and receive new information every day.
2. Whether respondents use the Internet regularly: This variable equals one if the answer is yes to the question, "Do you regularly use the Internet (or the World Wide Web) for sending and receiving e-mail or for any other purpose, such as making purchases, searching for information, or making travel reservations?" People with a high level of information acquisition and processing ability would be more likely to seek and receive new information from the Internet and understand how to use the technology.
3. A measure of the respondent's financial sophistication: I use the financial literacy questions and answers developed in Lusardi et al. (2009), which is available in the 2008 and 2010 HRS survey waves. In this empirical study, I use all single individuals in 2008–2010 for this variable. The measure is the number of correct answers, standardized to have a mean of zero and a variance of one, to the following eight questions (answers are shown in

the brackets): (1) You should put all your money into the safest investment you can find and accept whatever return it pays [No]. (2) An employee of a company with publicly traded stock should have a lot of his or her retirement savings in the company's stock [No]. (3) It is best to avoid owning stocks of foreign companies [No]. (4) Even older retired people should hold some stocks [Yes]. (5) You should invest most of your money in a few good stocks that you select rather than in lots of stocks or in mutual funds [No]. (6) If the interest rate falls, bond prices will rise [Yes]. (7) If one is smart, it is easy to pick individual company stocks that will have better than average returns [No]. (8) There is no way to avoid people taking advantage of you, if you invest in the stock market [No]. People with a high level of information acquisition and processing ability would be more likely to have better understanding and knowledge because their information capacity would have helped them achieve a high level of financial sophistication.

4. Whether respondents follow the stock market: The survey question used for this is "How closely do you follow the stock market: very closely, somewhat, or not at all?" The variable is one if the respondent answers "very closely" or "somewhat." Following the stock market is a behavioral outcome that depends on both person and situation. Given a situation, people with a high level of information capacity would seek for and process stock market information more easily than those with a low level of information capacity. Although the chapter is agnostic about the exact mechanism, one possible mechanism is that the internal cost of the effort of information acquisition and processing is low for those with high levels of information capacity. I also consider the situational factor by studying the subsample of stock investors. People who currently invest in the stock market have a larger situational incentive to follow the stock market. I examine with stock investors whether investors with a high level of information capacity would be more likely to collect and process relevant information by closely following the stock market.

For the variables in 1, 2, and 4, I use logit models. For the logit models, the marginal effects and cluster-robust standard errors are calculated. For the variable in 3, I use a linear model:

$$Y_{it} = \beta_0 + \beta_1 O_i + \beta_2 C_i + \beta_3 A_i + \gamma' \mathbf{X}_{\text{it}} + \varepsilon_{it},$$

where Y_{it} is the outcome variable described above for an individual i at time t; and O_i, C_i, and A_i are Openness, Conscientiousness, and Agreeableness measures, respectively. $\mathbf{X}_{\text{it}}$ is a vector of control variables. Ordinary least squares

(OLS) is used for estimation. All specifications are estimated with robust standard errors clustered at the U.S. census regions × metro type level.

In addition to Openness, Conscientiousness, and Agreeableness, I control for the other Big Five traits (Extraversion and Neuroticism), demographic variables (age in a quadratic form, race, gender, marital status, employment status, and whether self-employed), presence of dependents, home ownership status, natural log of total medical spending, years of education, natural log of income (standardized to have a mean of zero and a variance of one), year fixed effects, and region (nine U.S. census regions), and metro type (highly urban, medium-size, and rural) fixed effects. I use income from the previous survey wave to reduce any concern of reverse causality.

I also include a measure of fluid intelligence which Jaeggi et al. (2008) define as "the ability to reason and to solve new problems independently of previously acquired knowledge." Fluid intelligence is designed to capture executive functioning and speed of response related to sequential, inductive, and quantitative reasoning, and is considered an important aspect of cognitive ability and general intelligence.[2] I consider fluid intelligence to be as important as noncognitive ability such as the Big Five personality traits in operationalizing information capacity. Therefore, I include the fluid intelligence measure in the empirical model.

In 2010 and 2012, the HRS conducted the number series test as a measure of fluid intelligence, which asks a respondent to fill in a blank in a progressive numerical sequence with a certain pattern. I standardize the test scores by subtracting the mean and dividing by the standard deviation of the 2010 sample.

In column 1 of Table 4.1, I examine the likelihood of reading books, magazines, or newspapers daily. The marginal effects of Openness and Conscientiousness are statistically significant at the 1%-level. A one-standard-deviation increase in Openness and Conscientiousness would increase the likelihood by 3.5 and 2.3 percentage points, respectively. These effects are large and are comparable to the effect of one additional year of education (2.6 percentage points). The effects are much larger than that of a one-standard-deviation increase in the log of income (0.2 percentage points).

In column 2 of the table, I find that Openness is statistically significant at the 1%-level and increases the likelihood of using the Internet daily by 4.8 percentage points. The marginal effect of Openness is larger than that of being male (−4.6 percentage points) or one additional year of education (3.1 percentage points).

[2] See Almlund et al. (2011, p. 39) for a brief introduction.

Table 4.1 Predictive Power of Openness, Conscientiousness, and Agreeableness for Information-Related Behaviors and Wealth Accumulation

	Outcomes Related to Information Capacity					Wealth and Hardship	
	(1)	(2)	(3)	(4)	(5)	(6)	(7)
Openness	0.035	0.048	0.102	0.057	0.050	0.006	0.034
	(0.009)	(0.006)	(0.029)	(0.007)	(0.014)	(0.009)	(0.026)
Conscientiousness	0.023	0.005	0.077	0.025	0.036	0.071	−0.118
	(0.009)	(0.005)	(0.035)	(0.007)	(0.015)	(0.010)	(0.022)
Extraversion	0.008	−0.018	−0.037	0.005	−0.011	0.043	−0.067
	(0.009)	(0.004)	(0.033)	(0.009)	(0.015)	(0.008)	(0.013)
Agreeableness	−0.001	0.007	−0.061	−0.030	−0.031	−0.062	0.110
	(0.009)	(0.006)	(0.052)	(0.007)	(0.017)	(0.010)	(0.020)
Neuroticism	−0.014	−0.015	0.021	0.006	−0.011	−0.014	0.178
	(0.007)	(0.005)	(0.027)	(0.005)	(0.012)	(0.007)	(0.017)
White	0.075	0.109	0.210	0.030	0.045	0.237	−0.158
	(0.016)	(0.015)	(0.054)	(0.012)	(0.030)	(0.028)	(0.048)
Male	−0.092	−0.046	0.126	0.128	0.153	0.072	−0.124
	(0.019)	(0.008)	(0.081)	(0.009)	(0.029)	(0.031)	(0.025)
Years of education	0.026	0.031	0.047	0.022	0.011	0.041	−0.007
	(0.003)	(0.002)	(0.011)	(0.003)	(0.005)	(0.004)	(0.007)
ln(income)	0.002	0.037	0.144	0.029	0.008	0.088	−0.074
	(0.009)	(0.005)	(0.038)	(0.006)	(0.011)	(0.010)	(0.017)
Fluid intelligence	0.052	0.280	0.355	0.089	0.059	0.151	−0.099
	(0.020)	(0.015)	(0.072)	(0.014)	(0.042)	(0.024)	(0.035)
R-squared*	0.10	0.32	0.20	0.08	0.08	0.47	0.23
Number of observations	4,780	11,024	1,509	10,670	1,670	11,051	4,734

* For the the logit models in columns 1, 2, 4, and 5, the marginal effects and pseudo R-squared are reported.

Notes: The sample is all single individuals in 2012–2014 for columns 1–2 and 4–7. For column 3, all single individuals in 2008–2010 are used. Dependent variables are the following: (1) read books, magazines, or newspapers daily, (2) daily use of web for e-mail. etc. (3) the standardized score of financial sophistication measure (0–8) to have a mean of zero and a variance of one, (4)–(5) closely follow the stock market, (6) total net wealth, and (7) a self-reported measure of difficulty in paying monthly bills (1–5) standardized to have a mean of zero and a variance of one. In column 5, only the individuals with stock are included. In column 6, wealth (in 10,000s) is transformed by the inverse hyperbolic sine function and standardized to have a mean of zero and a variance of one. The logit model is used for columns 1, 2, 4, and 5. The OLS estimation is used in columns 3 and 6–7. Standard errors, in parentheses, are clustered at the region × metro type (highly-urban, medium-size, and rural). Openness, Conscientiousness, Extraversion, Agreeableness, and Neuroticism are standardized to have a mean of zero and a variance of one. The natural log of income is lagged by two years and standardized to have a mean of zero and a variance of one. All specifications control for age, marital status, employment status, whether self-employed, presence of dependents, home ownership, natural log of medical spending, year fixed effects, and region and metro-type fixed effects.

I also find in column 3 that Openness and Conscientiousness have large effects on financial sophistication. A one-standard-deviation increase in Openness and Conscientiousness increases the financial sophistication score by 10.2 percent and 7.7 percent of one standard deviation of the financial sophistication measure, respectively. In columns 1–3, Agreeableness does not play a significant role. This is probably because the dependent variables in those columns are mostly related to the information acquisition and processing rather than to discerning incorrect information.

In columns 4–5, I find that Openness, Conscientiousness, and Agreeableness have large effects on the likelihood of following the stock market closely. In column 4, I use all individuals in the sample. The three traits increase the likelihood by 5.7, 2.5, and −3.0 percentage points, respectively. The marginal effects of the three straits are larger than those of race (3.0 percentage points) or one more year of education (2.2 percentage points). They are also comparable to the effect of a one-standard-deviation increase in the natural log of income (2.9 percentage points).

In column 5, I use only the individuals with stockholdings. These individuals have more incentives to follow the stock market. However, I still find strong heterogeneity across individuals. A one-standard-deviation increase in Openness, Conscientiousness, and Agreeableness increases the likelihood by 5.0, 3.6, and −3.1 percentage points, respectively. They have larger effects than education (1.1 percentage points) and income (0.8 percentage points). The result suggests that being curious (Openness) and organized and thorough (Conscientiousness) matters for following the stock market. Agreeableness has a negative effect, which indicates that distrustful and antagonistic people tend to follow the stock market closely.

It is noteworthy that fluid intelligence has a statistically significant effect at the 1%-level in all our regressions except for column 5. It has an especially large effect on the daily use of the Internet (column 2) and financial sophistication (column 3). The magnitude of its marginal effect in columns 2–3 is approximately three times larger than that of Openness or Conscientiousness. In columns 1 and 4, the effect of fluid intelligence is comparable to the sum of the marginal effects of Openness, Conscientiousness, and Agreeableness. In column 5, fluid intelligence is not statistically significant at the 10%-level. However, the magnitude of the coefficient is as large as that of Openness.

Overall, our results show that Openness, Conscientiousness, and Agreeableness are important noncognitive abilities in predicting activities and behavioral outcomes that are related to information capacity. In the next section, I use the three traits together as a proxy for information capacity.

3.3 Accumulation of Wealth and the Big Five Personality Traits

In this section, I study the long-term financial consequences of having different levels of Openness, Conscientiousness, and Agreeableness. In particular, I examine total net wealth and financial hardship. The results in the previous section suggest that the three personality traits are a proxy for information capacity. In this section, I use them together in our empirical model as a vector measurement of information capacity. Therefore, the evidence presented in this section would indirectly show the financial and economic value of information capacity.

Since information capacity would improve the quality of decision making, I hypothesize that Openness and Conscientiousness have a positive effect on wealth accumulation and a negative effect on economic hardship. On the other hand, Agreeableness would have a negative effect on wealth accumulation and a positive effect on economic hardship.

I first examine total net wealth, including all financial, business, housing, and other physical assets. The net value of total wealth is calculated as the sum of all financial, business, housing, and other physical assets such as vehicles less the sum of all debt including mortgages, home loans, credit card debts, and other debts. Financial assets include stocks, bonds, CDs, T-bills, checking, savings, and money market accounts. To deal with negative net wealth, I transform net wealth in $10,000s with the inverse hyperbolic sine function rather than the natural log function. After the transformation, I standardized it to have a mean of zero and a variance of one using the 2010 sample distribution.

In column 6 of Table 4.1, I include the personality traits and demographic variables (age in a quadratic form, race, gender, years of education, marital status, employment status, presence of dependents), year fixed effects, and region × metro-type fixed effects, fluid intelligence, and other financial variables, including the z-score of the natural log of income from the previous wave, home ownership status, and the natural log of total medical spending. The effects of Conscientiousness and Agreeableness are significant at the 1%-level. A one-standard-deviation increase in Conscientiousness and Agreeableness changes the net wealth by 7.1 percent and −6.2 percent of one standard deviation, respectively. The magnitude of these effects is comparable to that of being male (7.2 percent) and larger than that of one additional year of education (4.1 percent). The result suggests that more conscientious and disagreeable people have more wealth. This is consistent with our hypothesis that the information capacity may influence wealth accumulation positively.

A one-standard-deviation increase in Openness increases the standardized net wealth by 0.6 percent of one standard deviation, which is not statistically significant at the 10%-level. The effect of Openness is relatively smaller than that of Conscientiousness and Agreeableness. Although Openness may be positively correlated with the information acquisition ability, it may also increase spending because Openness to new experience may lead to more costly hobbies and activities, such as traveling or buying new technology gadgets, which in turn decrease net wealth.

In column 7 of Table 4.1, the dependent variable is a subjective measure of economic hardship, which is also standardized to have a mean of zero and a variance of one. The survey question for this measure is "How difficult is it for (you/your family) to meet monthly payments on (your/your family's) bills?" and scaled from 1 ("not at all") to 5 ("completely"). I find that Conscientiousness and Agreeableness have negative and positive effects, respectively. The magnitude of these effects is comparable to that of being male or being white. This is consistent with our hypothesis that information capacity improves an individual's economic well-being.

Fluid intelligence shows large effects in both columns 6 and 7. For the net wealth (column 6), the marginal effect of fluid intelligence is more than twice of that of Conscientiousness and Agreeableness. In column 7, the effect of fluid intelligence is similar to that of Conscientiousness or Agreeableness. As fluid intelligence is an important part of information capacity, the large effects of fluid intelligence further support our hypothesis that information capacity plays an important role in wealth accumulation and economic well-being.

Overall, our results in this section suggest that information capacity has positive long-term consequences on wealth accumulation and economic hardship even after race, gender, education, and income are accounted for.

4. Conclusion

During the past century, psychology and economics have developed theories that have taken very different paths. However, in the last couple of decades, attempts have been made in behavioral economics to merge psychological ideas into economics. The information-theoretic perspective given in this chapter provides a common foundation to model human ability for both psychology and economics.

I have defined information capacity to measure the amount of entropy reduced by human ability. I have proposed to approximate information capacity by operationalizing with the Big Five personality traits. To model information capacity, incorporating Openness, Conscientiousness, and Agreeableness in neoclassical economic models would be a challenging but very rewarding task. In particular, modeling the ability to discern incorrect information would be a fruitful endeavor. I conjecture that Agreeableness would provide grounds for developing the underlying mechanisms for discerning incorrect information.

Acknowledgments

I thank Amos Golan, Aman Ullah, and the editors for the careful review. I am also grateful to Ron Laschever, Alan Adelman, David Slichter, Sol Polachek, and Karim Chalak for helpful discussions and comments.

References

Almlund, M., Duckworth, A. L., Heckman, J., and Kautz, T. (2011). "Personality Psychology and Economics." In Hanushek, E. A., Machin, S. J., and Woessmann, L. (eds.), *Handbook of the Economics of Education*, vol. 4, chap. 1 (pp. 1–181). North-Holland: Elsevier.

Benjamin, B. V., Cesarini, D. D., Chabris, C. F., Glaeser, E. L., D. Laibson, D., Gunason, V., et al. (2012). "The Promises and Pitfalls of Genoeconomics," *Annual Review of Economics*, 4: 627–662.

Camerer, C. F., Loewenstein, G., and Rabin, M. (2011). *Advances in Behavioral Economics*. Princeton, NJ: University Press.

Caplin, A., and Dean, M. (2015). "Revealed Preference, Rational Inattention, and Costly Information Acquisition." *American Economic Review*, 105: 2183–2203.

Cover, T., and Thomas, J. (2006). *Elements of Information Theory*, 2nd ed. New York: John Wiley.

Heckman, J., and Kautz, T. (2012). "Hard Evidence on Soft Skills." *Labour Economics*, 19: 451–464.

HRS. (2012). "Health and Retirement Study, 2006–2012 Public Use Datasets." Produced and distributed by the University of Michigan with funding from the National Institute on Aging (grant number NIA U01AG009740). Ann Arbor, MI.

Jaeggi, S. M., Buschkuehl, M., Jonides, J., and Perrig, W. J. (2008). "Improving Fluid Intelligence with Training on Working Memory." *Proceedings of the National Academy of Sciences*, 105, 6829–6833.

Jaynes, E. T. (1957a). "Information Theory and Statistical Mechanics." *Physical Review*, 106: 620–630.

Jaynes, E. T. (1957b). "Information Theory and Statistical Mechanics. II." *Physical Review*, 108, 171–190.

John, O. P., Naumann, L. P., and Soto, C. J. (2008). "Paradigm Shift to the Integrative Big Five Trait Taxonomy: History, Measurement, and Conceptual Issues." In John, O. P., Robins, R. W., and Pervin, L. A. (eds.), *Handbook of Personality: Theory and Research*, 3rd ed. (pp. 114–158). New York: Guilford Press.

Kahneman, D. (1973). *Attention and Effort*. Englewood Cliffs, NJ: Prentice-Hall.

Kahneman, D. (2003). "Maps of Bounded Rationality: Psychology for Behavioral Economics." *American Economic Review*, 93: 1449–1475.

Lusardi, A., Mitchell, O. S. and Curto, V. (2009). "Financial Literacy and Financial Sophistication in the Older Population: Evidence from the 2008 HRS." Working Paper *WP 2009-216*. Michigan Retirement Research Center, University of Michigan.

Matějka, F., and McKay, A. (2014). "Rational Inattention to Discrete Choices: A New Foundation for the Multinomial Logit Model." *American Economic Review*, 105: 272–298.

McCrae, R. R., and Costa, P. T. (1986). "Personality, Coping, and Coping Effectiveness in an Adult Sample." *Journal of Personality*, 54: 385–404.

McCrae, R. R., and Costa, P. T. (1987). "Validation of the Five-Factor Model of Personality across Instruments and Observers." *Journal of Personality and Social Psychology*, 52: 81–90.

McCrae, R. R., and Costa, P. T. (1999): "A Five-Factor Theory of Personality." In Pervin, L. A. and John, O. P. (eds.), *Handbook of Personality: Theory and Research*, 2nd ed. (pp. 139–153). New York: Guilford Press.

Mroczek, D. K., and Kolarz, C. M. (1998). "The Effect of Age on Positive and Negative Affect: A Developmental Perspective on Happiness," *Journal of Personality and Social Psychology*, 75, 1333–1349.

Prenda, K. M., and Lachman, M. E. (2001). "Planning for the Future: A Life Management Strategy for Increasing Control and Life Satisfaction in Adulthood," *Psychology and Aging*, 16: 206–216.

Roberts, B. W. (2006). "Personality Development and Organizational Behavior." *Research in Organizational Behavior*, 27: 1–40.

Roberts, B. W. (2009). "Back to the Future: Personality and Assessment and Personality Development." *Journal of Research in Personality*, 43: 137–145.

Roberts, B. W., and DelVecchio, W. F. (2000). "The Rank-Order Consistency of Personality Traits from Childhood to Old Age: A Quantitative Review of Longitudinal Studies." *Psychological Bulletin*, 126: 3–25.

Shannon, C. E. (1948). "A Mathematical Theory of Communication." *Bell System Technical Journal*, 27: 379–423.

Simon, H. A. (1955). "A Behavioral Model of Rational Choice." *Quarterly Journal of Economics*, 69: 99–118.

Simon, H. A. (1956)."Rational Choice and the Structure of the Environment." *Psychological Review*, 63: 129–138.

Simon, H. A. (1978). "Rationality as Process and as Product of Thought." *American Economic Review*, 68: 1–16.

Sims, C. A. (2003). "Implications of Rational Inattention." *Journal of Monetary Economics*, 50: 665–690.

Sims, C. A. (2006). "Rational Inattention: A Research Agenda." Working Paper, Princeton University.

Stewart, N. (2009). "The Cost of Anchoring on Credit-Card Minimum Repayments." *Psychological Science*, 20: 39–41.

Stigler, G. J., and Becker, G. S. (1977). "De gustibus non est disputandum." *American Economic Review*, 67: 76–90.

Thaler, R. H. (2015). *Misbehaving: The Making of Behavioral Economics*. New York: W. W. Norton.

Thaler, R. H., and Sunstein, C. R. (2008). *Nudge: Improving Decisions about Health, Wealth, and Happiness*. New Haven, CT: Yale University Press.

Tversky, A., and Kahneman, D. (1974). "Judgment under Uncertainty: Heuristics and Biases." *Science*, 185: 1124–1131.

5

Information Recovery Related to Adaptive Economic Behavior and Choice

George Judge

1. Introduction

The outcome of economic-behavioral processes and systems involves the rich ingredients of complexity, uncertainty, volatility, and ambiguity. The statistical complexity of information recovery emerges because economic systems are dynamic and seldom in equilibrium and there is no unique time-invariant econometric model. Most conceptual micro- and macroeconomic models are single valued-equilibrium in nature, and in practice, processes and systems are stochastic and seldom, if ever, in equilibrium. Although there is only one sampling distribution consistent with an economic system in equilibrium, there are a large number of possible ways an economic process-system may be out of equilibrium. In this situation, it seems inappropriate to ask what the model is, but, rather, one should ask what it could be. For many econometric problems and data sets, the natural solution may not be a fixed distribution but a set of distributions, each with its own probability. In this situation, the value of an econometric prediction is related to how far the system is out of equilibrium.

The resulting uncertainty about existing conditions and the dynamics of the process create problems for econometric model specification and make it difficult, using traditional direct econometric methods, to capture the underlying hidden structure or key dynamic functions of the system. Although economic processes are, in general, simple in nature, the underlying dynamics are complicated and not well understood. The result is a family of economic models, each incorrectly specified and containing inadequacies that provide an imperfect link to the indirect noisy observational data. This type of data makes it impossible to distinguish between mutual influence and causal influence and does not contain dynamic or directional information. Even introducing a lag in the mutual observations fails to distinguish information that is actually exchanged from information that is shared and does not support time causality.

George Judge, *Information Recovery Related to Adaptive Economic Behavior and Choice* In: *Advances in Info-Metrics: Information and Information Processing across Disciplines*. Edited by: Min Chen, J. Michael Dunn, Amos Golan, and Aman Ullah, Oxford University Press (2021).
DOI: 10.1093/oso/9780190636685.003.0005

The indirect noisy observations used in an attempt to identify the underlying dynamic system and to measure causal influence require the solution of a stochastic inverse problem that is the inverse of the conventional forward problem, which relates the model parameters to the observed data. Thus, the data are in the effects domain, and our interest lies in the causal domain. If the number of measurements-data points are smaller than the number of unknown parameters to be estimated, the stochastic inverse problem is in addition ill posed. Without a large number of assumptions, the resulting stochastic ill-posed, underdetermined inverse problem cannot be solved by traditional estimation and inference methods. As a result, conventional semi- and parametric estimation and inference methods are fragile under this type of model and data uncertainty and in general are not applicable for answering causal influence dynamic economic system questions. Application of traditional econometric models and methods that are not suited to the ill-posed inverse information recovery task has led to what Caballero (2010) has called a pretense of knowledge syndrome.

Given imperfect reductionist and often toy economic-econometric models and indirect noisy effects data, with information recovery in mind, we recognize the inadequate nature of traditional economic-econometric methods in terms of measuring evidence and making inferences. As a basis for solving the stochastic information recovery problem and providing a behavior-related optimizing-criterion measure, we recognize the connection between adaptive behavior and causal entropy maximization and use an information-theoretic family of entropic functions as a basis for linking the data and the unknown and unobservable behavioral model parameters. Consistent with this objective, we present several econometric models, suggest possible applications, and note some economic-econometric implications of the information-theoretic approach. We end with a question concerning the use of traditional estimation and inference criteria that do not have a connection to economic behavior and choice data.

2. Causal Entropy Maximization

In this chapter, we recognize that economic data comes from systems with dynamic adaptive behavior that are nondeterministic, involve uncertainties that can be quantified by the notion of information and are driven toward a certain stationary state associated with a functional and hierarchical structure (Georgescu-Roegen, 1971; Raine et al., 2006; Annila and Salthe, 2009). Contrary to economic theories in which a utility function is usually maximized, it is far from definitive whether a self-organized stochastic dynamic system is optimal with respect to any scalar functional that is known a priori. As a result, dynamic economic systems involve myriad interdependent micro components that give rise to a nearly instantaneous feedback adaptive behavior world that seldom, if ever, is in equilibrium.

As we seek new ways to think about the causal adaptive behavior of large complex and dynamic economic systems, we use entropy as the systems status, measure-optimizing criterion. This permits us to recast economic-behavioral systems in terms of path microstates where entropy reflects the number of ways a macrostate can evolve along a path of possible microstates and the more diverse the number of microstates, the larger the causal path entropy. A uniform unstructured distribution of the microstates corresponds to a macrostate with maximum entropy and minimum information. From an economic information recovery standpoint, we use the connection between adaptive intelligent behavior, causal entropy maximization (AIB-CEM), and self-organized equilibrium-seeking behavior in an open dynamic economic system (Wissner-Gross and Freer, 2013). In this context, economic systems are equilibrium-stationary state seeking on average but may not be in a particular equilibrium. Although only one stationary state is consistent with an economic system in equilibrium, there are a large number of ways an economic path-dependent competitive, interacting process-system may be out of equilibrium. Thus, in the behavioral area, causal entropy maximization is a link that leads us to believe that an economic-behavioral system with a large number of agents, interacting locally and in finite time, is in fact optimizing itself. In this setting, things such as preferences, a commodity, economic value, optimal resource allocation, income and causal path entropy are behavior related and represent essentially the same thing. In this adaptive interdependent behavioral system, it is useful to be reminded that data does not behave, people behave. Thus, systems may be measured in terms of outcomes such as income distribution that provides an informational-probability link to the underlying self- organizing system. It is useful to note that in areas such as computer science, maximum causal inference has been used with dynamically sequential revealed information from interacting processes (see Ziebart et al., 2010, 2013).

The connection between causal adaptive behavior and entropy maximization, based on a causal generalization of entropic forces, suggests that economic social systems do not evolve in a deterministic or a random way, but tend to adapt behavior in line with an optimizing principle. This is a natural process in an effective behavioral working system. One reason for seeking an entropy-based adaptive behavior causal framework is that it permits the interpretation of adaptive economic behavior in terms of entropic functions and thereby the use of information-theoretic econometric methods. Information-theoretic methods are increasingly being used to analyze self-organizing processes in both analytical and numerical research, and this consistency of the econometric model, the data, and the estimation and inference process has potential for turning economics from a descriptive science to a predictive or at least a comprehensive and behavior-related quantitative one. It also raises questions about the use of traditional reductionist economic models and econometric estimation and inference methods.

3. An Information Recovery Framework

In section 1, we noted the connection between adaptive intelligent behavior and entropy maximization. In the context of dynamic economic systems information recovery, this connection suggests a basis for establishing a causal influence-econometric model link to the data. With this behavioral-entropy connection, a natural solution is to make use of information-theoretic estimation and inference methods that are designed to deal with the nature of economic-econometric models and data, and the resulting stochastic inverse problems. In this context, the Cressie and Read (1984), Read and Cressie (1988) (CR) family of entropic functions provides a basis for linking the data and the unknown and unobservable behavioral model parameters. This permits the researcher to exploit the statistical machinery of information theory to gain insights relative to the adaptive behavior underlying the causal behavior of a dynamic process from a system that may not be in equilibrium. Thus, in developing an information-theoretic econometric approach to estimation and inference, the CR single-parameter family represents a way to link the likelihood-entropic behavior informational functions with the underlying sample of data to recover estimates of the unknown parameters of the sampling model of the process. Information-entropic functions of this type have an intuitive interpretation that reflects uncertainty as it relates to a model of the adaptive behavior for processes.

In identifying estimation and inference measures that may be used as a basis for characterizing the data-sampling process for indirect noisy observed data outcomes, we begin with the CR multiparametric convex family of entropic functional-power divergence measures:

$$I(\mathbf{p},\mathbf{q},\gamma) = \frac{1}{\gamma(\gamma+1)} \sum_{i=1}^{n} p_i \left[\left(\frac{p_i}{q_i} \right)^{\gamma} - 1 \right]. \tag{5.1}$$

In Eq. (5.1), γ is a parameter that indexes members of the CR family, p_i's represent the subject probabilities, and the q_i's are interpreted as reference probabilities. Being probabilities, the usual probability distribution characteristics of $p_i, q_i \in [0,1] \forall i$, $\sum_{i=1}^{n} p_i = 1$, and $\sum_{i=1}^{n} q_i = 1$ are assumed to hold. In (5.1), as γ varies, the resulting CR family of estimators that minimize power divergence exhibits qualitatively different sampling behavior that includes Shannon's entropy, the Kullback–Leibler measure, and in a binary context, the logistic distribution divergence.

The CR family of power divergences is defined through a class of additive convex functions that encompasses a broad family of test statistics and represents *a broad family of likelihood functional relationships* within a moments-based estimation context. In addition, the CR measure exhibits proper convexity in $\mathbf{p}$, for all values of γ and $\mathbf{q}$, and embodies the required probability system characteristics, such as additivity and invariance with respect to a monotonic transformation of the divergence measures. In the context of extremum metrics, the general CR family of power divergence statistics represents a flexible family of pseudo-distance measures from which to derive empirical probabilities. The CR single-index family of divergence measures can be interpreted as encompassing a wide array of empirical goodness-of-fit and estimation criteria. As γ varies, the resulting estimators that minimize power divergence (MPD) exhibit qualitatively different sampling behavior.

To place the CR family of power divergence statistics in a relative entropy perspective, we note that there are corresponding Renyi (1961, 1970) and Tsallis (1988) families of entropy functionals-divergence measures. As demonstrated by Gorban, Gorban, and Judge et al. (2010), over defined ranges of the divergence measures, the CR and entropy families are equivalent. Relative to Renyi (1961 and 1970) and Tsallis (1988), the CR family has a more convenient normalization factor $1/(\gamma(\gamma+1))$ and has proper convexity for all powers, both positive and negative. The CR family allows for separation of variables in optimization, when the underlying variables belong to stochastically independent subsystems (Gorban et al., 2010). This separation of variables permits the partitioning of the state space and is valid for divergences in the form of a convex function.

3.1 Examples of Two Information-Theoretic Behavioral Models

As a first example, assume a stochastic economic-econometric model of behavioral equations that involve endogenous and exogenous variables. Data consistent with the econometric model may be reflected in terms of empirical sample moments-constraints such as $\mathbf{h}(\mathbf{Y},\mathbf{X},\mathbf{Z};\boldsymbol{\beta}) = n^{-1}[\mathbf{Z}'(\mathbf{Y}-\mathbf{X}\boldsymbol{\beta})] \xrightarrow{p} 0$, where $\mathbf{Y},\mathbf{X},$ and $\mathbf{Z}$ are, respectively, an $n\times 1$, $n\times k$, $n\times m$ vector/matrix of dependent variables, explanatory variables, and instruments, and the parameter vector $\boldsymbol{\beta}$ is the objective of estimation. A solution to the stochastic inverse problem, in the context of Eq. (5.1) and based on the optimized value of $I(\mathbf{p},\mathbf{q},\gamma)$, is one basis for representing a range of data-sampling processes and likelihood functions. As γ varies, the resulting estimators that minimize power divergence exhibit qualitatively different sampling behavior. Using empirical sample moments, a solution to the stochastic inverse problem, for any given choice of the γ parameter, may be formulated as the following extremum-type estimator for β:

$$\hat{\beta}(\gamma) = \arg\min_{\beta \in \mathrm{B}} \left[\min_{\mathbf{p}} \left\{ I(\mathbf{p}, \mathbf{q}, \gamma) \,|\, \sum_{i=1}^{n} p_i \mathbf{z}_i{}' (y_i - \mathbf{x}_i \beta) = 0, \sum_{i=1}^{n} p_i = 1, p_i \geq 0 \forall i \right\} \right], \tag{5.2}$$

where $\mathbf{q}$ is usually taken as a uniform distribution. This class of estimation procedures is referred to as minimum power divergence (MPD) estimation (see Gorban et al., 2010; Judge and Mittelhammer, 2012a, 2012b).

As a second example, consider an analysis of binary response data models (BRMs), which include the important analysis of discrete choice models (see, e.g., Judge and Mittelhammer, 2012). *The objective is to predict probabilities that are unobserved and unobservable from indirect noisy observations.* Traditionally, the estimation and inference methods, used in empirical analyses of binary response models, convert this fundamentally *ill-posed stochastic inverse problem* into a well-posed one that can be analyzed via conventional parametric statistical methods. This is accomplished by imposing a parametric functional form on the underlying data-sampling distribution and followed by maximum likelihood estimation.

Seeking to minimize the use of unknown information concerning model components, we begin by characterizing the $n \times 1$ vector of Bernoulli random variables, $\mathbf{Y}$, by the universally applicable stochastic representation

$$\mathrm{Y} = \mathbf{p} + \varepsilon, \text{where } \mathrm{E}(\varepsilon) = 0 \text{ and } \mathbf{p} \in \mathop{\times}_{i=1}^{n} (0,1). \tag{5.3}$$

The specification in (5.3) implies only that the expectation of the random vector $\mathbf{Y}$ is some mean vector of Bernoulli probabilities $\mathbf{p}$ and that the outcomes of Y are decomposed into means and noise terms. The Bernoulli probabilities in (5.3) are allowed to depend on the values of explanatory variables contained in the $(n \times k)$ matrix $\mathbf{X}$, where the conditional orthogonality condition, $\mathrm{E}[\mathbf{X}'(\mathrm{Y} - \mathbf{p}(\mathbf{X})|\mathbf{X})] = 0$, is implied. Given sampled binary outcomes, if we use $k < n$ empirical moment representations of the orthogonality conditions, $n^{-1}\mathbf{x}'(\mathbf{y} - \mathbf{p}) = 0$, to connect the data space to the unknown-unobservable probabilities, this constraint is the empirical implementation of the moment condition, $\mathrm{E}(\mathbf{X}'(\mathrm{Y} - \mathbf{p})) = 0$.

If we specify this as an extremum problem, we may characterize the conditional Bernoulli probabilities by minimizing, some member of the family of generalized Cressie-Read (CR) power divergence measures. In this context, under reference distribution $q_i = .5$, the multinomial specification of the minimum power divergence estimation problem in Lagrange form is

$$L(\mathbf{p},\lambda) = \sum_{i=1}^{n}\sum_{i=1}^{n}\left(\frac{1}{\gamma(\gamma+1)}\sum_{j=1}^{m} p_{ij}\left[\left(\frac{p_{ij}}{q_{ij}}\right)^{\gamma} - 1\right]\right) + \sum_{j=1}^{m}\lambda_j'\mathbf{x}'\left(y_j - p_j\right)$$
$$+ \sum_{i=1}^{n}\eta_i\left(\sum_{j=1}^{n} p_{ik} - 1\right). \tag{5.4}$$

Solving first-order conditions with respect to the $p_{ij}'s$ leads to the standard multivariate logistic distribution, when the reference distributions are uniform (see Mittelhammer and Judge, 2011).

3.2 Convex Entropic Divergences

In choosing a member of the CR family of likelihood-divergence functions (5.1), one might follow Gorban (1984) and Gorban and Karlin (2003) and consider in the context of the above examples a parametric family of convex information divergences which satisfy additivity and trace conditions. Convex combinations of CR(0) and CR(−1) span an important part of the probability space and produce a remarkable family of divergences-distributions. Using the CR divergence measures, this parametric family is essentially the linear convex combination of the cases where $\gamma = 0$ and $\gamma = -1$. This family is tractable analytically and provides a basis for joining (combining) statistically independent subsystems.

When the base measure of the reference distribution $\mathbf{q}$ is taken to be a uniform probability density function, we arrive at a one-parameter family of additive convex dynamic Lyapunov functions. From the standpoint of extremum minimization with respect to $\mathbf{p}$, the generalized divergence family, under uniform $\mathbf{q}$, reduces to

$$S_{\alpha}^{*} = \sum_{i=1}^{n}\left(-(1-\alpha)p_i \ln\left(p_i\right) + \alpha \ln\left(p_i\right)\right), 0 \le \alpha \le 1. \tag{5.5}$$

In the limit, as $\alpha \to 0$, the Kullback–Leibler or minimum I divergence of the probability mass function $\mathbf{p}$, with respect to $\mathbf{q}$, is recovered. As $\alpha \to 1$, the maximum empirical likelihood (MEL) solution is recovered. *This generalized family of divergence measures permits a broadening of the canonical distribution functions and provides a framework for developing a loss-minimizing estimation rule (Jeffreys, 1983).*

4. Further Examples-Applications

In the following subsections, the connection between adaptive behavior and causal entropy maximization is illustrated in the form of a state-space Markov model, a weighted and binary behavioral network model, and a time

4.1 A Stochastic State-Space Framework

Information-theoretic dynamic economic models appear naturally and can be given a directional meaning in a conditional adaptive behavior Markov framework, when state spaces and transition probabilities are introduced in the CR-MPD framework (Miller and Judge, 2015). Looking at the scene in terms of a stochastic process lets us get a peek at some of the hidden dynamics of the system and introduce the role of time and probabilistic causality. The condition of probabilistic causality introduces an arrow of time restriction, and the Markov process satisfies this restriction. For example, if the decision outcomes exhibit first-order Markov character, the dynamic behavior of the agents may be represented by conditional transition probabilities $p(j,k,t)$, which represent the probability that agent i moves from state $j = 1, 2, \ldots, K$, to state k at a time t. Given observations on the micro behavior $Y(i,k,t)$, the conditional discrete Markov decision process framework may be used to model the agent-specific dynamic economic behavior which varies with t. In the conditional case, we have $T\times(K-1)\times(K-1)$ unknown transition probabilities in the $(K-1)\times(T-1)$ estimating equations. Information recovery goes from the data to the parameters; thus, this is an ill-posed inverse estimation problem, and traditional estimation methods are not applicable.

Linking the sample analog of the Markov process to the indirect noisy observations

$$Y(k,t) = \sum_{j=1}^{K} Y(j,t-1)p(j,k,t) + e(k,t), \tag{5.6}$$

leads to a new class of conditional Markov models that is based on a set of estimating equations–moment equations

$$E\left[z_t' e(k,t)\right] = 0, \tag{5.7}$$

where $\mathbf{z}_t$ is an appropriate set of instrumental system or intervention variables. By substituting (5.6) into (5.7), we form a set of estimating equations expressed in terms of the unknown transition probabilities. As a basis for identifying parametric data sampling distributions and likelihood functions in the form of distances in probability space, we again use the CR family of power divergence measures

$$I(\mathbf{p},\mathbf{q},\gamma) = \frac{2}{\gamma(1+\gamma)} \sum_{t=1}^{T} \sum_{j=1}^{K} \sum_{k=1}^{K} p(j,k,t) \left[\left(\frac{p(j,k,t)}{q(j,k,t)} \right)^{\gamma} - 1 \right], \tag{5.8}$$

to provide access to a rich set of distribution functions that encompasses a family of estimation objective functions indexed by discrete probability distributions that are convex in p. Formally, the MPD problem may be solved by choosing transition probabilities p to minimize $I(\mathbf{p},\mathbf{q},\gamma)$, subject to the sample analog of (5.2), which is

$$\sum_{t=1}^{T} \mathbf{z}_t' \left(Y(k,t) - \sum_{j=1}^{K} Y(j,t-1) p(j,k,t) \right) = 0, \tag{5.9}$$

for each $j = 2, \ldots, K$ and the row-sum constraint

$$\sum_{k=1}^{K} p(j,k,t) = 1, \tag{5.10}$$

over all j and t. Given a sample of indirect noisy observations and corresponding moment conditions, the parametric family of convex entropic divergences defined in Eq. (5.5) may be used to choose an optimum member of the CR family.

4.2 Network Behavior Recovery

A new network-based paradigm that goes beyond traditional overly simplified modeling and mathematical anomalies is developing under the name of Network Science (e.g., see Willinger et al., (2009) and Barabasi (2012) and the references contained therein). It is based on observed adaptive behavior data sets that are indirect, incomplete, and noisy. Several things make this approach attractive for information recovery in economics and the other social sciences. In the economic-behavioral sciences, everything seems to depend on everything else, and this fits right into the interconnectedness–simultaneity of the nonlinear dynamic network paradigm. There is also a close link between evolving network structures and the equilibrium or disequilibrium of economic-behavioral systems and entropy maximization. Finally, in terms of a methodology, network problems appear to be consistent with the information-theoretic approach to information recovery.

To indicate the applicability of information-theoretic approach, an example may be useful. In an economic-behavioral network, the efficiency of information flow is predicated on discovering or designing protocols that efficiently route information. In many ways, this is like a transportation network where the emphasis is on design and efficiency in routing the traffic flows (e.g., see Castro et al. (2004) and the references therein). To carry this information flow analogy a bit farther, consider the problem of determining least-time point-to-point traffic flows between subnetworks, when only aggregate origin-destination traffic volumes are known. Given information about the network protocol in the form of a matrix A_{ij} that is composed of binary elements, traffic flows may be estimated from the noisy aggregate traffic data. Since the origin to destination routes-unknowns are much larger than the origin–destination data, this results in an ill-posed linear inverse problem of the type first introduced in section 3. If we write the inverse problem as

$$y_i = \sum_{j=1}^{N} A_{ij} p_j, \tag{5.11}$$

where y_i and A_{ij} are known and p_j is unknown and $\sum_{j=1}^{N} p_j = 1$, we may make use of the CR family of entropic divergence measures (5.1) and write the problem as the following constrained conditionally optimization problem:

$$I(\mathbf{p}, \mathbf{q}, \gamma) \mid \sum_{j=1}^{n} \lambda_i \left(\sum_{j=1}^{N} \left(A_{ij} p_j - y_i \right), \sum_{j=1}^{n} p_j = 1, p_j \geq 0 \right). \tag{5.12}$$

This is just the solution to a standard problem when a function must be inferred from insufficient sample-data information. Thus, network inference and monitoring problems have a strong resemblance to an inverse problem in which key aspects of a system are not directly observable. For an application of information-theoretic entropic methods to this type of network information flow problem, see Cho and Judge (2006, 2013) and Squartini (2015). It is interesting that network theory presents a model for producing scale-free patterns that are manifestations in physics of free-energy consumption. In other words, if economic systems did not consume energy in the least time, these patterns would not be present. We should recognize that economic systems work toward states of increasing entropy (consuming free energy), as well as a basis for describing these systems in relation to network theory. Finally, it is worth emphasizing that in actual networks, the flows from one node to another will themselves affect node-to-node capacities, making the evolution of a network an intractable problem beyond deterministic or statistical predictions (Hartonen et al., 2012).

4.3 Unlocking the Dynamic Content of Time Series

At this point, we should note that over the centuries economists have been interested in the informational behavioral content of time series data. In spite of many efforts in this area, because of a disconnect between the temporal information from the dynamic behavioral environment and the traditional estimation methods used, the temporal patterns underlying the time-dated outcomes have, in general, remained hidden. The function of dynamics in this process is to connect to system outcomes later on in time. To connect the underlying dynamic behavioral data with our corresponding entropy maximization formulation, we make use of the permutation entropy principle (Bandt and Pompe (B&P), 2002). The B&P concept provides a basis for analysis of ordinal nonlinear time series and permits a way of describing, in probability distribution form, the underlying dynamic state of the system. The objective is to extract from a nonlinear time series qualitative information in the form of temporal dynamics, and the idea is based on permutation patterns-ordinal relations among values of a time series. To provide a probability distribution of the temporal dynamics that is linked to the sample space, B&B proposed a method that takes time causality into account by comparing time-related observations-ordinal patterns in a time series. They consider the order relation between time series instead of the individual values. Permutation patterns-partitions-vectors are developed by comparing the order of neighboring observations; from this it is possible to create a probability distribution whose elements are the frequencies associated with this ith permutation pattern, $I = 1,2, \ldots, D!$. Using, for example, the gamma ≥ 0 member of the CR family with a uniform reference distribution, we may define the information content of such a distribution for the $D!$ distinct assessable states as $PE = -\sum_{i=1}^{D!} p_i \ln p_i$, with normalized version $PE_{\text{norm}} = -\frac{1}{\ln D!}\sum_{i=1}^{D!} p_i \ln p_i$. For a review and applications of the permutation entropy method, see Zanin et al., (2012) and Kowalski et al. (2012).

5. Summary Comments

In this chapter, we have (1) exhibited a connection between adaptive economic behavior and causal entropy maximization in equilibrium-seeking dynamic economic systems, and (2) made use of a broad family of entropic functionals and behavior-related information in the form of moment conditions to provide a basis for econometric information recovery. The applicability of this stationary state seeking-optimizing idea has been designated by behavioral and discrete choice econometric models, a Markov state-space econometric model, an empirical partial equilibrium binary network, and time series models and applications. The general approach is applicable for settings where data are generated from interacting dynamic processes. The next step is to extend the general applicability of this adaptive–optimizing behavior concept to a range of econometric settings and large, open, dynamic economic systems.

Finally, in section 2 we noted the statistical implications of using imperfect economic-econometric models and data, and for solution purposes the need to solve a stochastic inverse problem. As a start toward mitigating these problems, we suggested in section 2, as a criterion-status measure, the connection between adaptive behavior and causal entropy maximization, and in section 3 we suggested an information-theoretic estimation and inference framework. In contrast, we noted that traditional data-based methods such as maximum likelihood and variations of least squares are disconnected from the underlying dynamic behavioral process. Given the importance of recovering dynamic economic behavioral information, a natural question arises as to the use of traditional estimation and inference methods-criteria as a solution basis for stochastic inverse information recovery problems. The connection between adaptive dynamic economic behavior and causal entropy maximization provides a behavioral-based optimization criterion. This in conjunction with information-theoretic methods appears to be one way to move economics from a descriptive to a predictive science.

References

Annila, A., and Salthe, S. (2009). "Economies Evolve by Energy Dispersal." *Entropy*, 11: 606–633.

Bandt, C., and Pompe, B. (2002). "Permutation Entropy: A Natural Complexity Measure for Time Series." *Physical Review Letters*, 88(17): 174102-1–174102-4.

Barabasi, A-L. 2012. "The Network Takeover." *Nature Physics*, 8: 14–16.

Caballero, Ricardo J. 2010. "Macroeconomics after the Crisis: Time to Deal with the Pretense-of-Knowledge Syndrome." Journal of Economic Perspectives 24 (4), : 85–102.

Castro, R. M., Coates, G. Laing, Nowak, R., and Yu, B. 2004. "Network Tomography: Recent Developments." *Statistical Science*, 19: 499–517.

Cho, W., and Judge, G. (2006). "Information Theoretic Solutions for Correlated Bivariate Processes." *Economic Letters*, 7: 201–207.

Cho, W., and Judge, G. (2013). "An Information Theoretic Approach to Network Tomography." Working Paper, Uuniversity of California at Berkeley.

Cressie, N., and Read, T. (1984). "Multinomial Goodness of Fit Tests." *Journal of Royal Statistical Society of Series B*, 46: 440–464.

Georgescu-Roegen, N. (1971). *The Entropy Law and the Economic Process.* Cambridge, MA: Harvard University Press.

Gorban, A. (1984). "Equilibrium Encircling: Equations of Chemical Kinetics and Their Thermodynamic Analysis." Nauka: Novosibirsk.

Gorban, A., Gorban, P., and Judge, G. (2010). "Entropy: The Markov Ordering Approach." *Entropy*, 12: 1145–1193.

Gorban, A., and Karlin, I. (2003). "Family of Additive Entropy Functions out of Thermodynamic Limit." *Physical Review E*, 67: 016104.

Hartonen, T., and Annila, A. (2012). "Natural Networks As Thermodynamic Systems." *Complexity*, 18: 53–62.

Jeffreys, H. (1983). *Theory of Probability.* 3rd ed. Oxford: Clarendon Press.

Judge, G., and Mittelhammer, R. (2012a). *An Information Theoretic Approach to Econometrics.* Cambridge: Cambridge University Press.

Judge, G., and Mittelhammer, R. (2012b). "Implications of the Cressie-Read Family of Additive Divergences for Information Recovery." *Entropy*, 14: 2427–2438.

Kowalski, A., Martin, M., Plastino, A., and G. Judge, G. (2012). "On Extracting Probability Distribution Information from Time Series." *Entropy*, 14: 1829–1841.

Miller, D., and Judge, G. (2015). "Information Recovery in a Dynamic Statistical Model." *Econometrics*, 3:187–198.

Mittelhammer, R., and Judge, G. (2011). "A Family of Empirical Likelihood Functions and Estimators for the Binary Response Model." *Journal of Econometrics*, 164: 207–217.

Raine, A., Foster, J., and Potts, J. (2006). "The New Entropy Law and the Economic Process." *Ecological Complexity*, 3: 354–360.

Read, T., and Cressie, N. (1988). *Goodness of Fit Statistics for Discrete Multivariate Data.* New York: Springer Verlag.

Renyi, A. (1961). "On Measures of Entropy and Information." In Proceedings of the 4th Berkeley Symposium on Mathematics, Statistics and Probability, Berkeley.

Renyi, A. (1970). *Probability Theory.* Amerstam, North-Holland: Elsevier.

Squartini, T., Seri-Giacomi, E., Garlaschelli, D., and Judge, G. (2015). *Information Recovery in Behavioral Networks, Plos One*, 10(5): e0125077.

Tsallis, C. (1988). "Possible Generalization of Boltzmann-Gibbs Statistics." *Journal of Statistical Physics*, 52: 479–487.

Willinger, W., D. Alderson, D., and Doyle, J. (2009). "Mathematics and the Internet: A Source of Enormous Confusion and Great Potential." *Journal of the American Mathematical Society*, 56: 586–599.

Wissner-Gross, A., and Freer, C. (2013). "Causal Entropic Forces." *Physical Review Letters PRL*, 110: 168702-1-5.

Zanin, M., Zunino, L., Rosso, O., and Papo, D. (2012). "Permutation Entropy and Its Main Biomedical and Econophysics Applications: A Review." *Entropy*, 14: 1553–1577.

Ziebart, B., Bagnell, J., and Dey, A. (2010). "Modeling interaction via principle of maximum entropy." Proceedings, International Conference on Machine Learning, Haifa, Israel.

Ziebart, B. D, Bagnell, J. A. and Dey, A. K. (2013). "The Principle of Maximum Causal Entropy for Estimating Interacting Processes." *IEEE Transactions on Information Theory*, 59(4): 1966–1980.

PART III

INFO-METRICS AND THEORY CONSTRUCTION

This part demonstrates the use of info-metrics for basic theory construction. This is not a trivial task and is essential for every model construction. It applies the principle of maximum entropy to model or theory building within a constrained optimization framework. The constraints capture all the information used. These are symmetry conditions of the system studied or are often conceptualized as conservations rules-rules that govern the behavior of the system. These rules are defined over the entities of interest. In Chapter 6, Harte develops a unified theory of ecology. The chapter demonstrates in an elegant way how one can achieve two goals under the info-metrics framework. The first goal is to establish general laws and patterns, and the second is to document the diversity of idiosyncratic organismal forms, adaptations, and behaviors. The focus of this chapter is on progress to date in constructing such a unification on a foundation of information theory.

In Chapter 7, Caticha develops and discusses the notion of entropic dynamics. The notion of dynamics here emerges directly via info-metrics and the principle of maximal entropy. Simply stated, entropic dynamics is a framework in which dynamical laws such as those that arise in physics are derived as an application of information-theoretic inference. No underlying laws of motion, such as those of classical mechanics, are postulated. Instead, the dynamics is driven by entropy subject to constraints that reflect the information that is relevant to the problem at hand. The chapter provides a review of the derivation of three forms of mechanics: standard diffusion, Hamiltonian mechanics, and quantum mechanics.

Overall, these two chapters provide a coherent method for constructing theories and models using the info-metrics framework. They also demonstrate a consistent way for thinking about information and constraint specification for real-world problems.

6

Maximum Entropy

A Foundation for a Unified Theory of Ecology

John Harte

1. Introduction

Two seemingly irreconcilable goals characterize the field of ecology: the search for uniqueness and the search for generality.

On the one hand, many ecologists devote their career to the study of a particular habitat, and possibly even a specific location and a particular species. The unique aspects of the chosen system of study are often what motivate such intensely focused work and endow the findings with significance. It was the idiosyncratic that first drew many of us ecologists to the study of natural history; it was exciting to astound others with a fascinating description of a seemingly unique phenomenon we would observe in a nearby woods or pond. Journals like *Nature* and *Science* tend to look favorably upon submitted manuscripts that describe an organism or a community of organisms that exhibits some novel trait or in some other respect deviates from prior expectations.

On the other hand, many ecologists, being scientists, seek general laws—or at least patterns that are ubiquitous in space and persistent in time. An example in ecology where generality seems to hold is a pattern called the species-area relationship (SAR), describing how the number of species increases as one counts the species in ever-larger areas within an ecosystem. We return to this later in the chapter. Claims of a general law or a ubiquitous pattern in ecology are, however, sometimes construed by colleagues as being dismissive of their unique findings about particular organisms or species under particular conditions.

As examples of unique phenomena accumulate, seekers of generality sometimes discover that they have to revisit and revise both their claims of general patterns and their theories that purport to explain such patterns. The question naturally arises then: is there convergence toward general theory that is consistent with the idiosyncrasies of nature, or should we expect the ever-mounting accumulation of idiosyncratic ecological phenomena to preclude the possibility of general theory? In other fields, such as physics, virtually no one would doubt the answer: generality wins in the end. In ecology, it is an open question.

John Harte, *Maximum Entropy: A Foundation for a Unified Theory of Ecology* In: *Advances in Info-Metrics: Information and Information Processing across Disciplines.* Edited by: Min Chen, J. Michael Dunn, Amos Golan, and Aman Ullah, Oxford University Press (2021).
DOI: 10.1093/oso/9780190636685.003.0006

The work described here is premised on the compatibility of the two goals: seeking general laws and patterns, and documenting the diversity of idiosyncratic organismal forms, adaptations, and behaviors. Although I have spent many years as an ecologist studying and reveling in the unique dimensions of natural history, my recent work has attempted to devise a general unified theory of ecology, one that can accurately predict the patterns, such as the species-area relationship and very much more that ecologists observe. The focus of this chapter is on progress to date in constructing, upon a foundation of information theory, such as unification.

2. Ecological Theory

Ecological models and theories that can make quantitative predictions are often construed to lie along a spectrum ranging from purely statistical to purely mechanistic. A purely statistical approach to ecology might be based on no more than persistently observed correlations between measured quantities. Or it might be based on an ad hoc statistical assumption (e.g., nature is fractal, which might then lead to predicted power laws). If a handwaving rationale, based on a plausible mechanism, is used to motivate the observed correlation, then we are entering the realm of hybrid explanation. Moving further along the spectrum, suppose we seek to formulate a truly mechanistically based theory of ecology, capable say of predicting spatial distributions of individuals within species, abundance and metabolic rate distributions across species, the relationship between the abundance of a species and the average metabolic rate of its individuals, food-web architecture, species-area relationships, and the structure of taxonomic and phylogenetic trees. What mechanisms might we select as ingredients? Now we confront the central dilemma faced by ecologists seeking mechanistic, predictive theory.

2.1 The Ecologist's Dilemma

A mechanistic and quantitative theory of ecology would presumably include some sort of mathematical description of the mechanisms by which species and individuals reproduce, partition biotic and abiotic resources, and interact with each other in doing so. The mechanisms that ecologists generally agree operate in nature include predation, competition, mutualism, commensalism, birth, death, migration, dispersal, speciation, disease and resistance to disease, effects of crowding, effects of rarity on reproductive pairing and genetic diversity, effects of aggregation of individuals on predation, an assortment of reproductive strategies ranging from lekking behavior to broadcast insemination to male sparring to flower enticement of pollinators, as well as an equally wide range of organism behaviors and social dynamics somewhat distinct from reproduction, including harvesting strategies, eusociality, decoying, predator–escape strategies, plant signaling, nutrient cycling, resource substitution, hibernation, plant architectures to achieve structural goals, and lif-cycle adaptations to fluctuations in weather and resource supply.

The preceding is a long, yet quite incomplete, list of the many mechanisms that are observed to govern the interactions of millions of species composed of billions of locally adapted populations. These interactions play out on a stage with abiotic conditions that vary unpredictably in both space and time. Those abiotic conditions are, in turn, influenced by the organisms. And just in case the above list of ingredients to mechanistic theory does not offer enough choice, we could include trait and genetic differences between species, between populations within species, and between individuals within populations.

This morass of mechanistic, behavioral, and trait complexity, historical contingency, and environmental stochasticity generates the ecological phenomena we observe in nature. Few, if any, of the mechanisms are sufficiently understood that they can be incorporated, quantitatively, into ecological models or theory without adjustable parameters. As a consequence, mechanistic models and theories are difficult to falsify, and thus they do not, or at least should not, inspire confidence. Moreover, ecologists, like economists, must confront the fact that controlled experimentation and the isolation of causal mechanisms are extremely difficult; that system boundaries, beyond which mechanistic influence can be ignored, are porous and often ill defined; and that the very objects of study in ecology are disappearing every day as the human population and the scale of the human economy grow, resulting in the destruction of habitat and the extinction of species.

All of this constitutes the ecologist's dilemma, and it is daunting. In attempts to retain mechanism, parameterized ecological models have been constructed to provide insight into a relatively narrow set of questions. For example, from variations on the Lotka-Volterra equations (Lotka, 1956; Volterra, 1926) describing species interactions, a very small subset of the above-listed mechanisms—such as predation, competition, birth, and death—can be modeled. But the variety of possible functional forms that can be used to describe species interactions is large, and they usually contain many unmeasured adjustable parameters. Hence, while such models have shed light on a few aspects of stability and coexistence of diverse species (e.g., May, 1972), they are a long way from constituting a sound basis for building widely applicable, unified, predictive theory. So let us turn to an alternative approach.

2.2 Nonmechanistic Ecological Theory

In light of this dilemma, we are led to ask whether it is possible to construct quantitative predictive ecological theory that incorporates no explicit mechanisms. Could such theory accurately predict the observed patterns of variation within and between species? In other words, could it predict how body sizes or metabolic rates vary across individuals, how abundances vary across species, how abundance relates to metabolism, how species and individuals share space, and even more? Could it even predict how birth and death rates differ across species? The answer to all these questions is yes, but to understand how and why, we first review some very basic ideas about inference.

2.3 The Logic of Inference

Least-biased inference lies at the core of science. Science advances by gathering together prior knowledge about a system and then using that knowledge as a springboard to infer new insights. Often the new insights we seek can be expressed in the form of probability distributions. A fundamental tenet of science is that the inference procedure used to obtain these distributions should be as free from bias as possible. One way to minimize bias is to ensure that the inferred distributions satisfy the constraints imposed by all the prior knowledge but embody no additional and therefore unwarranted assumptions.

Remarkably, a rigorous mathematical procedure exists for inferring least biased probability distributions that are consistent with prior knowledge and otherwise embody no additional assumptions about the system being investigated. That procedure is called MaxEnt, or maximization of information entropy. This inferential procedure has been successfully applied in a variety of fields, including economics (Golan, Judge, and Miller, 1996), forensics (Roussev, 2010), medical imaging (Frieden, 1972; Skilling, 1984; Gull and Newton, 1986), neural signaling and bird flocking (Bialek, 2015), protein folding (Steinbach et al., 2002, Seno et al., 2008), genomics (Salamon and Konopka, 1992), and ecology (Phillips et al., 2006; Phillips, 2008; Elith et al., 2011; Harte et al., 2008; Harte 2011). Most impressively, in physics, it has been shown that the laws of thermodynamics and statistical mechanics can be derived using the MaxEnt procedure (Jaynes, 1957, 1982).

In applications of MaxEnt, information at the macroscale is usually in the form of values of the "state variables," which are used as constraints in order to derive least-biased probability distributions at the microscale. Thus, for a thermodynamic system the state variables might be chosen to be pressure, volume, and number of molecules of the bulk system, and then the distribution of molecular velocities, a phenomenon at the microscale, can be inferred. A parallel construction can be used in ecology, and it has led to the construction of a theory denoted by METE (maximum entropy theory of ecology), to which I now turn.

3. The Maximum Entropy Theory of Ecology: Basics and a Simple Model Realization

The goal of METE (pronounced mēt and appropriately meaning "to allocate") is to predict the forms of numerous metrics describing how, within an ecosystem, metabolism is allocated over individuals, how individuals are allocated over species and over space, how species are allocated over space and over higher taxonomic categories such as genera or families, as well as relationships between, for example, metabolic rate and abundance of individuals within species or within higher categories. It seeks to do this, at any spatial scale and for any broad taxonomic group of organisms, such as mammals or plants or birds or arthropods or microorganisms. A further goal is to do this for any type of habitat, such as desert, forest, meadow, or even the human gut.

The metrics referred to here can be static or dynamic. In ecosystems disturbed by human activity, the shapes of the patterns are frequently observed to change over time, and on a longer timescale, they may be changing as a consequence of natural successional processes and even evolution and diversification. The first model realization of METE was only applicable to systems in steady state, and it was constructed paralleling the method used to derive the laws of classical equilibrium statistical mechanics and thermodynamics by Jaynes (1957). In this model realization (denoted ASNE) of METE, the state variables used to predict the static metrics of macroecology include the area, A_0, of an ecosystem at any spatial scale at which census data exist, the total number of species S_0, censused in that area, the total number of individuals, N_0, across the S_0 species, and the summed metabolic rate, E_0, of the N_0 individuals. With the constraints that arise from the ratios of the state variables A_0, S_0, N_0, E_0, the maximum information entropy condition predicts the mathematical forms of many macroecological metrics.

Two distributions are at the core of the ASNE model of METE. The first distribution is a joint conditional distribution (the ecosystem structure function) over abundance, n, and metabolic rate, ε: $R(n, \varepsilon|S_0, N_0, E_0)$. $R{\cdot}d\varepsilon$ is the probability that if a species is picked from the species pool, then it has abundance n, and if an individual is picked at random from that species, then its metabolic energy requirement is in the interval $(\varepsilon, \varepsilon + d\varepsilon)$. The second is a species-level spatial distribution $\Pi(n|A, n_0, A_0)$. If a species has n_0 individuals in an area A_0, then

Π is the probability that it has n individuals in an area A within A_0. MaxEnt determines the forms of those two core functions. Figure 6.1 (from Harte and Newman, 2014) is a flow chart summarizing how the predicted forms of the macroecological metrics derive from R and Π in the ASNE model.

For the species-abundance distribution (SAD), the ASNE version of METE predicts a truncated Fisher logseries function ($\Phi(n)$ in Figure 6.1). In a comprehensive comparative review of SADs, White et al. (2012) demonstrated that across spatial scales, taxonomic groups, and habitats, the truncated logseries form of the SAD predicted between 83 and 93 percent of the observed variation in the rank abundance relationships of species across 15,848 globally distributed communities, including birds, mammals, plants, and butterflies, and outperformed the lognormal distribution. Some specific validations are shown in Figure 6.2.

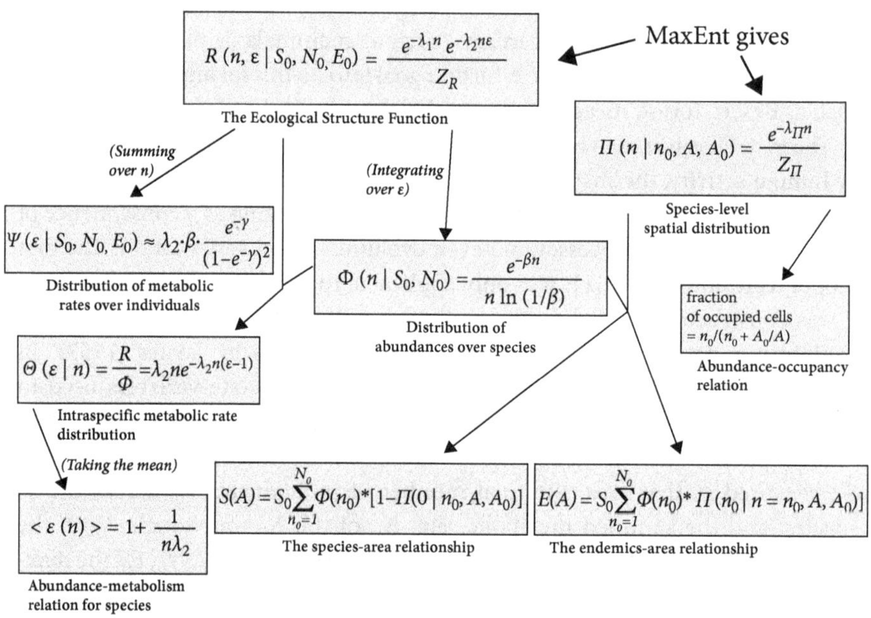

Figure 6.1 The structure of static ASNE.

Notes: The flow chart shows how the forms of the macroecological metrics are derived in the ASNE version of METE. The ecological structure function R and the spatial distribution function Π defined in the text are derived directly using the maximum entropy principle, implemented with the method of Lagrange multipliers. λ_1, λ_2, and λ_π are Lagrange multipliers that are uniquely determined by the values of the state variables; Z_R and Z_Π are partition functions that are determined from the Lagrange multipliers; $\beta = \lambda_1 + \lambda_2$; $\gamma = \lambda_1 + \lambda_2 \varepsilon$. The predicted forms of the metrics of macroecology, including the species abundance distribution, Φ, and the distribution of metabolic rates over all individuals, Ψ, are derived from R by appropriate summation over abundance or metabolic rate. The species-area and endemics-area relationships are derived, as indicated in the figure, by appropriately combining the functions R and Π. Complete details of derivations can be found in Harte (2011).

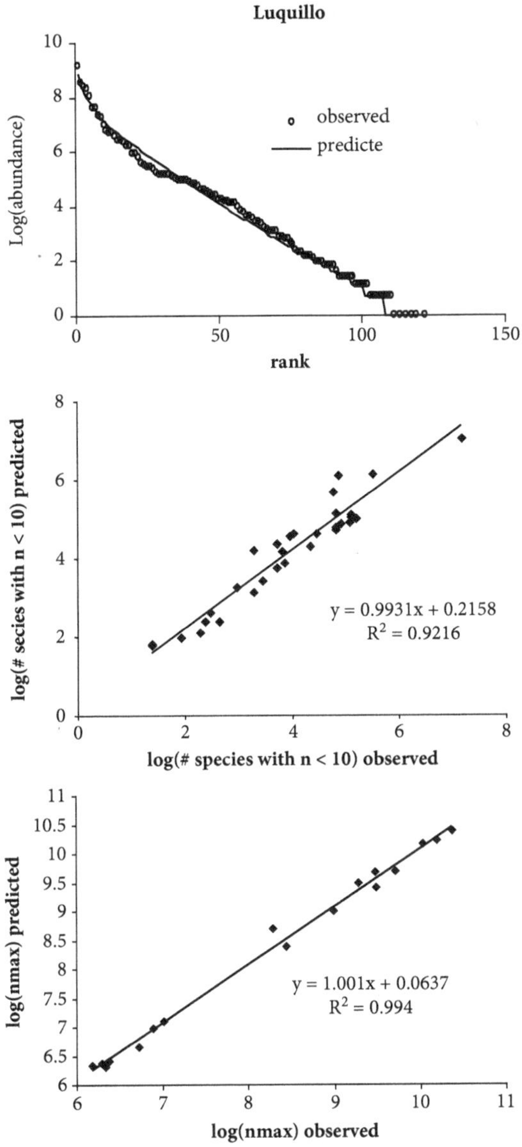

Figure 6.2 a. The species abundance distribution: Comparison of data and theory for the distribution of tree species abundances in the Luquillo tropical forest plot in Puerto Rico.

b. Test of the METE prediction (Eq. 7.40) for the number of rare species (here taken to be $n \leq 10$) for twenty-five distinct ecosystems. Only the values of N_0 and S_0 are used to make the predictions.

c. Test of the METE prediction (Eq. 7.41) of the most abundant species at the same twenty-five ecosystems. Only the values of N_0 and S_0 are used to make the predictions.

All comparisons are from Harte (2011). In Figure 6.2 and throughout, the abbreviation "log" refers to the natural logarithm.

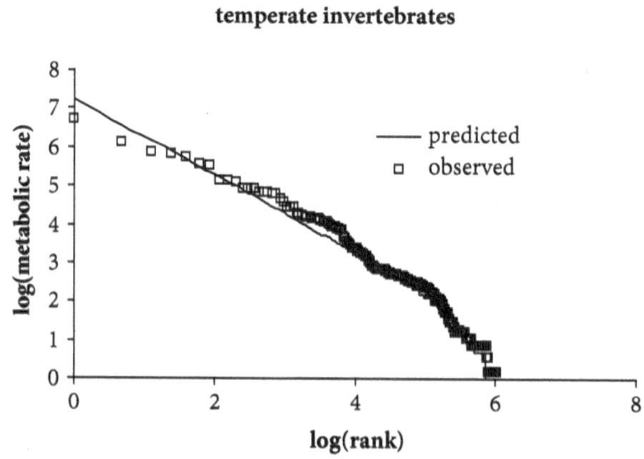

Figure 6.3 The distribution of metabolic rates over individuals: comparison of data and theory for the arthropods in a terrestrial ecosystem (Harte 2011).

The ASNE version of METE also predicts the distribution of metabolic rates across individuals ($\psi(\varepsilon)$ in Figure 6.1). Although the predicted distribution is not a commonly encountered function in ecology, it is approximately constant at very small values of metabolic rate, approximately power law at intermediate values, and exponentially decreasing at very large values. Again, the Ethan White group (Xiao et al., 2015) carried out the most comprehensive test of this prediction; comparisons of this prediction with sixty globally distributed forest communities, including over 300,000 individuals and nearly 2,000 species, revealed very good agreement for most communities. Some specific validations are shown in Figure 6.3.

The species-area relationship (SAR) is another example of an accurately predicted metric in the ASNE model of METE. Here, the theory makes a surprising prediction (Harte et al., 2009): all SARs should collapse onto a predicted universal curve if the SAR is re-plotted using appropriate scaling variables. This is illustrated in Figure 6.4; the details of the rescaling are described in the caption.

In contrast to these successes of the ASNE model of METE, Newman et al. (2014) and Xiao et al. (2015) show that the intraspecific distribution of metabolic rates ($\Theta(\varepsilon|n)$ in Figure 6.1) is not accurately predicted by ASNE/METE. We return to that and a related failure in the next section.

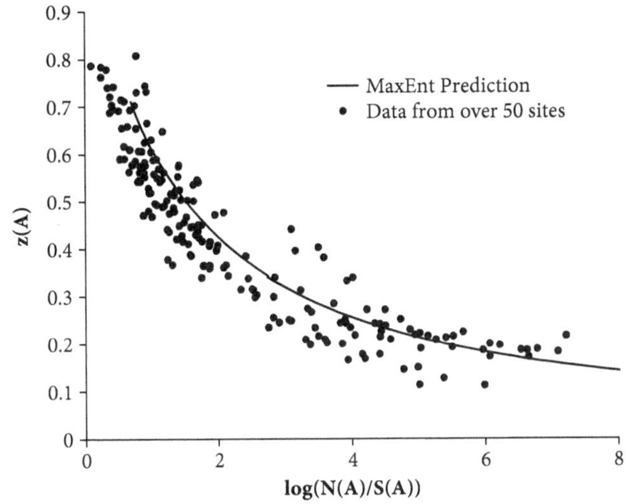

Figure 6.4 Universal scale collapse of species-area relationships (SARs) in the maximum entropy theory of ecology.

Predicted and observed values of the slope of an SAR at every spatial scale are plotted as a function of the value of the ratio of abundance to species richness at each scale. SAR data are derived from a variety of ecosystems, taxa, and spatial scales from multiple sources. For derivation of the theoretical prediction from MaxEnt and for sources of data, see Harte et al. (2009). Note that the often-assumed power-law form of the SAR with slope $z = 1/4$ would correspond to all the data points lying on a horizontal line, with intercept at 0.25.

4. Failures of the Static ASNE Model of METE

4.1 Energy Equivalence

Although many of the predictions are relatively accurate, two apparent types of failure emerged from numerous empirical tests. The first type of failure concerns the predicted form of the intraspecific distribution of metabolic rates ($\Theta(\varepsilon|n)$ in Figure 6.1) and the closely related "energy equivalence principle," which can be derived from $\Theta(\varepsilon|n)$. The latter prediction was originally considered a success, for the predicted inverse relationship between metabolic rate and abundance, or between (body size)$^{3/4}$ and abundance, has been widely claimed to be valid (Enquist et al., 1998; Damuth, 1981). Yet, census data reveal numerous exceptions to this ASNE/METE prediction (see, for example, White et al., 2007; Newman et al., 2014; Xiao et al., 2015).

The failure of the ASNE model prediction of energy equivalence appears remediable within the MaxEnt framework. By extending METE beyond the species level, including as state variables the numbers of higher-order entries in the taxonomic hierarchy (number of genera, number of families, etc.), we have shown that the predicted form of $\Theta(\varepsilon)$ and the energy equivalence are systematically altered (Harte et al., 2015). Instead of predicting that all the data should fall on a single straight line with slope −1 when log(metabolic rate) is plotted against log(abundance), now MaxEnt predicts that the data will fall on a series of such lines with different intercepts that depend on the species richness of the genus, the genus richness of the family and so on for each species. Thus, the data should partially fill a triangle rather than fall on a straight line. The essential idea here is that an additional state variable corresponding to a taxonomic classification at a level higher in the taxonomy than species alters the allocation of metabolism in such a way as to impose energy equivalence at the higher level, thus diluting the metabolism allocated to species that happen to be in species-rich higher taxonomic categories.

Analysis (Harte et al., 2015) of intertidal invertebrate data (Marquet et al., 1990) indicates that the taxonomically extended theory accurately predicts the pattern of failure of energy equivalence, and eliminates much of the scatter in a plot of log(abundance) versus log(metabolic rate). An additional success of the taxonomically extended METE derives from several publications that reveal patterns in macroecology exhibiting systematic dependence on the species richness of higher taxonomic categories. For example, ter Steege et al. (2013) show that the most abundant Amazonian tree species belong to genera that contain relatively few species. Schwartz and Simberloff (2001) and Lozano and Schwartz (2005) show that rare vascular plant species are overrepresented in species-rich families. Smith et al. (2004) show that species of mammals with the largest body sizes, and therefore the largest metabolic rates of individuals, belong to genera with the fewest species. Moreover, the variance of body size across species is also greatest in mammalian genera with the fewest species (Smith et al., 2004). Remarkably, these general patterns are all predicted in the extended theory (Harte et al., 2015).

The extended theory also predicts quite accurately (Figure 6.5) the distribution of species richness values across genera or families (Harte et al., 2015). The important role played by the entire taxonomic tree in the extended theory suggests the possibility that MaxEnt can provide new insight into the evolutionary history and phylogeny of extant communities.

Importantly, the taxonomically extended theory does not alter the successful ASNE predictions of the SAD, the SAR and the distribution of metabolic rates over individuals.

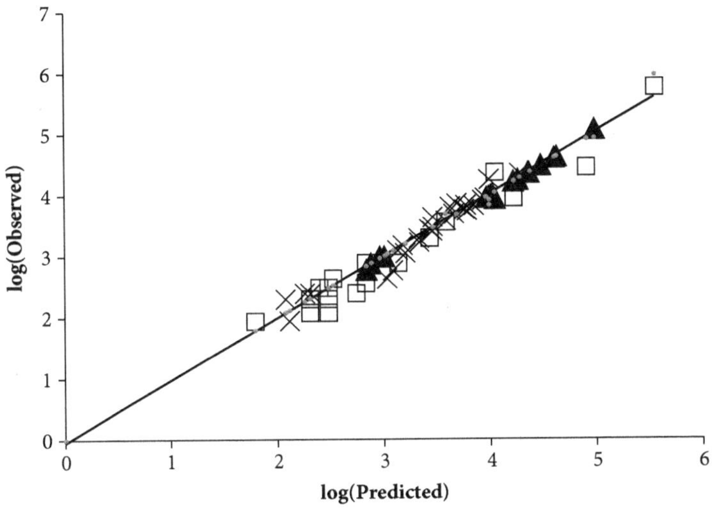

Figure 6.5 Comparison of observed and predicted distributions of species, $\Gamma(m)$, across families for arthropods, plants, and microorganisms and across genera for birds.

For each of the thirty data sets graphed, three types of data are plotted: number of families (or genera) with one species; number of families (or genera) with fewer than ten species; and number of species in the most species-rich family (or genus). For those three types of data, theory predicts 94, 99, and 95 percent of the variance, respectively. Regression slopes of observed against predicted are equal to 1, and intercepts are equal to 0 within 95% confidence intervals (Harte et al., 2015).

BOX: A Note on Theory Evaluation. A recent comparison (Xiao et al., 2016) of a version of the taxonomically extended METE versus a mechanistic model, which made a more limited set of predictions, concluded that the mechanistic model better described plant metabolic data. Aside from the fact that their analysis only looked at the extension to genera (and not families, which we showed in Harte et al. (2015) is the necessary extension to describe plant data), the comparison also highlights some very general issues concerning model or theory evaluation.

In the case of METE, the information that constrains entropy maximization and thus determines the predicted metrics is the numerical values of the state variables. There are no adjustable parameters once the measured values of the state variables are known, and hence the MaxEnt predictions can be pinned down before ever even looking at the shapes of the patterns to be predicted. In contrast, most mechanistic models in ecology do contain adjustable parameters. If one model or theory contains more adjustable parameters than another, that is easily accounted for using the Akaike information criterion (AIC; Akaike, 1973) to single out the "better" model.

But if a parameterized model predicts the shape of just a few metrics (say the species-abundance distribution and the metabolic rate distribution) with great accuracy, while a theory with no adjustable parameters predicts the species-abundance distribution and the metabolic rate distribution with a little less accuracy than does the model, but it also predicts fairly accurately a number of other metrics of macroecology such as the species-area relationship, the family-area relationship, and the structure of taxonomic trees, then which should be considered "better"? There is no agreed-upon answer to that question. When such comparisons are made in the literature, there is sometimes a preference for the model that best predicts the metrics that both theories predict, thereby ignoring the fact that the model that emerged on top by such a comparison may actually have far narrower explanatory power than the theory. There is no agreed upon criteria, however, for comparing models and theories that differ both in the number of things they predict and in the accuracy of their predictions.

4.2 METE Fails in Rapidly Changing Systems

The second type of failure of the original METE motivates a much more fundamental theoretical revision. Both the ASNE model of METE and its extensions to higher taxonomic levels fail to accurately predict empirical species-abundance distributions and species-area curves in ecosystems undergoing relatively rapid change (Harte, 2011; Newman et al., 2014, 2020; Rominger et al., 2016). Examples are rapidly diversifying ecosystems on newly formed islands, or ecosystems undergoing recent disturbance such as drought or experiencing relatively rapid succession, as, for example, in the aftermath of fire in fire-adapted systems or in the aftermath of the semi-isolation of an ecosystem from its source pool of immigrants as a consequence of land-use change.

Figure 6.6 shows one example, illustrating close agreement with the static METE SAR prediction (Figure 6.4) in an undisturbed Bishop pine stand (Mount Vision) and poor agreement in a successional, relatively rapidly changing, site (Bayview).

Another example, to be discussed in more detail later in this chapter, concerns the 50-ha tropical forest plot on Barro Colorado Island (BCI) in Panama, part of a network of plots managed by the Smithsonian Institution. The small island was formed when Gatun Lake was artificially created in 1912 for the construction of the Panama Canal, thereby reducing the immigrant seed source at the site of the plot. As a consequence (E. Leigh, personal communication), tree-species

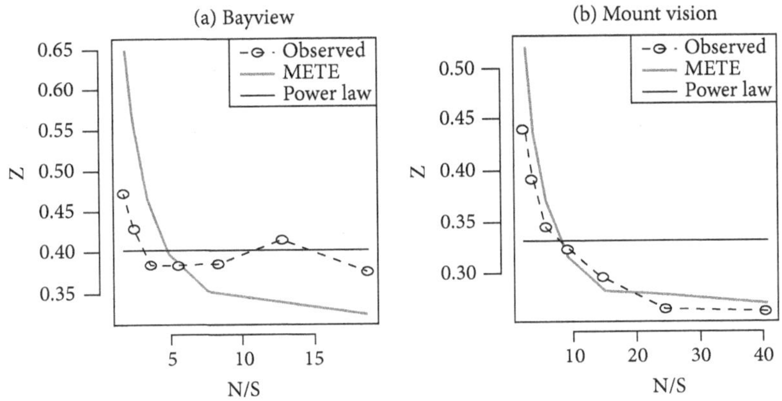

Figure 6.6 Empirical and METE-predicted SAR scale collapse curves for vegetation in two fire-adapted *Pinus muricata* (Bishop pine) stands at Point Reyes National Seashore:
(a) Bayview is a recently disturbed successional "dog-hair" site, having experienced a major fire in 1995.
(b) Mount Vision is a mature Bishop pine stand of trees with a more diverse understory that has not experienced fire in at least seventy-five years. Data from Newman et al. (2016).

richness censused every five years in the plot has slowly declined over three decades. When viewed as a log(abundance) versus rank graph, the shape of the species-abundance distribution is currently intermediate between a Fisher logseries distribution and a lognormal distribution, and is fairly well described by the zero-sum multinomial distribution (Hubbell, 2001). In contrast, other Smithsonian tropical forest plots that are less disturbed, such as Sherman, Cocoli, and Bukit Timah, show closer agreement with the logseries distribution predicted by the static theory (Harte, 2011).

Additional evidence that the predicted logseries SAD fails in disturbed, relatively rapidly changing ecosystems was reported for moth populations at relatively undisturbed and recently disturbed Rothamsted plots (Kempton and Taylor, 1974), where recently altered plots exhibited a shift from a logseries to a lognormal SAD.

And finally, arthropod species-abundance distributions from island sites of varying age in Hawaii reveal departures from the static ASNE/METE predictions at young sites that are plausibly most rapidly diversifying (Rominger et al., 2016; data from D. Gruner).

5. Hybrid Vigor in Ecological Theory

Failure of static METE to describe macroecological patterns in rapidly changing ecosystems suggests that with a mechanistic model of how state variables change in time, we might be able to conceive of a complete theory of macroecology. It would be a hybrid theory in the sense that a model incorporating explicit mechanisms would predict the values of the state variables, which then provide the constraints that allow the information entropy maximization procedure to predict the shapes of the probability distributions and functional relationships that describe patterns in macroecology.

The strategy behind the hybrid theory of macroecology that is sketched here is to combine the predictive power of the MaxEnt method with a mechanistic approach to predicting time-dependent state variables, which will then provide the MaxEnt constraints in a fully dynamic and self-contained theory. If our attempt is successful, we will have remedied the failure of METE to make accurate predictions when ecosystems are changing rapidly.

5.1 DynaMETE: A Natural Extension of the Static Theory

To remedy the failure of static MaxEnt theory, we propose an extension of the static METE to the stochastic dynamic realm (DynaMETE). Because now the state variables are changing in time, two things are needed: a mechanistic model of how the state variables are changing and a procedure for incorporating time-dependent constraints into the MaxEnt inference calculations.

We introduce a time-dependent probability distribution $P(S,N,E,t)$ for the state variables S, N, and E, which obeys the time-dependent master equation (van Kampen, 1981):

$$P(S,N,E,t+1)-P(S,N,E,t)=\sum_i\{-r_i(S,N,E)P(S,N,E,t) \\ +r_i(S',N',E')P(S',N',E',t)\} \qquad (6.1)$$

The first term on the right-hand side of Eq. (6.1), $\Sigma_i\ -r_i(S, N, E)\cdot P(S, N, E)$, accounts for all transitions out of the state (S, N, E), and the second term accounts for all transitions to the state (S, N, E) from a different state (S',N',E') during a unit time interval. The primed state variables are either equal to the unprimed variables or are displaced from them by an amount that depends on the transition process (i.e., the value of i). For example, consider a term in the sum over i corresponding to the process of birth; because birth only influences N and in fact raises it by 1, the contribution to the summations from that process are

$b(N) \cdot P(S, N, E)$ in the first term on the right-hand side and $b(N-1) \cdot P(S, N-1, E)$ in the second term on the right-hand side.

The choice of a first-order Markov process to describe the dynamics stems from a deliberate decision to avoid having to introduce the additional parameters that would be necessary if we assumed deeper-in-time historical dependence. We are assuming that the influence of evolutionary and ecological history on the dynamics of the system at the present time, t, is adequately captured by knowledge of the state of the system at time $t-1$. This is a frequently made assumption in ecology, but we cannot give a compelling justification for it from first principles.

In this first exploration of DynaMETE (1.0), we tentatively have chosen six transition processes (Table 6.1) and have assumed simplified forms for the dependencies of these transition rates on the state variables as shown in the table. In ongoing work, we are exploring more realistic dependencies of the transition rates on the state variables. This includes, for example, using insights from metabolic scaling theory (Brown et al., 2004; West et al., 2001; von Bertalanffy, 1957), which suggest modified forms for the dependence on state variables of the first three transitions listed in Table 6.1.

To incorporate stochastic and time-independent constraints into the MaxEnt inference procedure, we use the Bayesian superstatistics approach (Beck and Cohen, 2003). In particular, from the probability distribution for the state variables, $P(S,N,E,t)$, we infer the stochastic time-dependent structure function, $D(n,\varepsilon,t)$, using the expression:

$$D(n,\varepsilon,t) = \sum_S \sum_N \sum_E R(n,\varepsilon|S,N,E) \cdot P(S,N,E,t). \tag{6.2}$$

Table 6.1 The Transition Processes That Drive DynaMETE 1.0.

Transition Process	Effect of Transition	Assumed form in DynaMETE 1.0
Ontogenic Growth	$E \to E+1$	$g = g_0{*}E - g1{*}E^2$
Birth	$N \to N+1$	$b = b_0{*}N$
Death	$N \to N-1$	$d = d_0{*}N + d1{*}N^2$
Speciation	$S \to S+1$	$\lambda = \lambda_0{*}S$ or $\lambda = \lambda_0{*}N$
Immigration	$N \to N+1, S \to S$	$m = m_0{*}S/Smeta)$
	or	or
	$N \to N+1, S \to S+1$	$m = m_0{*}(1-S/Smeta)$
Local extinction	$S \to S-1, N \to N-1$	$\sigma = d(1){*}S{*}\ \Phi(1\|S, N)$

The species abundance distribution, $\Phi(n, t)$, in the expression for the extinction rate is determined from $D(n,\varepsilon,t\text{-}1)$ in Eq. (6.2).

This application of the superstatistics approach is no more nor less than an application of the rule of conditional probability. From $D(n,\varepsilon,t)$ the stochastic, time-dependent metrics of macroecology can be derived in the same way as sketched in Figure 6.1, with D replacing R.

To complete DynaMETE, we also need to know the dynamical evolution of the spatial distribution function Π. A master equation for the that probability distribution can be written using the same birth and death transition rates in Table 6.1 used to derive the stochastic distribution of state variables. The immigration term now has to be replaced with a dispersal kernel that informs us how far from a parent a newborn is located. Under the simplest plausible assumption, offspring disperse uniformly into a disk centered at the parent. With that assumption and with one additional parameter (the radius of the disk) introduced, the model is solvable and yields plausible outcomes (Harte, 2007).

Initial explorations of the predicted dynamical evolution of SAD under disturbance are yielding promising results. In one example, we examined the consequences of initiating an ecosystem in a steady-state solution of DynaMETE and then reducing the immigration rate parameter, m_0. In response to the perturbation, the output SAD changes from a Fisher logseries toward a lognormal distribution. This may be the origin of the SAD observed at BCI, where the 50-ha plot was partially cut off from an immigration source pool with the construction of Gatun Lake, and species richness may be slowly declining. There the observed SAD is intermediate in shape between the logseries and the lognormal distribution (Hubbell, 2001), in at least rough agreement with the preliminary results from DynaMETE.

We have also just begun to carry out trial runs to simulate differing and as yet untested assumptions about the cause of diversification on newly emerged islands. Two processes, speciation and immigration, can both contribute to diversification, and their relative importance is largely untested. Moreover, speciation rates may depend on the number of individuals on the island, on the number of species, or on something entirely different. The model output for the expectation value, $<S(t)>$, and for the dynamic SAD are different in form depending on what is assumed about the diversification mechanism.

Thus, the theory, along with diversification data from age transects in HI (Rominger et al., 2016), can allow determination of the relative importance of two mechanisms of diversification, immigration versus speciation, as newly formed islands become more species rich over time. It may provide a resolution to the long-standing puzzle in evolutionary biology: does species richness or total species abundance exert more influence on the rate of speciation within an ecosystem?

DynaMETE may also shed some light on the distinction between ecosystems that are changing in response to anthropogenic disturbances, such as clear cutting or the introduction of exotic species, versus the disturbances that ecosystems have experienced repeatedly in the past, such as in fire- or storm-adapted habitats. Although it would limit the eventual applicability of the theory, it would nevertheless be exciting if it successfully described ecological dynamics under natural disturbance regimes but not under anthropogenic disturbance.

Much still remains to be worked out, of course. DynaMETE is a work in progress. What is described here resembles an architectural footprint, street view, and structural framing; the detailed interiors have yet to be worked out, and they undoubtedly will evolve as we gain experience with the preliminary version described here.

If successful, this hybrid theory of macroecology will predict the shape of abundance and body size distributions, as well as species area relationships in disturbed ecosystems such as those shown in Figure 6.6. It will shed light on the drivers of diversification and explain the pattern of diversification observed in the fossil record subsequent to major extinction episodes. But even if it is successful, there are still reasons to seek a more fundamental theory, and that is the topic of the next section.

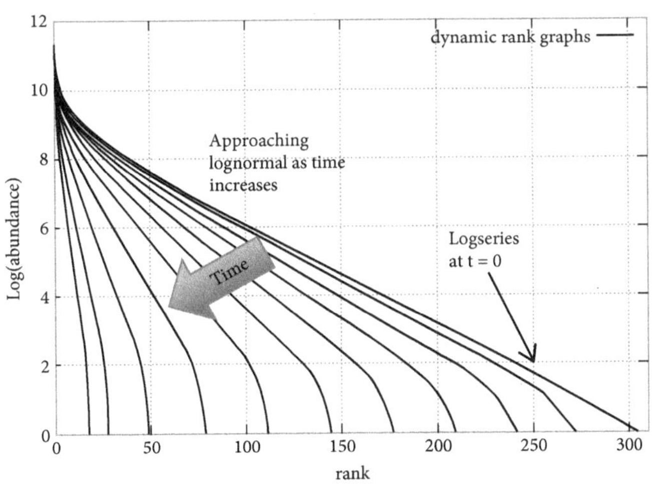

Figure 6.7 The dynamic species abundance distribution.

An ecosystem initially in steady state and with a logseries SAD is perturbed by setting the immigration rate constant to zero at $t = 0$. In response, the system slowly loses species, as expected, and the SAD shifts from the logseries toward the lognormal. The number of individuals, N, drops much more slowly than does the number of species. Using parameters from the BCI tropical forest plot, we find that each time step above (separate lines) corresponds to ~100 y.

6. The Ultimate Goal

Our hybrid theory, illustrated in Figure 6.8, and indeed any hybrid approach, still suffers from the difficulties raised in section 2.1. The six mechanisms that drive the dynamics were arbitrarily selected and are described by somewhat arbitrary, parameterized, mathematical representations. Far more satisfying than a hybrid theory would be a theory that was entirely based on maximum information entropy inference procedures. Is it conceivable that the mechanistic component of our hybrid theory could be replaced with MaxEnt-based theory? Some preliminary research suggests the answer is yes.

In an approach toward that goal, Zhang and Harte (2015) showed that per capita birth and death rates, as well as ontogenic growth, can be predicted from Boltzmann entropy maximization. Moreover, the conditions under which a collection of competing species can coexist are derived. The basic strategy in that approach is to maximize log(W), where W is the number of microstates compatible with a specified demographic macrostate.

Oversimplifying slightly, Zhang and Harte (2015) defined microstates as specific assignments of a resource to the individuals in a collection of species, with the specific amount of resource going to an individual determining whether that individual dies, survives for one time interval but does not reproduce, or survives and reproduces. Macrostates are specified sets of birth and death rates for each of the coexisting species. A generalization to multiple resources is straightforward. The degree of distinguishability of individuals within species compared to their distinguishability between species determines the combinatorial expression for W, and as a consequence exerts a strong influence on determining demographic the macrostate that corresponds to the maximum number of microstates.

To determine the relative distinguishability of individuals, we can use trait distributions (e.g., the distributions of metabolic rates of individuals within

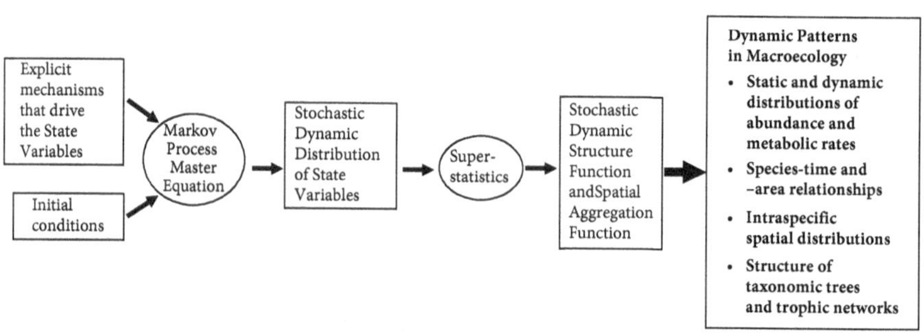

Figure 6.8 Architecture of a hybrid (MaxEnt + mechanism) version of DynaMETE.

The heavy arrow is implemented using the mathematics shown schematically in Figure 6.1.

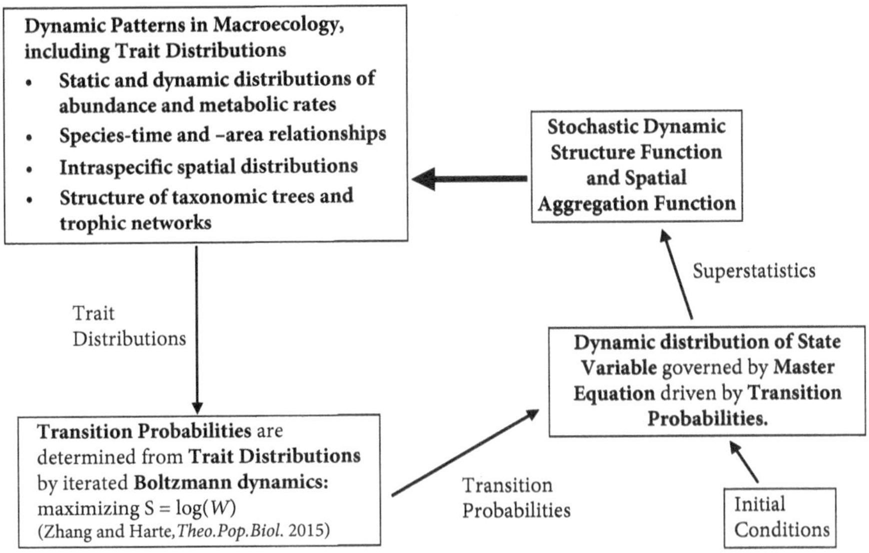

Figure 6.9 Architecture of a purely statistical (MaxEnt) version of DynaMETE. The heavy arrow is implemented using the mathematics shown schematically in Figure 6.1. If the box in the lower left of the figure is removed, along with the arrows leading in and out of it, then Figure 6.8 results.

and between species if the resource requirements of individuals are determined by their body size or metabolic rate). These trait distributions, in turn, can be inferred from the shapes of the macroecological metrics. This suggests a kind of bootstrap theory in which the state variables (or distributions of state variables) that are input to METE's predictions of the form of the metrics are themselves derived from a master equation that is driven by transition rates that are derived from entropy maximization. But the input to that last maximization step (maximization of W) is information contained in the ecological metrics! This circular (in a good sense) theoretical framework is sketched in Figure 6.9. The underlying assumption is that nature sits at the sweet spot, which, we hope, is a unique, self-consistent solution to the theory. It is a completely MaxEnt-based dynamic theory of macroecology. Whether prior specification of which traits are driving the system is necessary, or whether the theory can actually determine what the important traits are, remains to be seen, although I suspect that the former option is more likely.

We are just at the beginning of a journey into the theoretical wilderness of mechanism-less ecological theory. The only trail guide we possess is provided by the successes of theory in physics, where explicit mechanisms play an increasingly diminishing role, as in Jaynes's (1957; 1982) derivation of thermodynamics from MaxEnt, or in quantum "mechanics," where explicit mechanisms that result in quantum behavior do not appear to exist. It promises to be an exciting journey.

Appendix Some Epistemological Considerations

Here I review briefly some very general criticisms and philosophical issues that have been raised about MaxEnt applications to theory construction.

Ambiguity? There is a degree of flexibility in the way METE is constructed. In particular, different choices of the state variables will result in different predictions. For example, we might have chosen total biomass rather than total metabolic rate as a state variable. Is this a weakness of the theory? If it is, it is also a weakness of thermodynamics and other very successful theories. Thus, in formulating the thermodynamics of a container of gas, one might have chosen the interior surface area of the container rather than the volume, and then tried to formulate a relationship, analogous to $PV = nRT$, among area, pressure, temperature, and number of moles of gas. Such an attempt would lead nowhere; volume, not area, is the correct state variable. Similarly, metabolic rate, not biomass, is the correct state variable in METE. Indeed, all theories have seemingly arbitrary *ab initio* assumptions that define their framing concepts.

Regression Is the Alternative to Mechanism? There is a line of reasoning (Peters, 1991) that originates from the same concerns about mechanistic theory that I expressed in section 6.2, but it leads to conclusions that I cannot accept. The complexity of ecosystems led Peters to conclude that mechanism-based theory is unattainable and that its replacement must therefore be data-driven regression analysis. Predictions certainly can be made on the basis of regression, but there is no evidence to suggest that such predictions will be the best we can hope to achieve. The success of MaxEnt-based theory suggests that there are other alternatives, besides regression analysis, to mechanistically based models and theory.

Inevitable Trade-offs? Another possible concern stems from an influential and classic article that was written for ecologists and intended to apply to modeling efforts quite generally. Levins (1966) stated three desiderata of ecological models–generality, precision, and realism–and argued that there are necessarily trade-offs among them. He argued, for example, that if realism is increased by adding more mechanisms to the model, then precision and/or generality will be reduced. More generally, any attempt to enhance precision or generality will reduce the other two desiderata. A similar set of trade-offs, Levins argues, would apply to ecological theories.

Within that triad of desiderata, in METE or any other ecological theory built on the framework of MaxEnt, precision might be considered to be sacrificed because the theory only predicts probability distributions. But that is carping; the form of, and the parameters describing, the distributions are precisely predicted. The scope or generality of METE is certainly broad and so, by Levins's argument, that leaves realism as the sacrificial lamb. In fact, by the definition of realism intended by Levins, theories such as METE with no explicit mechanisms have zero realism. But that raises a bigger question: Why should one care about how many realistic processes found in nature are explicitly incorporated into a theory? If a theory is general and precise, and its many predictions are quite accurate, is that not good enough? What is gained by loading up that theory with a large number of processes that we know occur in nature?

One answer might be that, even though precision will be lost in the process because of the proliferation of parameters, we have nevertheless gained understanding by incorporating realistic mechanisms. Such understanding, however, is obtainable only if the theory remains valid. Yet, as we load more and more processes into a theory, thereby making it more realistic in Levins's sense of the term, the ensuing loss of precision makes it increasingly difficult to know if the theory is in fact valid. Its predictions might still be accurate but only because of the flexibility afforded by the many adjustable parameters. And as confidence in the validity of a theory is reduced, so goes our understanding.

Moreover, even if we define "understanding" to be the identification of the major processes that govern patterns in nature, theories like METE that lack explicit mechanism can nevertheless achieve that kind of understanding. As discussed next, we can sometimes infer mechanism from mechanism-less theory.

Mechanism without Mechanism. A persistent criticism of attempts to construct ecological theory purely on a foundation of MaxEnt is summarized by a phrase that once gained fame in U.S. politics: "Where's the beef?" (or more appropriately in our case, "where's the meat in METE?"). Where are the forces, the drivers, and the mechanisms that actually *explain* ecological phenomena? In the theory envisioned in outline form in Figure 6.9, we can't hide behind the response that it is the state variables that embody the drivers because there even the state variables derive from MaxEnt. If the set of state variables used in building a MaxEnt-based theory is insufficient to make accurate predictions, the nature of the discrepancy can sometimes point the way to identifying important mechanisms. In thermodynamics, for example, the failure of the ideal gas law, $PV = nRT$, under extreme values of the state variables led to the discovery of van der Waals's forces between molecules. In ecology, our earlier example of the need to include higher taxonomic categories to bring the form of the energy-equivalence prediction into close agreement with observations (Harte et al., 2015) suggests that evolutionary history, as reflected in the structure of taxonomic trees, influences the relationship between abundance and metabolism. It illustrates how a failure of any particular version of a MaxEnt-based theory can both aid in identifying relevant mechanisms that influence ecosystems and at the same time lead to improved theory. It also illustrates that one cannot separate the ecological theatre from the evolutionary play (Hutchinson, 1965).

For another example, if one adds to the ASNE state variables a new resource variable (call it W for water), the resulting ASNEW version of METE will produce a SAD that differs from the Fisher logseries predicted by ASNE, with the $(1/n)$ term altered to $(1/n^2)$ (Harte, 2011). In general, with a total of r resources (including metabolic rate E) to be allocated, the predicted SAD is a product of an exponential and a term $1/n^{r+1}$. This modification increases the predicted fraction of species that are rare. Intuitively, this makes sense; a greater number of limiting resources might either provide more specialized opportunities for rare species to survive or induce rarity in species that would in other contexts be abundant. Thus, METE provides a framework in which the degree of rarity in a community can be related to the number of resources driving macroecological patterns. METE, a nonmechanistic theory, can provide insight into what mechanisms might be driving macroecological patterns.

Zhang and Harte (2015) provide yet another example. In their approach to finding the macrostate with the most microstates in ecology, the analysis identifies several traits that exert the greatest influence on microstate density. In particular, they find that body size (or the closely related metabolic rate) and the ratio of the degree of distinguishability of individuals within a species to the degree of distinguishability of individuals between two species governs the likelihood that the two species can coexist and even their relative abundances. Thus, theory without any explicit mechanisms can inform us about what mechanisms actually govern the patterns we observe in macroecology.

References

Akaike, H. (1973). "Information Theory and an Extension of the Maximum Likelihood Principle." In B. N. Petrov and F. Csáki, 2nd International Symposium on Information Theory, Tsahkadsor, Armenia, USSR, September 2–8, 1971. Budapest: Akadémiai Kiadó, pp. 267–281.

Beck, C., and Cohen, E. G. D. (2003). "Superstatistics." *Physica A*, 322: 267–275.

Bialek, W. (2015). "Perspectives on Theory at the Interface of Physics and Biology." arXIV: 1512.08954v1 [physics.bio-ph].

Brown, J., Gillooly, J., Allen, A., Savage, V., and West, G. (2004). "Toward a Metabolic Theory of Ecology." *Ecology*, 85(7): 1771–1789.

Damuth, J. (1981). "Population Density and Body Size in Mammals." *Nature*. 290: 699–703.

Elith, J., et al. (2011). "A Statistical Explanation of MaxEnt for Ecologists." *Diversity Distributions* 17, 43–57.

Enquist, B. J., Brown, J. H., and West, G. B. (1998). "Allometric Scaling of Plant Energetics and Population Density." *Nature*, 395: 163–165.

Frieden, B. R. (1972). "Restoring with Maximum Likelihood and Maximum Entropy." *Journal of the Optical Society of America*, 62: 511–518.

Golan, A., Judge, G., and Miller, D. (1996). *Maximum Entropy Econometrics: Robust Estimation with Limited Data*. New York: John Wiley.

Gull, S. F., and Newton, T. J. (1986). "Maximum Entropy Tomography." *Applied Optics*, 25: 156–160.

Harte, J. (2007). "Toward a Mechanistic Basis for a Unified Theory of Spatial Structure in Ecological Communities at Multiple Spatial Scales." In D. Storch, P. Marquet, and J. Brown (eds.), *Scaling Biodiversity*, Chapter 6. Cambridge: Cambridge University Press.

Harte, J. (2011). *Maximum Entropy and Ecology: A Theory of Abundance, Distribution, and Energetics*. Oxford: Oxford University Press.

Harte, J., and Newman, E. (2014). "Maximum Entropy as a Foundation for Ecological Theory: Trends in Ecology and Eevolution." *BioScience*, 29(7): 384–389.

Harte, J., Rominger, A., and Zhang, Y. (2015). "Extending the Maximum Entropy Theory of Ecology to Higher Taxonomic Levels." *Ecology Letters*, 18: 1068–1077.

Harte, J., Smith, A., and Storch, D. (2009). "Biodiversity Scales from Plots to Biomes with a Universal Species-Area Curve." *Ecology Letters*, 12: 789–797.

Harte, J., Zillio, T., Conlisk, E., and Smith, A. (2008). "Maximum Entropy and the State Variable Approach to Macroecology." *Ecology*, 89: 2700–2711.

Hubbell, S. P. (2001). *The Unified Neutral Theory of Biodiversity and Biogeography*. Princeton, NJ: Princeton University Press.

Hutchinson, G. E. (1965). *The Ecological Theater and the Evolutionary Play*. New Haven, CT: Yale University Press.

Jaynes, E. (1957). "Information Theory and Statistical Mechanics." *Physical Review*, 106: 620–630.

Jaynes, E. T. (1982). "On the Rationale of maximum entropy methods." Proceedings of the the IEEE, 70: 939–952.

Kempton, R. A., and Taylor, L. R. (1974). "Diversity Discriminants for the Lepidoptera." *Journal of Animal Ecology*, 43: 381–399.

Levins, R. (1966). "The Strategy of Model Building in Population Biology." *American Scientist*, 54: 421–431.

Lotka, A. J. (1956). *Elements of Mathematical Biology*. New York: Dover Publications.

Lozano, F. and Schwartz, M. (2005). "Patterns of rarity and taxonomic group size in plants." Biological Conservation 126: 146–15.

Marquet, P. A., Navarrete, S. N., and Castilla, J. C. (1990). "Scaling Population Density to Body Aize in Rocky Intertidal Communities." *Science*, 250: 1125–1127.

May, R. M. (1972). "Will a Large Complex System Be Stable." *Nature*, 238: 413–414.

Newman, E., Harte, M., Lowell, N., Wilber, M., and Harte, J. (2014). "Empirical Tests of Within- and Across-Species Energetics in a Diverse Plant Community." *Ecology*, 95(10): 2815–2825.

Newman, E., Wilber, M., Kopper, K., Moritz, M., Falk, D., McKenzie, D. and Harte, J. (2020). "Disturbance Macroecology: A Comparative Study of Community Structure in a High-Severity Disturbance Regime", *Ecosphere*, 11(1):e03022. 10.1002/ecs2.3022.

Peters, R. H. (1991). *A Critique for Ecology*. Cambridge: Cambridge University Press.

Phillips, S. J., et al. (2006). "Maximum Entropy Modeling of Species Geographic Distributions." *Ecological Modeling*, 190: 231–259.

Phillips, S. J. (2007). "Transferability, Sample Selection Bias and Background Data in Presence-Only Modelling: A Response to Peterson et·al." *Ecography*, 31: 272–278.

Rominger, A. J., et al. (2016). "Community Assembly on Isolated Islands: Macroecology Meets Evolution." *Global Ecology and Biogeography*, 25: 769–780.

Roussev, V. (2010). "Data Fingerprinting with Similarity Digests." In K-P. Chow and S. Shenoi (eds.), *Advances in Digital Forensics*, (VI, pp. 207–226). New York: Springer.

Salamon, P., and Konopka, A. (1992). "A Maximum Entropy Principle for the Distribution of Local Complexity in Naturally Occurring Nucleotide Sequences." *Computers and Chemistry*, 16: 117–124.

Schwartz, M., and Simberloff, D. (2001). "Taxon Size Predicts Rates of Rarity in Vascular plants." *Ecology Letters*. 4, 464–469.

Seno, F., Trovato, A., Banavar, J., and Maritan, A. (2008). "Maximum Entropy Approach for Deducing Amino Acid Interactions in Proteins." *Physical Review Letters*, 100, 078102.

Skilling, J. (1984). "Theory of Maximum Entropy Image Reconstruction." In J. H. Justice (ed.), *Maximum Entropy and Bayesian Methods in Applied Statistics*, (pp. 156–178). Cambridge: Cambridge University Press

Smith, F., Brown, J., Haskell, J., Lyons, K., Alroy, J., Charnov, E., Da yan, T., et al. (2004). "Similarity of Mammalian Body Size Across The Taxonomic Hierarchy and Across Space And Time." *The American Naturalist*, 163(5), 672–691.

Steinbach, P., Ionescu, R., and Matthews, C. R. (2002). "Analysis of Kinetics Using a Hybrid Maximum-Entropy/Nonlinear-Least-Squares Method: Application to Protein Folding." *Biophysical Journal*, 82: 2244–2255.

Ter Steege, H., Pitman, N., Sabatier, D., Baraloto, C., Salomão, R., Guevara, J., Phillips, O., et al. (2013). "Hyperdominance in the Amazonian Tree flora." *Science*, 342: 325–333.

van Kampen, N. G. (1981). *Stochastic Processes in Physics and Chemistry*. Amsterdam.
Volterra, V. (1926). "Variazioni e fluttuazioni del numero d'individui in specie a)nimali conviventi." *Biophysical Journal*, Roma, 2: 31–113.
von Bertalanffy, L. (1957). "Quantitative Laws in Metabolism and Growth." *Quarterly Review of Biology*, 32: 217–231.
West, G. B., Brown, J. H., and Enquist, B. J. (2001). "A General Model dor Ontogenic Growth." *Nature*, 413: 628–631.
White, E., Ernest, S., Kerkhoff, A., and Enquist, B. (2007). "Relationships between Body Size and Abundance in eEcology." *Trends in Ecology and Evolution*, 22: 323–330.
White, E. P. Thibault, K., and Xiao, X. (2012). "Characterizing Species Abundance Distributions across Taxa and Ecosystems Using a Simple Maximum Entropy Model." *Ecology*, 93: 1772–1778.
Xiao, X., McGlinn, D., and White, E. (2015). "A Strong Test of the Maximum Entropy Theory of Ecology." *The American Naturalist*, 185(3): e-article DOI: 10.1086/679576.
Xiao, X., O'Dwyer, J., and White, E. (2016). "Comparing Process-Based and Constraint-Based Approaches for Modeling Macroecological Patterns." *Ecology*, 97: 1228–1238.
Zhang, Y., and Harte, J. (2015). "Population Dynamics and Competitive Outcome Derive from Resource Allocation Statistics: The Governing Influence of the Distinguishability of Individuals." *Theoretical Population Biology*, 105: 52–63.

7
Entropic Dynamics
Mechanics without Mechanism

Ariel Caticha

1. Introduction

The drive to explain nature has always led us to seek the mechanisms hidden behind the phenomena. Descartes, for example, claimed to explain the motion of planets as being swept along in the flow of some vortices. The model did not work very well but at least it gave the illusion of a mechanical explanation and thereby satisfied a deep psychological need. Newton's theory fared much better. He took the important step of postulating that gravity was a universal force acting at a distance but he abstained from offering any mechanical explanation—a stroke of genius immortalized in his famous "*hypotheses non fingo*." At first there were objections. Huygens, for instance, recognized the undeniable value of Newton's achievement but was nevertheless deeply disappointed: the theory *works* but it does not *explain*. And Newton agreed: any action at a distance would represent "*so great an absurdity... that no man who has in philosophical matters a competent faculty of thinking can ever fall into it*" (Isaac Newton's Third Letter, 1693).

Over the following eighteenth century, however, impressed by the many successes of Newtonian mechanics, people started to downplay and then even forget their qualms about the absurdity of action at a distance. Mechanical explanations were, of course, still required but the very meaning of what counted as "mechanical" suffered a gradual but irreversible shift. It no longer meant "caused by contact forces" but rather "described according to Newton's laws." Over time Newtonian forces, including those mysterious actions at a distance, became "real," which qualified them to count as the causes behind phenomena.

But this did not last too long. With Lagrange, Hamilton, and Maxwell, the notion of force started to lose some of its fundamental status. Indeed, after Maxwell succeeded in extending the principles of dynamics to include the electromagnetic field, the meaning of "mechanical explanation" changed once again. It no longer meant identifying the Newtonian forces but rather finding the right equations of evolution—which is done by identifying the right "Hamiltonian"

Ariel Caticha, *Entropic Dynamics: Mechanics without Mechanism* In: *Advances in Info-Metrics: Information and Information Processing across Disciplines*. Edited by: Min Chen, J. Michael Dunn, Amos Golan, and Aman Ullah, Oxford University Press (2021). © Oxford University Press.
DOI: 10.1093/oso/9780190636685.003.0007

for the theory.[1] Thus, today gravity is no longer explained through a force but through the curvature of spacetime. And the concept of force finds no place in quantum mechanics where interactions are described as the evolution of vectors in an abstract Hilbert space.

Our goal in this chapter is to derive useful dynamical models without invoking underlying mechanisms. This does not mean that such mechanisms do not exist; for all we know they might. It is just that useful models can be constructed without having to go through the trouble of keeping track of myriad microscopic details that often turn out to be ultimately irrelevant. The idea can be illustrated by contrasting the two very different ways in which the theory of Brownian motion was originally derived by Smoluchowski and by Einstein. In Smoluchowski's approach one keeps track of the microscopic details of molecular collisions through a stochastic Langevin equation and a macroscopic effective theory is then derived by taking suitable averages. In Einstein's approach, however, one focuses directly on those pieces of information that turn out to be relevant for the prediction of macroscopic effects. The advantage of Einstein's approach is the simplicity that arises from not having to keep track of irrelevant details that are eventually washed out when taking the averages. The challenge, of course, is to identify those pieces of information that happen to be relevant.

Our argument proceeds by stages. We will exhibit three explicit examples. Starting with the simplest, we first tackle a standard diffusion, which allows us to introduce and discuss the nontrivial concept of time. In the next stage we discuss the derivation of a Hamiltonian dynamics as a nondissipative type of diffusion. Finally, in a further elaboration invoking notions of information geometry, we identify the particular version of Hamiltonian dynamics that leads to quantum mechanics.

Ever since its origin in 1925 the interpretation of quantum mechanics has been a notoriously controversial subject. The central question revolves around the interpretation of the quantum state, the wave function: does it represent the actual real state of the system (its *ontic* state), or does it represent a state of knowledge about the system (an *epistemic* state)? Excellent reviews with extended references to the literature are given in (Stapp, 1972; Schlösshauer, 2004; Jaeger, 2009; Leifer, 2014). However, the urge to seek underlying mecha-

[1] Note for non-physicists: It is a remarkable and wonderful fact of physics that the fundamental equations of motion for particles and/or fields can all be expressed in terms of a "principle of least action." For our present purpose, the technical details are not crucial; it suffices to note that the action is written in terms of a "Hamiltonian" function, which usually, but not always, coincides with the energy of the system. The successor to Newtonian dynamics is thus commonly called Hamiltonian dynamics, and the corresponding equations of motion are called Hamilton's equations.

nisms has motivated a huge effort to identify hidden variables and subquantum levels of reality from which an effective quantum dynamics might emerge. (See, e.g., Nelson, 1985; de la Peña and Cetto, 1996; Smolin, 2006; Hooft, 2002, 2007; Adler, 2004; Grössing, 2008, Grössing *et al.*, 2012, 2016. In this work we adopt the point of view that quantum theory itself provides us with the most important clue: the goal of quantum mechanics is to calculate probabilities. Therefore, by its very nature, quantum mechanics must be an instance of Bayesian and entropic inference.

Entropic dynamics (ED) provides a framework for deriving dynamical laws as an application of entropic methods (Caticha, 2011, 2014, 2015; Caticha, Bartolomeo, and Reginatto, 2015). The use of entropy as a tool for inference can be traced to E. T. Jaynes (1957, 1983, 2003). For a pedagogical overview and further references see Caticha (2012).

The literature on the attempts to reconstruct quantum mechanics is vast and there are several approaches that, like ED, are also based on information theory (see, e.g., Wootters, 1981; Caticha, 1998a, 1998b, 2000; Brukner and Zeilinger 2003; Spekkens, 2007; Goyal, Knuth, and Skilling, 2010; Hardy, 2006; Hall and Reginatto, 2002; Reginatto, 2013; Chiribella, D'Ariano, and Perinotti, 2011; D'Ariano, 2017). Since all these approaches must sooner or later converge to the same fundamental equation of quantum theory—the Schrödinger equation—it is inevitable that they will show similarities, but there are important differences. What distinguishes ED is a strict adherence to Bayesian and entropic methods without invoking mechanisms operating at some deeper subquantum level.

This chapter is basically a self-contained review of material developed in Caticha (2014, 2015) and Caticha et al. (2015), but with one significant change. The gist of entropic dynamics is that the system of interest (e.g., particles or fields) undergoes a diffusion with a special systematic drift that is ultimately responsible for such quintessential quantum phenomena as interference and entanglement. A central idea introduced in Caticha (2014), but not fully developed in later publications, is that this peculiar drift is itself of entropic origin. More precisely, the drift follows the gradient of the entropy of some extra variables that remain hidden or at least inaccessible. Here we intend to pursue this fully entropic interpretation of the drift while taking advantage of later insights that exploit ideas from information geometry.

An important feature of ED is a central concern with the nature of time. In ED "entropic" time is *designed* to keep track of change. The construction of entropic time involves several ingredients. One must define the notion of "instants", show that these instants happen to be ordered, and finally, one must define a convenient measure of the duration or interval between the successive instants. As one might have expected, in an entropic approach to time it turns out that an arrow of time is generated automatically. Entropic time is intrinsically directional.

Here we will focus on the derivation of the Schrödinger equation but the ED approach has been applied to a variety of other topics in quantum mechanics that will not be reviewed here. These include the quantum measurement problem (Johnson and Caticha, 2012; Vanslette and Caticha, 2017); momentum and uncertainty relations (Nawaz and Caticha, 2012); the Bohmian limit (Bartolomeo and Caticha, 2016a, 2016b)) and the classical limit (Demme and Caticha (2017); the extensions to curved spaces (Nawaz, Abedi, and Caticha, 2016) and to relativistic fields (Ipek and Caticha, 2015; Ipek, Abedi, and Caticha, 2017); and the derivation of symplectic and complex structures from information geometry (Caticha, 2018).

2. The Statistical Model

As in any inference problem we must first decide on the subject matter—what are we talking about? We consider N particles living in flat Euclidean space $\mathbf{X}$ with metric δ_{ab}. Here is our first assumption: *The particles have definite positions x_n^a, collectively denoted by x, and it is their unknown values that we wish to infer.* (The index $n = 1 \ldots N$ labels the particles and $a = 1, 2, 3$ its spatial coordinates.) For N particles the configuration space is $\mathbf{X}_N = \mathbf{X} \times \ldots \times \mathbf{X}$. The fact that positions are unknown implies that the probability distribution $\rho(x)$ will be the focus of our attention.

Incidentally, this already addresses that old question of determinism versus indeterminism that has troubled quantum mechanics from the outset. In the standard view quantum theory is considered to be an extension of classical mechanics and therefore deviations from causality demand an explanation. In contrast, according to the entropic view, quantum mechanics is an example of entropic inference, a framework designed to handle insufficient information. From this entropic perspective indeterminism requires no explanation; uncertainty and probabilities are the expected norm. It is certainty and determinism that demand explanations (Demme and Caticha, 2017).

The assumption that the particles have definite positions that happen to be unknown is not an innocent statement. It represents a major shift away from the standard Copenhagen interpretation. Such a departure is not in itself a problem: our goal here is not to justify any particular interpretation of quantum mechanics but to reproduce its empirical success. According to the Copenhagen interpretation observable quantities, such as positions, do not in general have definite values; they can become definite but only as the immediate result of a measurement. ED differs in that positions play a very special role: at all times particles have definite positions that define the ontic state of the system. The wave function, on the other hand, is a purely epistemic notion; it defines our epistemic state about the system. In ED—just as in the Copenhagen interpretation—other observables such as energy or momentum do not in general have definite values; their values are created by the act of measurement. These other quantities are epistemic in that they reflect properties not of the particles but of the wave function. Thus, positions are ontic, while energies and momenta are epistemic. This point deserves to be emphasized: in the ED description of the double-slit experiment, we might not know which slit the particle goes through, but it definitely goes through one or the other.

The second assumption is that *in addition to the particles there also exist some other variables denoted* y. The assumption does not appear to be extreme.[2] First, the world definitely contains more stuff than the N particles we happen to be currently interested in and, second, it turns out that our main conclusions are very robust in that they turn out to be largely independent of most particular details about these y variables. We only need to assume that the y variables are themselves uncertain and that their uncertainty, described by some probability distribution $p(y|x)$, depends on the location x of the particles. As we shall soon see, it is the entropy of the distribution $p(y|x)$ that plays a significant role in defining the dynamics of x. Other details of the distribution $p(y|x)$ turn out to be irrelevant. For later reference, the entropy $S(x)$ of $p(y|x)$ relative to an underlying measure $q(y)$ on the space of y variables is

$$S(x) = -\int dy\, p(y|x) \log \frac{p(y|x)}{q(y)}. \tag{7.1}$$

For notational simplicity in multidimensional integrals such as (7.1), we will write dy instead of $d^n y$. Note also that $S(x)$ is a scalar function on the configuration space $\mathbf{X}_N$.

Once the microstates $(x,y) \in \mathbf{X}_N \times \mathbf{Y}$ are identified we proceed to tackle the dynamics. Our goal is to find the probability density $P(x'|x)$ for a change from an initial position $x \in \mathbf{X}_N$ to a new neighboring $x' \in \mathbf{X}_N$. Since both x' and the associated y' are unknown the relevant space is not just $\mathbf{X}_N$ but $\mathbf{X}_N \times \mathbf{Y}$ and,

[2] As we shall discuss in more detail in the concluding remarks, the y variables, though hidden from our view, do not at all play the technical role usually ascribed to "hidden variables."

therefore, the distribution we seek is the joint distribution $P(x',y'|x,y)$. This is found by maximizing the appropriate entropy,

$$\mathcal{S}[P,Q] = -\int dx'dy'\,P(x',y'|x,y)\log\frac{P(x',y'|x,y)}{Q(x',y'|x,y)}, \tag{7.2}$$

relative to the joint prior $Q(x',y'|x,y)$ and subject to the appropriate constraints.

2.1 Choosing the Prior

In Eq. (7.2) the prior $Q(x',y'|x,y)$ expresses our beliefs—or more precisely, the beliefs of an ideally rational agent—*before* any information about the specific motion is taken into account. We adopt a prior that represents a state of considerable ignorance: knowledge of x' and x tells us nothing about y', and knowledge of y' and y tells us nothing about x'. This is represented by a product,

$$Q(x',y'|x,y) = Q(x'|x)\,Q(y'|y). \tag{7.3}$$

The prior $Q(y'|y)$ for the y variables is chosen to be a uniform distribution. If the volume element in the space Y is proportional to $q(y)dy$, then $Q(y'|y) \propto q(y')$. The measure $q(y)$ need not be specified further. The prior $Q(x'|x)$ for the x variables is considerably more informative.

In a "mechanics without a mechanism" one does not explain why motion happens. The goal instead is to produce an estimate of what kind of motion one might reasonably expect. *The central piece of dynamical information is that the particles follow trajectories that are continuous.* This represents yet another significant deviation from the early historical development of quantum mechanics which stressed discreteness and discontinuity. Fortunately, in modern versions of quantum mechanics these two aspects are deemphasized. Ever since Schrödinger's pioneering work the discreteness of quantum numbers is not mysterious—certainly not more mysterious than the discrete resonances of classical vibrations. And the discontinuities implicit in abrupt quantum jumps, a relic from Bohr's old quantum theory, just do not exist; they were eventually recognized as unnecessary and effectively discarded as soon as Schrödinger formulated his equation. Incidentally, there is another source of discontinuity—

the so-called wave function collapse. Its mystery disappears the moment one recognizes the epistemic nature of the wave function: the collapse is not physical but merely an instance of probability updating (Johnson and Caticha, 2012; Vanslette and Caticha, 2017).

The assumption of continuity represents a tremendous simplification because it implies that motion can be analyzed as the accumulation of many infinitesimally short steps. We adopt a prior $Q(x'|x)$ that incorporates the information that the particles take steps that are infinitesimally short and that reflects translational and rotational invariance but is otherwise maximally ignorant about any correlations. Such a prior can itself be derived from the principle of maximum entropy. Indeed, maximize

$$S[Q] = -\int dx'\, Q(\Delta x) \log \frac{Q(\Delta x)}{\mu}, \tag{7.4}$$

where $\Delta x_n^a = x'^a_n - x_n^a$, relative to the uniform measure $\mu =$ constant, subject to normalization, and N independent constraints that enforce short steps and rotational invariance,

$$\langle \delta_{ab} \Delta x_n^a \Delta x_n^b \rangle = \kappa_n, \quad (n = 1 \dots N), \tag{7.5}$$

where κ_n are small constants. The result is a product of Gaussians,

$$Q(x'|x) \propto \exp -\frac{1}{2} \sum\nolimits_n \alpha_n \delta_{ab} \Delta x_n^a \Delta x_n^b. \tag{7.6}$$

The Lagrange multipliers α_n are constants that may depend on the index n in order to describe nonidentical particles. They will eventually be taken to infinity in order to enforce infinitesimally short steps.

The choice of a Gaussian prior turns out to be natural, not just because it arises in an informational context, but also because, as described by the central limit theorem, it arises whenever a "macroscopic" effect is the result of the accumulation of a large number of "microscopic" contributions.

Since proportionality constants have no effect on the entropy maximization, our choice for the joint prior is

$$Q(x', y'|x, y) = q(y') \exp -\frac{1}{2} \sum\nolimits_n \alpha_n \delta_{ab} \Delta x_n^a \Delta x_n^b. \tag{7.7}$$

Now we are ready to specify the constraints.

2.2 The Constraints

We first write the posterior as a product,

$$P(x',y'|x,y) = P(x'|x,y)\,P(y'|x',x,y), \tag{7.8}$$

and we specify the two factors separately. We require that the new x' depends only on x, so we set $P(x'|x,y) = P(x'|x)$, which is the transition probability we wish to find. The new x' is independent of the actual values of y or y', but, as we shall soon see, it does depend on their entropies.

As for the second factor in (7.8), we require that $P(y'|x',x,y) = p(y'|x')$—the uncertainty in y' depends only on x'. Therefore, the first constraint is that the joint posterior be of the form

$$P(x',y'|x) = P(x'|x)\,p(y'|x'). \tag{7.9}$$

To implement this constraint, substitute (7.9) into the joint entropy (7.2) to get

$$\mathcal{S}[P,Q] = -\int dx'\,P(x'|x)\log\frac{P(x'|x)}{Q(x'|x)} + \int dx'\,P(x'|x)\,S(x'), \tag{7.10}$$

where $S(x)$ is given in Eq. (7.1).

Having specified the prior and the constraints, the ME method takes over. We vary $P(x'|x)$ to maximize (7.10) subject to normalization. The result is

$$P(x'|x) = \frac{1}{\zeta}\exp\left[S(x') - \frac{1}{2}\sum\nolimits_n \alpha_n \delta_{ab}\Delta x_n^a \Delta x_n^b\right], \tag{7.11}$$

where ζ is a normalization constant. In Eq. (7.11) it is clear that infinitesimally short steps are obtained in the limit $\alpha_n \to \infty$. It is therefore useful to Taylor expand,

$$S(x') = S(x) + \sum\nolimits_n \Delta x_n^a \frac{\partial S}{\partial x_n^a} + \ldots \tag{7.12}$$

and rewrite $P(x'|x)$ as

$$P(x'|x) = \frac{1}{Z}\exp\left[-\frac{1}{2}\sum\nolimits_n \alpha_n\,\delta_{ab}\left(\Delta x_n^a - \langle\Delta x_n^a\rangle\right)\left(\Delta x_n^b - \langle\Delta x_n^b\rangle\right)\right], \tag{7.13}$$

where Z is the new normalization constant. A generic displacement $\Delta x_n^a = x_n'^a - x_n^a$ can be expressed as an expected drift plus a fluctuation,

$$\Delta x_n^a = \langle \Delta x_n^a \rangle + \Delta w_n^a, \quad \text{where} \quad \langle \Delta x_n^a \rangle = \frac{1}{\alpha_n} \delta^{ab} \frac{\partial S}{\partial x_n^b}, \tag{7.14}$$

$$\langle \Delta w_n^a \rangle = 0 \quad \text{and} \quad \left\langle \Delta w_n^a \Delta w_n^b \right\rangle = \frac{1}{\alpha_n} \delta^{ab}. \tag{7.15}$$

These equations show that as $\alpha_n \to \infty$, for very short steps, the dynamics is dominated by fluctuations Δw_n^a which are of order $O(\alpha_n^{-1/2})$, while the drift $\langle \Delta x_n^a \rangle$ is much smaller, only of order $O(\alpha_n^{-1})$. Thus, just as in the mathematical models of Brownian motion, the trajectory is continuous but not differentiable. In ED particles have definite positions but their velocities are completely undefined. Notice also that the particle fluctuations are isotropic and independent of each other. The directionality of the motion and correlations among the particles are introduced by a systematic drift along the gradient of the entropy of the y variables.

The introduction of the auxiliary y variables, which has played a central role of conveying relevant information about the drift, deserves a comment. It is important to realize that the same information can be conveyed through other means, which demonstrates that quantum mechanics, as an effective theory, turns out to be fairly robust[3] (Bartolemeo and Caticha, 2016a, 2016b). It is possible, for example, to avoid the y variables altogether and invoke a drift potential directly (see, e.g., Caticha, 2014; Caticha, Bartolomeo, and Reginatto, 2015). The advantage of an explicit reference to some vaguely defined y variables is that their mere existence may be sufficient to account for drift effects.

3. Entropic Time

Having obtained a prediction, given by Eq. (7.13), for what motion to expect in one infinitesimally short step we now consider motion over finite distances. This requires the introduction of a parameter t, to be called *time*, as a bookkeeping tool to keep track of the accumulation of successive short steps. Since the rules of inference are silent on the subject of time we need to justify why the parameter t deserves to be called time.

The construction of time involves three ingredients. First, we must identify something that one might call an "instant"; second, it must be shown that these instants are ordered; and finally, one must introduce a measure of separation

[3] In the case of physics, the y variables might, for example, describe the microscopic structure of spacetime. In the case of an economy, they may describe the hidden neural processes of various agents. It is a strength of the ED formulation that details about these auxiliary variables are not needed.

between these successive instants—one must define "duration." Since the foundation for any theory of time is the theory of change—that is, dynamics—the notion of time constructed in the next section will reflect the inferential nature of entropic dynamics. We call such a construction *entropic time.*

3.1 Time as an Ordered Sequence of Instants

Entropic dynamics is given by a succession of short steps described by $P(x'|x)$, Eq. (7.13). Consider, for example, the ith step which takes the system from $x = x_{i-1}$ to $x' = x_i$. Integrating the joint probability, $P(x_i, x_{i-1})$, over x_{i-1} gives

$$P(x_i) = \int dx_{i-1} P(x_i, x_{i-1}) = \int dx_{i-1} P(x_i|x_{i-1}) P(x_{i-1}). \tag{7.16}$$

This equation follows directly from the laws of probability; it involves no assumptions and, therefore, it is sort of empty. To make it useful something else must be added. Suppose we interpret $P(x_{i-1})$ as the probability of different values of x_{i-1} *at one "instant" labeled* t, then we can interpret $P(x_i)$ as the probability of values of x_i *at the next "instant,"* which we will label t'. Writing $P(x_{i-1}) = \rho(x,t)$ and $P(x_i) = \rho(x',t')$, we have

$$\rho(x',t') = \int dx \, P(x'|x) \rho(x,t). \tag{7.17}$$

Nothing in the laws of probability leading to Eq. (7.16) forces the interpretation (7.17) on us; this is the additional ingredient that allows us to construct time and dynamics in our model. Equation (7.17) defines the notion of "instant": If the distribution $\rho(x,t)$ refers to one instant t, then the distribution $\rho(x',t')$ generated by $P(x'|x)$ defines what we mean by the "next" instant t'. Iterating this process defines the dynamics. Entropic time is constructed instant by instant: $\rho(t')$ is constructed from $\rho(t)$, $\rho(t'')$ is constructed from $\rho(t')$, and so on.

The construction is intimately related to information and inference. An "instant" is an informational state that is complete in the sense that it is specified by the information—codified into the distributions $\rho(x,t)$ and $P(x'|x)$—that is sufficient for predicting the next instant. To put it briefly: *the instant we call the present is such that the future, given the present, is independent of the past.*

In the ED framework the notions of instant and of simultaneity are intimately related to the distribution $\rho(x)$. It is instructive to discuss this further. When we consider a single particle at a position $\vec{x} = (x^1, x^2, x^3)$ in three-dimensional space it is implicit in the notation that the three coordinates x^1, x^2, and x^3

occur simultaneously—no surprises here. Things get a bit more interesting when we describe a system of N particles by a single point $x = (\vec{x}_1, \vec{x}_2, \ldots \vec{x}_N)$ in $3N$-dimensional configuration space. Here it is also implicitly assumed that all the $3N$ coordinate values refer to the same instant; we take them to be simultaneous. But this is no longer so trivial because the particles are located at different places—we do not mean that particle 1 is at $\vec{x}_1$ now and that particle 2 is at $\vec{x}_2$ later. There is an implicit assumption linking the very idea of a configuration space with that of simultaneity, and something similar occurs when we introduce probabilities. Whether we talk about one particle or about N particles, a distribution such as $\rho(x)$ describes our uncertainty about the possible configurations x of the system at a given instant. The different values of x refer to the same instant; they are meant to be simultaneous. And therefore, in ED, a probability distribution $\rho(x)$ provides a criterion of simultaneity.

In a relativistic theory there is a greater freedom in the choice of instants and this translates into a greater flexibility with the notion of simultaneity. Conversely, as we have shown elsewhere, the requirement that these different notions of simultaneity be consistent with each other places strict constraints on the allowed forms of ED (Ipek, Abedi, and Caticha, 2017).

It is common to use equations such as (7.17) to define a special kind of dynamics, called Markovian, that unfolds in a time defined by some external clocks. In such a Markovian dynamics, the specification of the state at one instant is sufficient to determine its evolution into the future.[4] It is important to recognize that in ED we are not making a Markovian assumption. Although Eq. (7.17) is formally identical to the Chapman–Kolmogorov equation, it is used for a very different purpose. We do not use (7.17) to define a (Markovian) dynamics in a preexisting background time because in ED there are no external clocks. The system is its own clock, and (7.17) is used both to define the dynamics and to construct time itself.

3.2 The Arrow of Entropic Time

The notion of time constructed according to Eq. (7.17) is intrinsically directional. There is an absolute sense in which $\rho(x, t)$ is prior and $\rho(x', t')$ is posterior. If we wanted to construct a time-reversed evolution, we would write

[4] The application of entropic dynamics to continuous motion is, of course, heavily motivated by the application to physics; for all we know space and time form a continuum. The derivation of discrete Markov models as a form of entropic dynamics is a subject for future research. But even in the case of an essentially discrete dynamics (such as the daily updates of the stock market), the idealization of continuous evolution is still useful as an approximation over somewhat longer time scales.

$$\rho(x,t) = \int dx' \, P(x|x')\rho(x',t'), \tag{7.18}$$

where according to the rules of probability theory $P(x|x')$ is related to $P(x'|x)$ in Eq. (7.13) by Bayes' theorem,

$$P(x|x') = \frac{\rho(x,t)}{\rho(x',t')} P(x'|x). \tag{7.19}$$

This is not, however, a mere exchange of primed and unprimed quantities. The distribution $P(x'|x)$, Eq. (7.13), is a Gaussian derived from the maximum entropy method. In contrast, the time-reversed $P(x|x')$ is given by Bayes' theorem, Eq. (7.19), and is not in general Gaussian. The asymmetry between the inferential past and the inferential future is traced to the asymmetry between priors and posteriors.

The puzzle of the arrow of time (see, e.g., (Price, 1996; Zeh, 2002)) has been how to explain the asymmetric arrow from underlying symmetric laws. The solution offered by ED is that there are no underlying laws whether symmetric or not. The time asymmetry is the inevitable consequence of entropic inference. From the point of view of ED the challenge is not to explain the arrow of time but the reverse: how to explain the emergence of symmetric laws within an entropic framework that is intrinsically asymmetric. As we shall see in the next section, some laws of physics derived from ED, such as the Schrödinger equation, are indeed time-reversible even though entropic time itself only flows forward.

3.3 Duration: A Convenient Time Scale

To complete the construction of entropic time we need to specify the interval Δt between successive instants. The basic criterion is convenience: *duration is defined so that motion looks simple.* We saw in Eqs. (7.14) and (7.15) that for short steps (large α_n) the motion is largely dominated by fluctuations. Therefore, specifying Δt amounts to specifying the multipliers α_n in terms of Δt.

The description of motion is simplest when it reflects the symmetry of translations in space and time. In a flat spacetime this leads us to an entropic time that resembles Newtonian time in that it flows "equably everywhere and everywhen." Thus, we choose α_n to be independent of x and t, and we choose Δt so that $\alpha_n \propto 1/\Delta t$. Furthermore, it is convenient to express the proportionality constants in terms of some particle-specific constants m_n and an overall constant η that fixes the units of the m_ns relative to the units of time. The result is

$$\alpha_n = \frac{m_n}{\eta} \frac{1}{\Delta t}. \tag{7.20}$$

The constants m_n will eventually be identified with the particle masses and the constant η will be regraduated into $\hbar$.

4. The Information Metric of Configuration Space

Before we proceed to study the dynamics defined Eq. (7.13) with its corresponding notion of entropic time it is useful to consider the geometry of the N-particle configuration space, $\mathbf{X}_N$. We have assumed that the geometry of the single-particle spaces $\mathbf{X}$ is described by the Euclidean metric δ_{ab}. We can expect that the N-particle configuration space, $\mathbf{X}_N = \mathbf{X} \times \cdots \times \mathbf{X}$, will also be flat, but a question remains about the relative scales associated with each $\mathbf{X}$ factor. Information geometry provides the answer.

To each point $x \in \mathbf{X}_N$ there corresponds a probability distribution $P(x'|x)$. This means that $\mathbf{X}_N$ is a statistical manifold, the geometry of which is uniquely determined (up to an overall scale factor) by the information metric,

$$\gamma_{AB} = C \int dx' \, P(x'|x) \frac{\partial \log P(x'|x)}{\partial x^A} \frac{\partial \log P(x'|x)}{\partial x^B}. \tag{7.21}$$

Here the upper-case indices label both the particle and its coordinate, $x^A = x_n^a$, and C is an arbitrary positive constant (see, e.g., Caticha, 2012; Amari, 1985). Substituting Eqs. (7.13) and (7.20) into (7.21) in the limit of short steps ($\alpha_n \to \infty$) yields

$$\gamma_{AB} = \frac{Cm_n}{\eta \Delta t} \delta_{nn'} \delta_{ab} = \frac{Cm_n}{\eta \Delta t} \delta_{AB}. \tag{7.22}$$

Note the divergence as $\Delta t \to 0$. Indeed, as $\Delta t \to 0$, the distributions $P(x'|x)$ and $P(x'|x + \Delta x)$ become more sharply peaked and increasingly easier to distinguish. This leads to an increasing information distance, $\gamma_{AB} \to \infty$. To define a distance that remains useful for arbitrarily small Δt, we choose $C \propto \Delta t$. In fact, since γ_{AB} will always appear in the combination $\gamma_{AB} \Delta t / C$, it is best to define the "mass" tensor,

$$m_{AB} = \frac{\eta \Delta t}{C} \gamma_{AB} = m_n \delta_{AB}. \tag{7.23}$$

and its inverse,

$$m^{AB} = \frac{C}{\eta\Delta t}\gamma^{AB} = \frac{1}{m_n}\delta^{AB}. \tag{7.24}$$

Thus, up to overall constants, the metric of configuration space is the mass tensor. This result may be surprising. Ever since the work of Heinrich Hertz in 1894 (Lanczos, 1986), it has been standard practice to describe the motion of systems with many particles as the motion of a single point in an abstract space—the configuration space. The choice of geometry for this configuration space had so far been justified as being dictated by convenience—the metric is suggested by an examination of the kinetic energy of the system. In contrast, in ED there is no room for choice. Up to a trivial global scale, the metric is uniquely determined by information geometry.

To recap our results so far: with the multipliers α_n chosen according to (7.20), the dynamics given by $P(x'|x)$ in (7.13) is a standard Wiener process. A generic displacement, Eq. (7.14), is

$$\Delta x^A = b^A \Delta t + \Delta w^A, \tag{7.25}$$

where $b^A(x)$ is the drift velocity,

$$\langle \Delta x^A \rangle = b^A \Delta t \quad \text{with} \quad b^A = \eta m^{AB}\partial_B S, \tag{7.26}$$

and the uncertainty Δw^A is given by

$$\langle \Delta w^A \rangle = 0 \quad \text{and} \quad \langle \Delta w^A \Delta w^B \rangle = \eta m^{AB} \Delta t. \tag{7.27}$$

I finish this section with two remarks. The first is on *the nature of clocks*: In Newtonian mechanics time is defined to simplify the motion of free particles. The prototype of a clock is a free particle: it moves equal distances in equal times. In ED time is also defined to simplify the dynamics of free particles. The prototype of a clock is a free particle too: as we see in (7.27), *free particles* (because for sufficiently short times all particles are free) *undergo equal fluctuations in equal times*.

The second remark is on *the nature of mass*. As we shall soon see, the constants m_n will be identified with the particles' masses. Then, Eq. (7.27) provides an interpretation of what "mass" is: mass is an inverse measure of fluctuations. It is not unusual to treat the concept of mass as an unexplained primitive concept that measures the amount of stuff (or perhaps the amount of energy) and then state that quantum fluctuations are inversely proportional to mass. In an inference scheme such as ED the presence of fluctuations does not require an explanation. They reflect uncertainty, the natural consequence of incomplete information. This opens the door to an entropic explanation of mass: *mass is just a measure of uncertainty about the expected motion.*

5. Diffusive Dynamics

The dynamics of $\rho(x,t)$, given by the integral equation (7.17), is more conveniently rewritten in a differential form known as the Fokker–Planck (FP) equation (Reif, 1965),

$$\partial_t \rho = -\partial_A \left(b^A \rho \right) + \frac{1}{2} \eta m^{AB} \partial_A \partial_B \rho . \tag{7.28}$$

(For the algebraic details, see, e.g. Caticha, 2012.) The FP equation can also be written as a continuity equation,

$$\partial_t \rho = -\partial_A \left(\rho v^A \right) . \tag{7.29}$$

The product ρv^A in (7.29) represents the probability current, and v^A is interpreted as the velocity of the probability flow—it is called the *current velocity*. From (7.28) the current velocity in (7.29) is the sum of two separate contributions,

$$v^A = b^A + u^A . \tag{7.30}$$

The first term is the drift velocity $b^A \propto \partial^A S$ in (7.26) and describes the tendency of the distribution $\rho(x)$ to evolve so as to increase the entropy of the y variables. One could adopt a language that resembles a causal mechanism: *it is as if* the system were pushed by an entropic force. But, of course, such a mechanistic language should not be taken literally. Strictly speaking, entropic forces cannot be causal agents because they are purely epistemic. They might influence our beliefs and expectations, but they are not physically capable of pushing the system—the "real causes," if any, lie elsewhere. Or perhaps there are no real causes. In classical mechanics, for example, there is nothing out there that causes a free particle to persist in its uniform motion in a straight line. Similarly, in ED we are not required to identify what makes changes happen the way they do.

The second term in (7.30) is the *osmotic velocity*,

$$u^A = -\eta m^{AB} \partial_B \log \rho^{1/2} . \tag{7.31}$$

It represents diffusion, the tendency for probability to flow down the density gradient. Indeed, one might note that the osmotic or diffusive component of the current, ρu^A, obeys a version of Fick's law in configuration space,

$$\rho u^A = -\frac{1}{2}\eta m^{AB}\partial_B\rho = -D^{AB}\partial_B\rho \tag{7.32}$$

where $D^{AB} = \eta m^{AB}/2$ is the diffusion tensor (Reif, 1965). Here, too, one might be tempted to adopt a causal mechanistic language and say that the diffusion is driven by the fluctuations expressed in Eq. (7.27). *It is as if* the system were bombarded by some underlying random field. But Eq. (7.27) need not represent actual fluctuations; it represents our uncertainty about where the particle will be found after Δt. The mere fact that we happen to be uncertain about the position of the particles does not imply that something must be shaking them.[5]

Next we note that since both b^A and u^A are gradients, their sum—the current velocity—is a gradient too,

$$v^A = m^{AB}\partial_B\Phi \quad \text{where} \quad \Phi = \eta\left(S - \log\rho^{1/2}\right). \tag{7.33}$$

The FP equation,

$$\partial_t\rho = -\partial_A\left(\rho m^{AB}\partial_B\Phi\right), \tag{7.34}$$

can be conveniently rewritten in yet another equivalent but very suggestive form involving functional derivatives. For some suitably chosen functional $\tilde{H}[\rho,\Phi]$, we have

$$\partial_t\rho(x) = \frac{\delta\tilde{H}}{\delta\Phi(x)}. \tag{7.35}$$

It is easy to check that the appropriate functional $\tilde{H}$ is

$$\tilde{H}[\rho,\Phi] = \int dx\,\frac{1}{2}\rho m^{AB}\partial_A\Phi\partial_B\Phi + F[\rho], \tag{7.36}$$

where the integration constant $F[\rho]$ is some unspecified functional of ρ.

We have just exhibited our first example of a mechanics without mechanism: a standard diffusion derived from principles of entropic inference. Next we turn our attention to the derivation of quantum mechanics.

[5] E. T. Jaynes warned us that "the fact that our information is able to determine [a quantity] F to 5 percent accuracy, is not enough to make it fluctuate by 5 percent!" He called this mistake the Mind Projection Fallacy (Jaynes, 1990).

6. Hamiltonian Dynamics

The previous discussion has led us to a *standard* diffusion. It involves a single dynamical field, the probability density $\rho(x)$, which evolves in response to a fixed nondynamical "potential" given by the entropy of the y variables, $S(x)$. In contrast, a *quantum* dynamics consists in the coupled evolution of two dynamical fields: the density $\rho(x)$ and the phase of the wave function. This second field can be naturally introduced into ED by allowing the entropy $S(x)$ to become dynamical: the entropy S guides the evolution of ρ, and in return, the evolving ρ reacts back and induces a change in S. This amounts to an entropic dynamics in which each short step is constrained by a slightly different drift potential; the constraint Eq. (7.9) is continuously updated at each instant in time.

Clearly, different updating rules lead to different forms of ED. The rule that turns out be particularly useful in physics is inspired by an idea of Nelson's (Nelson, 1979). We require that S be updated in such a way that a certain functional, later to be called "energy," remains constant. Such a rule may appear to be natural—how could it be otherwise? Could we possibly imagine physics without a conserved energy? But this naturalness is deceptive because ED is not at all like a classical mechanics. Indeed, the classical interpretation of a Langevin equation such as (7.25) is that of a Brownian motion in the limit of infinite friction. This means that, in order to provide a classical explanation of quantum behavior, we would need to assume that the particles were subjected to *infinite* friction while undergoing *zero* dissipation. Such a strange dynamics could hardly be called "classical"; the relevant information that is captured by our choice of constraints cannot be modeled by invoking some underlying classical mechanism—this is mechanics without a mechanism. Furthermore, while it is true that an updating rule based on the notion of conserved total energy happens to capture the relevant constraints for a wide variety of physics problems, its applicability is limited even within physics. For example, in the curved spacetimes used to model gravity, it is not possible to even define global energy, much less require its global conservation. (For the updating rules that apply to curved spaces see Ipek, Abedi, and Caticha, 2017.)

6.1 The Ensemble Hamiltonian

In the standard approach to mechanics the conservation of energy is derived from an action principle plus the requirement of a symmetry under time translations—today's laws of physics are the same as yesterday's. Our derivation proceeds in the opposite direction: we first identify energy conservation as the relevant piece of information and from it we derive the equations of motion (Hamilton's equations) and their associated action principle.

It turns out that the empirically successful energy functionals are of the form (7.36). We impose that, irrespective of the initial conditions, the entropy S or, equivalently, the potential Φ in (7.33) will be updated in such a way that the functional $\tilde{H}[\rho, \Phi]$ in (7.36) is always conserved,

$$\tilde{H}[\rho + \delta\rho, \Phi + \delta\Phi] = \tilde{H}[\rho, \Phi], \tag{7.37}$$

or

$$\frac{d\tilde{H}}{dt} = \int dx \left[\frac{\delta \tilde{H}}{\delta \Phi} \partial_t \Phi + \frac{\delta \tilde{H}}{\delta \rho} \partial_t \rho \right] = 0. \tag{7.38}$$

Using Eq. (7.35), we get

$$\frac{d\tilde{H}}{dt} = \int dx \left[\partial_t \Phi + \frac{\delta \tilde{H}}{\delta \rho} \right] \partial_t \rho = 0. \tag{7.39}$$

We want $d\tilde{H}/dt = 0$ to hold for arbitrary choices of the initial values of ρ and Φ. When using Eq. (7.34) this translates into requiring $d\tilde{H}/dt = 0$ for arbitrary choices of $\partial_t \rho$. Therefore, Φ must be updated according to

$$\partial_t \Phi = -\frac{\delta \tilde{H}}{\delta \rho}. \tag{7.40}$$

Equations (7.35) and (7.40) are recognized as a conjugate pair of Hamilton's equations and the conserved functional $\tilde{H}[\rho, \Phi]$ in Eq. (7.36) is then called the *ensemble Hamiltonian.*

Thus, the form of the ensemble Hamiltonian $\tilde{H}$ is chosen so that the first Hamilton equation (7.35) is the FP Eq. (7.29), and then the second Hamilton equation (7.40) becomes a generalized form of the Hamilton–Jacobi equation,

$$\partial_t \Phi = -\frac{\delta \tilde{H}}{\delta \rho} = -\frac{1}{2} m^{AB} \partial_A \Phi \partial_B \Phi - \frac{\delta F}{\delta \rho}. \tag{7.41}$$

This is our second example of a mechanics without mechanism: *a nondissipative ED leads to Hamiltonian dynamics.*

6.2 The Action

We have just seen that the field $\rho(x)$ is a "generalized coordinate" and $\Phi(x)$ is its "canonical momentum." Now that we have derived Hamilton's equations, (7.35) and (7.40), it is straightforward to invert the usual procedure and *construct* an action principle from which they can be derived. Just define the differential

$$\delta A = \int dt \int dx \left[\left(\partial_t \rho - \frac{\delta \tilde{H}}{\delta \Phi} \right) \delta \Phi - \left(\partial_t \Phi + \frac{\delta \tilde{H}}{\delta \rho} \right) \delta \rho \right] \tag{7.42}$$

and then integrate to get an "action,"

$$A[\rho, \Phi] = \int dt \left(\int dx \Phi \dot{\rho} - \tilde{H}[\rho, \Phi] \right), \tag{7.43}$$

with "Lagrangian,"

$$L = \int dx \Phi \dot{\rho} - \tilde{H}[\rho, \Phi]. \tag{7.44}$$

Therefore, by construction, imposing $\delta A = 0$ for arbitrary choices of $\delta \Phi$ and $\delta \rho$ leads to (7.35) and (7.40). Thus, in the ED approach, the action $A[\rho, \Phi]$ is not particularly fundamental; it is just a clever way to summarize the dynamics in a very condensed form.

7. Information Geometry and the Quantum Potential

Different choices of the functional $F[\rho]$ in Eq. (7.36) lead to different dynamics. Earlier we used information geometry, Eq. (7.21), to define the metric m_{AB} of configuration space. Here we use information geometry once again to motivate the particular choice of the functional $F[\rho]$ that leads to quantum theory. (For an alternative derivation based on the preservation of symplectic and metric structures, see Caticha (2019).)

The special role played by the particle positions leads us to consider the family of distributions $\rho(x|\theta)$ that are generated from a distribution $\rho(x)$ by translations in configuration space by a vector θ^A, $\rho(x|\theta) = \rho(x - \theta)$. The extent to which $\rho(x|\theta)$ can be distinguished from the slightly displaced $\rho(x|\theta + d\theta)$ or, equivalently, the information distance between θ^A and $\theta^A + d\theta^A$, is given by

$$d\ell^2 = g_{AB} d\theta^A d\theta^B \tag{7.45}$$

where

$$g_{AB}(\theta) = \int dx \frac{1}{\rho(x-\theta)} \frac{\partial \rho(x-\theta)}{\partial \theta^A} \frac{\partial \rho(x-\theta)}{\partial \theta^B}. \tag{7.46}$$

Changing variables $x - \theta \to x$ yields

$$g_{AB}(\theta) = \int dx \frac{1}{\rho(x)} \frac{\partial \rho(x)}{\partial x^A} \frac{\partial \rho(x)}{\partial x^B} = I_{AB}[\rho]. \tag{7.47}$$

Note that these are translations in configuration space. They are not translations in which the system is displaced as a whole in three-dimensional space by the same constant amount, $(\vec{x}_1, \vec{x}_2, \ldots \vec{x}_N) \to (\vec{x}_1 + \vec{\varepsilon}, \vec{x}_2 + \vec{\varepsilon}, \ldots \vec{x}_N + \vec{\varepsilon})$. The metric (7.47) measures the extent to which a distribution $\rho(x)$ can be distinguished from another distribution $\rho'(x)$ in which *just one particle* has been slightly shifted while all others remain untouched, for example, $(\vec{x}_1, \vec{x}_2, \ldots \vec{x}_N) \to (\vec{x}_1 + \vec{\varepsilon}, \vec{x}_2, \ldots \vec{x}_N)$.

The simplest choice of functional $F[\rho]$ is linear in ρ, $F[\rho] = \int dx \rho(x) V(x)$, and the function $V(x)$ will play the role of the familiar scalar potential. In an *entropic* dynamics one might also expect contributions that are of a purely informational nature. Information geometry provides us with two tensors: one is the metric of configuration space $\gamma_{AB} \propto m_{AB}$, and the other is $I_{AB}[\rho]$. The simplest nontrivial scalar that can be constructed from them is the trace $m^{AB} I_{AB}$. This suggests

$$F[\rho] = \xi m^{AB} I_{AB}[\rho] + \int dx \rho(x) V(x), \tag{7.48}$$

where $\xi > 0$ is a constant that controls the relative strength of the two contributions. The case $\xi < 0$ leads to instabilities and is therefore excluded. (From Eq. (7.47) we see that $m^{AB} I_{AB}$ is a contribution to the energy such that those states that are more smoothly spread out tend to have lower energy.) The case $\xi = 0$ leads to a qualitatively different theory—a hybrid dynamics that is both indeterministic and yet classical (Bartolomeo and Caticha, 2016). The term $m^{AB} I_{AB}$ is usually called the "quantum" potential or the "osmotic" potential.[6] It is the crucial term that accounts for all quintessentially "quantum" effects—superposition, entanglement, wave packet expansion, tunneling, and so on.

With the choice (7.48) for $F[\rho]$, the generalized Hamilton–Jacobi equation (7.41) becomes

$$-\partial_t \Phi = \frac{1}{2} m^{AB} \partial_A \Phi \partial_B \Phi + V - 4\xi m^{AB} \frac{\partial_A \partial_B \rho^{1/2}}{\rho^{1/2}}. \tag{7.49}$$

[6] To my knowledge, the relation between the quantum potential and the Fisher information was first pointed out in (Reginatto, 1998).

8. The Schrödinger Equation

Once we have the coupled equations (7.34) and (7.49) we are done—it may not yet be obvious yet, but this is quantum mechanics. Purely for the sake of convenience, it is useful to combine ρ and Φ into a single complex function

$$\Psi_k = \rho^{1/2} \exp(i\Phi/k), \tag{7.50}$$

where k is an arbitrary positive constant, which amounts to rescaling the constant η in Eq. (7.33). Then the two equations (7.34) and (7.49) can be written into a single complex equation,

$$ik\partial_t \Psi_k = -\frac{k^2}{2} m^{AB} \partial_A \partial_B \Psi_k + V\Psi_k + \left(\frac{k^2}{2} - 4\xi\right) m^{AB} \frac{\partial_A \partial_B \mid \Psi_k \mid}{\mid \Psi_k \mid} \Psi_k, \tag{7.51}$$

which is quite simple and elegant except for the last nonlinear term. It is at this point that we can take advantage of our freedom in the choice of k. Since the dynamics is fully specified through ρ and Φ, the different choices of k in Ψ_k all lead to different versions of the same theory. Among all these equivalent descriptions, it is clearly to our advantage to pick the k that is most *convenient*—a process sometimes known as "regraduation."[7] The optimal choice, $k_{\text{opt}} = (8\xi)^{1/2}$, is such that the nonlinear term drops out and is identified with Planck's constant,

$$\hbar = (8\xi)^{1/2}. \tag{7.52}$$

Then Eq. (7.51) becomes the Schrödinger equation,

$$i\hbar\partial_t \Psi = -\frac{\hbar^2}{2} m^{AB} \partial_A \partial_B \Psi + V\Psi = \sum_n \frac{-\hbar^2}{2m_n} \nabla_n^2 \Psi + V\Psi, \tag{7.53}$$

where the wave function is

$$\Psi = \rho e^{i\Phi/\hbar}. \tag{7.54}$$

The constant $\xi = \hbar^2/8$ in Eq. (7.48) turned out to play a crucial role: it defines the numerical value of what we call Planck's constant, and it sets the scale that separates quantum from classical regimes. The conclusion is that for any positive

[7] Other notable examples of regraduation include the Kelvin definition of absolute temperature and the Cox derivation of the sum and product rule for probabilities (Caticha, 2012).

value of the constant ξ it is always possible to combine ρ and Φ to a physically equivalent but more convenient description where the Schrödinger equation is linear.

9. Some Final Comments

A theory such as ED can lead to many questions. Here are a few.

9.1 Is ED Equivalent to Quantum Mechanics?

Are the Fokker–Planck Eq. (7.34) and the generalized Hamilton–Jacobi Eq. (7.49) fully equivalent to the Schrödinger equation? This question, first raised by Wallstrom (1994) in the context of Nelson's stochastic mechanics (Nelson, 1985), is concerned with the issue of whether the phase $\Phi(x)$ and the wave function $\Psi(x)$ are single-valued or multivalued functions of x. Briefly stated, Wallstrom's objection is that Nelson's stochastic mechanics leads to phases Φ and wave functions Ψ that are either both multivalued or both singlevalued. The problem is that both alternatives are unsatisfactory. On one hand, quantum mechanics forbids multivalued wave functions. On the other hand, singlevalued phases can exclude physically relevant states such as states with nonzero angular momentum. We will not discuss this issue except to note that the objection does not apply once the relation between the quantum phase and gauge symmetry is properly taken into account (Caticha, 2019; Cararra and Caticha, 2017) and particle spin is incorporated into ED (Caticha and Carrara, 2020). A similar argument was developed by Takabayasi (1983) in the very different context of his hydrodynamical approach to quantum theory.

9.2 Is ED a Hidden-Variable Model?

Let us return to these mysterious auxiliary y variables. Should we think of them as hidden variables? There is a trivial sense in which the y variables are "hidden": they are not directly observable.[8] But being unobserved is not sufficient to qualify as a hidden variable. The original motivation behind attempts to construct hidden variable models was to explain or at least ameliorate certain

[8] The y variables are not observable *at the current stage of development of the theory*. It may very well happen that once we learn where to look we will find that they have been staring us in the face all along.

aspects of quantum mechanics that clash with our classical preconceptions. But the y variables address none of these problems. Let us mention a few of them:

1. *Indeterminism*: Is ultimate reality random? Do the gods really play dice? In the standard view quantum theory is considered an extension of classical mechanics—indeed, the subject is called quantum *mechanics*—and therefore deviations from causality demand an explanation. In the entropic view, on the other hand, *quantum theory is not mechanics; it is inference*—entropic inference is a framework designed to handle insufficient information. From the entropic perspective indeterminism requires no explanation. Uncertainty and probabilities are the norm; it is the certainty and determinism of the classical limit that demand an explanation (Demme and Caticha, 2017).
2. *Nonclassical mechanics*: A common motivation for hidden variables is that a subquantum world will eventually be discovered where nature obeys essentially classical laws. But in ED there is no underlying classical dynamics—both quantum mechanics and its classical limit are derived. The peculiar nonclassical effects associated with the wave-particle duality arise not so much from the y variables themselves but rather from the specific nondissipative diffusion that leads to a Schrödinger equation. The important breakthrough here was Nelson's realization that diffusion phenomena could be much richer than previously expected—nondissipative diffusions can account for wave and interference effects.
3. *Nonclassical probabilities*: It is often argued that classical probability fails to describe the double-slit experiment; this is not true (see, e.g., Section 2.5 in [Caticha, 2012]). It is the whole entropic framework—and not just the y variables—that is incompatible with the notion of quantum probabilities. From the entropic perspective it makes just as little sense to distinguish quantum from classical probabilities as it would be to talk about economic or medical probabilities.
4. *Nonlocality*: Realistic interpretations of the wave function often lead to such paradoxes as the wave function collapse and the nonlocal Einstein–Podolski–Rosen (EPR) correlations (Schlösshauer, 2004; Jaeger, 2009). Since in the ED approach the particles have definite positions and we have introduced auxiliary y variables that might resemble hidden variables, it is inevitable that one should ask whether this theory violates Bell inequalities. Or, to phrase the question differently: where precisely is nonlocality introduced? The answer is that the theory has been formulated directly in $3N$ dimensional configuration space and the Hamiltonian has been chosen to include the highly nonlocal quantum potential, Eq. (7.48). So, yes, the

ED model developed here properly describes the highly nonlocal effects that lead to EPR correlations and to violations of Bell inequalities.

9.3 On Interpretation

We have derived quantum theory as an example of entropic inference. The problem of interpretation of quantum mechanics is solved because instead of starting with the mathematical formalism and then seeking an interpretation that can be consistently attached to it, one starts with a unique interpretation and then one builds the formalism. This allows a clear separation between the ontic and the epistemic elements. In ED there is no risk of confusing which is which. "Reality" is represented through the positions of the particles, and our "limited information about reality" is represented by probabilities as they are updated to reflect the physically relevant constraints. In ED all other quantities, including the wave function, are purely epistemic tools. Even energy and momentum and all other so-called observables are epistemic; they are properties not of the particles but of the wave functions (Johnson and Caticha, 2012; Vanslette and Caticha, 2017.

To reiterate a point we made earlier: since "quantum" probabilities were never mentioned, one might think that entropic dynamics is a classical theory. But this is misleading: in ED probabilities are neither classical nor quantum; rather, they are tools for inference. All those nonclassical phenomena, such as the nonlocal effects that arise in double-slit interference experiments, or the entanglement that leads to nonlocal Einstein–Podolski–Rosen correlations, are the natural result of the linearity that follows from including the quantum potential term in the ensemble Hamiltonian.

The derivation of laws of physics as examples of inference has led us to discuss the concept of time. The notion of entropic time was introduced to keep track of the accumulation of changes. It includes assumptions about the concept of instant, of simultaneity, of ordering, and of duration. A question that is bound to be raised is whether and how entropic time is related to the actual, real, "physical" time. In a similar vein, to quantify the uncertainties in the motion—the fluctuations—to each particle we associated one constant m_n. We are naturally led to ask: How are these constants related to the masses of the particles?

The answers are provided by the dynamics itself: by deriving the Schrödinger equation from which we can obtain its classical limit (Newton's equation, $F = ma$ [Demme and Caticha, 2017]) we have shown that the t that appears in the laws of physics is entropic time and the constants m_n are masses. The argument is very simple: it is the Schrödinger equation and its classical limit that are used to design and calibrate our clocks and our mass-measuring devices. We conclude that by their very design, *the time measured by clocks is entropic time, and what mass measurements yield are the constants m_n*. No notion of time that is in any way more "real" or more "physical" is needed. Most interestingly, even though the dynamics is time-reversal invariant, entropic time is not. The model automatically includes an arrow of time.

Finally, here we have focused on the derivation of examples of dynamics that are relevant to physics, but the fact that ED is based on inference methods that are of universal applicability and, in particular, the fact that in the entropic dynamics framework one deliberately abstains from framing hypotheses about underlying mechanisms suggests that it may be possible to adapt these methods to fields other than physics (Abedi, Bartolomeo, 2019).

Acknowledgments

My views on this subject have benefited from discussions with many students and collaborators including M. Abedi, D. Bartolomeo, C. Cafaro, N. Caticha, A. Demme, S. DiFranzo, A. Giffin, S. Ipek, D. T. Johnson, K. Knuth, S. Nawaz, P. Pessoa, M. Reginatto, C. Rodriguez, and K. Vanslette.

References

Abedi, M., D. Bartolomeo, and A. Caticha. (2019). "Entropic Dynamics of Exchange Rates and Options" *Entropy,* 21: 586.

Adler, S. (2004). *Quantum Theory as an Emergent Phenomenon.* Cambridge: Cambridge University Press.

Amari, S. (1985). *Differential-Geometrical Methods in Statistics.* Berlin: Springer-Verlag.

Bartolomeo, D., and A. Caticha. (2016). "Trading Drift and Fluctuations in Entropic Dynamics: Quantum Dynamics as an Emergent Universality Class." *Journal of Physics Conference Series,* 701, 012009; arXiv:1603.08469.

Bartolomeo, D., and A. Caticha. (2016). "Entropic Dynamics: The Schrödinger Equation and Its Bohmian Limit." In A. Giffin and K. Knuth (eds.), *Bayesian Inference and Maximum Entropy Methods in Science and Engineering.* AIP Conference Proceedings 1757: 030002; arXiv:1512.09084.

Brukner, C., and A. Zeilinger. (2003). "Information and Fundamental Elements of the Structure of Quantum Theory." In L. Castell and O. Ischebeck (eds.), *Time, Quantum, Information.* New York: Springer; arXiv:quant-ph/0212084.

Carrara, N., and A. Caticha. (2017). "Quantum Phases in Entropic Dynamics." Presented at MaxEnt 2017, 37th International Workshop on Bayesian Inference and Maximum Entropy Methods in Science and Engineering, July 9–14, 2017, Jarinu, Brazil; arXiv:1708.08977.

Caticha, A. (1998a). "Consistency and Linearity in Quantum Theory." *Physics Letters A,* 244: 13–17; (1998b). "Consistency, Amplitudes, and Probabilities in Quantum

Theory," *Physical Review A*, 57: 1572–1582); (2000). "Insufficient Reason and Entropy in Quantum Theory." *Foundations of Physics*, 30: 227–251.

Caticha, A. (2011). "Entropic Dynamics, Time, and Quantum Theory." *Journal of Physics A: Mathematical and Theoretical*, 44: 225303; arXiv.org/abs/1005.2357.

Caticha, A. (2014). "Entropic Dynamics: An Inference Approach to Quantum Theory, Time and Measurement." *Journal of Physics: Conference Series*, 504: 012009; arXiv:1403.3822.

Caticha, A. (2019). "The Entropic Dynamics approach to Quantum Mechanics" *Entropy*, 21:943; arXiv:1908.04693.

Caticha, A. (2012). *Entropic Inference and the Foundations of Physics* (EBEB 2012, São Paulo, Brazil). https://www.albany.edu/physics/faculty/ariel-caticha.

Caticha, A., and N. Carrara. (2020). "The Entropic Dynamics of Spin," arXiv:2007.15719.

Caticha, A., D. Bartolomeo, and M. Reginatto. (2015). "Entropic Dynamics: From Entropy and Information Geometry to Hamiltonians and Quantum Mechanics." In A. Mohammad-Djafari and F. Barbaresco (eds.), *Bayesian Inference and Maximum Entropy Methods in Science and Engineering*, AIP Conference Proceedings 1641: 155; arXiv:1412.5629.

Caticha, A. (2015). "Entropic Dynamics." *Entropy*, 17: 6110; arXiv:1509.03222.

Chiribella, G., G. M. D'Ariano, and P. Perinotti. (2011). "Informational derivation of quantum theory," *Physical Review A*, 84: 012311.

D'Ariano, G. M. (2017). "Physics without Physics: The Power of Information-Theoretical Principles." *International Journal of Theoretical Physics:* 56: 97.

de la Peña, L., and A. M. Cetto. (1996). *The Quantum Dice, an Introduction to Stochastic Electrodynamics*. Dordrecht, Holland: Kluwer.

Demme, A., and A. Caticha. (2017). "The Classical Limit of Entropic Quantum Dynamics." In G. Verdoolaege (ed.), *Bayesian Inference and Maximum Entropy Methods in Science and Engineering*. AIP Conference Proceedings 1853: 090001; arXiv.org:1612.01905.

E. T. Jaynes: *Papers on Probability, Statistics and Statistical Physics*. (1983). Ed. by R. D. Rosenkrantz. Dordrecht: Reidel.

Goyal, P., K. Knuth, and J. Skilling. (2010). "Origin of Complex Quantum Amplitudes and Feynman's Rules." *Physical Review A*, 81: 022109.

Grössing, G. (2008). "The Vacuum Fluctuation Theorem: Exact Schrödinger Equation via Nonequilibrium Thermodynamics." *Physics Letters A*, 372: 4556.

Grössing, G. et al. (2012), "The Quantum as an Emergent System," *Journal of Physics: Conference Series*, 361: 012008.

Grössing, G. et al., ed. (2016), *EmQm15: Emergent Quantum Mechanics 2015*, *Journal of Physics: Conference Series*, 701. http://iopscience.iop.org/issue/1742-6596/701/1.

Hall, M. J. W., and M. Reginatto. (2002). "Schrödinger Equation from an Exact Uncertainty Principle," *Journal of Physics A* 35: 3289–3299; "Quantum Mechanics from a Heisenberg-type Inequality," *Fortschritte der Physik*, 50: 646–656.

Hardy, L. (2011). "Reformulating and Reconstructing Quantum Theory," arXiv:1104.2066.

Hooft, G. 't. (2002). "Determinism beneath Quantum Mechanics," presented at "Quo vadis Quantum Mechanics?", Temple University, Philadelphia, September 25, 2002; arXiv:quant-ph/0212095;

Hooft, G. 't. (2007). "Emergent Quantum Mechanics and Emergent Symmetries," presented at the 13th International Symposium on Particles, Strings and Cosmology, PASCOS, Imperial College, London, July 6, 2007; arXiv:hep-th/0707.4568.

Ipek, S., and A. Caticha (2015). "Entropic Quantization of Scalar Fields." In A. Mohammad-Djafari and F. Barbaresco (eds.), *Bayesian Inference and Maximum Entropy Methods in Science and Engineering*. AIP Conference Proceedings 1641: 345; arXiv:1412.5637.

Ipek, S., M. Abedi, and A. Caticha. (2017). "Entropic Dynamics of Scalar Quantum Fields: A Manifestly Covariant Approach." In G. Verdoolaege (ed.). *Bayesian Inference and Maximum Entropy Methods in Science and Engineering*. AIP Conference Proceedings 1853: 090002.

Isaac Newton's third letter to Bentley, February 25, 1693. In I. B. Cohen (ed.), *Isaac Newton's Papers and Letters on Natural Philosophy and Related Documents*, 1958 (p. 302). Cambridge: Cambridge University Press,

Jaeger, G. (2009). *Entanglement, Information, and the Interpretation of Quantum Mechanics*, Berlin: Springer-Verlag.

Jaynes, E. T. (1957). "Information Theory and Statistical Mechanics, I and II," *Physical Review* 106: 620 and 108: 171.

Jaynes, E. T. (1990). "Probability in Quantum Theory." In W. H. Zurek (ed.), *Complexity, Entropy and the Physics of Information*. Reading MA: Addison-Welsey. Online at http://bayes.wustl.edu.

Jaynes, E. T. (2003). *Probability Theory: The Logic of Science*. Edited by G. L. Bretthorst. Cambridge: Cambridge University Press.

Johnson, D. T., and A. Caticha. (2012). Entropic Dynamics and the Quantum Measurement Problem." In K. Knuth et al. (eds.), *Bayesian Inference and Maximum Entropy Methods in Science and Engineering*. AIP Conference Proceedings 1443, 104; arXiv:1108.2550

Lanczos, C. (1986). *The Variational Principles of Mechanics*. New York: Dover.

Leifer, M. S. (2014). "Is the Quantum State Real? An Extended Review of ψ-ontology Theorems." *Quanta 3*: 67; arXiv:1409.1570.

Nawaz, S., and A. Caticha. (2012). "Momentum and Uncertainty Relations in the Entropic Approach to Quantum Theory." In K. Knuth et al. (eds.), *Bayesian Inference and Maximum Entropy Methods in Science and Engineering*. AIP Conference Proceedings 1443: 112; arXiv:1108.2629.

Nawaz, S., M. Abedi, and A. Caticha. (2016). "Entropic Dynamics in Curved Spaces." In A. Giffin and K. Knuth (eds.), *Bayesian Inference and Maximum Entropy Methods in Science and Engineering*. AIP Conference Proceedings 1757: 030004; arXiv:1601.01708.

Nelson, E. (1979). "Connection between Brownian Motion and Quantum Mechanics." *Einstein Symposium Berlin*, Lecture Notes on Physics 100, p. 168. Berlin: Springer-Verlag.

Nelson, E. (1985). *Quantum Fluctuations*. Princeton, NJ: Princeton University Press.

Price, H. (1996). *Time's Arrow and Archimedes' Point*. New York: Oxford University Press.

Reginatto, M. (1998). "Derivation of the Equations of Nonrelativistic Quantum Mechanics Using the Principle of Minimum Fisher Information." *Physics Reviews A*, 58: 1775.

Reginatto, M. (2013). "From Information to Quanta: A Derivation of the Geometric Formulation of Quantum Theory from Information Geometry." arXiv:1312.0429.

Reif, F. (1965). *Fundamentals of Statistical and Thermal Physics*. New York: McGraw-Hill.

Schlösshauer, M. (2004). "Decoherence, the Measurement Problem, and Interpretations of Quantum Mechanics," *Reviews of Modern Physics*, 76: 1267.

Smolin, L. (2006). "Could Quantum Mechanics Be an Approximation to Another Theory?" arXiv:quant-ph/0609109.

Spekkens, R. (2007). "Evidence for the Epistemic View of Quantum States: A Toy Theory." *Physical Review A*, 75: 032110.

Stapp, H. P. (1972). "The Copenhagen Interpretation." *American Journal of Physics*, 40: 1098.

Takabayasi, T. (1983). "Vortex, Spin, and Triad for Quantum Mechanics of Spinning Particle," *Progress of Theoretical Physics*, 70: 1–17.

Vanslette, K., and A. Caticha. (2017). "Quantum Measurement and Weak Values in Entropic Quantum Dynamics." In G. Verdoolaege, *Bayesian Inference and Maximum Entropy Methods in Science and Engineering*. AIP Conference Proceedings 1853: 090003; arXiv:1701.00781.

Wallstrom, T. C. (1989). "On the Derivation of the Schrödinger Equation from Stochastic Mechanics." *Foundations of Physics Letters*, 2: 11.

Wallstrom, T. C. (1994). "The Inequivalence between the Schrödinger Equation and the Madelung Hydrodynamic Equations." *Physical Review A*, 49: 1613.

Wootters, W. K. (1981). "Statistical Distance and Hilbert Space." *Physical Review D*, 23: 357–362.

Zeh, H. D., (2002). *The Physical Basis of the Direction of Time*. Berlin: Springer-Verlag.

PART IV

INFO-METRICS IN ACTION I: PREDICTION AND FORECASTS

This is the first of two parts on info-metrics in action. It demonstrates the versatility of the info-metrics framework for inference in problems across the sciences. In Chapter 8, Kravchenko-Balasha develops a new info-metrics way for understanding cancer systems. It uses an approach known as information-theoretic surprisal analysis for identifying proteins that are associated with cancer. Unlike the chapters in the previous parts, here info-metrics is connected directly with thermodynamic equilibrium. Simply stated, this chapter shows how thermodynamic-based information-theoretic concepts and experimental cancer biology can be integrated for the purpose of translating biological molecular information into a predictive theory.

In Chapter 9, Bernardini Papalia and Fernandez-Vazquez apply info-metrics to spatial disaggregation of socioeconomic data. This is a common problem of disaggregating statistical information. Such disaggregation means that the problem is underdetermined, and one must resort to info-metrics in order to do the disaggregation in the least biased way. The spatial disaggregation problem, discussed in this chapter, is especially difficult when dealing with socioeconomic data. This is due to the inherent spatial dependence and spatial heterogeneity. This chapter develops an entropy-based spatial forecast disaggregation method for count areal data that uses all available information at each level of aggregation.

In Chapter 10, Foster and Stutzer use a large deviation approach to study the performance and risk aversion of different financial funds. Large deviation theory is used for estimating low-probability events, given certain information. It is directly connected to information theory and info-metrics. This chapter uses large deviations to rank mutual funds' probabilities of outperforming a benchmark portfolio. It also shows that ranking fund performance in this way is identical to ranking each fund's portfolio with a generalized entropy measure and is equivalent to an expected generalized power utility index that uses a risk-aversion coefficient specific to that fund.

In Chapter 11, Lahiri and Wang apply concepts from info-metrics to estimate macroeconomic uncertainty and discord. The focus of this chapter is on estimating forecast uncertainty emerging from the Survey of Professional Forecasters. This is an important problem in the macroeconomic and forecasting literature. This chapter also contrasts the info-metrics forecasts with that of the more conventional moment-based estimates, and it uses info-metrics to measure ex ante "news" or "uncertainty shocks" in real time.

In Chapter 12, Nelson develops a new way to assess probabilistic forecasts. Rather than using Shannon information, that quantity is translated to the probability domain. This is an interesting idea. Translation reveals that the negative logarithmic score and the geometric mean are equivalent measures of the accuracy of a probabilistic inference. The chapter also generalizes these transformations and ideas to other familiar information and entropy measures, such as the Rényi and Tsallis entropies.

Overall, this part demonstrates much of the strength and usefulness of info-metrics for solving problems, and assessing these solutions, across various disciplines. In each case, it uses the observed information together with minimally needed structure or assumptions in order to predict, or forecast, the entity of interest. When applicable, the authors compare the info-metrics predictions and forecasts with other methods.

8

Toward Deciphering of Cancer Imbalances: Using Information-Theoretic Surprisal Analysis for Understanding of Cancer Systems

Nataly Kravchenko-Balasha

1. Background

Biological systems are complex systems controlled by protein networks. A biological network is a web with molecular nodes, or subunits (e.g., proteins) that are linked to one another via activating or inhibiting interactions. Those interactions generate a flow of biological information, and the direction of this flow is highly dependent on the concentrations of the molecules (e.g., proteins). The subunits process input information (e.g., external signals, such as nutrients) and transduce it to other specific interaction partners, thereby generating *signaling networks*. Eventually, the protein–protein interactions have phenotypic outcomes (observable characteristics), such as cell migration, cell proliferation, and tissue architecture.

Protein networks may change significantly when pathological processes take place. Cancer results from the acquisition of a wide range of genetic alterations that may lead to the extensive reorganization of protein–protein interaction networks and abnormal phenotypic outcomes that are hard to predict. Even though the genetic impairments precede and in many cases underlie the protein network rearrangement, it is the oncoproteins that eventually are responsible for cancer progression. Furthermore, tumors eventually become dependent on several key oncoproteins that are responsible for cancer growth and survival, a phenomenon known as "oncogene addiction" (Weinstein, 2002). This has led to the idea that the inhibition of components of oncoprotein-activated signaling pathways would be an effective therapeutic strategy for cancer, and this has proven to be the case. Selective drugs have been developed that demonstrate significant antitumor activity. However, this effort has been hindered by a

Nataly Kravchenko-Balasha, *Toward Deciphering of Cancer Imbalances: Using Information-Theoretic Surprisal Analysis for Understanding of Cancer Systems* In: *Advances in Info-Metrics: Information and Information Processing across Disciplines.* Edited by: Min Chen, J. Michael Dunn, Amos Golan, and Aman Ullah, Oxford University Press (2021).

DOI: 10.1093/oso/9780190636685.003.0008

number of factors, including molecular variations between tumors (intertumor heterogeneity) and between cells inside the tumor (intratumor heterogeneity); and therapy-induced physiological changes in host tissues that can reduce or even nullify the desired antitumor effects of therapy (Shaked, 2016). These factors are often *tumor-specific* (Hood and Flores; Robin et al., 2013; Zhu and Wong, 2013) and lead to the consequent difficulty in predicting how patient-specific altered protein networks will respond to cancer therapy, highlighting the urgent need for personalized treatment (Hood and Flores, 2012; Drake et al., 2016).

A series of quantitative techniques addressing the complex nature of protein networks have been developed, such as reverse-engineering algorithms, based on chemical kinetic-like differential equations (Young, Tegnér, Collins, Tegnér, and Collins, 2002; Bansal and Bernardo, 2007); Bayesian methods, based on elucidating the relationships between a few genes at a time (Friedman, Linial, Nchman, and Pe'er, 2000); and multivariate statistical methods that include clustering methods (Mar, Wells, and Quackenbush, 2011), principal component analysis (Jolliffe, 2002), singular value decomposition (Alter, Brown, and Botstein, 2000), meta-analysis (Rhodes et al., 2004), and machine learning (van Dijk, Lähdesmäki, de Ridder, and Rousu, 2016; see Creixell et al., 2015; Kravchenko-Balasha, 2012 for additional discussion). However, despite the considerable progress in the field of data analysis (Creixell, 2015) and encouraging attempts to implement these approaches in the field of personalized medicine (Alyass, Turcotte, and Meyre, 2015), it is still the case that aggressive tumors do not respond well to the currently available therapeutics. This suggests that our understanding of patient-specific rewiring and the alterations of cancer signaling networks is far from completion. A number of experimental strategies are being developed, aiming to test the drug responsiveness of cancer tissues to diverse combinations of FDA-approved drugs in a patient-specific manner (Pemovska et al., 2013). For example, Pemovska et al. showed that anticancer agents, already in clinical use for various types of cancer, can be effective against patient-derived acute myeloid leukaemia samples (Pemovska et al., 2013). However, our ability to integrate and interpret in an individualized manner "omics" data, data which characterizes and quantifies large pools of biological molecules, remains an important limiting factor in the transition to personalized medicine (Alyass, Turcotte, and Meyre, 2015; Koomen et al., 2008; Leyens et al., 2014).

An approach based on physical rules may overcome this limitation by providing a rigorous and quantitative *theory that is valid for every single patient.* Such an approach differs from straight statistical analyses, which regard a sub-population as representative of the entire population. Information-theoretical approaches based on physical rules enable the investigation of patient-specific *reorganization* of molecular signaling, thereby gaining a critical understanding of the network structure of signaling imbalances in every analyzed tissue. These structures can guide us in the choice and application of treatment in the future.

This chapter shows how thermodynamic-based information-theoretical theory and experimental cancer biology can be integrated for the purpose of translating biological molecular information into a predictive theory.

2. Information-Theoretic Approaches in Biology

The conversion of omics molecular data, obtained from different tumors/cancer samples into knowledge, such as patient-oriented treatments, requires detailed understanding of how molecular signals are integrated and processed and how tumor networks are "rewired" during oncogenesis (Lee et al., 2012). Information-theoretical approaches, though not common in biological research, have been successfully applied to analysis of biological networks in a number of cases (see, e.g., Nykter et al., 2008; Waltermann and Klipp, 2011; Levchenko and Nemenman, 2014). These approaches allow analysis and interpretation of large and sometimes noisy data through discovery of mathematical laws that govern its behavior. The approaches usually rely on the concept of statistical entropy. Statistical entropy is a measure of the amount of uncertainty an observer possesses regarding the distribution of states of a system. If the observed/calculated entropy is high, then the distribution is uniform, and no one state is considerably more probable than its fellow states. When the entropy is low, one or a few states stand out, yielding information about the system. This information is actually the knowledge (e.g., knowledge about protein–protein correlations) enabling the state of a system to be distinguished from numerous available potential states.

In order to analyze nonequilibrium biological systems *based on physico-chemical rules*, we utilize a thermodynamic-based information-theoretical surprisal analysis, which has been previously applied to nonequilibrium systems in chemistry and physics (see, e.g., Levine, 2005; Levine and Bernstein, 1974; Ben-Shaul, Levine, and Bernstein, 1972). Surprisal analysis characterizes the probability of a system to be in different states. In our case, these states may have a rich internal protein–protein network structure. The analysis is based on the assumption that the entropy of a system is at its global maximum when the system is unconstrained. Under conditions of constant temperature and pressure, *an entropy maximum* is equivalent *to a free-energy minimum*, and the state of lowest free energy in biology is the steady state (McQuarrie, 2000). The steady state is unchanging over time, and all the processes are balanced (Nelson, Lehninge, and Cox, 2008). Thus, this state can also be named "balanced state." Upon application of constraints to the system (see elaboration in the following sections), the system deviates from this basal, balanced state, and reaches a new constrained state. Surprisal analysis identifies the balanced state of the system, as well as the number of constraints that operate on the system and deviate it from the balanced state. The analysis determines how each and every constraint affects the different components of the system. Furthermore, the analysis uncovers the arrangement of the system's components and how these components are interconnected. Thus, the analysis achieves a comprehensive map of the state and structure of the system.

Surprisal analysis begins by assuming that a small set of constraints is known. If the assumed constraints do not reproduce the experimental molecular distributions, one is *surprised* and therefore must search for additional constraints. Thus "*surprisal*" is the sum over the constraints and the quantification of the deviation from the steady, balanced state.

A simple example for a constraint from physics is a drifted Brownian motion of electrons in a conductor. In an uncharged conductor, electrons move in a random Brownian motion. Upon application of the electric field a constraint in the system is generated, thereby increasing the probability of an electron moving in the direction opposite to the field. Thus the motion of electrons is drifted.

An aggressively growing tumor is apparently not at the balanced system. Surprisal analysis considers that environmental and genomic constraints preclude the tumor cells from reaching the balanced state. The analysis takes the experimental expression levels of molecules as input (e.g., proteins, mRNA, DNA) and extracts the expected distribution of the expression levels in the balanced state, as well as deviations of the expression levels from this state. The analysis determines how each constraint affects each molecule and to what extent its expression level deviates from the level expected in the balanced state. The network structure, as well as the organization of the molecules into independent subnetworks, is determined.

Importantly, it is not trivial to identify the balanced state in biological systems, let alone complex systems such as cancer. The ability to deduce the distribution of protein expression levels in the balanced state from the experimental data is a unique aspect of surprisal analysis.

We have implemented surprisal analysis in several cancer systems and have provided encouraging experimental validation of the approach (see the following sections). We have recently demonstrated that, as in chemistry and physics, the direction of change, as, for example, the direction of cell–cell movement in cancer systems, can be predicted and thereby experimentally controlled (Kravchenko-Balasha, Shin, Sutherland, Levine, and Heath, 2016; Kravchenko-Balasha, Wang, Remacle, Levine, and Heath, 2014).

The approach can be applied to different types of biological data, such as systems evolving in time (Flashner-Abramson, Abramson, White, and Kravchenko-Balasha, (2018), transitions from normal to cancerous phenotypes

(Poovathingal, Kravchenko-Balasha, Shin, Levine, and Heath, 2016; Kravchenko-Balasha, Simon, Levine, Remacle, and Exman, 2014; Kravchenko-Balasha et al., 2011), or from mesenchymal to epithelial phenotypes (Zadran, Arumugam, Herschman, Phelps, and Levine, 2014), and data sets that include multiple samples (Kravchenko-Balasha, Levitzki, Goldstein, Rotter, Gross, Remacle et al., 2012; Kravchenko-Balasha, Johnson, White, Heath, and Levine, 2016; Zadran, Remacle, and Levine, 2013; Vasudevan et al., 2018; Vasudevan, Flashner-Abramson, Remacle, Levine, and Kravchenko-Balasha, 2018; Flashner-Abramson, Vasudevan, Adejumobi, Sonnenblick, and Kravchenko-Balasha, 2019), as well as single-cell studies (Kravchenko-Balasha, Shin, Sutherland, Levine, and Heath, 2016; Kravchenko-Balasha, Wang, Remacle, Levine, and Heath, 2014; Poovathingal et al., Heath, 2016).

3. Theory of Surprisal Analysis

This section provides a brief presentation of the theory of surprisal analysis. For more details, see Kravchenko-Balasha, 2012; Kravchenko-Balasha et al., 2011; Vasudevan, Flashner-Abramson et al, 2018; Remacle, Kravchenko-Balasha, Levitzki, and Levine, 2010.

To identify the balanced state and deviations thereof in every biological sample (e.g., cancer tissue) we utilize experimental protein expression levels. To understand biological states, additional types of molecules can be used, such as mRNA (molecules that transfer genetic information from DNA to the machinery that synthesizes proteins) or metabolites (the intermediate products of metabolic reactions). Proteins are large, complex molecules that perform the main functions in the cell and constitute the direct targets of the targeted cancer therapies used in the clinics today. Therefore, in this chapter we focus mainly on the protein molecules.

As explained earlier, the analysis breaks down the experimental expression levels of the proteins to the expression level expected in the balanced state, and the sum of the deviations from this value due to different constraints that operate in the system. For each protein, the expected level in the balanced state and the deviations from this state sum up to the experimental value.

To perform a search for the maximum entropy subject to constraints, we use the method of Lagrange undetermined multipliers:

$$\pounds = \text{Entropy} - \sum_{\alpha} \lambda_{\alpha} \text{Constraint}_{\alpha} \tag{8.1}$$

where α is the index of the constraint, and λ_{α} are the amplitudes of the constraints. A given constraint α can have a large amplitude (λ_{α}) in one sample but can have effectively a negligible amplitude in other samples. The amplitudes, λ_α, are the Lagrange undetermined multipliers. There may be a set of Lagrange multipliers in each biological sample k.

Seeking the maximum of £ is the solution of the problem of maximization of the entropy subject to constraints. Using the Shannon grouping property (Golan and Lumsdaine, 2016; Shannon, 1948), we can present the entropy as $S = -\sum_i X_i [\ln(X_i) - 1] + \sum_i X_i S_i$ (Remacle et al., 2010). The first term, $-\sum_i X_i [\ln(X_i) - 1]$, is the mixing term, representing the entropy of a mixture of protein species X_i. The second term, $\sum_i X_i S_i$, represents the weighted sum over the entropies of each species i (the detailed theoretical and computational background is presented in (Remacle et al., 2010). The grouping property enables assigning a weight to each protein, denoting the contribution of each molecule to the balanced state, and then identifying the deviation from the global maximum due to the constraints (see Remacle et al., 2010). Constraints are presented as: $\langle G_\alpha \rangle = \sum_i G_{i\alpha} X_i$. This equation means that the protein concentration X_i for each protein i in a given sample is limited by the quantity $\langle G_\alpha \rangle$, which represents a biological process responding to the constraint α. $G_{i\alpha}$ is the weight of protein i in the biological process corresponding to the constraint α (Remacle et al., 2010).

The maximization of function (8.1) leads to the exponential form (a detailed description of the steps leading from (8.1) to (8.2) is presented in Remacle et al., 2010):

$$X_i(k) = X_i^o \exp\left(-\sum_{\alpha=1} G_{i\alpha} \lambda_\alpha(k)\right) \tag{8.2}$$

$X_i(k)$ represents the experimental expression level of protein i in sample k. X_i^o is the expression level of the protein at the global maximum of entropy (the balanced state without constraints). Using X_i^o values, we obtain the distribution of the protein levels at the global maximum of the entropy and thus the balanced state of the system. The exponent term in Eq. (8.2), $\exp\left(-\sum_{\alpha=1} G_{i\alpha} \lambda_\alpha(k)\right)$, describes the deviation of the expression level of protein i from the global entropy maximum due to the constraints $\alpha = 1, 2, 3 \ldots$. By taking a natural logarithm of Eq. (8.2), we obtain the following:

$$\ln X_i(k) = \underbrace{\ln X_i^0(k)}_{\text{basal level}} - \underbrace{\sum_{\alpha=1}^{n} G_{i\alpha} \lambda_\alpha(k)}_{\substack{\text{deviations from the steady state} \\ \text{due to constraints}}} \tag{8.3}$$

Equation (8.3) is the working equation that we utilize to describe the states in biological systems.

In practical terms, a matrix is constructed, containing protein expression levels (rows) in the different samples (columns). To determine the amplitudes of the constraints, $\lambda_\alpha(k)$, and the weight of each protein i in the constraint α, $G_{i\alpha}$, the matrix is diagonalized. This is achieved by utilizing SVD (singular value decomposition (Alter et al., 2000) as a mathematical tool.

In order to find the actual number of constraints in the system, we define the minimal number of constraints required to reproduce the experimental protein levels in the data set. In other words, we compute the sum $\ln X_i^0(k) - \sum_{\alpha=1}^{n} G_{i\alpha}\lambda_\alpha(k)$ for escalating values of n (i.e., for escalating numbers of constraints) and compare the result to $\ln X_i(k)$. We also calculate error limits (using standard deviations of the experimental data) as described in (Vasudevan, Flashner-Abramson et al, 2018; Gross and Levine, 2013), allowing us to identify constraints whose amplitudes exceed the threshold values.

To understand the biological nature of the constraints, proteins with significant deviations from the steady states are grouped into the protein–protein networks using the String database (Szklarczyk et al., 2011). This database utilizes previous knowledge, based on experimental observations, and assigns to each pair of proteins a probability to have a real functional connectivity. Thereby we relate each constraint to the unbalanced biological network, which we often term as an "unbalanced process" that deviates the system from the steady state. We propose that through the manipulation of the proteins (e.g., inhibiting) that contribute the most to the deviations, we can test our predictions on how the states of the system can be changed.

4. Using Surprisal Analysis to Understand Intertumor Heterogeneity

Targeted cancer therapies (therapies that act on specific molecular targets, usually proteins, associated with cancer) have been developed and demonstrated significant antitumor activity. Molecular variations between tumors (intertumor heterogeneity) impaired our ability to predict how patient-specific altered protein networks will respond to targeted cancer therapy. This notion led to the idea that targeted therapy should be based on patient-specific molecular targets within the tumor.

The field of personalized medicine has accelerated in recent years, and a massive amount of protein measurements for different types of cancer is accumulating. We suggest that incorporating experimental measurements obtained from cancer tissues into the thermodynamic-based framework might be advantageous: not only does it provide an in-depth understanding of the patient-specific altered protein-signaling networks, but it also allows predictions about how to interrupt these unbalanced networks as a means to terminate the disease.

Using surprisal analysis, we identify the unbalanced protein networks (= processes) operating in the entire population of patients as well as the subset of these networks that operate in every patient. The analysis assumes that tumors with different functional properties—as measured, for example, by proteomics (large-scale study of protein expression levels in the sample)—are subject to the influence *of different constraints.* Using the surprisal analysis procedure, these constraints are related to the tumor/patient-specific changes in the protein/phosphoprotein expression levels. As described earlier, proteins that undergo coordinated changes in the expression levels relative to their balanced expression levels are organized into groups, allowing us to identify unbalanced processes operating in each tumor (Figure 8.1). Molecular comprehension of those processes may suggest how the cellular states can be changed in order to stop the disease (Kravchenko-Balasha et al., 2016; Flashner-Abramson et al., 2019). Each unbalanced process can influence *several* protein pathways. *Several distinct unbalanced processes* may operate in each tumor, and each protein can participate in several unbalanced processes due to the nonlinearity of biological networks (Kravchenko-Balasha et al., 2016). Surprisal analysis identifies these processes for every *single protein and every patient tissue through decomposition of the protein expression level,* as shown in Eq. (8.3). Recently, we have shown that different patients may display similar biomarker expression levels, albeit carrying biologically distinct tumors, characterized by different sets of unbalanced molecular processes (Vasudevan et al., 2018). We suggest that in order to rationally design

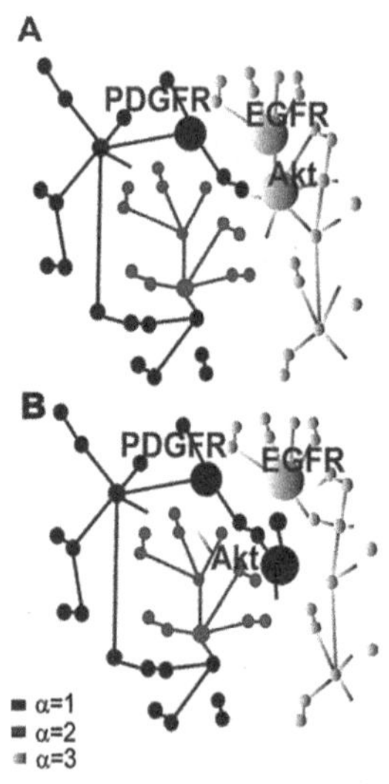

Figure 8.1 Scheme – Patient-specific rewiring of signaling networks. Using surprisal analysis, a *set* of distinct unbalanced processes in each tumor is identified. In this scheme, 3 processes, $\alpha = 1,2,3$, are presented. Central oncoproteins, Akt and EGFR, can be involved in the same process, as in the process $\alpha = 3$ in patient (**A**). Alternatively, signaling rewiring can lead to decoupling between Akt and EGFR (Logue, J. S., and Morrison, D. K. 2012), as in patient (**B**), where they appear in distinct unbalanced processes. Inhibition of EGFR in patient (**B**) will not affect the proteins involved in processes $a = 1,2$, including Akt.

patient-specific combined therapies, an accurate resolution of the patient-specific unbalanced protein network structures is required. A simple example on how oncogenic factors can cause patient-specific rewiring of protein-signaling networks is provided in Figure 8.1. Rewiring in protein networks can lead to considerable differences in patient responses to the same therapy.

In the following subsection, I provide an example of how surprisal analysis can be applied to proteomic data sets to characterize tumor variability.

4.1 A Thermodynamic-Based Interpretation of Protein Expression Heterogeneity in Glioblastoma Multiforme Tumors Identifies Tumor-Specific Unbalanced Processes

The analysis was performed on the proteomics (protein expression) data collected from a panel of eight human brain tumors (glioblastoma, GBM (Kravchenko-Balasha et al., 2016). Tumors labeled GBM 10 and 12 expressed wild-type (*a typical*, wt) epidermal growth factor receptor (EGFR), which plays a significant role in tumor development in many types of cancer (Szklarczyk et al., 2011). GBM 1, 15, and 26 overexpressed wtEGFR, and GBM 6, 39, and 59 overexpressed the mutant EGFR variant III (EGFRvIII), constitutively active, oncoprotein. GBM tumors exhibit high intertumor protein expression variability, which is a consequence of both patient-specific genetic backgrounds and various tumor-specific driver or passenger mutations. Using the measured expression levels of over 1,000 proteins and hundreds of phosphorylation sites—that provide measurement of the functional activity of the proteins—we decomposed the experimental levels of every measured protein according to the unbalanced processes as described earlier.

Identification of the Most Stable State. Utilizing *basal*, steady-state levels of every protein *i*, as calculated using surprisal analysis, we built a histogram that represents the distribution of protein expression levels of *different protein species* at the steady state. An example for such a histogram is shown for the tumor GBM59 (black histogram, Figure 8.2A). The long tail in this histogram is composed of the proteins that had the highest intensities during the experiment. In the experiments, in which molecular intensities correspond to the molecular concentrations, these groups of molecules usually comprise the steady-state "core"—the group of molecules that contribute the most to the steady-state network (Kravchenko-Balasha et al., 2012).

The gray histogram (Figure 8.2A) is composed of the values of the deviations from the steady state as calculated for every measured protein. It clearly shows that only a limited number of proteins have significant $\sum_{\alpha=1}^{n} G_{i\alpha}\lambda_{\alpha}$ values. These

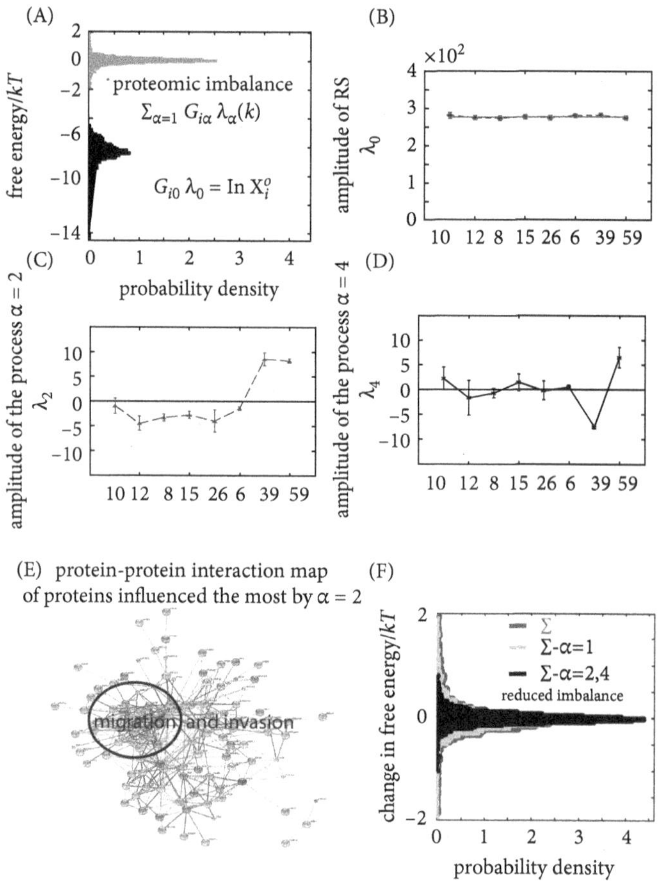

Figure 8.2 Surprisal analysis of 8 different GBM tumors, adapted from (Kravchenko-Balasha et al., 2016). (**A**) Histogram of the protein intensities at the steady state (black) and deviations thereof (grey). (**B**) The amplitude, $\lambda_0(k)$, of the steady state (reference state, RS) for every GBM tumor k is shown. To within the small error bars the steady state is found to be invariant across all tumors. (**C-D**) Amplitudes of the unbalanced processes $\alpha = 2, 4$ as represented by $\lambda_2(k)$ (**C**) *and* $\lambda_4(k)$ (**D**). (**E**) 140 of the proteins with the most significant values of G_{i2} (induced due to the process $\alpha = 2$) were used as an input for generation of the protein-protein network. Only 115 *connected* proteins are shown. (**F**) The effect of a 4-fold decrease in the weights of the unbalanced processes ($\alpha = 2, 4$) operating in the GBM59 on the free energy. The notation Σ in the figure means $\sum_{\alpha=1} G_{i\alpha}\lambda_a(k)$ and is the same data shown in the upper part of panel (**A**). The notation $\Sigma - \alpha = 2{,}4$ in the figure means the sum $\sum_{\alpha=1}^{7} G_{i\alpha}\lambda_a(k)$ with λ_2 and λ_4 decreased by 4-fold.

proteins need to be examined in order to understand the biological nature of the constraints. The majority of the values are centered around zero.

We indeed found a large group of proteins (more than 500) that contribute significantly to the steady state and do not contribute to the deviations from the steady state. These proteins participate in homeostatic functions of the cell, such as protein and RNA metabolism. The amplitude of the steady state remains invariant for different GBM tumors (Figure 8.2B); thus, the description of the stable steady state is robust and shared across all eight tumors. This result corroborates the previous data, where surprisal analysis of transcriptomic data sets (in which mRNA levels were measured instead of the protein levels) revealed that the most stable and *invariant* state of different organisms, cell lines, and healthy/diseased individuals was associated with groups of transcripts involved in cellular homeostasis (Kravchenko-Balasha et al., 2011, 2012; Vasudevan, 2018).

Unbalanced Processes Operating in GBM Tumors. Beyond the commonality of the steady state, we find that different constraints are operating in different tumors, meaning the tumors vary significantly from each other. Examples for two unbalanced processes, $\alpha = 2$ and $\alpha = 4$, are shown in Figures 8.2C and 8.2D. Figure 8.2C,D shows that the unbalanced processes $\alpha = 2$ and $\alpha = 4$ play a significant role in the tumors GBM59 and GBM39, since the amplitudes of the processes differ significantly from 0. These tumors are similar in the process $\alpha = 2$ (**Figure 8.2C**) and share the common molecular processes associated with cell migration. However, they differ in the $\alpha = 4$ constraint (**Figure 8.2D**).

In order to understand the biological meaning of the processes, the proteins participating in these unbalanced processes are assembled into a functional protein–protein network according to STRING database (Szklarczyk et al., 2011). As mentioned earlier, this database integrates all known experimental evidence about protein–protein interactions and assigns for each pair of proteins a probability to have a real functional interaction. Functional network is shown for the unbalanced process $\alpha = 2$ (Figure 8.2E). Pairs of proteins with known functional interactions are connected with a line. The width of the line marks the degree of confidence for protein–protein interaction (according to the STRING database). The unbalanced process $\alpha = 2$ includes a big group of the proteins involved in the process of cell migration and invasion– one of the central processes responsible for the diffusive infiltration of the GBM cells in the brain that leads to therapeutic failure (Holland, 2000).

Molecular mining of the constraint 4 revealed that the GBM 39 tumor has an additional induced migration module with oncoprotein PDGFR, as a potential drugable target, while analysis of the GBM59 tumor points to the enhancement of glycolysis (the cellular process that transforms glucose to energy) through pPKM2 protein.

Figure 8.2F suggests that inhibiting the unbalanced processes $\alpha = 2, 4$, through inhibition of, for example, central proteins participating in those processes, could effectively reduce the signaling imbalance in the GBM59 tumor. Without processes $\alpha = 2, 4$, the tumor imbalance reduced to 0 (black distribution in Figure 8.2F). The same analysis is performed for every tumor individually (Kravchenko-Balasha, 2016).

In a broader sense, we show that surprisal analysis can provide a structure of tumor unbalanced networks by providing a subset of unbalanced processes for every tumor. In our example, tumors GBM59 and GBM39 had a *network signature* composed of the processes $\alpha = 2, 4$. Process 4 has an opposite behavior in these tumors. Moreover, these processes did not appear in any other tumor from the data set. Other tumors were assigned a different structure (= signature) of unbalanced processes, signature that may help to design patient-specific drug combinations in the future (Kravchenko-Balasha, 2016).

Network structure, as resolved by surprisal analysis, resembles a computer's hardware and software. While there is a common "hardware" to different tumors and tissues, represented by the shared invariant protein networks, the intertumor heterogeneity is represented by cellular "software," or unbalanced molecular processes, which significantly vary in protein composition (Kravchenko-Balasha et al., 2012, 2016). Some of the unbalanced processes are shared by several tumors, but some of them constitute unique classifiers of tumor subgroups (Kravchenko-Balasha et al., 2016). This finding is highly significant because it suggests that the search for drug targets should focus on the unbalanced processes, reducing the number of possible targets. These processes represent intertumor heterogeneity and provide insights for designing patient-specific combination therapies.

Comparison of the results of our theoretical interpretation of statistical multivariate methods, commonly used to analyze large data sets, such as PCA (principal component analysis) and *k*-means clustering (a clustering method that partitions *n* observations into *k* clusters in which each observation belongs to the cluster with the nearest mean, serving as a representative of the cluster), revealed that surprisal analysis provides a superior level of resolution of patient-specific alterations in protein expression levels (Kravchenko-Balasha et al., 2016).

We have recently validated the approach by demonstrating that a strategy based on surprisal analysis allows the rational design of patient-specific drug combinations (Flashner-Abramson et al, 2019) and prediction of the response of cells to drug treatment (Flashner-Abramson et al., 2018). We experimentally show our ability to dissect patient-specific signaling signatures with high resolution by designing efficient combinations of targeted drugs against different cancer cell lines. We show that our predictions of therapies against personalized signaling signatures were more effective in killing the cancer cells than were clinically prescribed treatments (for more details, see Flashner-Abramson et al., 2019).

5. Toward Understanding Intratumor Heterogeneity

Cellular intratumor heterogeneity in cancer can have a crucial impact on therapeutic success (Fisher, Pusztai, and Swanton, 2013). Small subpopulations of cancer stem cells (cells that are able to give rise to all cell types found in a particular tumor) or drug-resistant cells might be masked in bulk assays and may eventually form a drug-resistant malignancy. Moreover, the tumor contains nonmalignant cells, such as stromal, hematopoietic, and endothelial cells that can support the tumor growth and influence the molecular processes in the tumor. Thus, detailed single-cell analysis can help identify the heterogeneous *intratumor* signaling processes through an accurate resolution of the intratumor protein correlation networks (Poovathingal et al., 2016; Shi et al., 2011, 2012; Wei et al., 2016). Single-cell measurements and analysis may have clear advantages when only a limited number of samples/proteins is available (Poovathingal et al., 2016 and Figure 8.3).

In this subsection, I show how surprisal analysis can be extended to quantify the changes of protein copy numbers in individual cells (Poovathingal et al., 2016; Shin et al., 2011). A single-cell quantitative framework of surprisal analysis allows collecting the fluctuations of protein expression levels from *single cells* to the thermodynamics-based theory. In such an approach, each sampled cell is regarded as a typical representative of a large number of cells from the bulk culture (Shin et al., 2016; Poovathingal et al., 2016).

Here we discuss an example of such an analysis performed on the nontransformed breast MCF10a cells that have been treated with the Benzo[a]pyren (B[a]P) for ninety-six days. B[a]P is a carcinogen present in cigarette smoke. The B[a]P treatment resulted in transition of the cells from a nontransformed to a transformed, cancer-like phenotype within three months. Eleven signaling oncoproteins were measured on the single-cell level at eight time points for a period of three months (Figure 8.4).

The thermodynamic-based interpretation of the cellular transformation from the nontransformed to the cancer-like phenotype allowed us to identify a time point in which this transition occurred. Interestingly, this transition had classical hallmarks of phase transitions, familiar to us from the physical world, such as the ice-to-water transition. The hallmarks of these types of transitions include phase coexistence (as coexistence of ice and water phases) and loss of degrees of freedom (e.g., example, when heat is added to the sample without an increase in the temperature of water). In this chapter, I give an example of how surprisal analysis leads to the discovery of the phase coexistence in the MCF10a cells treated with B[a]P. For more details, see Poovathingal, et al., 2016, which provides additional discussion on the phase transition in MCF10a cells, including the hallmark of loss of degrees of freedom.

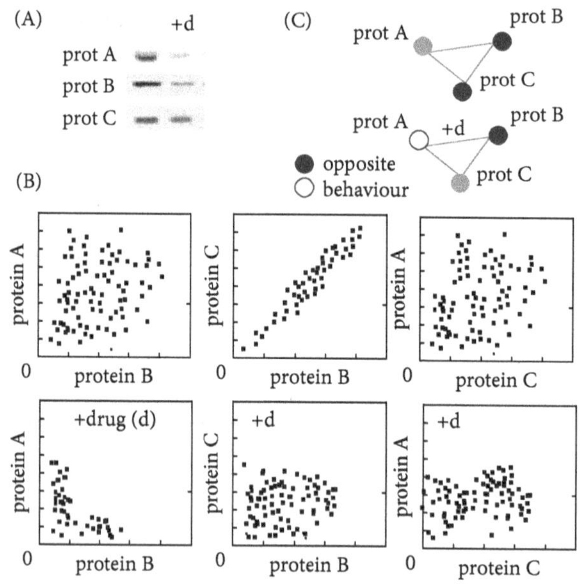

Figure 8.3 Significance of single cell measurements, adapted from (Poovathingal et al., 2016). Measurements of several proteins from the same single cells yield quantitative protein-protein correlations/anti-correlations. This is a uniquely single cell measurement. (**A**) Consider the levels of the three hypothetical phosphoproteins (prot A, prot B, and prot C) representing a small signaling network within a bulk culture. Drug treatment (+d) of the cells may collectively repress all protein levels as shown in the bulk assays. (**B**) Single cell analysis presented in the two-dimensional scatter plots reveals a deeper picture. Note that in the plots for the untreated cells, all phosphoprotein levels are high, but only protein C and protein B are strongly correlated.

Upon treatment, all proteins are repressed, but protein A and protein B are anti-correlated while the others are non-correlated. (**C**) This inferred correlation network before and after treatment (+d) is shown in the network graphic. This network, generated at the single cell level, can provide a rich set of testable hypotheses.

We analyzed potential phase coexistence in the cell population of MSF10A cells by taking a cue from the ice/water transition system. We can distinguish between two phases (water and ice) based on different physical attributes such as the different index of light reflection or differences in the shear modulus. In other words, we looked for an observable parameter that differs in value between the two phases (= two subpopulations). We used the amplitude, λ_1, of the constraint $\alpha = 1$ (the most significant constraint) in every cell as a distinguishing observable parameter for the cellular transition. Using this parameter, we can identify the clear coexistence of two subpopulations at days 12 and 28. Scatter-

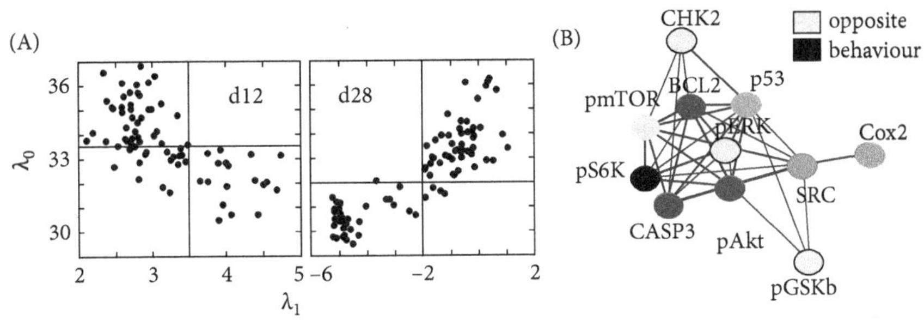

Figure 8.4 Single cell analysis of MCF10A breast immortalized cells, treated with B[a]P using microfluidic-based single cell barcode chip (Poovathingal et al., 2016). (**A**) Each dot represents a value of the amplitude λ_1 calculated for every single cell. The amplitude λ_1 of the process $a = 1$, which is illustrated in (**B**), has different values in different cells and thereby divides the cell population into two distinct subpopulations: At day 12 (d12) there is a significant subpopulation (in lower right corner) in which pERK is induced and pS6K is reduced (note anti-correlation between pERK and pS6K as presented in (**B**)); in the second subpopulation (upper left corner) the process is insignificant. At day 28 (d28) there is a subpopulation in which pERK is reduced and pS6K is significantly induced (lower left corner). This process in insignificant in the cells in the upper right corner.

plots of λ_1 versus λ_0 for days 12 and 28 are shown in Figure 8.4A. These plots show that two cellular subpopulations coexist at those time points and differ in the significance of the first unbalanced process. A biological interpretation of the process $\alpha = 1$ was provided by analyzing the proteins participating most in process $\alpha = 1$ (**Figure 8.4B**). Figure 8.4B shows that the process $\alpha = 1$ is largely characterized by an anticorrelation of the oncoproteins pS6K and pERK. Every subpopulation had a different protein signaling signature, since the significance of this process was different in different cells (**Figure 8.4A**).

In addition to the physical understanding of the process, two significant biological insights could be derived from the analysis:

1. Actual transition (phase transition) significantly preceded the emergence of the traditional precancerous phenotype, which appeared only after ~3 months.

Thus, early identification of the transition on the molecular levels might have important implications in early diagnostics.

2. At days 12 and 28, the activity of oncoprotein pS6K varies significantly in different cellular subpopulations. Thus, inhibition of this oncoprotein may

influence only one subpopulation of the cells in a tumor, whereas the other subpopulation may continue to proliferate.

The approach can be generally applied to other transitions, such as transitions to an invasive and/or a metastatic phenotype, or transitions to drug-resistant phenotypes. The open question is whether it can be used to anticipate those transitions. Furthermore, these results suggest that the approach can be used to characterize the intratumor heterogeneity via accurate resolution of the cellular subpopulations and the unbalanced processes they harbor (**Figure 8.4**)—the course that we are currently pursuing.

6. Using Surprisal Analysis to Predict a Direction of Change in Biological Processes

This section is dedicated to a series of studies that show how a thermodynamic-based interpretation of biological processes enables us to predict cellular behaviors (Kravchenko-Balasha et al., 2014; Shin et al., 2011).

For example, Kravchenko-Balasha et al. (2014) shows how a thermodynamic-based interpretation of cell–cell interactions via the cell–cell protein signaling reveals a role for those interactions in determining cellular superstructures and allows the prediction of cell–cell spatial distributions.

The architecture of tumors plays an important role in tumor development. Understanding these structures and the forces that drive their formation is of high importance in cancer research and therapy. For example, diffuse architectures of the aggressive brain tumors, glioblastomas, in which cells infiltrate within the brain, are the main reason for the therapeutic failure of these tumors (Holland, 2000).

To help understand cell-to-cell spatial organization, we developed a methodology (Kravchenko-Balasha et al., 2014) that combines *two-cell* functional proteomics (in which levels of different proteins are measured in isolated pairs of the cells) and surprisal analysis. We found that signal transduction in two interacting glioblastoma (GBM) cancer cells depends on the cell–cell separation distance. Using thermodynamic-based analysis of the protein concentrations as a function of cell–cell distance in two interacting cells, we were able to identify the cell–cell separation distance that corresponded to the steady state of the cell–cell protein signaling. In other words, we were able to identify the cell–cell distance range in which constraints were equal to 0. Thus, we predicted that this distance range should occur with higher probability in multicellular populations (when many cells are grown together in the same environment). Indeed, this separation distance was found to be the dominant cell separation distance range in multicellular population. We therefore predicted that aggressive GBM cells would exhibit a scattered distribution, whereas less aggressive GBM cells would closely pack (Kravchenko-Balasha et al., 2014), consistent with experimental observations of others *in vivo* (Inda et al., 2010).

Furthermore, we have recently demonstrated that the thermodynamic-based characterization of the cell–cell signaling imbalances permits the accurate prediction and control of the direction of movement of invasive cancer cells (Kravchenko-Balasha, Shin et al., 2016). Here I provide a brief description of this study.

Controlling cell migration is important in tissue engineering and medicine. Cell motility depends on factors such as nutrient concentration gradients and cell–cell signaling via soluble factors.

We sought a physical-based approach, which allows controlling the direction of cell–cell movement in a predictive fashion. We hypothesized that the time-course measurements of proteins involved in cell–cell communication in isolated pairs of GBM cells as a function of cell–cell distance and a parallel detection of the cell–cell relative motion over time will give us a clue about how this motion can be controlled.

A single-cell barcode chip (SCBC, described in **Figure 8.5A**) was used to experimentally interrogate several secreted proteins (proteins secreted by a cell to mediate cell–cell communication) in hundreds of isolated glioblastoma brain

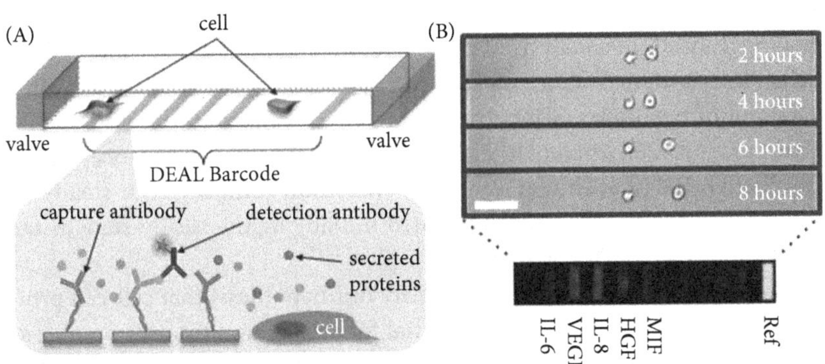

Figure 8.5 Schematics of the experimental details, adapted from (Kravchenko-Balasha, Shin et al., 2016). **(A)** Drawing of a single cell microchamber that captures a pair of GBM cells. Each microchamber includes a fluorescent-antibody based technology that allows to measure the levels of different proteins in pairs of GBM cells (Bottom). Five assayed proteins were measured at the end of the experiment (8h). For each pair of the cells we measured the cell-cell distance every 2 hour **(B)**. A representative time-lapse image of a two-cell chamber over 8 h is shown. (Scale bar: 100 μm)

cancer cell pairs and to monitor their relative motion over time. Following a two-hour acclimation period, cell movements within the individual microchambers were tracked using microscopy imaging over a period of six hours (**Figure 8.5B**). At the end of this period, specific secreted proteins were captured on designated elements of miniaturized *antibody arrays* that were patterned within each microchamber (**Figure 8.5A**, lower panel). This antibody array allows us to measure expression levels of the proteins of interest. The levels of five secreted proteins, involved in cell–cell signaling and communication (interleukin (IL)-6, IL-8, vascular endothelial growth factor [VEGF], hepatocyte growth factor [HGF], and macrophage migration inhibitory factor [MIF]) were measured as a function of cell–cell distance. We used surprisal analysis of *the protein levels* to determine the most stable separation distance between two cells, which corresponds to the steady state of cell–cell signaling (Kravchenko-Balasha, Shin et al., 2016). This distance is the most stable because a change of the cell–cell separation distance in either direction would lead to an increase in the free energy of cell–cell signaling. Thus, the direction of cellular motion is predicted to be the direction that reduces the influence of the constraints operating on the system, leading to a more stable cell–cell separation. Using experimental protein expression levels as a function of cell–cell distance, we resolved two distance-dependent constraints. The amplitudes of the constraints were significantly reduced at a cell–cell distance of ~200 μm (Kravchenko-Balasha, Shin et al., 2016). This distance is predicted to be the most probable cell–cell separation distance, since it yields the most stable cell–cell protein signaling. Indeed, the analysis of cell trajectories indicated that more cell pairs reached a separation distance of ~200 μm after eight hours relative to the initial time point of two hours of incubation. Moreover, cells that had an initial cell–cell distance of ~200 μm (at 2h) did not change their cell–cell distance over time (Kravchenko-Balasha, Shin et al., 2016). This result confirmed our steady-state prediction and shows that cells separated initially by the steady-state distance have the lowest cell-cell potential.

Potential of the cell movement. To verify our hypothesis that cell-cell protein signaling is what determines the potential for cell motion towards a stable separation distance, we examined changes in cell–cell separation distances (Δcell–cell r) in two-hour intervals (**Figure 8.6B**), and compared the results to those obtained for isolated single cells. Cells initially located at separations of < 200 μm showed close to Gaussian distribution of cell–cell displacements for $\Delta t = 2$h (**Figure 8.6B**). A clear development of a tail toward higher positive values is evident in the time intervals $\Delta t = 4$h and $\Delta t = 6$h (**Figure 8.6B**). On the other hand, cell pairs initially located at > 200 μm exhibited a tendency to reduce the cell–cell distance (i.e., negative Δcell–cell r) over time (Kravchenko-Balasha, Shin et al., 2016). This suggests that the cell–cell interaction provides a potential

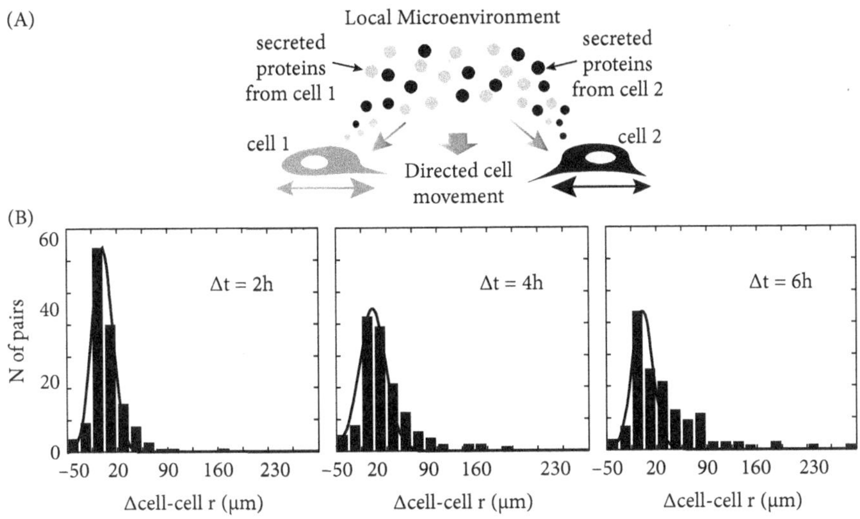

Figure 8.6 Evidence for the role of cell-cell potential, adapted from (Kravchenko-Balasha, Shin et al., 2016). (**A**) Directed cell movement results from the influence of proteins secreted to the local microenvironment on cells. (**B**) Histograms of changes in the cell-cell separation distance relative to the initial value (Δcell-cell r). The initial measurement was taken 2 hours following cell adaptation to the chip. Shown are results for time intervals of $\Delta t = 2h$, 4h and 6h for cell pairs that were initially separated by distances smaller than our predicted steady state distance, 200 μm. The histograms were fitted to a Gaussian distribution, aiming to highlight the deviations with time. The fit to a Gaussian was acceptable at the shortest time point, ($R^2 = 0.95$ for $\Delta t = 2h$) and deteriorated at longer time points ($R^2 = 0.89$ for $\Delta t = 4h$; $R^2 = 0.7$ for $\Delta t = 6h$). The strong development of asymmetry with time is evidence for the role of a cell-cell potential.

gradient that *directs* cell–cell motion. Our results suggested that the cell–cell motion could be described as Brownian (random) motion (as represented by close to Gaussian distribution) biased by cell–cell potential (as represented by the deviation from the Gaussian distribution; see **Figure 8.6B**).

To verify that hypothesis, we solved the Langevin equation for 5,000 simulated cell–cell trajectories. The Langevin equation describes heavily damped motion (as in our case, the motion of the cells) that includes two parts: one part describes the random motion, and the second part the motion biased by the potential (in our case, cell–cell potential). In other words, we proposed that the equations of motions that are used to simulate the general case of high-friction motion apply to cell–cell movement force, which takes the following form (Kravchenko-Balasha, Shin et al., 2016):

$$\frac{dr}{dt} = \underbrace{-\frac{1}{\gamma}\frac{dU(r)}{dr}}_{\substack{\text{movement due to}\\ \text{cell–cell potential}}} - \underbrace{\sqrt{2D}R(t)}_{\text{random motion}}$$

$r(t)$ is the cell–cell distance as a function of time, $U(r)$ is the cell–cell potential as determined by the signaling, and γ is the friction coefficient. The higher the friction, the less effective is the potential $U(r)$. The random motion is described by the isotropic random force $R(t)$, and it is the only force term when the cell is isolated. The movement under a random force alone leads to a Gaussian distributed displacements.

Surprisal analysis of the protein data set revealed the existence of a free-energy gradient, with an energy minimum at a cell separation of 200 μm. Taking this distance range as the range with the minimal cell–cell potential, we solved the Langevin equation for 5,000 simulated cell pairs that were allowed to move 8 hours as in the experiment. The theoretical computations (**Figure 8.7**) accurately reproduced the experimental measurements (**Figure 8.6**), providing strong evidence that the movement of *interacting* cells follows the rules of Brownian dynamics: away from the stable separation distance, the signaling is unbalanced and the cell motion is described by two tendencies, random motility and directed movement due to a cell–cell potential gradient. Gradients in protein secretion levels create the cell–cell potential. In contrast, examination of the motion of isolated single cells reveals that the motion of noninteracting cells is subject to a nondirectional random force (Brownian-like motion, Kravchenko-Balasha, Shin et al., 2016).

Restraining the unbalanced signaling inhibits the directed motion. To verify that unbalanced signaling induces *directed* motion, we inhibited the two proteins

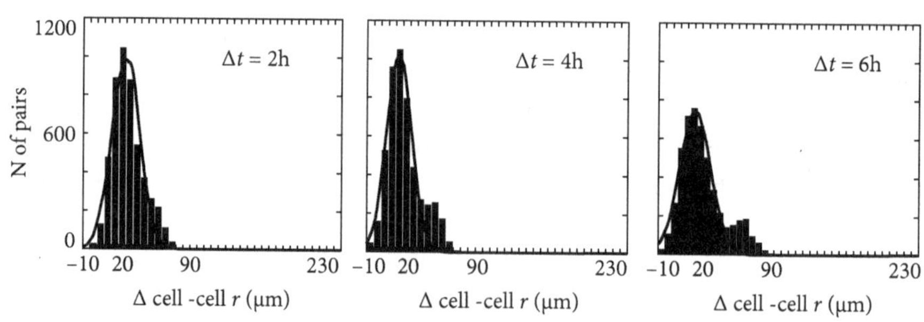

Figure 8.7 Theoretical histograms of cell–cell displacements, generated using 5,000 trajectories calculated using Langevin equation, adapted from (Kravchenko-Balasha, Shin et al., 2016). The cell-cell potential was determined by the observed cell–cell signaling vs. distance. The potential used was harmonic about the steady-state separation of 200 μm.

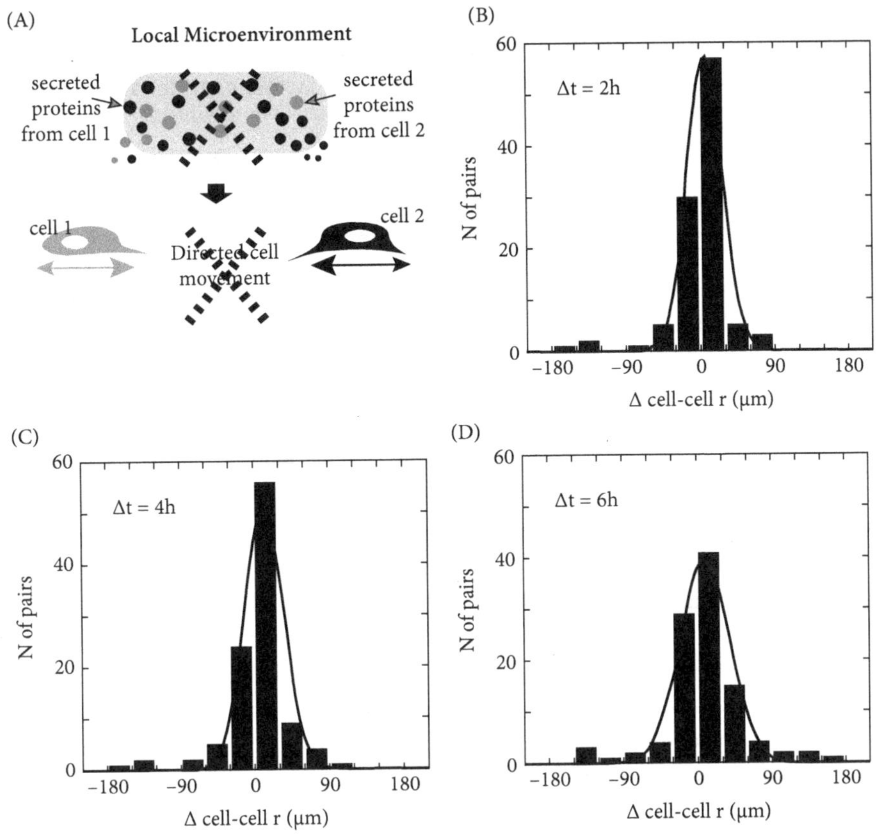

Figure 8.8 Cell–cell motion after treatment with neutralizing antibodies against IL-6 and HGF, adapted from (Kravchenko-Balasha, Shin et al., 2016). (**A**) Schematic illustration of the local microenvironmental condition. (**B–D**) Following the inhibition of signaling, the changes in the distribution of the cell–cell displacements (Δcell–cell r) after $\Delta t = 2$ h (**B**), $\Delta t = 4$ h (**C**), and $\Delta t = 6$ h (**D**). The results shown are for about 150 U87EGFR cell pairs that were initially separated by less than 200 μm. The measured displacements (Δcell–cell r) were binned into histograms. The histograms were fitted to Gaussian distribution ($R^2 > 0.95$). Even after a delay of 6 h, the histogram of the cell–cell distances could be well fitted by a Gaussian distribution ($R^2 = 0.97$).

most responsible for the free-energy gradient, IL-6 and HGF (**Figure 8.8**). This signaling inhibition resulted in the complete loss of directed movement, such that the cells moved in a Brownian-like manner (**Figure 8.8**), similar to the movement of single isolated cells. In other words, we have determined a cell–cell potential that defines *the direction and extent of motion* in a manner that is similar to other two-body interacting systems in physics and chemistry.

In summary, we showed that soluble factor signaling between two cells can define a free-energy gradient, which, in turn, directs the relative cell motion. Experimental control of the levels of these soluble factors provides a handle for controlling cellular motion in a predictive fashion.

7. Summary

A series of studies discussed in this chapter provide an example of how the molecular alterations at multiple levels (bulk and single-cell levels) can be interpreted using high-throughput technologies and information-theoretic surprisal analysis to resolve signaling imbalances in cancer systems. Use of the thermodynamic-based information theory to interpret biological systems can be powerful because, as in chemistry and physics, it enables accurate prediction of biological behaviors. We demonstrate that, similar to nonequilibrium systems in chemistry and physics, the direction of change in cancer processes can be predicted and that this knowledge can be used to manipuate cancer phenotypes (Kravchenko-Balasha, Shin et al., 2016; Flashner-Abramson et al, 2018).

The analysis provides an accurate, concise structure of network imbalances, through identification of a unique set of distinct unbalanced processes in each sample. An example for patient-specific unbalanced protein network structures was provided in the subsection "Using Surprisal Analysis to Understand Intertumor Heterogeneity." We believe that this approach can be utilized to rationally design patient-specific combination therapies, a direction that we are actively pursuing (Flashner-Abramson et al, 2019).

In general, this multidisciplinary approach can provide a predictive experimental-theoretical framework for studying complex, nonequilibrium, healthy and diseased systems. Reduction of multiple variable data sets to a small number of physical parameters by modeling biological processes using fundamental physicochemical laws allows us to accurately identify stable and unbalanced states in the system. Using this knowledge, we can generate rationalized strategies for the manipulation of those states, and thus for the manipulation of pathological phenotypes.

References

Alter, O., Brown, P. O., and Botstein, D. (2000). "Singular Value Decomposition for Genome-Wide Expression Data Processing and Modeling." *Proceedings of the National Academy of Sciences USA* 97(18):10101–10106.

Alyass, A., Turcotte, M., and Meyre, D. (2015). "From Big Data Analysis to Personalized Medicine for All: Challenges and Opportunities." *BMC Medical Genomics* 8(1): 33.

Bansal, M., and Bernardo, D. (2007). "Inference of Gene Networks from Temporal Gene Expression Profiles." *IET Systems Biology* 1(5): 306–312.

Ben-Shaul, A., Levine, R. D., and Bernstein, R. B. (1972). "Entropy and Chemical Change. II. Analysis of Product Energy Distributions: Temperature and Entropy Deficiency." *Journal of Chemical Physics* 57(12): 5427.

Creixell, P., et al. (2015). "Pathway and Network Analysis of Cancer Genomes." *Naturemethods* 12(7): 615–621.

Drake, J. M., et al. (2016). "Phosphoproteome Integration Reveals Patient-Specific Networks in Prostate Cancer." *Cell* 166(4): 1041–1054.

Fisher, R., Pu.sztai, L., and Swanton, C. (2013). "Cancer Heterogeneity: Implications for Targeted Therapeutics." *British Journal of Cancer* 108(3): 479–485.

Flashner-Abramson, E., Abramson, J., White, F. M., and Kravchenko-Balasha, N. (2018). "A Thermodynamic-Based Approach for the Resolution and Prediction of Protein Network Structures." *Chemical Physics* 514: 20–30.

Flashner-Abramson, E., Vasudevan, S., Adejumobi, I. A., Sonnenblick, A., and Kravchenko-Balasha, N. (2019). "Decoding Cancer Heterogeneity: Studying Patient-Specific Signaling Signatures Towards Personalized Cancer Therapy." *Theranostics* 9(18): 5149–5165.

Friedman, N., Linial, M., Nachman, I., and Pe'er, D. (2000). "Using Bayesian Networks to Analyze Expression Data." *Journal of Computational Biology* 7(3–4): 601–620.

Golan, A., and Lumsdaine, R. L. (2016). "On the Construction of Prior Information—An Info-Metrics Approach." In Gloria Gonzalez-Rivera, R. Carter Hill, and Tae-Hwy Lee (eds.), *Essays in Honor of Aman Ullah (Advances in Econometrics, Volume 36)*. Emerald Group Publishing Limited, pp. 277–314.

Gross, A., and Levine, R. D. (2013). "Surprisal Analysis of Transcripts Expression Levels in the Presence of Noise: A Reliable Determination of the Onset of a Tumor Phenotype." *PLoS One* 8(4): e61554.

Holland, E. C. (2000). "Glioblastoma Multiforme: The Terminator." *Proceedings of the National Academy of Sciences USA* 97(12): 6242–6244.

Hood, L., and Flores, M. (2012). "A Personal View on Systems Medicine and the Emergence of Proactive P4 Medicine: Predictive, Preventive, Personalized and Participatory." *NewBiotechnology* 29(6): 613–624.

Inda, M. M., et al. (2010). "Tumor Heterogeneity Is an Active Process Maintained by a Mutant EGFR-Induced Cytokine Circuit in Glioblastoma." *Genes and Development* 24(16): 1731–1745.

Jolliffe, I. T. (2002). *Principal Component Analysis*. 2nd ed. New York: Springer. Available at: http://www.loc.gov/catdir/enhancements/fy0817/2002019560-t.html.

Koomen, J. M., et al. (2008). "Proteomic Contributions to Personalized Cancer Care." *Molecular and Cellular Proteomics* 7(10): 1780–1794.

Kravchenko-Balasha, N., et al. (2011). "Convergence of Logic of Cellular Regulation in Different Premalignant Cells by an Information Theoretic Approach." *BMC Systems Biology* 5: 42.

Kravchenko-Balasha, N., et al. (2012). "On a Fundamental Structure of Gene Networks in Living Cells." *Proceedings of the National Academy of Sciences USA* 109(12): 4702–4707.

Kravchenko-Balasha, N., Johnson, H., White, F. M., Heath, J. R., and Levine, R. D. (2016). "A Thermodynamic Based Interpretation of Protein Expression Heterogeneity in Different GBM Tumors Identifies Tumor Specific Unbalanced Processes." *Journal of Chemistry and Physics* 120(26): 5990–5997.

Kravchenko-Balasha, N., Shin, Y. S., Sutherland, A., Levine, R. D., and Heath, J. R. (2016). "Intercellular Signaling through Secreted Proteins Induces Free-Energy Gradient-Directed Cell Movement." *Proceedings of the National Academy of Sciences USA* 113(20): 5520–5525.

Kravchenko-Balasha, N., Simon, S., Levine, R. D., Remacle, F., and Exman, I. (2014) "Computational Surprisal Analysis Speeds-Up Genomic Characterization of Cancer Processes." *PLoS One* 9(11): e108549.

Kravchenko-Balasha, N., Wang, J., Remacle, F., Levine, R. D., and Heath, J. R. (2014). "Glioblastoma Cellular Architectures Are Predicted through the Characterization of Two-Cell Interactions." *Proceedings of the National Academy of Sciences USA* 111(17): 6521–6526.

Lee, M. J., et al. (2012). "Sequential Application of Anticancer Drugs Enhances Cell Death by Rewiring Apoptotic Signaling Networks." *Cell* 149(4): 780–794.

Levchenko, A., and Nemenman, I. (2014). "Cellular Noise and Information Transmission." *Current Opinion in Biotechnology* 28: 156–164.

Levine, R. D. (2005). *Molecular Reaction Dynamics*. Cambridge: The University Press.

Levine, R. D., and Bernstein, R. B. (1974). "Energy Disposal and Energy Consumption in Elementary Chemical Reactions: Information Theoretic Approach." *Accounts of Chemical Research* 7(12): 393–400.

Leyens, L., et al. (2014). "Quarterly of the European Observatory on Health Systems and Policies." *Eurohealth International Eurohealth Incorporating Euro Observer* 20(3): 41–44.

Logue, J. S., and Morrison, D. K. (2012). "Complexity in the Signaling Network: Insights from the Use of Targeted Inhibitors in Cancer Therapy." *Genes and Development* 26(7): 641–650.

Mar, J. C., Wells, C. A., and Quackenbush, J. (2011). "Defining an Informativeness Metric for Clustering Gene Expression Data." *Bioinformatics* 27(8): 1094–1100.

McQuarrie, D. A. (2000). *Statistical Mechanics* (University Science Books). 1st ed. New York: Harper&Row Publishers.

Nelson, D. L., Albert, L. Lehninger, A. L., and Cox, M. M. (2008). *Lehninger Principles of Biochemistry*, 5th ed. New York: Macmillan.

Nykter, M., et al. (2008). "Critical Networks Exhibit Maximal Information Diversity in Structure-Dynamics Relationships." *Physical Review Letters* 100(5): 058702.

Pemovska, T., et al. (2013). "Individualized Systems Medicine Strategy to Tailor Treatments for Patients with Chemorefractory Acute Myeloid Leukemia." *Cancer Discovery* 3(12): 1416–1429.

Poovathingal, S. K., Kravchenko-Balasha, N., Shin, Y. S., Levine, R. D., and Heath, J. R. (2016). "Critical Points in Tumorigenesis: A Carcinogen-Initiated Phase Transition Analyzed via Single-Cell Proteomics." *Small* 12(11):1425–1431.

Remacle, F., Kravchenko-Balasha, N., Levitzki, A., and Levine, R. D. (2010). "Information-Theoretic Analysis of Phenotype Changes in Early Stages of Carcinogenesis." *Proceedings of the National Academy of Science USA* 107(22): 10324–10329.

Rhodes, D. R., et al. (2004). "Large-Scale Meta-Analysis of Cancer Microarray Data Identifies Common Transcriptional Profiles of Neoplastic Transformation and Progression." *Proceedings of the National Academy of Science USA* 101(25): 9309–9314.

Robin, X., et al. (2013). "Personalized Network-Based Treatments in Oncology." *Clinical Pharmacology and Therapeuticsr* 94(6): 646–650.

Shaked, Y. (2016). "Balancing Efficacy of and Host Immune Responses to Cancer Therapy: The Yin and Yang Effects." *Nature Reviews. Clinical Oncology* 13(10): 611–626.

Shannon, C. E. (1948). "A Mathematical Theory of Communication." *Bell Labs Systems Technical Journal* 27(July 1928): 379–423.

Shi, Q., et al. (2012). "Single-Cell Proteomic Chip for Profiling Intracellular Signaling Pathways in Single Tumor Cells." *Proceedings of the National Academy of Sciences USA* 109(2): 419–424.

Shin, Y. S., et al. (2011). "Protein Signaling Networks from Single Cell Fluctuations and Information Theory Profiling." *Biophysics Journal* 100(10): 2378–2386.

Szklarczyk, D., et al. (2011). "The STRING Database in 2011: Functional Interaction Networks of Proteins, Globally Integrated and Scored." *Nucleic Acids Research* 39(Database issue): D561–8.

van Dijk, A. D. J., Lähdesmäki, H., de Ridder, D., and Rousu, J. (2016). "Selected Proceedings of Machine Learning in Systems Biology: MLSB 2016." *BMC Bioinformatics* 17(S16): 51–52.

Vasudevan, S., Flashner-Abramson, E., Remacle, F., Levine, R. D., and Kravchenko-Balasha, N. (2018). "Personalized Disease Signatures through Information-Theoretic Compaction of Big Cancer Data." *Proceedings of the National Academy of Sciences USA* 115(30): 7694–7699.

Waltermann, C., and Klipp, E. (2011). "Information Theory Based Approaches to Cellular Signaling." *Biochimica Biophysica Acta* 1810(10): 924–932.

Wei, W., et al. (2016). "Single-Cell Phosphoproteomics Resolves Adaptive Signaling Dynamics and Informs Targeted Combination Therapy in Glioblastoma." *Cancer Cell* 29(4): 563–573.

Weinstein, I. B. (2002). "Cancer. Addiction to Oncogenes—The Achilles Heal of Cancer." *Science* 297(5578): 63–64.

Yarden, Y., and Sliwkowski, M. X. (2001). "Untangling the Erbb Signalling Network." *Nature Reviews Molecular Cell Biology* 2(2):127–137.

Yeung, M. K. S., Tegnér, J., Collins, J. J., Tegner, J., and Collins, J. J. (2002). "Reverse Engineering Gene Networks Using Singular Value Decomposition and Robust Regression." *Proceedings of the National Academy of Sciences USA* 99(9): 6163–6168.

Zadran, S., Arumugam, R., Herschman, H., Phelps, M. E., and Levine, R.D. (2014). "Surprisal Analysis Characterizes the Free Energy Time Course of Cancer Cells Undergoing Epithelial-to-Mesenchymal Transition." *Proceedings of the National Academy of Sciences USA* 111(36): 13235–13240.

Zadran, S., Remacle, F., and Levine, R. D. (2013). "miRNA and mRNA Cancer Signatures Determined by Analysis of Expression Levels in Large Cohorts of Patients." *Proceedings of the National Academy of Science USA* 110(47): 19160–19165.

Zhu, J-J, and Wong, E. T. (2013). Personalized medicine for glioblastoma: Current challenges and future opportunities. *Current Molecular Medicine* 13(3): 358–367.

9

Forecasting Socioeconomic Distributions on Small-Area Spatial Domains for Count Data

Rosa Bernardini Papalia and Esteban Fernandez-Vazquez

1. Introduction

Statistical information for empirical analysis is frequently available at a higher level of aggregation than is desired. The spatial disaggregation of socioeconomic data is considered complex owing to the inherent spatial properties and relationships of the spatial data, namely, spatial dependence and spatial heterogeneity. Spatial disaggregation procedures are proposed in the literature as useful tools to help researchers make the best use of aggregated data in studies of land use and agriculture (Chakir, 2009). These studies are similar to those that have been used in statistics for analyzing contingency tables with known marginals or in procedures based on moment restrictions often used in cases where data from a small, potentially nonrepresentative data set can be supplemented with auxiliary information from another data set that may be larger and/or more representative of the target population (Imbens and Hellerstein, 2010). The moment restrictions yield weights for each observation that can subsequently be used in weighted regression analysis.

When sample data refers to areas or locations, two scenarios have to be addressed: spatial autocorrelation between observations and spatial heterogeneity in relations. It follows that the standard assumptions of traditional statistical methods—that data values are derived from independent observations or that a single relationship with constant variance exists across the sample data—are no longer guaranteed (Anselin, 1988, 1990; LeSage, 1999). An adequate procedure is to implement spatial models that allow assessing the magnitude of space influences by introducing a specific weighting scheme in which relationships among spatial areas are specified (Anselin, 2010). The spatial pattern of the data is defined according to the choice of a spatial weights or contiguity matrix where the underlying structure is generally defined by 0–1 values, with the value 1

Rosa Bernardini Papalia and Esteban Fernandez-Vazquez, *Forecasting Socioeconomic Distributions on Small-Area Spatial Domains for Count Data* In: *Advances in Info-Metrics: Information and Information Processing across Disciplines.* Edited by: Min Chen, J. Michael Dunn, Amos Golan, and Aman Ullah, Oxford University Press (2021).

DOI: 10.1093/oso/9780190636685.003.0009

assigned to spatial units having a common border or within a critical cutoff distance (Fischer, 2006).

In this chapter, we focus on model specifications for areal spatial cross-sectional data, typically count data. In the literature, two main alternatives for count data are commonly used. The first one refers to use of the traditional spatial econometric models for continuous data requiring count data transformation to meet the model's assumptions. The second alternative refers to use of adequate model specifications that incorporate spatial lag autocorrelation in modeling counts (Lambert, Brown, and Florax, 2010). The currently available spatial models are based on the normality assumption and on specific prior distributions within Bayesian hierarchical models, which are sometimes inappropriate under skew data or when the sample sizes in the areas are not large enough to rely on the Central Limit Theorem.

This chapter seeks to illustrate the usefulness and flexibility of the information- theoretic (IT)-based estimation methods in relation to all sources of uncertainties and in the presence of researchers' ignorance about data-sampling processes, model functional forms, and error correlation structures (Bhati, 2008). The idea is to illustrate the use of IT methods in estimating small-area spatial models for count areal data that do not force the analyst to commit to any parametric distributional assumptions. Our specific objective is to propose information–theoretic-based disaggregation methods using in-sample and out-of-sample data for estimating count data at a detailed spatial scale, generally characterized by small sample sizes. We consider, additionally, the potential effect of data affected by spatial dependencies or spatial heterogeneities. The objective is to illustrate, develop, and implement IT-based estimation methods (Golan et al., 1996; Golan, 2008, 2018) to produce "reliable" estimates and to forecast socioeconomic distributions for small geographic areas and/or small demographic domains when the parameter of interest is a small-area count. Simulation studies are carried out in order to validate the estimation method and to test the properties of the proposed estimators. The Monte Carlo experiment will include assessment of the performance of some ancillary/auxiliary data (from administrative records or surveys) in producing the desired estimates.

The rest of this chapter is organized as follows. The next section summarizes the main spatial perspectives for areal data. Spatial models for count data are introduced in section 3 by reviewing the relevant literature. The IT model formulation and estimation are described in section 4. In section 5, the results of a series of Monte Carlo experiments are presented. In section 6, we introduce an empirical application finalized to the estimation of unemployment levels of immigrants at municipal scale in Madrid in 2011. section 7 concludes with a brief discussion of the fi ndings and enumerates promising potential directions for future work.

2. Spatial Perspectives for Areal Data

Spatial data reflect geographical information that can take into account the effect of spatial influences in explanation of the phenomenon of interest. Observations for which the spatial arrangement, the absolute location, and/or the relative position is explicitly taken into account are termed spatial data. Three main classes of spatial data can be distinguished: (1) Geostatistical or spatially continuous data, that is, observations associated with a continuous variation measure over space (given observed values at fixed sampling points, (2) areal or lattice data related to the discrete measured attribute over space, and (3) spatial point patterns, where the objects are point locations at which events of interest have occurred (Cressie 1993; Besag 1974).

When sample data have a location component, two scenarios have to be considered: spatial autocorrelation between observations and spatial heterogeneity in relations. Spatial structures generally associated with absolute location effects refer to the impact on each unit—the effect of being located at a particular point in space—while those structures associated with relative location effects consider relevant the position of one unit relative to other units.

The substantive implications of properly modeling spatial heterogeneity and dependence are linked with methodological issues. First, estimation of models incorporating spatial heterogeneity poses both identification and collinearity problems (due to the correlation between unobserved individual specific effects and explanatory variables). Several solutions have been proposed to deal with this kind of problem in the continuous setting case (Bernardini Papalia and Fernandez Vazquez, 2018; Peeters and Chasco, 2006).

2.1 Spatial Dependence

Spatial dependence (relative location effects) is traditionally modeled by assuming a given structure in the form on which spatial spillovers are produced. The most commonly assumed spatial structures are given by a spatial autoregressive process in the error term and/or a spatially lagged dependent variable, or a combination of both. A spatial error model (SEM) specification assumes that the spatial autocorrelation is modeled by a spatial autoregressive process in the error terms: spatial effects are assumed to be identical within each unit, but all the units are still interacting spatially through a spatial weight matrix. The presence of spatial dependence is then associated with random shocks (due to the joint effect of misspecification, omitted variables, and spatial autocorrelation). Alternatively, a spatial autoregressive (SAR) model specification—also called a spatial lag model—assumes that all spatial dependence effects are captured by the lagged term. The spatial autocorrelation is then modeled by including a spatially lagged dependent variable. Global and local measures of spatial autocorrelation are computed to determine whether the data exhibit spatial dependence, and a series of test statistics based on the Lagrange Multiplier (LM) or Rao Score (RS) principle are used to determine whether the variables in the model sufficiently capture the spatial dependence in the data. If the variables do not fully model the dependence, the diagnostics indicate whether the researcher should estimate a model with a spatially lagged dependent variable, a spatially lagged error term, or both. The LM/RS principle can also be extended to more complex spatial alternatives, such as higher-order processes and spatial error components models.

Spatial association (spatial autocorrelation), corresponds to situations where observations or spatial units are nonindependent over space. To identify this association, a scatterplot where each value is plotted against the mean of neighboring areas—the Moran's scatterplot or a spatial autocorrelation statistic such as Moran's I or Geary's C—can be used. Moran's I is a measure of global spatial autocorrelation, while Geary's C is more sensitive to local spatial autocorrelation. Both of these statistics require the choice of a spatial weights or contiguity matrix, W, that represents our understanding of spatial association among all area units (Fischer 2006). Usually, $w_{ii} = 0$, $i = 1, \ldots, n$, where n is the number of spatial units, but for $i = j$, the association measure between area i and area j, w_{ij}, can be defined in many different ways, being the most usual definition the minimum distance between areas. In order to capture dependencies across spatial units, spatial variables are introduced in the model specification. These spatial variables can be weighted averages of neighboring dependent variables; the weighted average of neighboring explanatory variables; or errors where the definition of neighbors is carried out through the specification of the spatial weights matrix W (Anselin, 2010; Arbia, 2006; Le Sage, 1999).

2.2 Spatial Heterogeneity

The unobserved spatial heterogeneity (absolute location effects) can be introduced by assuming (1) slope heterogeneity across spatial units, implying that parameters are not homogeneous over space but vary over different geographical locations and (2), the presence of cross-sectional correlation due to the presence of some common immeasurable or omitted factors.

The term *spatial heterogeneity* refers to variation in relations over space, due to the structural instability or nonstationarity of relationships (Le Sage, 1999; Le Sage, Pace 2009). Heterogeneity can be related to the spatial structure or to the spatial process generating data. In this perspective, additional information that may be provided by the spatial structure, such as heteroscedasticity, spatially varying coefficients, random coefficients, and spatial structural change, can be introduced (Anselin, 2010). Spatial heterogeneity can be identified by detecting (1) structural instability (by using the Chow test) and (2) heteroscedasticity (by using the Breusch–Pagan test; Anselin 2010).

2.3 The Role of the Map

The most distinctive element of spatial models is the existence of a map. In regional sciences and spatial econometrics, the map is typically geographical in nature, indicating where the units of interest are located. Spatial models typically assume that proximity on the map implies high correlation in the response variables. The map is multidimensional (two or more dimensions) and can imply a rich variety of spatial relationships.

The selection of appropriate spatial weights is of crucial importance in spatial modeling and implies an assumption about the variables that determine the relative similarities of units. Estimation results may depend on the choice of this matrix. There are several approaches to define spatial relations between two locations or spatial units, but they are usually classified into two main groups: the spatial contiguity approach and the distance-based approach. Typical types of neighboring matrices for the spatial contiguity approach are the linear, the rook, the bishop, and the queen contiguity matrices $\boldsymbol{W}$. Regarding the distance approach, we have, for example, the k-nearest neighbors or the critical cutoff neighborhood matrices (Le Sage, 1999). Following is a summary of the main characteristics of the criteria enumerated before to define the $\boldsymbol{W}$ matrix:

- Contiguity Matrix: Represents an $n \times n$ symmetric matrix, where $w_{ij} = 1$, when i and j are neighbors and 0 when they are not. By convention, the diagonal elements are set to zero. W is usually standardized so that all rows sum to one.
  - Linear contiguity: Define $w_{ij} = 1$ for regions that share a common edge to the immediate right or left to the region of interest;
  - Rook contiguity: Two regions are considered neighbors if they share a common border, and only for these regions w_{ij} is set to 1.
  - Bishop contiguity: Define $w_{ij} = 1$ only for regions that share a common vertex.
  - Queen contiguity: Regions that share a common border or a vertex are considered neighbors, and for these $w_{ij} = 1$.
- Distance Approach: makes direct use of the latitude-longitude coordinates associated with spatial data observations (Arbia, 2006).

- Critical Cutoff Neighborhood: Two regions , i and j,are considered neighbors if $0 \leq d_{ij} < d^*$, being d_{ij} the appropriate distance adopted between regions, and d^* representing the critical cutoff or threshold distance, beyond which no direct spatial influence between spatial units is considered.
- k-Nearest Neighbor: Given the centroid distances from each spatial unit i to all units $j = i$ ranked as $d_{ij}(1) < d_{ij}\ (2) < \cdots < d_{ij}\ (n-1)$, for each $k = 1, 2, \ldots, n-1$, the set $Nk(i) = \{j(1), j(2), \ldots, j(k)\}$ contains the k closest units to i, and for each given k, the k-nearest neighbor matrix, has the form: $w_{ij} = 1, j\ (i) \in Nk(i), i \in 1, \ldots, n$, and is zero otherwise.

3. Spatial Models for Count Data

Spatial models can represent three different types of spatial patterns. First, spatial models can capture spatial lags: the idea that the units are directly affected by other units is present in models in which units are economic agents which are known to interact during the choice process. Second, spatial models can capture spatially correlated errors, the idea that important latent variables that drive behavior of the unit can be inferred from unit proximity on the map (when omitted variables induce a form of spatial dependence). In this case, there exists a statistical adjustment for missing variables that determine the response variable. Third, spatial models can capture spatial drift, the idea that model parameters are a function of a unit's location on the map. Models can be regarded as a representation of unobserved heterogeneity in which the parameters follow a spatial process.

For studying the spatial patterns in count data, where the dependent variable consists in a count of positive integers (Billé and Arbia, 2016), several types of spatial models may be employed by taking into account the asymmetric distributions and a potential high portion of zeros that characterize this type of data.

For modeling count data, one approach proposed in the literature is to convert the count parameter of interest into an approximately continuous variable and to use spatial econometric models for continuous data, as a spatial autoregressive (SAR) model, or a spatial error (SEM) model, or mixed models with both spatial lag and spatial error considered simultaneously. As a consequence, this approach demands count data transformation to meet the standard model's assumptions necessary for estimation and for use of the standard statistical tests. This is the case of the transformation of a Poisson loglinear model to a linear log model, given a Poisson distribution for the number of count outcomes observed in each spatial unit. In such a case, the most used transformation is the log transformation by taking into account that the dependent variable needs to be converted into a rate variable by inclusion of an offset. These models have many applications, not only to the analysis of counts of events, but also in the context of models for contingency tables, and to the analysis of survival data. A linear model is then fitted to the transformed data, and the conversion into a rate variable allows comparisons between the results obtained through the Poisson loglinear model (fitted before the covariates also had to be transformed) and the spatial econometrics models. This solution is motivated by the complexity and computational burden required by the direct estimation of spatial discrete choice models (multidimensional integration problems).

Recent developments in spatial econometric approaches for count data have been proposed, suggesting a count estimator that models the response variable as a function of neighboring counts such as a spatial autoregressive Poisson model (SAR-Poisson) (Lambert, Brown, and Florax, 2010). In the presence of overdispersion, the Poisson model underestimated the actual zero frequency and the right-tail values of the distribution, and the negative binomial (NB) models are usually adopted (Cameron and Trivedi, 1986; Haining, Griffith, and Law, 2009).

When the presence of zeros does not depend on the unobserved heterogeneity, the most used formulations are (1) the zero-inflated models, which consist of a mixed formulation that gives more weight to the probability of observing a zero or (2) hurdle models, which are characterized by two independent probability processes, which are generally a binary process and a Poisson process. Alternatively, modified hurdle models have been developed by Winkelmann (2004) and Deb and Trivedi (2002).

For count data defined into spatial units of the lattice, some alternative models can be specified, where the spatial dependency structure is defined conditionally. The large-scale variation data is normally integrated in the model through a regression component, which is added to the structure of the mean of observations. Part of the spatial autocorrelation can be modeled by including known covariate risk factors in a regression model, but it is common for spatial structures to remain in the residuals after accounting for these covariate effects. For modeling the residual autocorrelation, the most common feature is to expand the linear predictor with a set of spatially correlated random effects as part of a hierarchical model (Banerjee, Carlin, and Gelfand, 2004). The random effects are usually represented by a conditional autoregressive model that includes a priori spatial autocorrelation through the contiguity structure of the spatial units. A recent review of literature, including different methodological solutions in spatial discrete models and several applied fields, is included in Billé and Arbia (2016).

4. Information-Theoretic Methods for Spatial Count Data Models: GCE Area Level Estimators

This section proposes IT estimators of count data based on direct—that is, sample counts—observations of the counts of interest. It is partially based on the discussion of previous work by Bhati (2008) and on more recent research on small-area model estimation of means presented in Bernardini-Papalia and Fernandez-Vazquez (2018) and in Golan (2018). Our discussion here focuses on the problem of the estimation of a small-area count y_i that represents the unknown parameter of interest (at the population level), but for which an estimate in the sample is available.

Let us assume that the population sizes N_i are known and that we have D samples for a set of $i = 1,..,D$ small areas and direct estimates are observable and denoted as c_i. These counts indicate the number of units sampled that hold a given characteristic of interest in sample i.

Let us also assume that some aggregate information for the whole set of D areas is available as well (this represents the available ancillary information's contribution). For example, this is the typical case where some external source provides reliable estimates of the aggregate count $Y = \sum_{i=1}^{D} y_i$, even when the counts for each one of the populations i are unknown.

The typical estimates $\hat{y}_i$ for the population counts y_i are based in the sample counts c_i and are calculated as:

$$\hat{y}_i = c_i\left(\frac{N_i}{n_i}\right) \tag{9.1}$$

where n_i represents the number of units sampled relative to the population i.

Note that these estimates do not guarantee consistency with the observable information available at an aggregate level, Y. A generalized cross-entropy (GCE)-based estimator (Golan, Judge, and Miller, 1996) can be applied in this context as an effective way of adjusting the estimates considered in (9.1) to make them consistent with the additional out-of-sample information contained in Y. We proceed by adding a random term to each sample count c_i that accounts for the possible discrepancies between the information observed in the samples and the aggregate. Denoting this term by e_i, we assume it as the realization of a random variable that can take a range of M possible values contained in a vector $b'_i = \left[b_{i1},\ldots,b_i^*,\ldots,b_{iM}\right]$. Each support vector $\boldsymbol{b}_i$ will be assumed to be different for each area i, and it will account for a potential adjustment in our initial estimate.

In order to illustrate this idea, let us consider the simplest case with $M = 2$ values. In this case, the support vector for e_i will be given by $\boldsymbol{b}'_i = [b_{i1}, b_{i2}]$, where b_{i1} indicates the minimum possible value of this adjustment, while b_{i2} represents the maximum value for a correction in c_i. Given the uncertainty about the process under study and the nature of count of the variable of interest, we can set values for these bounds in an intuitive way as $b_{i1} = -c_i$ and $b_{i2} = n_i - c_i$. These limits represent an adjustment in c_i that leads to its minimum (0) and maximum (n_i) possible realizations in the sample. Once these possible realizations for each e_i term have been specified, a probability distribution $\boldsymbol{p}'_i = [p_{i1}, p_{i2}]$ should be assigned to them and estimated.

In this context, the GCE estimator of y_i will be given by the expression:

$$\hat{y}_i^{GCE} = [c_i + \hat{e}_i]\left(\frac{N_i}{n_i}\right), \tag{9.2}$$

where

$$\hat{e}_i = \sum_{m=1}^{M} \hat{p}_{im} b_{im} \tag{9.3}$$

The additional information represented by Y can be easily included in a GCE estimation as follows:

$$\underset{p}{Min}\, KL(p/q) = -\sum_{i=1}^{D}\sum_{m=1}^{M} p_{im} \ln(p_{im}/q_{im}) \tag{9.4}$$

subject to:

$$\sum_{m=1}^{M} p_{im} = 1; i = 1, \ldots, D \tag{9.5}$$

$$\sum_{i=1}^{D}\left[c_i + \sum_{m=1}^{M} p_{im} b_{im}\right]\left(\frac{N_i}{n_i}\right) = Y \tag{9.6}$$

Constraints (9.5) are normalization constraints, while (9.6) contains the additional aggregate information. The objective function to minimize is the Kullback–Leibler divergence between our target probability distributions $\boldsymbol{p}_i$, and some a priori distribution $\boldsymbol{q}_i.\boldsymbol{q}_i$ should contain the prior value we would assign to the term e_i.

To recover the probability vectors $\boldsymbol{p}_i$, the Lagrangian function would be:

$$L=\sum_{i=1}^{D}\sum_{m=1}^{M} p_{im}\ln\left(p_{im}/q_{im}\right)+\lambda\left[Y-\sum_{i=1}^{D}\left[c_i+\sum_{m=1}^{M} p_{im}b_{im}\right]\left(\frac{N_i}{n_i}\right)\right]$$
$$+\mu_i\left[1-\sum_{m=1}^{M} p_{im}\right] \tag{9.7}$$

With the first-order conditions:

$$\frac{\partial L}{\partial p_{im}}=\ln\left(p_{im}\right)-ln\left(q_{im}\right)+1-\lambda b_{im}\left(\frac{N_i}{n_i}\right)-\mu_i=0; \tag{9.8}$$
$$i=1,\dots,D; m=1,\dots,M$$

$$\frac{\partial L}{\partial \lambda}=Y-\sum_{i=1}^{D}\left[c_i+\sum_{m=1}^{M} p_{im}b_{im}\right]\left(\frac{N_i}{n_i}\right)=0 \tag{9.9}$$

$$\frac{\partial L}{\partial \mu_i}=1-\sum_{m=1}^{M} p_{im}=0; i=1,\dots,D \tag{9.10}$$

The solution to this system of equations and parameters yields the following:

$$\hat{p}_{im}=\frac{q_{im}\exp\left[\lambda b_{im}\left(\frac{N_i}{n_i}\right)\right]}{\Omega\left(\hat{\lambda}\right)}=\frac{q_{im}\exp\left[\hat{\lambda} b_{im}\left(\frac{N_i}{n_i}\right)\right]}{\sum_{m=1}^{M} q_{im}\exp\left[\hat{\lambda} b_{im}\left(\frac{N_i}{n_i}\right)\right]} \tag{9.11}$$

where $\Omega\left(\hat{\lambda}\right)=\sum_{m=1}^{M} q_{im}\exp\left[\hat{\lambda} b_{im}\left(\frac{N_i}{n_i}\right)\right]$ is a normalization factor and $\hat{\lambda}$ is the estimate of the Lagrange multiplier associated with constraint (9.6), which contains the additional information. As a consequence, the constrained optimization problem depicted in Eqs. (9.4) to (9.6) can be formulated in terms of the unconstrained- dual $L(\lambda)=\lambda Y-\ln\Omega(\lambda)$, depending only on the parameter λ. In the absence of any additional information as the aggregate in (9.6), there are no reasons to correct the sample counts c_i at all, and the estimates produced by this GCE solution are equivalent to the estimates in (9.1). Consequently, the elements of $\boldsymbol{q}_i$ should be specified in a way that results in prior values of e_i equal to zero. For illustration purposes but without loss of generality, let us again consider the simplest case with $M=2$ values with $b_{i1}=-c_i$ and$b_{i2}=n_i-c_i$.

In this case, it is straightforward to find that the elements in the a priori distribution $\boldsymbol{q}'_i = [q_{i1}, q_{i2}]$ that verify this condition should be calculated as

$$q_{i1} = 1 - (c_i/n_i) \text{ and } q_{i2} = c_i/n_i. \tag{9.12}$$

5. Simulation Experiments

We illustrate how the GCE estimator works, taking as our reference the artificial count data generated for a set of forty-seven spatial units. In our simulation, we randomly generate the population size on each unit i as $N_i \sim U(500, 2000)$, while the units sampled in each case will be proportional to the original population size. More specifically, the sample sizes for each area of interest were generated as $n_i = N_i \times N(0.1, 0.01)$, which produces sample sizes approximately equal to 10 percent of the original population size.[1] The counts at the population level y_i (to be estimated) were generated from a binomial process as $y_i \sim B(N_i, p)$, where the parameter p will vary depending on the specific scenario. The N_i, n_i and y_i elements were set as constant through the simulation draws. On each sample n_i, the sample counts were simulated to distribute as their equivalent elements in the population, being the c_i counts generated as $c_i \sim B(n_i, p)$ and varying on each draw of the simulation.

The population counts y_i are unknown, but the population sizes N_i and the aggregate Y for the forty-seven small areas of interest are assumed as known. Based on this sample and on out-of-sample information, the GCE estimator defined in Eqs. (9.4) to (9.6) is computed. In all experiments, we opted for the simplest case depicted before, and we have set support vectors for the e_i terms with $M = 2$ points as $\boldsymbol{b}'_i = [-c_i, n_i - c_i]$, with corresponding prior distributions as in (9.12). As an indicator of accuracy, we employed the average empirical mean squared error (AEMSE), which is defined as the mean of the squared differences between actual population counts and their corresponding estimates.[2] We consider different scenarios that account for different levels of uncertainty in the data-generation process, as well as spatial heterogeneity and spatial dependence. On each scenario considered, we generate 1,000 draws

[1] Given that this procedure will produce noninteger sample sizes, these were conveniently rounded to the closest integer.

[2] In order to assess the quality of the estimates provided by each estimator, we have considered the AEMSE, which is obtained by taking the mean value of all the EMSEs obtained for all areas: The AEMSE provides a global measure of the quality of the estimates, and the results for different estimators can be compared: the lower the AEMSE, the better the estimates fit to the real values.

(samples) and compare the AEMSE of the estimates based on sample counts and the GCE estimators.[3]

5.1 Scenario 1: GCE in a Spatial Homogeneous Process

In this simple case, we generate the population counts y_i as explained before, and we assume that the parameter p in the binomial distribution is constant across the $D = 47$ spatial units. In order to account for different levels of uncertainty in the process, we set different values of this parameter and calculate the corresponding AEMSE for the two estimators under evaluation. Table 9.1 summarizes the outcomes of the simulation.

Not surprisingly, values of parameter p closer to the case of maximum uncertainty—that is, $p = 1 - p = 0.5$—increase the deviation indicators for both estimators. Nevertheless, the results of the experiment under any of the values considered for parameter p suggest that the count estimates obtained by the

Table 9.1 Average Empirical Mean-Squared Error (1,000 draws)

		$\hat{y}_i$	$\hat{y}_i^{GCE}$
p	**0.1**	1325.555	1269.501
	0.2	2379.582	2335.308
	0.3	2731.347	2676.778
	0.4	3139.169	3085.364
	0.5	3338.338	3270.762
	0.6	3421.277	3359.648
	0.7	2973.667	2899.258
	0.8	2197.155	2131.774
	0.9	1314.188	1271.260

Note: p denotes the parameter in the binomial distribution $y_i \sim B(N_i, p)$; the column $\hat{y}_i$ reports the AEMSE corresponding to the direct estimator from sample counts; and $\hat{y}_i^{GCE}$ shows the AEMSE for the estimates obtained by GCE.

[3] In our exercise, the GCE estimator is compared only with a direct estimation based on sample counts. Other indirect estimators that connect small areas through a linking model based on auxiliary information are not considered for the sake of simplicity. This type of estimator requires additional information (predictors of the variable of interest) and larger sample sizes.

proposed GCE estimator are able to correct the initial estimates based only on sample information, and the inclusion of the out-of-sample data represented by the information at an aggregate level, Y, produces, on average, lower deviation measures in all the cases.

5.2 Scenario 2: GCE in a Spatial Heterogeneous Process

The scenario depicted earlier represents a situation where the data-generating process is homogeneous across space; that is, in all the D spatial units under study the number of counts in the population distribute as $y_i \sim B(N_i, p)$. In order to introduce some heterogeneity in this stochastic process, in this subsection we will assume that each area i has its own idiosyncratic behavior and the populations counts are generated as $y_i \sim B(N_i, p_i)$. This heterogeneity in the parameters p_i is present in the sample counts as well, now distributing the corresponding $c_i's$ as $c_i \sim B(n_i, p_i)$. More specifically, we allow the p_i parameters to range from 0.3 to 0.7, distributing uniformly in this range across the $D = 47$ spatial units. The remaining conditions of the numerical experiment stay unchanged with respect to the first scenario.

Again, we will compare the accuracy of the estimates of the sample counts and the GCE-based estimates. Given that the sample counts c_i play an important role in both the support vectors $\boldsymbol{b}_i$ and the prior distributions $\boldsymbol{q}_i$, the spatial heterogeneity that we introduce in the process should affect the conditions of the GCE estimates. Table 9.2 summarizes the results of this second experiment.

Table 9.2 reports the value of the p parameter, together with the population and sample sizes, which are kept constant through the simulation draws, together with the AEMSE value for each spatial unit. The patterns found in the first scenario of spatial homogeneity are not affected, generally speaking, and we found that the GCE-based estimator produces on average lower deviation indicators than the estimates based only on sample counts. The average value found for the AEMSE is similar to the errors found in the previous experiments when parameter p took values from 0.3 to 0.7 homogeneously across all the spatial units. The flexibility of the proposed GCE estimator seems to incorporate the spatial heterogeneity into the support vectors and the prior distribution and produces AMESEs that are on average lower. Even when the variability is a bit higher than the estimates based only on sample counts (1,352 for the GCE and 1,347 for the estimates), the AEMSE obtained by applying the GCE estimator is lower for thirty-eight out of the forty-seven populations than that produced by estimates based solely on the sample information.

Table 9.2 Average Empirical Mean-Squared Error by Spatial Unit (1,000 draws)

Spatial unit	p_i	N_i	n_i	$\hat{y}_i^{GCE}$	$\hat{y}_i$
1	0.509	1308	101	2851.917	2951.309
2	0.465	1094	89	2865.957	3012.416
3	0.444	1321	109	3057.352	3111.748
4	0.370	1690	147	3965.466	4029.287
5	0.442	1286	107	3204.479	3221.457
6	0.570	1971	158	7590.878	7236.469
7	0.341	1035	89	2732.591	2711.916
8	0.413	1829	104	4359.303	4377.284
9	0.637	753	203	1620.058	1630.628
10	0.433	1897	162	4684.542	4799.325
11	0.657	881	157	2667.785	2664.069
12	0.495	1407	195	3136.77	3242.236
13	0.609	1167	164	2764.587	2918.711
14	0.688	1673	62	3437.704	3586.458
15	0.603	1963	167	4559.996	4528.275
16	0.352	1135	152	2366.113	2472.41
17	0.541	1871	124	4461.646	5057.875
18	0.403	1745	85	5818.447	5685.72
19	0.368	1787	127	4058.534	4004.35
20	0.408	1847	236	4015.085	4356.79
21	0.474	1403	103	4230.106	4480.261
22	0.451	692	84	1682.54	1708.855
23	0.444	1806	151	4401.089	4545.794
24	0.536	1897	188	4777.641	4853.952
25	0.360	724	131	1662.863	1677.531
26	0.471	552	220	1343.095	1355.315
27	0.591	562	164	1241.99	1251.116
28	0.608	1041	167	2947.621	3068.559
29	0.321	1232	67	2359.833	2432.632
30	0.497	729	104	1744.571	1756.664
31	0.557	1264	177	3240.022	3269.468
32	0.366	743	73	1666.211	1674.28
33	0.450	1723	81	4327.463	4355.646
34	0.708	1642	82	3497.716	3781.644
35	0.453	611	64	1538.825	1569.874
36	0.382	1858	109	5289.252	5217.836
37	0.441	1835	61	4136.085	4454.779
38	0.458	1680	125	5088.123	4959.327
39	0.355	1244	96	2746.501	2800.39
40	0.655	1176	190	2901.515	2949.412
41	0.509	850	111	1908.633	1949.568
42	0.614	549	109	1472.844	1497.382
43	0.421	1084	105	2925.693	2926.714
44	0.468	1500	183	3559.88	3809.476
45	0.508	620	102	1383.17	1397.464
46	0.604	1226	67	2906.771	2952.254
47	0.630	640	150	2082.582	2075.864
Total		**60,543**	**6002**	**3218.763**	**3284.485**

Note: p_i denotes the parameter in the binomial distributions $y_i \sim B(N_i, p)$ and $c_i \sim B(n_i, p_i)$;N_i shows the population sizes and n_i the sample size on i, respectively. As in Table 9.1, the column $\hat{y}_i$ reports the AEMSE corresponding to the direct estimator and $\hat{y}_i^{GCE}$ shows the AEMSE for the estimates obtained by GCE.

5.3 Scenario 3: GCE in a Process of Spatial Dependence

In this last scenario, we simulate the potential effect of a process of spatial dependence on the data generation. For this purpose, we specify a spatial structure on the set of $D = 47$ spatial units used in the experiment. In particular, we assume that these areas correspond to the spatial division of Spain in NUTS-3 regions—Spanish provinces, excluding the islands. The conditions for simulating population and sample sizes, N_i and n_i, respectively, are the same as those in the previous scenarios. The counts at the population level y_i to be estimated, however, are generated now in a process that includes spatial heterogeneity and spatial dependence.

More specifically, the population counts y_i are assumed to be affected by a SAR process, where the counts in one area i are conditioned by the counts in neighbor areas. The data-generating process is assumed to be:

$$y_i = \rho \sum_{j \neq i}^{D} w_{ij} y_j + B(N_i, p_i) = \rho \sum_{j \neq i}^{D} w_{ij} y_j + \varepsilon_i \qquad (9.13)$$

or, in matrix terms:

$$y = \rho W y + \varepsilon = [I - \rho W]^{-1} \varepsilon. \qquad (9.14)$$

where ε_i stands for a stochastic process that generates counts in the population i following a binomial distribution, but that now represents only part of the counts. The other part of the values of y_i is given as a weighted average of the counts in the neighbor areas $\sum_{j \neq i}^{D} w_{ij} y_j$ multiplied by a parameter ρ that measures the strength of this spatial effect.[4]

As explained before, w_{ij} is the typical element of a matrix of spatial contiguity W with null cell elements in its main diagonal ($w_{ii} = 0$). For our experiment, we have defined two alternative W matrices according to two different contiguity criteria for the $D = 47$ Spanish NUTS-3 regions. One is based on the criterion of five-nearest neighbors, specifying that the spatial unit j is classified as a neighbor of i—and consequently $w_{ij} > 0$—if j is located within the five-closest regions to i, taking the Euclidean distance between their centroids. Otherwise, i and j are not considered as neighbors, and the term w_{ij} is set to zero. The second criterion is based on a distance decay function between regions, setting w_{ij} as $w_{ij} = d_{ij}^{-2}$, being d_{ij} the Euclidean distance between the centroid of regions i and j. Note

[4] Again, this procedure generates noninteger results, so the initial values simulated were rounded to the closest integer.

that the first definition of the spatial weighting matrix produces "sparse" W matrices, while the second criterion leads to "dense" W matrices. In both cases, the elements of these matrices have been conveniently row-standardized. The spatial parameter ρ is naturally bounded to be $|\rho| < 1$, and in this experiment we set values of −0.05, −0.10, −0.20, 0.05, 0.10, and 0.20 to simulate different intensities and types of correlation among regions in the SAR process.

Additionally, the sample counts are also affected by the spatial autocorrelation as well, and now they distribute as $c_i \sim B\left(n_i, p_i^s\right)$, where $p_i^s = y_i / N_i$. In other words, the binomial model that produces counts at the sample level is conditioned in the parameters p_i^s by the population counts y_i that are spatially autocorrelated. As in the previous cases, we compare the performance of the proposed GCE estimator with the traditional method of relying on the sample counts by calculating the AEMSE. The conditions regarding the availability of additional information are the same as before, assuming that the total count $Y = \sum_{i=1}^{D} y_i$ and the population sizes N_i are both observable.[5] Table 9.3 summarizes the outcomes of the experiment.

Table 9.3 reports the AEMSE under each one of the W matrices generated and for the different values of parameter ρ. Generally speaking, the results are not much different from the previous cases where the errors produced by

Table 9.3 Average Empirical Mean-Squared Error: Spatial Dependence Process (1,000 draws)

		$\hat{y}_i$	$\hat{y}_i^{GCE}$
$\rho = -0.20$	**W5-nearest**	16376.342	9668.563
	Winverse square distance	12917.346	8451.825
$\rho = -0.10$	**W5-nearest**	5491.971	4186.049
	Winverse square distance	6444.147	4286.229
$\rho = -0.05$	**W5-nearest**	4234.902	3399.808
	Winverse square distance	3838.230	3400.169
$\rho = 0.05$	**W5-nearest**	4519.476	3624.388
	Winverse square distance	3780.437	3192.515
$\rho = 0.10$	**W5-nearest**	7522.995	5247.854
	Winverse square distance	7838.294	4784.393
$\rho = 0.20$	**W5-nearest**	7658.885	5417.671
	Winverse square distance	28428.855	14853.592

Note: The values of parameter ρ measures the positive or negative spatial autocorrelation as in Eq. (9.13); the column $\hat{y}_i$ reports the AEMSE corresponding to the direct estimator based on sample counts, and $\hat{y}_i^{GCE}$ shows the AEMSE for the estimates obtained by GCE.

[5] Note that the total $Y = \sum_{i=1}^{D} y_i$ acting as a constraint is the sum of y_i now generated as in Eq. (9.13).

the GCE estimator were lower than a simple estimate based on sample counts. The presence of a spatial autocorrelation process increases the gains produced by including additional information that corrects the initial estimates and makes them consistent with this observable aggregate. Given that the presence of this type of spatial effects is likely to be happening when dealing with real data—the counts in one spatial unit are likely to be correlated with the counts in neighbor locations—these results show the usefulness of the GCE estimator that improves the accuracy of our estimates by accounting for the out-of-sample information.

6. An Empirical Application: Estimating Unemployment Levels of Immigrants at Municipal Scale in Madrid, 2011

This section illustrates how IT estimators of count data can be applied in the context of a real-world example. The small area count y_i to be estimated in this case is the number of unemployed immigrants living in the municipalities of the region of Madrid, Spain, in 2011. The only database containing information about the labor status and nationality status—distinguishing between immigrant and national individuals—at a detailed spatial scale is the Spanish Population Census (SPC) conducted in 2011. The sample corresponding to the province of Madrid in the SPC was formed by approximately 180,000 individuals who were surveyed about their socioeconomic characteristics.

The sample surveyed approximately 18,000 adult female and 14,000 adult male immigrants; roughly 5,000 of the female immigrants were classified as unemployed versus 3,200 male unemployed immigrants. Besides these regional figures, the heterogeneity across the municipalities of the regions was huge: the data in the sample of the SPC estimated that the capital city of Madrid concentrated more than 50 percent of the unemployed immigrants, while approximately 35 percent of the unemployed immigrants lived in thirty-two midsize municipalities larger than 20,000 inhabitants, and less than 15 percent lived in rural areas of less than 20,000 inhabitants. Because of data privacy issues, the microdata on the SPC identifies place of residence of the individuals only if they live in a municipality with a population larger than 20,000. This allows estimating counts of male and female unemployed immigrants for only 33 out of the 179 municipal areas of the Madrid region. For those municipalities smaller than 20,000 inhabitants, only average estimates for four population intervals are possible.[6]

[6] These population intervals are municipalities between 10,000 and 20,000; between 5,000 and 10,000; between 2,000 and 5,000; and, finally, those municipalities with populations smaller than 2,000 inhabitants.

We can approach the information contained in the SPC microdata as $D = 33 + 4 = 37$ samples for small areas, for which we have counts c_i of male and female-unemployed immigrants. The population sizes N_i—that is, total number of adult male or female immigrants—for the 179 municipalities are known from the annual municipal registers. Additionally, the Labor Force Survey (LFS) provides us with regional aggregates on the number of male and female unemployed immigrants for the region of Madrid.[7] While this database is the reference for studying labor market issues in Spain, the geographical detail that it allows is limited, and it is unable to get estimates for labor market indicators for spatial units more detailed than NUTS3 regions (provinces).

The information on the regional aggregates of the LFS, however, can be incorporated as constraints to a GCE estimator, as the one depicted in Eqs. (9.2) to (9.6). In this case, the LFS provides us with the aggregate count $Y = \sum_{i=1}^{D} y_i$, being y_i , the unobservable true municipal counts of male or female unemployed immigrants. The GCE estimator can be applied to adjust the initial estimates on the SPC, calculated as in (9.1), to make them consistent with the regional aggregates Y on the LFS. Continuing with the simplest case explained before with $M = 2$ values, the support vector for e_i will be given by the two points $b_{i1} = -c_i$ and $b_{i2} = n_i - c_i$, where c_i and n_i denote, respectively, the count estimates and the sample size for each municipality—or class of municipality—sampled in the SPC. With regard to the a priori distribution $\boldsymbol{q}_i$, it will be set as formed by the point probabilities $q_{i1} = 1 - (c_i/n_i)$ and $q_{i2} = c_i/n_i$, which reflect our initial guess about the absence of adjustment if no out-of-sample information is included in the GCE program in the form of an aggregate constraint as in Eq. (9.6).

These aggregate figures included in the right-hand side of Eq. (9.6) correspond to the estimates for the region of Madrid as contained in the LFS for 2011; the estimates were of 95,400 male unemployed immigrants and 69,800 female unemployed immigrants. Consequently, the count estimates c_i are adjusted by the GCE estimator, by estimating the probabilities of the term e_i, to make them consistent with these aggregates. The results of this process are summarized in Tables 9.4 and 9.5.

These tables contain the outcome of the estimation process, together with some characteristics of the thirty-three municipalities that can be uniquely

[7] See http://www.ine.es/en/inebaseDYN/epa30308/docs/resumetepa_en.pdf for more information on this survey.

Table 9.4 Estimates of Male Unemployed Immigrants y Municipality

Municipality	N_i	n_i	c_i	$\hat{y}_i$	$\hat{y}_i^{GCE}$
Alcalá de Henares	19,297	466	100	4,141	4,124
Alcobendas	6,746	197	46	1,575	1,573
Alcorcón	9,156	285	68	2,185	2,180
Algete	1,221	43	15	426	426
Aranjuez	3,504	106	21	694	694
Arganda del Rey	6,813	210	34	1,103	1,101
Arroyomolinos	442	11	3	121	121
Boadilla del Monte	1,500	84	27	482	482
Ciempozuelos	1,877	59	9	286	286
Colmenar Viejo	3,071	90	19	648	648
Collado Villalba	5,461	196	47	1,310	1,308
Coslada	9,697	206	41	1,930	1,926
Fuenlabrada	11,846	372	93	2,962	2,954
Galapagar	2,936	98	20	599	599
Getafe	11,592	332	62	2,165	2,159
Leganés	10,180	314	67	2,172	2,167
Madrid (Capital City)	232,665	7,317	1,639	52,117	49,627
Majadahonda	3,738	171	46	1,006	1,005
Mejorada del Campo	1,430	30	5	238	238
Móstoles	12,787	355	85	3,062	3,054
Navalcarnero	1,390	46	12	363	363
Parla	13,999	320	70	3,062	3,053
Pinto	2,648	110	16	385	385
Pozuelo de Alarcón	2,668	116	29	667	667
Rivas-Vaciamadrid	3,988	101	26	1,027	1,026
Rozas de Madrid, Las	3,443	167	34	701	700
San Fernando de Henares	3,209	60	10	535	534
San Sebastián de los Reyes	4,122	137	35	1,053	1,052
Torrejón de Ardoz	11,011	302	57	2,078	2,073
Torrelodones	729	43	14	237	237
Valdemoro	4,181	138	29	879	878
Villaviciosa de Odón	718	30	4	96	96
Tres Cantos	35	74	24	11	11
Between 10,000 and 20,000 (15 municipalities)	33,100	348	64	401	401
Between 5,000 and 10,000 (31 municipalities)	3,292	343	91	335	335
Between 2,000 and 5,000 (34 municipalities)	1,321	655	166	873	873
Less than 2,000 (66 municipalities)	1,807	509	113	6,087	6,043
Regional totals: male unemployed immigrants estimated				**98,011**	**95,400**

Note: N_i, n_i and c_i denotes, respectively the population sizes, the sample sizes and the sample counts on each municipality i. $\hat{y}_i$ shows the direct estimator based on sample counts and $\hat{y}_i^{GCE}$ reports the estimates obtained by GCE.

Table 9.5 Estimates of Female Unemployed Immigrants by Municipality

Municipality	N_i	n_i	c_i	$\hat{y}_i$	$\hat{y}_i^{GCE}$
Alcalá de Henares	16,820	515	156	5,095	4,568
Alcobendas	7,986	258	85	2,631	2,504
Alcorcón	8,919	349	106	2,709	2,558
Algete	1,322	39	15	508	505
Aranjuez	3,261	110	43	1,275	1,252
Arganda del Rey	5,680	211	52	1,400	1,346
Arroyomolinos	522	22	8	190	189
Boadilla del Monte	2,240	152	55	811	800
Ciempozuelos	1,771	85	26	542	536
Colmenar Viejo	3,055	143	43	919	901
Collado Villalba	5,414	229	74	1,750	1,692
Coslada	8,791	249	73	2,577	2,434
Fuenlabrada	10,448	419	142	3,541	3,322
Galapagar	3,080	142	39	846	829
Getafe	10,710	410	114	2,978	2,773
Leganés	9,376	350	113	3,027	2,855
Madrid (Capital City)	250,038	9,521	2,534	66,547	8,982
Majadahonda	5,277	243	64	1,390	1,341
Mejorada del Campo	1,398	45	16	497	493
Móstoles	11,587	432	135	3,621	3,364
Navalcarnero	1,238	50	18	446	442
Parla	11,678	372	131	4,112	3,834
Pinto	2,571	136	42	794	781
Pozuelo de Alarcón	4,236	241	72	1,266	1,232
Rivas-Vaciamadrid	3,914	133	33	971	945
Rozas de Madrid, Las	4,905	274	81	1,450	1,405
San Fernando de Henares	3,052	80	21	801	785
San Sebastián de los Reyes	4,787	196	60	1,465	1,421
Torrejón de Ardoz	10,002	347	122	3,517	3,312
Torrelodones	1,139	57	13	260	258
Valdemoro	4,170	176	44	1,043	1,013
Villaviciosa de Odón	933	32	12	350	348
Tres Cantos	44	94	23	11	11
Between 10,000 and 20,000 (15 municipalities)	32,016	440	149	479	474
Between 5,000 and 10,000 (31 municipalities)	3,078	431	153	365	363
Between 2,000 and 5,000 (34 municipalities)	1,076	778	264	1,093	1,073
Less than 2,000 (66 municipalities)	1,575	612	186	10,842	8,860
Regional totals: female unemployed immigrants estimated				**132,115**	**69,800**

Note: N_i, n_i, and c_i denote, respectively, the population sizes, sample sizes, and sample counts of each municipality i. $\hat{y}_i$ shows the direct estimator based on sample counts, and $\hat{y}_i^{GCE}$ reports the estimates obtained by GCE.

identified in the SPC sample and the four types of smaller areas that are grouped according to their population. For each spatial unit, the first column (N_i) reports the total number of immigrants according to the municipal registers. The second column (n_i) shows the number of individuals sampled in the SPC, while the column labeled c_i indicates the number of individuals in the sample who are classified as immigrant and unemployed simultaneously. The fourth column ($\hat{y}_i$) reports the estimates that can be calculated from the sample in the SPC, calculated as in Eq. (9.1). If these initial estimates are summed across the municipalities of the region, this sum is not consistent with the official aggregates published in the LFS for 2011. The differences in the case of the male unemployed immigrants are relatively small (less than 3 percent of deviation), but for the case of the female immigrants, these deviations are sizable: while the official figures contained in the LFS for the region of Madrid were of 69,800 female unemployed immigrants, the initial estimates almost double this aggregate and would indicate that this figure goes up to more than 130,000 individuals. Moreover, the estimate for the capital city of Madrid (66,547) virtually equals the regional aggregate contained in the LFS.

The GCE estimator proposed in this work allows for correcting these deviations by incorporating the out-of-sample information given in the form of the aggregates observable in the LFS. These GCE-based estimates are reported in the last column of both tables ($\hat{y}_i^{GCE}$). Note that these estimates are relatively similar to the estimates of the counts $\hat{y}_i$ in the case of the table for the male immigrants, but the adjustments are much more substantial in the case of the females. Clearly, this is a consequence of the large discrepancy between the regional figures published in the LFS and the estimates of the counts from the SPC for the female immigrants. The size of these adjustments are, however, highly heterogeneous across municipalities, and they are mainly concentrated on correcting the estimate for the capital city of Madrid. The size of this particular correction, as well as for the rest of the spatial units, depends on the choice made for the points included in the supporting vectors for e_i and the a priori probabilities of these points ($\mathbf{q}_i$). These choices can be naturally modified. It would be possible to allow for a reduced range between b_{i1} and b_{i2}, which will produce smaller adjustments of the initial estimates—if, for example, we had confidence on the information from the sample of SPC. Alternatively but complementarily, if we think that the initial estimates are upper or lower biased, we could also select a priori distributions $\mathbf{q}_i$ different from the initial point of departure that assumes a null correction. It would be possible, for example, to give more prior weight to positive (negative) adjustment if we suspect that the sample estimate underestimates (overestimates) the true counts.

7. Conclusions

This chapter focuses on count data models incorporating spatial structures used to produce small-area estimates in the presence of ancillary information at an aggregate level.

Within this framework, the general basic problem is that any functional transformation of the count data model in the presence of spatial structures leads to inconsistent estimates. This is because the expected value of the functional transformation of a random variable depends on higher-order moments of its distribution. Therefore, for example, if the errors are heteroskedastic, the transformed errors will be generally correlated with the covariates.

To address this kind of problem in estimating small-area count parameters of interest, we propose an IT-based estimation method and assess its performance using Monte Carlo simulations. The focus of the experiments presented here is on alternative scenarios representing different levels of uncertainty in the data-generation process that are: (1) a spatial homogeneous process, (2) a spatial heterogeneous process, and (3) a process of spatial dependence.

The results suggest that the IT methods produce in all cases more accurate estimates of the unknown small-area population count of interest, if compared to a simple estimator based only on sample counts.

We found that, in the presence of a spatial heterogeneous process, the IT method is robust to different patterns of heterogeneity in the data-generation process. In addition, in the presence of a spatial autocorrelation process, the gains derived, including additional information to the estimation that correct the initial estimates and make them consistent with this observable aggregate, seem to be increased. Finally, our Monte Carlo simulation results suggested that the proposed estimators should be used as a substitute for the standard methods in the presence of spatial structure in the count data. The empirical application that estimates the number of unemployed immigrants living in the municipalities of Madrid, Spain, in 2011 suggests the importance of accounting for the information on the regional aggregates of the Labour Force Survey, which is incorporated as constraints and allows for an adjustment of the initial sample estimates to make them consistent with these aggregates.

Future direction of our research includes the generalization of the IT model specification to deal with cross-sectional time series data by allowing time to impact the model components where the errors are correlated over time. These models, known as spatiotemporal models, provide considerable fl exibility in capturing different types of dependence. They also offer a promising direction for new work in empirical applications. Extended IT formulations that impose restrictions from theories by adding constraints based on some specific properties of the functions (Golan, Judge, Perloff, 1996; Wu, Sickles, 2018) represent an extension of our future work on economic applications."

References

Anselin, L. (1988). *Spatial Econometrics: Methods and Models.* Dordrecht, The Netherlands: Kluwer.

Anselin, L. (1990). "Spatial Dependence and Spatial Structural Instability in Applied Regression Analysis." *Journal of Regional Science*, 30: 185–207.

Anselin, L. (2010). "Thirty Years of Spatial Econometrics." *Papers in Regional Science*, 89: 3–26.

Arbia, G. (2006). *Spatial Econometrics, Statistical Foundations and Applications to Regional Convergence.* Berlin: Springer-Verlag.

Banerjee, S., Carlin B. P., and Gelfand, A. E. (2004). *Hierarchical Modeling and Analysis for Spatial Data.* Boca Raton, FL: Chapman and Hall/CRC.

Bernardini Papalia, R., and Fernandez Vazquez, E. (2018). "Information Theoretic Methods in Small Domain Estimation." *Econometrics Reviews*, 37(4): 347–359.

Besag, J. (1974). "Spatial Interaction and the Statistical Analysis of Lattice Systems (With Discussion)". *Journal of the Royal Statistical Society* B, 36(2): 192–236.

Bhati, A. (2008). "A Generalized Cross-Entropy Approach for Modeling Spatially Correlated Counts." *Econometric Reviews*, 27: 574–595.

Billé, A. G., and Arbia, G. (2016). "Spatial Discrete Choice and Spatial Limited Dependent Variable Models: A Review with an Emphasis on the Use in Regional Health Economics." Mimeo.

Chakir, R. (2009). "Spatial Downscaling of Agricultural Land-Use Data: An Econometric Approach Using Cross Entropy." *Land Economics*, 85: 238–251.

Cameron, A. C., and Trivedi, P. K. (1986). "Econometrics Models Based on Count Data: Comparisons and Applications of Some Estimators and Tests." *Journal of Applied Econometrics*, 1: 29–53.

Cressie, N. (1993). *Statistics for Spatial Data.* New York: John Wiley.

Deb, P., and Trivedi, P. K. (2002). "The Structure of Demand for Health Care: Latent Class versus Two-Part Models." *Journal of Health Economics*, 21: 601–625.

Fischer, M. (2006). *Spatial Analysis and Geo Computation.* New York: Springer.

Golan, A. (2008): *Information and Entropy Econometrics—A Review and Synthesis, Foundations and Trends in Econometrics.* Now Publishers.

Golan, A. (2018). *Foundation of Info-Metrics: Modeling, Inference, and Imperfect Information.* Oxford: Oxford University Press.

Golan, A., Judge, G., and Miller, D. (1996). *A Maximum Entropy Econometrics: Robust Estimation with Limited Data.* New York: John Wiley.

Haining, R., Griffith, D., and Law, J. (2009). "Modelling Small Area Counts in the Presence of Overdispersion and Spatial Autocorrelation." *Computational Statistics and Data Analysis*, 53(8): 2923–2937.

Imbens, G. W., and Hellerstein, J. K. (2010). "Imposing Moment Restrictions from Auxiliary Data by Weighting." NBER Working Paper No. t0202.

Lambert, D. M., Brown, J., and Florax, R. J. G. M. (2010). "A Two-Step Estimator for a Spatial Lag Model of Counts-Theory, Small Sample Performance and an Application." *Regional Science and Urban Economics*, 40: 241–252.

LeSage, J. (1999). *The Theory and Practice of Spatial Econometrics*. Toledo, Ohio: University of Toledo.

LeSage, J., and Pace, R. (2009). *Introduction to Spatial Econometrics*. Boca Raton, FL: CRC Press.

Peeters, L., and Chasco, C. (2006). "Ecological Inference and Spatial Heterogeneity: An Entropy-Based Distributionally Weighted Regression Approach." *Papers in Regional Sciences*, 85(2): 257–276.

Winkelmann, R. (2004). "Health Care Reform and the Number of Doctor Visits: An Econometric Analysis." *Journal of Applied Econometrics*, 19(4): 455–472.

Wu, X., and Sickles, R. (2018). "Semiparametric Estimation under Shape Constraints." Working Papers from Rice University.

10

Performance and Risk Aversion of Funds with Benchmarks

A Large Deviations Approach

F. Douglas Foster and Michael Stutzer

1. Introduction

As noted by Roll (1992, p.13), "Today's professional money manager is often judged by total return performance relative to a prespecified benchmark, usually a broadly diversified index of assets." He argues that "[t]his is a sensible approach because the sponsor's most direct alternative to an active manager is an index fund matching the benchmark." A typical example, of more than just professional interest to academic readers, is the following statement by the TIAA-CREF Trust Company:

> Different accounts have different benchmarks based on the client's overall objectives. . . . Accounts for clients who have growth objectives with an emphasis on equities will be benchmarked heavily toward the appropriate equity index—typically the S&P 500 index—whereas an account for a client whose main objective is income and safety of principal will be measured against a more balanced weighting of the S&P 500 and the Lehman Corporate/Government Bond Index (TIAA-CREF, p. 3).

How should plan sponsors and the investors they represent evaluate the performance of a fund like this? William Sharpe (1998, p. 32) asserts that

> [t]he key information an investor needs to evaluate a mutual fund is (i) the fund's likely future exposures to movements in major asset classes, (ii) the likely added (or subtracted) return over and above a benchmark with similar exposures, and (iii) the likely risk vis-á-vis the benchmark.

Procedures for implementing this recommendation will differ, depending on the quantitative framework used for measuring "return over and above a

F. Douglas Foster and Michael Stutzer, *Performance and Risk Aversion of Funds with Benchmarks: A Large Deviations Approach* In: *Advances in Info-Metrics: Information and Information Processing across Disciplines.* Edited by: Min Chen, J. Michael Dunn, Amos Golan, and Aman Ullah, Oxford University Press (2021).
DOI: 10.1093/oso/9780190636685.003.0010

benchmark" in (ii) and "risk vis-á-vis the benchmark" in (iii). Let $R_p - R_b$ denote a portfolio p's "return over and above a benchmark," b. The natural generalization of mean-variance efficiency *relative to a benchmark the investor wants to beat* is Roll's (1992) Tracking Error Variance (TEV)-efficiency, resulting from minimization of the tracking error variance $Var\left[R_p - R_b\right]$ subject to a constraint on the desired size of $E\left[R_p - R_b\right] > 0$. The tracking error variance measures the "risk vis-á-vis the benchmark." The most common scalar performance measure consistent with TEV efficiency is the Information Ratio (Goodwin, 1998), defined as:

$$E\left[R_p - R_b\right] / \sqrt{Var\left[R_p - R_b\right]} \tag{10.1}$$

Note that (1) reduces to the textbook Sharpe Ratio when the benchmark portfolio is a constant return risk-free asset.[1] A different scalar performance measure is the expected exponential (a.k.a. CARA) utility:

$$E\left[-e^{-\gamma\left(R_p - R_b\right)}\right]. \tag{10.2}$$

When $R_p - R_b$ is normally distributed, maximization of (10.2) yields a specific TEV-efficient portfolio that depends on the fixed $\gamma > 0$ used to evaluate portfolios in the opportunity set. These results have motivated Brennan (1993), Gomez and Zapatero (2003) and Becker et al. (1999) to assume that (10.2) is a fund manager's criterion when ranking normally distributed portfolios. Of course, only one value of γ will lead to the choice of the portfolio that maximizes the Information Ratio (10.1).

But several questions arise when considering the general legitimacy of using either (10.1) or (10.2) to rank portfolios relative to a designated benchmark:

1. The TEV logic underlying both (10.1) and (10.2) is single-period. Is there an appropriate performance measure for those wanting to beat the designated benchmark over a long horizon?
2. When differential returns are not normally distributed, should (10.1) or (10.2) be modified?
3. An advantage of (10.1) is the absence of critical, exogenous preference parameters, like γ in (10.2). When differential returns are not normally distributed, can a performance measure analogous to (10.1) be found that obviates the need to know exogenous parameters like γ in (10.2)?

[1] Note that subtracting gross returns in (10.1) gives the same number as the more common subtraction of net returns.

This chapter argues that all three questions can be affirmatively answered by ranking portfolios in accord with an index of the probability that they will outperform the benchmark over typical long-term investors' time horizons. Section 2 provides the answer to the first question by using the Gärtner–Ellis Large Deviations Theorem (Bucklew, 1990, Chap. 2) to show that the appropriate performance measure is the following function:

$$D_p \equiv \max_{\gamma} - \lim_{T\to\infty} \frac{1}{T} \log E\left[e^{-\gamma\left(\sum_{t=1}^{T}\left(\log R_{pt} - \log R_{bt}\right)\right)}\right]. \tag{10.3}$$

We will also show how (10.3) can be viewed as a generalized entropy. Section 2 also answers the second and third questions by showing that ranking a portfolio p in accord with the generalized entropy (10.3) is equivalent to ranking it in accord with the asymptotic expected *power* utility of the ratio of wealth invested in the portfolio to wealth that would be earned by alternatively investing in the benchmark. This utility has relative risk aversion equal to $1+\gamma_p$, where γ_p denotes the maximized value of γ in (10.3), and hence does not need to be exogenously specified. However, when approximation of time averaged log gross returns by arithmetic averaged net returns is reasonable (e.g., when $Var\left[R_p - R_b\right]$ is small) and when $R_p - R_b$ is independently and identically distributed (IID), section 10.2.3 shows that the single-period exponential, TEV-based criterion (10.2) does arise from our criterion *without* assuming normality, by substituting our maximizing fund-specific γ_p from (10.3) for the fixed γ in (10.2). We also show that that under the additional assumption of normality, this substitution of γ_p for γ in (10.2) reduces to the Information Ratio (10.1). *In this sense, the outperformance probability hypothesis nests the better-known criteria (10.1) and (10.2) as special cases, and hence is not subject to critiques commonly made of different probability-based criteria, for example, expected utility maximization subject to "safety-first," Value-At-Risk (VAR) constraints* [1]. Section 10.2.3 contains the appropriate modifications of the Information Ratio that arise under non-IID normality. Then, we develop some historical returns-based estimators of the general performance index in section 10.3. Section 10.3.1.1 applies them to rank the relatively few mutual fund portfolios that, according to standard hypothesis tests, could outperform the S&P 500 in the long run. Section 10.4 develops some consequences of the auxiliary hypothesis that fund managers act (either now or eventually) as if they maximized the outperformance probability. We argue that a recent and comprehensive study of an alternative fund manager behavioral hypothesis—expected exponential utility maximization with a fixed managerial risk-aversion coefficient—provides little empirical evidence in favor of that alternative. In light of this, we argue that the well-established scientific principles of Popperian falsifiability and Occam's Razor weigh in favor of the outperformance probability maximization hypothesis, unless and until future empirical evidence weighs in favor of something else, because the hypothesis eliminates the free risk-aversion parameter in conventional expected utility hypotheses. Finally, we show that if the outperformance probability maximization hypothesis is true, econometric estimates of managerial risk aversion in tests of (what would then be) the misspecified expected utility maximization hypothesis are subject to a Lucas Critique (Lucas, 1976). Section 5 concludes with several future research topics that are directly suggested by our findings.

While some other connections to the portfolio choice and asset pricing literatures are made in the following section, the most closely related papers use other outperformance probability criteria and are now summarized. Unlike our approach, Stutzer (2000) is not based on the probability that the fund's cumulative return should exceed the benchmarks and did not contain an empirical ranking of mutual funds. But that paper's method was adopted for a time by Morningstar, Inc. to produce its "Global Star Ratings" of mutual funds (Kaplan and Knowles, 2001). Our alternative approach should therefore be of interest to performance analysts like them, as well as the fund managers they rank. Our approach also permits a stochastic benchmark, generalizing the constant growth rate benchmark used in the constantly rebalanced portfolio choice model of Stutzer (2003). Pham (2003) used the framework of that portfolio choice paper to model optimal dynamic portfolio choice when a risky asset's returns are generated by the process adopted in Bielecki, Pliska, and Sherris (2000). Finally, Browne (1999, Sec. 4) formulated a related, but more specific, criterion for optimal dynamic portfolio choice. After imposing restrictive portfolio and benchmark parametric price process restrictions, Browne characterized the portfolio that maximized "the probability of beating the benchmark by some predetermined percentage, before going below it by some other predetermined percentage." Our analysis differs from his by not assuming that agents are constrained to use such floors and ceilings, specific time horizons, or specific parametric return processes.

2. An Index of Outperformance Probability

Ex-ante, wealth at some future time T arising from initial wealth W_0 invested in some portfolio strategy p will be denoted $W_T^p = W_0 \prod_{t=1}^{T} R_{pt}$, where R_{pt} denotes the random gross return from the strategy between times $t-1$ and t. Note that the validity of this expression does not depend on the length of the time inteval between $t-1$ and t, nor the particular times t at which the random gross returns are measured. Similarly, an alternative investment of W_0 in a different portfolio b, dubbed the "benchmark," yields $W_T^b = W_0 \prod_{t=1}^T R_{bt}$. Taking logs and subtracting shows that

$$\log W_T^p - \log W_T^b = \log \frac{W_T^p}{W_T^b} = \sum_{t=1}^T \log R_{pt} - \sum_{t=1}^T \log R_{bt}. \tag{10.4}$$

From (10.4), we see that the portfolio strategy p outperforms the benchmark b when and only when the sum of its log gross returns exceeds the benchmark's. Dividing both sides of (10.4) by T yields the following expression for the difference of the two continuously compounded growth rates of wealth:

$$\log W_T^p/T - \log W_T^b/T = \frac{1}{T} \log \frac{W_T^p}{W_T^b} = \frac{1}{T} \sum_{t=1}^T \left(\log R_{pt} - \log R_{bt} \right). \tag{10.5}$$

Suppose one wants to rank portfolios according to the rank ordering of their respective probabilities for the event that $W_T^p > W_T^b$. Using (10.5), we see that this *outperformance event* is the event that (10.5) is greater than zero—that is, the portfolio's continuously compounded growth rate of wealth exceeds that of the benchmark. Hence, one desires a rank ordering of the probabilities

$$Prob\left[\frac{1}{T} \sum_{t=1}^T \left(\log R_{pt} - \log R_{bt} \right) > 0 \right]. \tag{10.6}$$

which is equivalent to ordering the complementary probabilities from lowest to highest; that is, *we seek to rank a portfolio strategy inversely to its underperformance probability*:

$$Prob\left[\frac{1}{T} \sum_{t=1}^T \left(\log R_{pt} - \log R_{bt} \right) \le 0 \right]. \tag{10.7}$$

Of course, the rank ordering of portfolio strategies via (10.7) could depend on the exact value of the investor's horizon T. Because it is difficult for performance analysts to determine an exact value of an investor's horizon (when one exists), and because short-horizon investors may have different portfolio rankings than long-horizon investors, let us try to develop a ranking of (10.7) that will be valid for all T greater than a suitably large T. This is similar in spirit to the motivation behind choice of an infinite horizon for the investor's objective in most consumption-based asset pricing models (for a survey, see Kocherlakota (1996), or in many portfolio choice models (e.g., in Grossman and Vila, 1992).

Supporting evidence in Stutzer (2003) shows that portfolios with relatively low underperformance probabilities (10.7) for suitably large T often also have relatively low underperformance probabilities for small T (or even *all* T) as well. Hence, shorter-term investors may also make use of the results.

To produce this ranking, we use the following two-step procedure. First, reasonably assume that investors are not interested in portfolio strategies that (almost surely) will not even beat the benchmark when given an infinite amount of time to do so; in that event, they would prefer investing in the benchmark to investing in the portfolio. Hence, one should restrict the ranking to portfolios for which the underperformance probability (10.7) approaches zero as $T \to \infty$. More formally, one need only rank portfolio strategies p for which

$$\lim_{T\to\infty} \frac{1}{T} \sum_{t=1}^{T} \left(\log R_{pt} - \log R_{bt}\right) > 0. \tag{10.8}$$

Inequality (10.8) requires that the so-called *ergodic mean* of the log portfolio gross returns exceeds the benchmark's (a.e.). In the familiar special case of IID returns, (10.8) requires that the portfolio's expected log gross return exceeds the benchmark's. In fact, if one weren't worried about the probability of underperforming at finite horizons T, applying the law of large numbers to the limit in (10.8) shows that the highest performing fund would be the one with the highest expected log return, that is, the "growth optimal" fund that maximizes the expected log utility of wealth. It will almost surely *asymptotically* generate more wealth than the benchmark and all other funds will. But this "time diversification" argument in favor of log utility is irrelevant because, as Rubinstein (1991) so effectively demonstrated, significant underperformance probabilities persist over time spans well in excess of typical investors' retirement horizons. As he summarized: "The long run may be long indeed."

So we will now quantify this downside, and formulate alternative utility-based formulations that reflect it. To do so, the second step of our procedure seeks a rank-ordering index for the underperformance probability (10.7) of portfolios satisfying (10.8), that is, those portfolios whose underperformance probabilities decay to zero as $T \to \infty$. Fortunately, the powerful, yet simply stated, Gärtner–Ellis Large Deviations Theorem (Bucklew, Chap. 2) is tailor-made for this purpose. This theorem shows that those portfolios' underperformance probabilities (10.7) will decay to zero as $T \to \infty$, at a portfolio-dependent exponential rate. As a result, the underperformance probability of a portfolio with a higher decay rate will approach zero faster as $T \to \infty$, and hence its complementary outperformance probability will approach 1 faster, so the portfolio will have a higher probability of outperforming for suitably large T. Again, it is important to emphasize that while this is formally an asymptotic criterion, in practice it will produce a ranking that applies to much shorter investor horizons T as well (see Stutzer, 2003) for substantial evidence establishing this). In summary, *the underperformance probability's rate of decay to zero as $T \to \infty$ is our proposed ranking index for portfolios. A portfolio whose underperformance probability decays to zero at a higher rate will be ranked higher than a portfolio with a lower decay rate.*

Direct application of the Gärtner–Ellis Large Deviations Theorem (Bucklew, 1990, Chap. 2) shows that the decay rate of the underperformance probability (10.7) is

$$D_p \equiv \max_{\theta_p} - \lim_{T\to\infty} \frac{1}{T} \log E\left[e^{\theta_p \sum_{t=1}^{T} (\log R_{pt} - \log R_{bt})}\right] \tag{10.9}$$

Under the restriction (10.8), the maximizing θ_p in (10.9) will be negative (see Stutzer, 2003), so without loss of generality one may substitute a value $-\gamma_p$, where $\gamma_p > 0$. Hence, the rank-ordering index is the decay rate:

$$D_p = \max_{\gamma_p > 0} - \lim_{T\to\infty} \frac{1}{T} \log E\left[e^{-\gamma_p \sum_{t=1}^{T} (\log R_{pt} - \log R_{bt})}\right] \tag{10.10}$$

An expected utility interpretation of (10.10) is found by first substituting (10.4) into it and simplifying to yield:

$$D_p = \max_{\gamma_p > 0} \lim_{T\to\infty} - \frac{1}{T} \log E\left[\left(\frac{W_T^p}{W_T^b}\right)^{-\gamma_p}\right]. \tag{10.11}$$

There are two differences between the right-hand side of (10.11) and a power utility of wealth with an exogenous degree of relative risk aversion $1+\gamma$. First, the argument of the power function in (10.11) is not the wealth invested in the portfolio strategy, but instead is the ratio of it to the wealth that would have been earned if invested in the benchmark. This ratio is the state variable in the formulations of Browne (1999). It is analogous to the argument in the period utility of an "external habit formation," consumption-based criterion (Campbell, Lo, and MacKinlay, 1997, p. 327). Instead of our ratio of individual wealth to wealth created from an exogenous benchmark investment, its argument is the ratio of individual consumption to an exogenous benchmark function of aggregate (past) consumption ("keeping up with the Joneses"). In fact, when generalizing this argument to model other forms of this consumption externality, Gali (1994, footnote 2) noted that "such a hypothesis may be given an alternative interpretation: agents in the model can be thought of as professional "portfolio managers" whose performance is evaluated in terms of the return on their portfolio relative to the rest of managers and/or the market." While our infinite horizon criterion for pure investment (10.11) is perhaps more reminscent of the asymptotic growth of the expected utility criterion $J_p = \lim_{T\to\infty} \frac{1}{T} \log E\left[\left(W_T^p\right)^{\alpha}\right]^{1/\alpha}$ used by Grossman and Zhou (1993) and Bielecki, Pliska, and Sherris (2000), it should be possible to adapt the reasoning leading to (10.11) to analyze the consumption/investment problem with consumption externalities. The meaning of the degree of risk aversion in all these models is not the usual aversion to mean preserving spreads of distributions of wealth or consumption, but rather of distributions of wealth or consumption *relative to a benchmark.*

The second and more unusual difference between our criterion and all these others is that the curvature parameter γ_p on the right-hand side of (10.11) is determined by maximization of expected utility, and is hence dependent on the stochastic process for $\log R_{pt} - \log R_{bt}$. Hence, outperformance probability maximizing portfolio choice may be reconciled with expected utility maximizing portfolio choice by requiring that *both* the portfolio p *and* the degree of risk aversion $1 + \gamma_p$ be chosen to maximize D_p in (10.10) or (10.11).

Applicability of the Gärtner–Ellis Theorem requires that one maintain the assumptions that the limit in (10.10) exists (possibly as the extended real number $+\infty$) for all $\gamma > 0$ and is differentiable at any γ yielding a finite limit. Many log return processes adopted by analysts will satisfy these hypotheses, as will be demonstrated by example.

2.1 Entropic Interpretation

One can also show that (10.10) is a generalization of a suitably minimized value of the Kullback–Leibler Information Criterion or relative entropy $I(P \parallel \mu)$ between probability distributions P and μ. To see this, note that when the process $\log R_{pt} - \log R_{bt}$ is IID with the identical distribution of $\log R_p - \log R_b$ denoted by μ, (10.10) simplifies to $D_p \overset{IID}{=} \max_{\gamma_p} -\log E_\mu\left[e^{-\gamma_p(\log R_p - \log R_b)}\right]$. Kullback's Lemma (Kitamra and Stutzer, 2002, Proposition 2.2) shows that

$$D_p \overset{IID}{=} \max_{\gamma_p} -\log E_\mu\left[e^{-\gamma_p(\log R_p - \log R_b)}\right] = \min_{\{P: E_P[\log R_p - \log R_b]=0\}} I(P \parallel \mu) \quad (10.12)$$

Hence, the underperformance probability decay rate to zero is the constrained minimum value of the relative entropy, over the set of probability distributions that make the expected log portfolio return equal to the benchmark's expected log return. As such, (10.10) or (10.11) may be viewed as the value of a generalized entropy appropriate for general non-IID excess log returns.

2.2 Time-Varying Gaussian Log Returns

In order to both illustrate the calculation (10.10) and to relate it to the Information Ratio, suppose that for each time t, $\log R_{pt} - \log R_{bt}$ is normally distributed, so that each partial sum $\sum_{t=1}^{T}\left(\log R_{pt} - \log R_{bt}\right)$ in (10.10) is a normally distributed random variable with mean $\sum_t E\left[\log R_{pt} - \log R_{bt}\right]$ and variance $Var\left[\sum_t\left(\log R_{pt} - \log R_{bt}\right)\right]$. But (10.10) is just -1 times the time average of the log moment generating functions of these normally distributed random variables, evaluated at the maximizing $-\gamma_p$. Hence, in the Gaussian case, (10.10) is just the quadratic function

$$D_p = \max_{\gamma} \lim_{T\to\infty} \frac{\sum_{t=1}^{T} E\left[\log R_{pt} - \log R_{bt}\right]}{T}\gamma - \frac{1}{2}\frac{Var\left[\sum_{t=1}^{T}\left(\log R_{pt} - \log R_{bt}\right)\right]}{T}\gamma^2. \tag{10.13}$$

This first term in (10.13) will exist whenever the ergodic mean (10.8) exists, while the second term will exist whenever the analogous ergodic variance exists. These are standard assumptions to make in econometric estimation. Setting the first derivative of (10.13) equal to zero and solving for the maximizing γ yields:

$$\gamma_p = \frac{\lim_{T\to\infty}\sum_{t=1}^{T} E\left[\log R_{pt} - \log R_{bt}\right]/T}{\lim_{T\to\infty} Var\left[\sum_{t=1}^{T}\left(\log R_{pt} - \log R_{bt}\right)\right]/T} \tag{10.14}$$

which from assumption (10.8) is positive, as asserted earlier. Now substitute (10.14) back into (10.13) and rearrange to obtain the following underperformance probability decay rate in the Gaussian case:

$$D_p = \frac{1}{2}\left(\frac{\lim_{T\to\infty}\sum_{t=1}^{T} E\left[\log R_{pt} - \log R_{bt}\right]/T}{\sqrt{\lim_{T\to\infty} Var\left[\sum_{t=1}^{T}\log R_{pt} - \log R_{bt}\right]/T}}\right)^2. \tag{10.15}$$

Hence, the Gaussian performance index (10.15) depends on the ratio of the long-run mean excess log return to its long-run standard deviation and is hence a generalization of the usual Information Ratio. This differs from the Gaussian (i.e., second-order) approximation of the performance criterion in Grossman and Zhou (1993) and Bielecki, Pliska, and Sherris (2000), which (the latter paper shows) is the long-run mean minus an exogenous risk-aversion parameter times the long-run variance.

2.3 Familiar Performance Measures as Approximations

The single-period, exponential utility performance measure (10.2) has been rationalized by its consistency with Roll's (1992) TEV-efficiency when the difference in returns $R_{pt} - R_{bt}$ is IID normal. To obtain something akin to (10.2) by an approximation of our index (10.10), first approximate a log gross return $\log R$ by its net return $R - 1$; that is, substitute $R_{pt} - R_{bt}$ for $\log R_{pt} - \log R_{bt}$ in (10.10). Now under the restriction that the difference in the time series of equity returns is produced by a serially independent process, one obtains the following performance measure:

$$\max_{\gamma_p} - \lim_{T\to\infty} \frac{1}{T} \sum_{t=1}^{T} \log E\left[e^{-\gamma_p\left(R_{pt}-R_{bt}\right)}\right] \tag{10.16}$$

which under the additional restriction that the independent distributions are identically distributed reduces to the single-period performance measure:

$$-\log E\left[e^{-\gamma_p\left(R_p-R_b\right)}\right] \tag{10.17}$$

that rank-orders portfolios in the same way as the single-period expected exponential utility:

$$E\left[-e^{-\gamma_p\left(R_p-R_b\right)}\right]. \tag{10.18}$$

Note that (10.18) is based on the exponential function used in the TEV-based index (10.2), despite the fact that the argument for it made no use of normality. But (10.18) is not the same as (10.2), because (10.18) uses the portfolio-dependent, expected utility maximizing γ_p when ranking portfolio p, rather than some constant value of γ used for all p. However, if we impose the additional restriction that is used to rationalize the TEV hypothesis, that is, that $R_p - R_b$ is normally distributed, we can substitute the Gaussian (quadratic) log moment generating function into the equivalent problem (10.17) and solve to yield the following maximizing γ_p

$$\gamma_p = E\left[R_p - R_b\right] / Var\left[R_p - R_b\right]. \tag{10.19}$$

Substituting (10.19) back into that log moment generating function and rearranging yields

$$\frac{1}{2}\left(\frac{E\left[R_p - R_b\right]}{\sqrt{Var\left[R_p - R_b\right]}}\right)^2 \tag{10.20}$$

which is half the squared Information Ratio (10.1). Hence, we see that under the log return approximation and the IID normal process restriction used to rationalize the TEV hypothesis, use of the fund-specific maximizing γ_p (10.19) transforms the exponential utility (10.2) into a parameter-free performance measure (10.20) that ranks the funds in the same order as the Information Ratio (10.1)! In fact, when the differential return itself is normally distributed (rather than the differential *log gross* return), Stutzer (2003, Proposition 2) shows that the Information Ratio (10.1) ranks portfolios in accord with their respective returns' probabilities of outperforming the benchmark return *on average* over the horizon T, rather than their respective *cumulative* returns outperforming the benchmark cumulative returns *at* T. Hence, our general performance measure (10.10) or (10.11) may be viewed as a generalization of the TEV-based performance measures (10.1) and (10.2), to be used when the approximation of log returns by net returns, the IID assumption, and the normality assumption are unwarranted.

Finally, to see what happens when we maintain the IID assumption without either the approximation of log returns by net returns or the normality assumption, one can apply the IID restriction directly to the difference of log gross returns in (10.10), producing the following index:

$$\max_{\gamma} - \log E\left[\left(\frac{R_p}{R_b}\right)^{-\gamma}\right] \tag{10.21}$$

which yields the same rank ordering of portfolios as the expected power utility of the return ratio

$$E\left[-\left(\frac{R_p}{R_b}\right)^{-\gamma_p}\right], \tag{10.22}$$

instead of the expected exponential utility of the return difference (10.18).

3. Nonparametric Estimation of the Performance Measure

The simplest estimator of the performance index (10.10) arises when one makes the additional assumptions that the differential log gross return process $X_{pt} \equiv \log R_{pt} - \log R_{bt}$ is independently and identically distributed (IID). As argued earlier, one need only rank the fund portfolios p that almost surely will outperform the benchmark, that is, that satisfy (10.8). Under the IID assumption, (10.8) requires that $E\left[X_p\right] > 0$, and (10.10) reduces to

$$D_p = \max_{\gamma>0} - \log E\left[e^{-\gamma X_p}\right] \tag{10.23}$$

Given an observed time series of past observations, denoted $X_p(t)$, $t = 0, \ldots T$, one can consistently estimate (10.23) by substituting its sample analog, which is just:

$$\hat{D}_p^{IID} = \max_{\gamma>0} - \log \frac{1}{T} \sum_{t=1}^{T} e^{-\gamma X_p(t)}. \tag{10.24}$$

But what if X_{pt} is not IID, necessitating a good estimator for the general rate function (10.10) rather than its IID special case (10.23). An argument in Kitamura and Stutzer (2002, pp. 168–169) showed that the estimator in Kitamura and Stutzer (1997) provides the basis for an analog estimator of the general rate function (10.10). Specifically, one must select a "bandwidth" integer $K > 0$ used to smooth the series $X_p(t)$ (via a two-sided average of radius K about each $X_p(t)$), resulting in the general rate function estimator:

$$\hat{D}_p^{K} = \max_{\gamma>0} - \frac{1}{2K+1} \log \frac{1}{T-2K} \sum_{t=K+1}^{T-K} e^{-\gamma \frac{\sum_{j=-K}^{K} X_p(t-j)}{2K+1}} \tag{10.25}$$

Of course, an analyst who was truly confident that some specific parametric stochastic process generated $X_p(t)$ could attempt to either directly calculate a formula for (10.10) (as was done for Gaussian processes in section 2.2) and then estimate it, or could use the parametric process to construct a direct simulation estimator of (10.10).

3.1 Empirical Results

It is useful to compare our fund performance and risk-aversion coefficient estimates with results from other recent studies. To foster this comparison, we examined mutual funds during the 228 months starting in January 1976 and ending in December 1994. This coincides with the estimation period used in Becker et al. (1999) and is almost identical to the estimation period used in the mutual fund performance study of Wermers (2000). We also followed Wermers in using the CRSP Survival-Bias Free U.S. Mutual Fund Database, originated by Mark Carhart (1997). After describing our results and quantifying the sampling error present in them, we will use them to reexamine some of the claims in Wermers, which of course were based on different performance criteria. We will then compare our fund-specific risk-aversion coefficient estimates to the fund *manager* risk-aversion coefficient estimates reported in Becker et al. (1999). We follow them in adopting the most commonly cited mutual fund benchmark, the S&P 500 Index.[2]

3.1.1 Fund Performance

In accord with the rationale presented in section 10.2, one should only rank funds (if any) that would asymptotically outperform the S&P 500; that is, one should only rank funds that satisfy (10.8). The S&P 500 benchmark monthly returns $\log R_b(t)$ were subtracted from the corresponding monthly returns of each of the equity mutual funds returns $\log R_p(t)$ to produce the historical time series $X_p(t)$, and were used to conduct the usual *one-way* paired difference of means tests of the null hypothesis $H_0 : E\left[X_p\right] = 0$ versus the alternative hypothesis $H_1 : E\left[X_p\right] > 0$. The test statistic is $\overline{X_p} \equiv \sum_{t=1}^{228} X_p(t)/228$.

Performance analysts will not be surprised to find that the null hypothesis was rejected in favor of the alternative ($t > 1.65$) for only 32 of the 347 CRSP mutual funds whose returns persisted from January 1976 to December 1994. Now if one were testing an hypothesis about whether or not a *typical* fund beat the S&P 500, a survival bias would result from examining only those funds that survived over that entire period. But that hypothesis is *not* being tested here. Even though the procedure here examines only those 347 funds that were skillful or lucky enough to survive those 228 months, a standard hypothesis test concludes that only 32 of the 347 could asymptotically outperform the S&P 500, strengthening our conclusion that relatively few funds will do so.

Summary statistics and performance rankings for those 32 funds are reported in Table 10.1, ranked in order of their estimated performance index values $\hat{D}_p^{IID}\%$ given by (10.24). This estimator might be problematic for funds whose X_{pt} are serially correlated. A standard test of the hypothesis that the first through sixth autocorrelation correlation coefficients are all zero was rejected at the 5

[2] To obtain a total return benchmark, we used the CRSP value-weighted index, which includes distributions from the S&P 500 stocks.

percent level for only 8 of the 32 funds, and even those had low autocorrelation coefficients. Not surprisingly, use of the alternative estimator $\hat{D}_p^K$ given by (10.25) gave the exact same performance ranking of the funds, for each value of the "bandwidth" K tested (i.e. $K = 3$, 6, and 12), and hence are not reported in the table.[3] A (naive) analyst who is only concerned with the terminal wealth in the funds at the end of the ranking period, that is, the cumulative return over the ranking period, would *not* rank the funds by $\hat{D}_p\%$. Instead, that analyst would rank the funds in order of the average difference in log returns over the rating period (10.5), listed in the second column of Table 10.1 as $\overline{X}_p\%$. But $\overline{X}_p\%$ ranks the funds very differently than $\hat{D}_p\%$, which is used by analysts concerned with eventual underperformance that did not happen during the ranking period. However, the two rankings do agree on the top-ranked fund 36450. Perhaps not surprisingly, this is the Fidelity Magellan Fund. Our estimates show that its portfolio strategy resulted in the least probability (10.7) of underperforming the S&P 500 benchmark because the probability of underperforming it decays to zero as $T \to \infty$ at the highest estimated rate $\hat{D}_p^{IID} = 4.95\%$ per month. Using the well-known compounding "Rule of 72" approximation, the underperformance probability (10.7) will eventually be cut in half about every $72/4.95 \approx 15$ months. While the bottom-ranked fund 18050 should also eventually outperform the S&P 500 (due to rejection of its null $E\left[X_p\right] = 0$ with $t > 1.65$), its probability of underperformance is estimated to die off much slower; that is, it will eventually halve only every $72/0.60 = 120$ months.

We designed and conducted a bootstrap resampling study in order to examine the likely impact of sampling error on the stability of our nonparametrically estimated rankings. We resampled the 228 months (with replacement) 10,000 times to construct alternative possible $X_p(t)$ series for each of the 32 funds that could (on the basis of the nonparametric hypothesis test) outperform the S&P 500. After each of the 10,000 replications, we estimated the 32 funds performance index values (10.24) and ranked them. In each replication, we followed Brown and Goetzmann (1995) in classifying fund performance as either above ("high") or below ("low") the median performance for that replication. Figure 10.1 shows the nonparametric results for the funds listed in Table 10.1. The first panel of Figure 10.1 clearly shows a high degree of stability where it is most needed, that is, at the top end of the ranking. The highest rated fund (Fidelity Magellan) stayed above the median in virtually all replications. More detail is provided in the second panel of Figure 10.1, which lists each fund's four transition probabilities

[3] We also investigated the possibility that GARCH processes might have generated the excess log fund returns, in which case fitted GARCH models could be used to produce simulation estimates of (10.10). But stationarity testing of the popular GARCH(1,1) specification failed for all but six of the thirty-two funds.

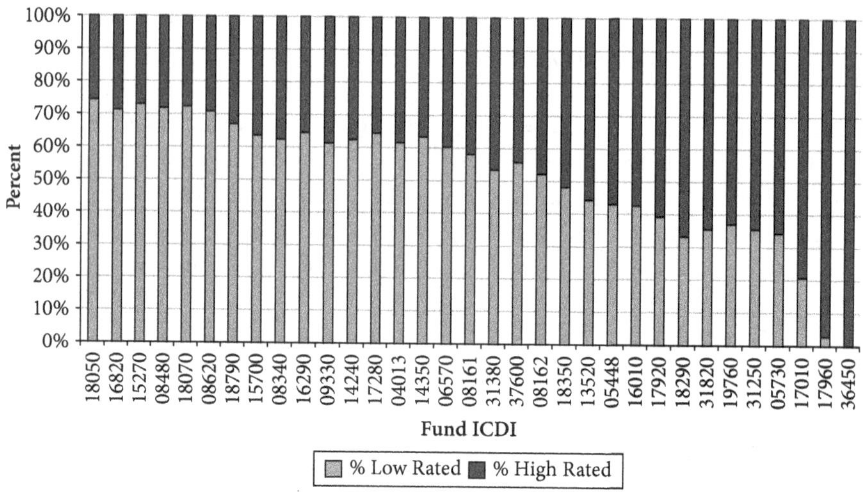

Figure 10.1 Fund Categorization.

of moving from above (below) the median to above (below) the median over successive replications. The panel confirms that it is more common for low funds to stay low and high funds to stay high in successive replications; as expected, the top- and bottom-ranked funds experience the most stability in this sense.

It is fruitful to reexamine one of the findings highlighted in Wermers's study (2000) of mutual fund performance, conducted over an almost identical period. Wermers (2000, p. 1686) asked the question: "Do higher levels of mutual fund trading result in higher levels of performance?" Wermers attempted to answer this question by constructing a hypothetical portfolio implemented by annually shifting money into funds that had relatively high turnover during the previous year. He concluded (p. 1690): "Although these high-turnover funds have negative (but insignificant) characteristic-adjusted net returns, their average unadjusted net return over our sample period significantly beats that of the Vanguard Index 500 fund."

Wermers's conclusion leaves two questions unanswered. First, we have shown that fund investors, who want their invested wealth to exceed that which would have accrued in the S&P 500 stocks, must not use the average net return as a performance measure. Rather, they must use the average log (1 + net return), akin to the geometric average, which can be significantly lower when returns are volatile. Second, Wermers examined a *strategy* of annually moving money from low-turnover funds to high-turnover funds. Significant load payments may pile up while doing this, but even if they did not, investors may also want to know whether or not it pays to buy-and-hold individual mutual funds that have relatively high average turnover.

As Wermers notes, annual turnover rates are reported for each fund. So we compiled the median of the nineteen years' turnover rates for each of the 32 funds that could (on the basis of the standard hypothesis test) asymptotically outperform the S&P 500. These ranged from a high of 181 percent to a low of 15 percent; half of these 32 funds had median annual turnover below 65 percent. Half of the 347 funds that had the full 228 monthly returns reported for our data period had median annual turnover below 43 percent, as reported in Table 10.2. So it is fair to say that the 32 outperforming funds had somewhat higher turnover than the rest.

Also, half of the 32 funds in Table 10.2 had median annual stock allocations that exceeded 85 percent, compared to 80 percent for the 347 funds. Moreover, the 32 funds had expense ratios that were similar to the 347 funds; half of each group had a median annual expense ratio below eighty-nine basis points. Hence, the outperformance of the 32 funds does not appear to be associated with atypical stock allocations or expense ratios.

The final column of Table 10.1 lists the estimates of the fund-specific coefficients of risk aversion $1+\gamma_p$ from (10.24), and their bootstrapped standard errors (in parentheses). No fund's coefficient is implausibly large, nor are the estimates imprecise; that is, the corresponding standard errors are considerably smaller than the estimates. Table 10.1 shows that there is a positive, but not perfect, correlation between a fund's degree of risk aversion and its outperformance probability ranking. To help see why, let us approximate the index by substituting the net return difference for the log return difference and assume that the net return difference is IID Gaussian. Then, Eq. (10.19) shows that the manager's degree of risk aversion will be high whenever the ratio $E\left[R_p - R_b\right]/Var\left[R_p - R_b\right]$ is high. In this case, the approximated performance index is (10.20), that is, half the squared or Information Ratio, which is half the ratio $E\left[R_p - R_b\right]^2/Var\left[R_p - R_b\right]$. So in this case, γ_p differs from D_p only because its numerator $E\left[R_p - R_b\right]$ is not squared. So there will be a *tendency* for a fund's performance and risk-aversion coefficient to be directly related, but the relationship is not perfect. The most extreme outperformance would occur when $R_p = R_b + c$ with $c > 0$, that is, when the portfolio return is perfectly correlated with the benchmark but is always higher by a constant amount. In this case, $Var\left[R_p - R_b\right] = 0$, so the degree of risk aversion is infinite, and shorting the benchmark to purchase the portfolio would then provide an arbitrage opportunity—the ultimate in outperformance! The top-ranked fund 36450 (Fidelity Magellan) certainly did not choose a portfolio that was that good, but its outstanding performance is a noisy signal that its portfolio management acted *as if* it was highly risk averse to the consequences of underperforming the S&P 500. Specifically, its estimated coefficient of risk aversion is $1+\gamma_p = 13.5$.

Table 10.1 The 32 out of 347 CRSP mutual funds that could (on the basis of a standard hypothesis test) statistically significantly outperform the S&P 500 Index during January 1976–December 1994.

Mutual Funds with $E\left[X_p\right] > 0$ CRSP #	$\overline{X_p}\%$	SDev %	Skew	Kurt	$\hat{D}_p\%$ (24)	$1+\hat{\gamma}_p$
36450	.80	2.4	−.66	5.94	4.95	13.5(5.5)
17960	.50	2.1	−.42	3.62	2.77	12.6(4.1)
17010	.33	1.7	−.26	0.64	1.81	12.8(4.2)
05730	.41	2.4	−.16	1.00	1.41	8.9(2.9)
31250	.47	2.9	.36	1.51	1.38	7.9(2.5)
19760	.42	2.6	.17	1.76	1.37	8.5(2.8)
31820	.29	1.8	−.09	0.71	1.34	10.9(3.8)
18290	.43	2.7	−.01	2.26	1.28	8.0(2.7)
17920	.35	2.3	.33	3.25	1.23	9.0(3.4)
16010	.25	1.6	−.02	2.31	1.19	11.4(4.6)
13520	.26	1.7	.06	0.64	1.15	10.9(4.1)
18350	.27	1.9	.92	4.15	1.07	10.2(3.9)
08162	.40	2.8	−.065	5.11	0.97	5.8(2.8)
37600	.33	2.4	.16	.80	0.95	7.8(2.9)
31380	.37	2.7	−.79	6.43	0.94	6.9(3.1)
08161	.55	4.1	−.72	5.40	0.86	5.1(1.9)
06570	.27	2.1	−.09	1.34	0.83	8.2(3.4)
14350	.26	2.0	−.37	2.30	0.82	8.3(3.6)
04013	.44	3.5	−.01	0.86	0.81	5.7(2.0)
17280	.03	2.3	−.30	1.63	0.81	7.4(3.0)
14240	.38	3.0	−.74	4.04	0.80	6.1(2.5)
09330	.47	3.7	−.51	1.52	0.78	5.3(1.9)
16290	.27	2.2	−.39	2.19	0.76	7.6(3.3)
08340	.24	2.0	.01	1.29	0.76	8.2(3.5)
15700	.23	1.9	−.57	5.07	0.74	8.4(4.1)
18790	.30	2.5	−.26	4.02	0.72	6.8(3.0)
08620	.29	2.4	−.45	2.35	0.72	6.9(3.0)
18070	.35	3.1	.19	3.45	0.66	5.8(2.4)
08480	.35	3.1	−.30	.40	0.65	5.7(2.2)
15270	.23	2.1	.42	3.35	0.64	8.3(3.4)
16820	.30	2.7	−.28	0.97	0.63	6.2(2.6)
18050	.33	3.0	−.12	1.27	0.60	5.6(2.3)

Rankings are in order of the nonparametrically estimated decay rate $\hat{D}_p$ from (10.24). No fund's implied degree of relative risk aversion $1+\hat{\gamma}_p$ from (10.24) is implausibly high, and all have standard errors (in parentheses) that are generally much lower than the estimates themselves, unlike the risk-aversion estimates reported in Becker, 1999.

4. Outperformance Probability Maximization as a Fund Manager Behavioral Hypothesis

Becker et al. (1999) used funds' returns from the same period we did in order to test the hypothesis that fund managers with fixed (but possibly different) coefficients of risk aversion acted as if they maximized (10.2). Their Table 5 (p. 139) reports summary statistics for the individual asset allocator funds' managers' estimated risk-aversion parameters γ. For the "asset allocation" category of funds that they (reasonably) believed to behave most in accord with their model, the mean value of the fund-specific risk-aversion parameters is an implausibly high $\gamma = 93.6$, while the median is actually negative, that is, $\gamma = -13.4$. If so, more than half the fund managers had negative risk aversion, resulting in a strictly *convex* utility (10.2). Hence, they would not have acted as if they solved the first-order conditions employed by Becker et al. in their model. Equally implausible and/or imprecise estimates of managerial risk aversion γ were reported for the other fund categories (1999, Table 3, pp. 135–136). They accurately conclude that "the risk aversion estimates are imprecise" (p. 145), and reported that the standard Hansen J-Statistic test of their model's GMM moment restrictions frequently failed.[4] In contrast, our estimates of fund-specific risk-aversion coefficients, listed in Table 10.1, are neither very high nor imprecisely estimated. So we now examine the possibility that fund managers might instead be trying to maximize the probability of outperforming their respective benchmarks.

4.1 Scientific Principles for Evaluating Managerial Behavioral Hypotheses

Suppose a researcher wants to construct an hypothesis of fund manager behavior based on maximization of some criterion and then test it using data from a group of managed funds. In order to produce an hypothesis that is both tractable and testable, suppose the researcher adopts the following (not atypical) maintained assumptions:

1. All fund managers in the group evaluate performance relative to the *same* benchmark portfolio.
2. All fund managers in the group perceive the *same* investment opportunity set.
3. All fund managers in the group have the technical ability to maximize the criterion.

Under these maintained assumptions, the outperformance probability maximizing hypothesis predicts that all fund managers in the test group would have

[4] For readers interested in a detailed analysis of the theoretical and econometric problems in that paper, we have written an appendix, available upon request.

chosen the *same* portfolio $p_{\max} \equiv \arg\max_p D_p$, which would be associated with the same coefficient of risk aversion $1 + \gamma_{p_{\max}}$—something that is unlikely to occur in practice. The same sort of counterfactual implication also arises in the standard mean-variance hypothesis of individual investor behavior. In the presence of a risk-free asset and the maintained assumptions listed above, the mean-variance hypothesis predicts that all investors in the test group will hold risky assets in the same proportions that the Sharpe Ratio maximizing "tangency" portfolio does, that is, in the same proportions as the market portfolio in the capital asset pricing model (CAPM).

Now consider the Becker et al. (1999) hypothesis that managers, each of whom has a fixed risk-aversion parameter γ, maximize (10.2). Under the three maintained assumptions listed earlier, the hypothesis's portfolio choice prediction still depends on each manager's unobservable risk-aversion parameter γ. The hypothesis would not predict that all the managers in the test group choose the same portfolio. Only managers with the same unobservable risk-aversion parameter would be predicted to choose the same portfolio. Hence, under the three maintained assumptions, the hypothesis need not make the prediction of homogeneous behavior that the outperformance probability maximization hypothesis of fund manager behavior would make (and with the additional presence of a riskless asset, the mean-variance hypothesis of individual behavior would make). But the ability to avoid this was obtained through flexibility arising from the hypothesis's introduction of an additional free parameter (i.e., a manager's value of γ). From a scientific viewpoint, is this a strength or a weakness?

Subsequent to Popper's seminal work (1977) scientists have considered a more potentially falsifiable hypothesis to be better than a less potentially falsifiable one, unless empirical evidence clearly favors the latter. This paper showed that the outperformance probability maximization hypothesis is equivalent to maximizing an expected utility over *both* a fund manager's risk-aversion "parameter" and the investment opportunity set. By endogenizing γ, it is eliminated as a free parameter, enabling the hypothesis to make a determinate portfolio choice prediction under the typical theoretical conditions (i)–(iii) above. This sharp prediction is of course more potentially falsifiable than the set of γ-dependent predictions made by the conventional hypothesis (10.2), and is thus favored by Popper's falsifiability principle for evaluating scientific hypotheses. A related principle—the Principle of Occam's Razor—favors the simplest hypothesis that can potentially make the right prediction. In addition to being a principle accepted by most scientists, Jefferys and Berger (1992) have found a Bayesian rationalization for it, summarized as follows:

Table 10.2 Comparison of the 32 Outperforming Funds' Characteristics to the Median Characteristics of All 347 CRSP Mutual Funds, January 1976–December 1994 The 32 outperforming funds had slightly higher turnover and fractional allocation to equities than the 347 funds did and similar expense ratios. Hence, the outperformance of the 32 funds does not appear to be associated with atypical stock allocations or expense ratios.

Mutual Funds with $E[X_p] > 0$						
CRSP #	**Fund**	**Turnover**	**Expense %**	**Stock %**	**Bond %**	**Cash %**
36450	Fidelity Magellan	1.26	1.08	92	2	1
17960	Fidelity Destiny I	0.75	0.68	NA	NA	NA
17010	NY Venture	0.59	1.02	91	2	5
05730	Mutual Shares	0.67	0.74	67	21	13
31250	Lindner Growth	0.21	0.89	80	6	14
19760	Sequoia	0.32	1.00	66	0	27.2
31820	Guardian Park Ave.	0.48	0.72	82	10	7
18290	Acorn Inv. Trust	0.32	0.82	85	3	9
17920	Van Kamp-Amer Pace/A	0.45	1.01	86	0	14
16010	Neub &Berm Partners Fund	1.81	0.95	73	2	25
13520	Fidelity Equity Income	1.07	0.72	68	26	5
18350	Pioneer II	0.26	0.81	90	1	8
08162	20th Cent.Select	0.95	1.01	99	0	1
37600	SteinRoe Special	0.61	0.96	84	1	9
31380	Evergreen Fund/Y	0.48	1.13	92	0	8
08161	20th Cent.Growth	1.05	1.01	99	0	1
06570	Templeton Growth/I	0.15	0.75	87	4	10
14350	AMCAP	0.18	0.73	84	0	14
04013	Spectra	NA	NA	NA	NA	NA
17280	Nicholas	0.27	0.86	89	0	11
14240	Weingarten Equity	1.09	1.19	96	0	4
09330	Loomis-Sayles Cap. Dev.	1.29	0.79	98.6	0	1.4
16290	IDS New Dimensions/A	0.70	0.74	87	0	13
08340	Windsor	0.34	0.53	84	2	10
15700	Van Kamp-Amer Comstock/A	0.61	0.79	85	0	11
18790	Janus	1.63	1.02	71	1	27
08620	Growth Fund of America	0.22	0.76	81	0	16
18070	Value Line Leveraged Growth	1.13	0.91	95	0	3
08480	St. Paul Growth	0.84	1.00	84	0	16
15270	Charter	1.06	1.25	77	0	23
16820	Putnam Voyager/A	0.65	1.10	93	0	6
18050	Van Kamp-Amer Emerg.Grwth/A	0.93	1.05	91	0	8.30
	Median: All 347 Funds	**0.43**	**0.89**	**80**	**0**	**6**

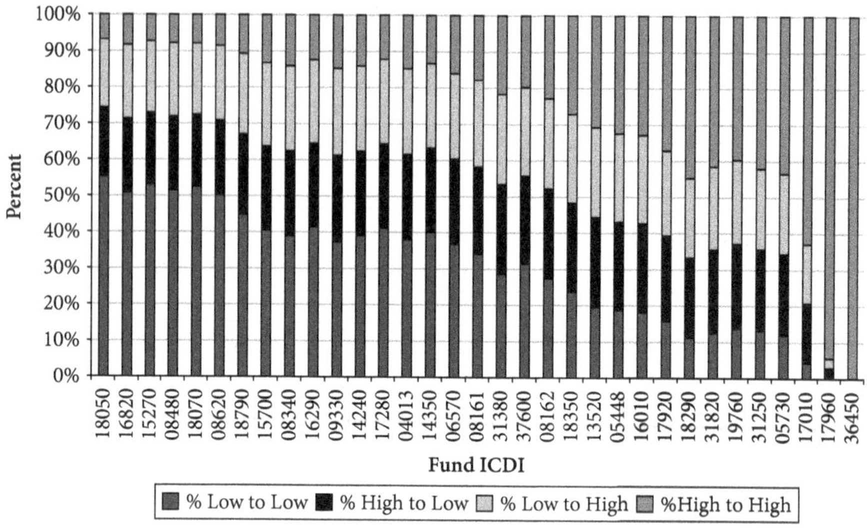

Figure 10.2 Persistence in Mutual Fund Rankings.

> Ockham's Razor, far from being merely an *ad-hoc* principle, can under many practical situations in science be justified as a consequence of Bayesian inference.... a hypothesis with fewer adjustable parameters has an enhanced posterior probability because the predictions it makes are sharp. [15]

Of course, these principles cannot be used to favor an hypothesis that is obviously counterfactual *about an absolutely critical fact* over an hypothesis that is not. But the homogeneity prediction of the outperformance probability maximization hypothesis was predicated on the three assumptions listed above, and it is highly unlikely that all three assumptions are valid in the real world. Currently, Morningstar separates domestic equity funds into twenty different categories, based on capitalization, style, and sector specializations. Morningstar ranks a fund against the others in its category, rather than against the entire fund universe. Hence, it is certainly possible that some of the 32 funds ranked here tried to beat benchmarks other than the S&P 500, perhaps benchmarks specific to their capitalization, style, or sector specialization, thus violating the first assumption listed above. In fact, Becker et al. also permitted funds in different categories to have different benchmarks; specifically, funds were assumed to have benchmarks that could be different (unobservable) h-weighted averages of the S&P 500 and T-Bills. The second assumption listed above is also unlikely, as it requires that the funds face the same restrictions (if any) on trading and that the fund managers agree on the forms and parameters of all portfolios' differential (log gross) return processes. The latter is particularly unlikely to be true in practice, due to differences in managers' opinions about the likely future performance of the individual stocks, bonds, and so on, that can be used to form their portfolios. Readers doubting this point should watch a randomly chosen episode of the public television show *Wall Street Week*, which would also cast doubt on the the realism of the third assumption, that is, that all fund managers have the technical skill to maximize a quantitative criterion function. Hence, even if the test group of 32 fund managers examined herein were trying to maximize the probability of outperforming the S&P 500, the likely violations of the second and third assumptions could explain why they chose different portfolios, with differing fund-specific outperformance probabilities and coefficients of risk-aversion evidenced in Table 10.1.

Furthermore, if our outperformance probability hypothesis is correct, the alternative hypothesis—that a manager with a fixed risk-aversion parameter γ maximizes (10.2)—is subject to the critique made in a justly celebrated paper by Robert E. Lucas (1976). Lucas criticized econometric analyses that incorrectly hold parameters of an optimizing agent's decision rule fixed when analyzing the effects of policy-induced changes in the decision-making environment. He showed that these parameters would change endogenously when agents optimized more thoroughly than was incorrectly assumed.

Econometric analyses that fix γ as a managerial preference "parameter" are subject to a similar critique. Our hypothesis implies that the optimal portfolio is associated with a portfolio-specific degree of risk aversion that depends on the decision-making environment, that is, the benchmark portfolio and investment opportunity set, via maximization of the performance measure (10.10) or (10.11). Plan sponsors and/or their investors who designate a fund benchmark for management to beat are analogous to Lucas's policymakers. Should they designate a tougher benchmark, they should anticipate that the outperformance probability maximizing manager will act as-if he or she had a lower degree of (endogenous) risk aversion. For example, if the Fidelity Magellan Fund actually found the portfolio strategy that maximized the probability of outperforming the S&P 500, then the first row in Table 10.1 shows that its degree of risk aversion was 13.5. But had investors insisted that Magellan designate an even tougher benchmark to beat than the S&P 500, its manager would have acted as-if he/she had a lower degree of risk aversion.

This critique is most starkly illustrated in the case of a single manager contracted to run two separate funds: one for a group of conservative investors who designated the three-month T-Bill as the benchmark portfolio, and the other for a group of investors who designated the S&P 500 as the benchmark portfolio.[5]

[5] This is not unrealistic; many fund managers run more than one fund.

The manager would quickly surmise that it is much easier to outperform a T-Bill benchmark than an S&P 500 benchmark and that he/she should choose a much more conservative portfolio when managing the former portfolio in order to maximize (minimize) the probability of outperforming (underperforming) it over finite investor horizons. A priori, there is no reason for theorists to rule out the possibility that the manager acted as if he/she used a higher coefficient of risk aversion when managing the former fund than he/she did when managing the latter. While this explanation for the different choices is unusual, it follows from our power utility criterion (10.11), which was *derived* from the deeper hypothesis of outperformance maximization. The conventional hypothesis that (10.2) is maximized *assumes rather than derives* the form of the utility (i.e., exponential) and then imposes the as yet empirically and experimentally unsupported *ad-hoc restriction* that its constant degree of risk aversion is completely exogenous to the manager's benchmark *and* the investment opportunity set used by the manager to beat it.

5. Conclusions

Mutual fund performance is often measured relative to a designated benchmark portfolio. This chapter provides performance analysts with a simple way of ranking funds in accord with their respective probabilities of outperforming a benchmark portfolio. We derived a closed form for the ranking index when funds' excess log returns (over the benchmark's) are generated by time-varying Gaussian processes. More generally, the outperformance probability index is (1) a generalization of the familiar constrained minimum value of Kullback–Leibler's relative entropy, and (2) an asymptotic expected generalized power utility, which differs in two ways from the familiar power utility of wealth. First, the argument of the utility function is the ratio of wealth earned in the fund to what would have otherwise been earned from investing in the benchmark. Second and more surprisingly, the curvature (i.e., risk-aversion) parameter value required to evaluate the expected power utility of a fund's portfolio is the value that maximizes the expected utility of that fund's portfolio! Hence, the fund performance ranking index uses a power utility whose coefficient of risk aversion varies endogenously from fund to fund.

In order to illustrate the feasibility and plausibility of this approach, we derived simple nonparametric estimators for the performance-ranking index and the fund-specific coefficients of risk aversion required to evaluate it. These were used to rank the performance of mutual funds that (based on standard hypothesis tests) could asymptotically outperform the S&P 500. We concluded that only 32 out of 347 funds will be able to asymptotically outperform the S&P 500, even though those 347 funds managed to survive the nineteen-year test period. Those that outperformed had overall equity allocations and expense ratios that were similar to those that did not. The fund-specifc coefficients of risk aversion required to evaluate the relative performance of those 32 funds ranged between 5.6 and 13.5. The highest ranked fund is Fidelity Magellan, which also had the highest coefficient of risk aversion.

These theoretical and empirical findings should benefit investors and performance analysts who want to rank funds in accord with their probabilities of outperforming a benchmark they want to beat. But for academic readers interested in fund manager *behavior*, we also formulated the hypothesis that a fund manager attempts to maximize the probability of outperforming a designated benchmark. We contrasted this hypothesis with the extant alternative hypothesis that a fund manager with a fixed coefficient of risk aversion attempts to maximize the expected utility of fund returns in excess of the benchmark. We argued that a recent empirical test of this alternative hypothesis provided little evidence in favor of it. In the absence of convincing empirical evidence favoring that alternative, the scientific desiderata of Popperian falsifiability and Occam's Razor weigh in favor of the outperformance probability maximization hypothesis, which eliminates the assumption that the fund manager has an unobservable, econometrically free curvature (i.e., risk aversion) parameter. Moreover, if the outperformance probability maximization hypothesis is valid, a manager's degree of risk aversion is endogenous and hence will change when the manager is faced with a different benchmark or investment opportunity set. An investor or performance analyst who misspecifies the manager's degree of risk aversion as an econometrically free parameter would then be subject to the celebrated Lucas Critique.

Acknowledgments

The authors wish to acknowledge recent comments from the editor (Aman Ullah) and from an anonymous referee. We also acknowledge earlier comments from Tom Smith, Richard Heaney, Juan-Pedro Gomez, J.C. Duan, Rama Cont, and seminar participants at the Information and Entropy Econometrics Conference in Washington, DC, University of Minnesota, University of Colorado, CIRANO Extremal Events Conference in Montreal, French Finance Association, European Financial Management Association, Ecole Polytechnique in Paris, Australian National University, University of Melbourne, University of Queensland, University of Auckland, and University of Otago. Foster acknowledges the support of the Australian Research Council.

References

Basak, S., and Shapiro, A. (2001). "Value-at-Risk Based Risk Management: Optimal Policies and Asset Prices." *Review of Financial Studies*, 14(2): 371–405.

Becker, C., Ferson, W., Myers, D. H., and Schill, M. J. (1999). "Conditional Market Timing with Benchmark Investors." *Journal of Financial Economics*, 52(1): 119–148.

Bielecki, T. R., Pliska, S. R., and Sherris, M. (2000). "Risk Sensitive Asset Allocation." *Journal of Economic Dynamics and Control*, 24: 1145–1177.

Brennan, M. (1993). "Agency and Asset Pricing. Finance Working Paper no. 6-93." Anderson Graduate School of Management, University of California, Los Angeles.

Brown, S. J., and Goetzmann, W. N. (1995). "Performance Persistence." *Journal of Finance*, 50(2): 679–698.

Browne, S. (1999). "Beating a Moving Target: Optimal Portfolio Strategies for Outperforming a Stochastic Benchmark." *Finance and Stochastics*, 3: 275–294.

Bucklew, J. A. (1990). *Large Deviation Techniques in Decision, Simulation, and Estimation*. New York: John Wiley.

Campbell, J. Y., Lo, A. W., and MacKinlay, C. A. (1997). *The Econometrics of Financial Markets*. Princeton, NJ: Princeton University Press.

Carhart, M. (1997). "On Persistence in Mutual Fund Performance." *Journal of Finance*, 52(1): 57–82.

Gali, J. (1994). "Keeping Up with the Joneses: Consumption Externalities, Portfolio Choice, and Asset Prices." *Journal of Money, Credit, and Banking*, 26(1): 1–8.

Gómez, J-P., and Zapatero, F. (2003). "Asset Pricing Implications of Benchmarking: A Two-Factor CAPM." *European Journal of Finance*, 9(4): 343–357.

Goodwin, T. H. (1998). "The Information Ratio." *Financial Analysts Journal*, 54(4): 34–43.

Grossman, S., and Vila, J-L. (1992). "Optimal Dynamic Trading with Leverage Constraints." *Journal of Financial and Quantitative Analysis*, 27(2): 151–168.

Grossman, S., and Zhou, Z. (1993). "Optimal Investment Strategies for Controlling Drawdowns." *Mathematical Finance*, 3(3): 241–276.

Jefferys, W. H., and Berger, J. O. (1992). "An Application of Robust Bayesian Analysis to Hypothesis Testing and Occam's Razor." *Journal of the Italian Statistical Society*, 1: 17–32.

Kaplan, P., and Knowles, J. (2001, March). *The Stutzer Performance Index: Summary of Mathematics, Rationale, and Behavior*. Quantitative Research Department, Morningstar, Chicago, IL.

Kitamura, Y., and Stutzer, M. (1997). "An Information-Theoretic Alternative to Generalized Method of Moments Estimation." *Econometrica*, 65(4): 861–874.

Kitamura, Y., and Stutzer, M. (2002). "Connections between Entropic and Linear Projections in Asset Pricing Estimation." *Journal of Econometrics*, 107: 159–174.

Kocherlakota, N. R. (1996). "The Equity Premium: It's Still a Puzzle. *Journal of Economic Literature*, 34(1): 42–71.

Lucas, R. E. (1976). "Econometric Policy Evaluation: A Critique." In Karl Brunner and Allan Meltzer (Eds.), *The Phillips Curve and Labor Markets: Volume 1, Carnegie-Rochester Conference Series on Public Policy*. Amsterdam: North-Holland, 1976.

Pham, H. (2003). "A Large Deviations Approach to Optimal Long Term Investment." *Finance and Stochastics*, 7: 169–195.

Popper, K. (1977). *The Logic of Scientific Discovery*. New York: Routledge. (14th Printing).

Roll, R. (1992). "A Mean/Variance Analysis of Tracking Error." *Journal of Portfolio Management*, 18(4): 13–22.

Rubinstein, M. (1991). "Continuously Rebalanced Investment Strategies." *Journal of Portfolio Management*, 18(1), 78–81.

Sharpe, W. (1998). "Morningstar's Risk-Adjusted Ratings." *Financial Analysts Journal*, 54(4): 21–33.

Stutzer, M. (2000). "A Portfolio Performance Index." *Financial Analysts Journal*, 56(3): 52–61.

Stutzer, M. (2003). "Fund Managers May Cause Their Benchmarks to Be Priced Risks." *Journal of Investment Management*, 1(3): 1–13.

Stutzer, M. (2003). "Portfolio Choice with Endogenous Utility: A Large Deviations Approach." *Journal of Econometrics*, 116: 365–386.

TIAA-CREF. (2000, September). TIAA-CREF Trust Company's personal touch. TIAA-CREF Investment Forum.

Wermers, R. (2000). "Mutual Fund Performance: An Empirical Decomposition into Stock-Picking Talent, Style, Transactions Costs, and Expenses." *Journal of Finance*, 55(4): 1655–1703.

11

Estimating Macroeconomic Uncertainty and Discord

Using Info-Metrics

Kajal Lahiri and Wuwei Wang

1. Introduction

For informed decision making in businesses and governments, forecasts as well as their uncertainties are important. Not surprisingly, policymakers in diverse fields are increasingly interested in the uncertainty surrounding point forecasts. A rich literature in economics and finance has developed models of *ex post* uncertainty by analyzing different functions of forecast errors in vector autoregressive and volatility models. However, the decision makers need *ex ante* uncertainty in real time before observing the outcome variable. Density forecasts obtained from surveys of experts are particularly suited to produce such information. Attempts have been made to compare these subjective uncertainties with those generated from estimated time series models.[1]

Before survey density forecasts became available, variance of point forecasts or disagreement was used as a convenient proxy for uncertainty. However, using density forecasts from the Survey of Professional Forecasters (SPF), Zarnowitz and Lambros (1987) and Lahiri et al. (1988) distinguished between aggregate forecast uncertainty and disagreement. This literature has attracted increasing attention in recent years.[2] Most recently, a number of interesting studies on the methodology of using info-metrics in economic forecasting have evolved.[3] Shoja and Soofi (2017) decompose entropy of the consensus density forecast into average uncertainty and disagreement, and use information divergence as a proxy for disagreement. In this chapter,

[1] See Lahiri and Liu (2006a), Clements (2014b), and Knüppel (2014). For various problems in generating *ex ante* forecast uncertainty from time series models, see Fresoli et al. (2015) and Mazzeu et al. (2018).

[2] See, for example, Lahiri and Sheng (2010a), Boero et al. (2008, 2013), and Abel et al. (2016).

[3] See Mitchell and Hall (2005), Lahiri and Liu (2006b, 2009), Mitchell and Wallis (2011), Rich and Tracy (2010), Kenny et al. (2015) and the references therein.

Kajal Lahiri and Wuwei Wang, *Estimating Macroeconomic Uncertainty and Discord: Using Info-Metrics*
In: *Advances in Info-Metrics: Information and Information Processing across Disciplines.* Edited by:
Min Chen, J. Michael Dunn, Amos Golan, and Aman Ullah, Oxford University Press (2021).
DOI: 10.1093/oso/9780190636685.003.0011

we utilize the info-metrics approach to estimate uncertainty, and we decompose it into components that include a disagreement term. We compare the conventional moment-based measures with info-metrics-based measures, including entropy and information divergence.

Our approach involves fitting continuous distributions to the histogram data. Many previous studies have utilized raw histograms, assuming that the probability mass is located on the midpoint of the bin. This approach essentially treats each forecast as a discrete distribution. For various reasons, the alternative of fitting continuous distributions to the histograms has become common in recent years. For output and inflation forecasts, it is reasonable to assume that the true underlying distributions are continuous, even though surveys are recorded in the form of histograms (i.e., probabilities attached to the bins) for convenience. However, changing from a discrete to a continuous system may involve trade-offs, especially when the variable of interest is uncertainty; cf. Golan (1991). Furthermore, there are a certain number of observations in the survey where forecasters attached probabilities to only a few bins. In these cases, assumption on distribution is more tenuous and may have an indeterminate impact at the individual level. These all remain challenges, though some of the issues do not affect the stylized facts after averaging over individuals. However, given that the target variables are continuous, we choose to adopt the method of fitting continuous distributions to the histograms. Following Giordani and Soderlind (2003), some researchers have assumed normal distributions or that the probability mass in a bin is uniformly distributed. The later assumption is inconsistent with distributions that are globally unimodal. Since many of the histograms are not symmetric and exhibit widely different shapes, we follow Engelberg et al. (2009) and use generalized beta as our preferred choice.

We are the first to apply continuous distributions to histograms while using info-metrics to estimate uncertainty, after approximating all individual histograms that have gone through occasional changes in survey bin sizes and number of bins. We further adjust for different forecast horizons to estimate a quarterly time series of aggregate uncertainty. The estimated information measures are then utilized successfully in several macroeconomic applications, including vector autoregression (VAR) models to estimate the effect of uncertainty on the macro economy. We use Jenson–Shannon Information to measure *ex ante* "news" or "uncertainty shocks," and we study its effect on uncertainty in real time.

The chapter is organized as follows: in section 2, we briefly introduce the data set and the key features of the recorded density forecasts. In section 3, we compare different measures of uncertainty by analyzing variances, and we confirm previous findings that disagreement regarding point forecast should not be used as a sole proxy for uncertainty. In section 4, we employ info-metrics to estimate uncertainty, and we study the difference between two approaches—the moment-based and the entropy approach. In section 5, we correct for horizons to get a time series of uncertainty, and we compare our measures to other popular uncertainty indexes. In section 6, we apply Jensen–Shannon Information divergence to individual densities to estimate "uncertainty shocks" or "news" from successive fixed-target forecasts, and we study the impact of these on uncertainty. In section 7, we evaluate the effects of uncertainty on macroeconomic variables using VAR. Finally, section 8 summarizes the main conclusions of the study.

2. The Data

The U.S. SPF, spearheaded by Victor Zarnowitz[4] in the late 1960s, is a unique data set that provides probability forecasts from a changing panel of professional forecasters. The forecasters are mostly from financial institutions, universities, research institutions, consulting firms, forecasting firms, and manufacturing. SPF was formerly conducted jointly by the American Statistical Association (ASA) and the National Bureau of Economic Research (NBER). It started in the fourth quarter of 1968 and was taken over by the Federal Reserve Bank of Philadelphia in the second quarter of 1990. Similar siblings of the survey, though considerably much younger, include the Bank of England Survey of External Forecasters and the European Central Bank's SPF. The timing of the survey has been planned carefully so that the forecasters will have the initial estimate of selected quarterly macroeconomic variables at the time of forecasting and at the same time issue forecasts in a timely manner. For instance, while providing forecasts in the middle of quarter 3, a forecaster has the information of the first estimate of the real GDP for the second quarter.

The survey asks professional forecasters about their predictions of a number of macro variables, including real GDP, nominal GDP, GDP deflator, and unemployment rate. Forecasters provide their point forecasts for these variables for the current year, the next year, the current quarter, and the next four quarters. Besides the point forecasts, they also provide density forecasts for output and price-level growth for the current and next year. The definition of the "output" variable has changed in the survey several times. It was defined as nominal GNP before 1981 Q3 and as real GNP from 1981 Q3 to 1991 Q4. After 1991 Q4, it is real GDP. The definition of the inflation variable "price of output" or GDP price deflator (PGDP) has changed several times as well. From 1968 Q4 to 1991

[4] See Zarnowitz (2008) for a remarkable autobiography.

Q4, it stood for the implicit GNP deflator. From 1992 Q1 to 1995 Q4, it was the implicit GDP deflator. From 1996 Q1 onward, it forecasts the GDP price index. The density forecasts for these two variables (output and price of GDP) are fixed-target probability forecasts of their annual growth rates in percentages for predefined intervals (bins) set by the survey. In this chapter, the numerical values of closed "bins" or "intervals" like "+1.0 to +1.9 percent" will be defined as [1, 2), and an open interval like "decline more than 2%" will be defined as an open bin (−∞, −2). We have utilized all density forecasts ever recorded in SPF during 1968 Q4 to 2017 Q3.[5]

3. Aggregate Uncertainty, Aggregate Variance and Their Components

In the absence of density forecasts, the early literature used variance of the point forecasts or disagreement as proxy for uncertainty. In this section, we calculate aggregate variance (i.e., the variance of the consensus distribution), average of the individual variances and forecaster discord (i.e., disagreement) as the variance of the mean forecasts to highlight the decomposition and their trends over last few decades. Unlike many previous studies where the recorded histograms are treated as discrete data or approximated by normal distributions, we fit generalized beta distributions or triangular distributions to the density forecast histograms following Engelberg, Manski, and Williams (2009). For density forecasts with more than two intervals, we fit generalized beta distributions. When the forecaster attaches probabilities to only one or two bins, we assume that the subjective distribution has the shape of an isosceles triangle. There are many different ways to generalize beta distributions to generate other distributions.[6] The generalized beta distribution we choose has four parameters, and its probability density function is defined as follows:

$$f(x;\alpha,\beta,l,r) = \frac{1}{B(\alpha,\beta)(r-l)^{\alpha+\beta-1}}(x-l)^{\alpha-1}(r-x)^{\beta-1}, l \le x \le r, \alpha > 0, \beta > 0 \tag{11.1}$$

[5] Initially, we had a total 24,011 forecast histograms, but our final sample is 23, 867 after dropping a few forecasts due to bimodality, confusion regarding target years and horizons, and so on.

The Fed is unsure about the correct horizon of the density forecasts for 1985 Q1 and 1986 Q1. In order to salvage these two rounds, we compared forecast uncertainty measures of each respondent from each survey to that of the same person in adjacent quarters, and we concluded that the "next year" forecasts in 1985 Q1 and 1986 Q1 were for horizon 4. That is, both were current-year forecasts. This enabled us to derive a continuous series for the variables of interest from the beginning of the survey until 2017 Q3.

[6] See, for instance, Gordy (1998) and Alexander et al. (2012).

where B is the beta function. We further restrict α and β to be greater than 1 to maintain unimodality of the fitted individual density distribution.

The two parameters α and β define the shape of the distribution, and the other two parameters l and r define the support. The generalized beta are the preferred forms of distributions for the recorded histograms for several reasons. First, when the histograms are treated as discrete distributions, with the usual assumption that the probability mass within an interval is concentrated at the midpoint of each interval, it does not reflect the expected continuity and unimodality of the true underlying distributions. Second, compared to the normal distribution, the generalized beta distribution is more flexible to accommodate different shapes in the histograms. The histograms often display excess skewness as well as different degrees of kurtosis. Finally, generalized beta distributions are truncated at both sides, while the normal distribution is defined over an open interval $(-\infty, +\infty)$, which is not true with most of the histograms and counterintuitive to the fact that the target variables have historical bounds.

We have adopted the triangular distributions when only one or two bins have positive probability masses (nearly 8 percent of our histograms). Normal distributions will shrink to a degenerate distribution in these cases. While the use of triangular distribution yields a unique solution for each observation, we should be cognizant of possible limitations of the assumption. The triangular distribution may exaggerate the spread and uncertainty embedded in the distribution. In the fitting process, we restrict the triangular distribution to be isosceles, and we allow the support to cover the whole bin, which has a probability of not less than 50 percent. There are triangular distributions we could fit with shorter supports and smaller variances if we changed the restrictions and assumptions. In fact, Liu and Sheng (2019) find that the triangular distribution tends to overestimate the associated uncertainty. However, to avoid multiple solutions for densities with only one or two bins and in the absence of additional information, we choose isosceles triangular distributions.

3.1 Fitting Continuous Distributions to Histogram Forecasts

Suppose forecaster i (i is forecaster ID number) makes a point forecast $y_{i,t,h}$ for the variable y_t, with a horizon of h quarters for the target year t and a probability forecast p_i (for convenience subscripts t and h are omitted for now), where p_i is a $k \times 1$ vector of probabilities (in percentage points) corresponding to k predefined intervals such that $\sum_{j=1}^{k} p_i(j) = 100$. The target variable y is the annual growth rate of output or price level in percentages. We fit a continuous distribution f_i to the histogram of p_i. The fitting process is to minimize the sum of squared differences between the cumulative probabilities at the nodes. As mentioned earlier, the choice between generalized beta and triangular distribution depends on whether there are more than two bins with positive probabilities. Histograms fitted to generalized beta are also divided into four different cases, depending on whether the open bin on either end of the support has positive probabilities. If all probabilities are attached to closed bins, then for the generalized beta distribution whose density is $f(x, \alpha, \beta, l, r)$, l and r are set to be the lower bound and upper bound of the bins that have positive probabilities. Then in the fitting process only the shape parameter α and β need to be solved in the minimization problem. If the left (right) open bin has positive probabilities, then l (r) needs to be solved in the minimization problem.

3.2 Uncertainty Decomposition—Decomposing Aggregate Variance

The mean and variance of f_i are μ_i and σ_i^2, respectively (subscripts t and h are omitted for convenience). For individual i, σ_i^2 is the forecast uncertainty. If we take the average of σ_i^2 over all forecasters, we get the average uncertainty $\overline{\sigma_i^2}$. If we take the average of all individual means, we get the mean consensus forecast $\overline{\mu_i}$. The variance of the means μ_i (N is the number of forecasters in that quarter) - $\sigma_{\overline{\mu}}^2 = \frac{1}{N}\sum_{i=1}^{N}\left(\mu_i - \overline{\mu_i}\right)^2$ is the disagreement about individual means. We also pool all individual distributions into a consensus distribution and obtain the aggregate variance as the variance of that consensus (aggregate) distribution. Lahiri, Teigland, and Zaporowski (1988) showed the following identity:

$$\textit{Aggregate Variance} = \textit{disagreement in means} + \overline{\sigma_i^2}. \tag{11.2}$$

Boero, Smith, and Wallis (2013) have corroborated the above identity by characterizing the aggregate density as a finite mixture.[7] Multiplying both sides by N, we get another representation of (11.2) in terms of total variation:

$$\sum_{i=1}^{N}\int f_i(x)\left(x-\overline{\mu_i}\right)^2 dx = \sum_{i=1}^{N}\left(\mu_i-\overline{\mu_i}\right)^2 + \sum_{i=1}^{N}\sigma_i^2 \tag{11.3}$$

The left-hand side of (11.3) is the total variation, which is the sum of the variation over bins across all forecasters from the consensus mean. On the right-hand side, we have the sum of squares of mean deviations and the sum of individual

[7] Interestingly, the mixture model formulation and this decomposition is well known in the Bayesian forecast combination literature for some time, see Draper (1995).

variances, which are N times of disagreement and average uncertainty, respectively. Obviously, Eqs. (11.2) and (11.3) are equivalent and show that the variance of the aggregate distribution is the sum of the forecaster discord regarding their means and average of the individual variances. Therefore, disagreement is only one component of the aggregate uncertainty and may not be a good proxy for uncertainty if the average variance turns out to be the dominant component during a period and varies over time (cf. Lahiri and Sheng, 2010a). We now proceed to calculate these components from SPF density forecasts and look at their variations and relative importance.

3.3 Estimation of Variance

To compute the aggregate variance, we could follow two approaches. In approach 1, we can get the consensus histogram $p_c = \frac{1}{N}\sum_{i=1}^{N} p_i$ first and then fit a generalized beta distribution f_{p_c} to the consensus histogram, and obtain the variance of f_{p_c} denoted by $\sigma^2_{f_{p_c}}$. In approach 2, we utilize the individual generalized beta and triangular distributions that we have fitted to the individual histograms. We take the average of them to get the consensus distribution $f_c = \frac{1}{N}\sum_{i=1}^{N} f_i$, and we compute the variance of f_c, denoted by $\sigma^2_{f_c}$. The order of completing these two procedures will yield different results. Since the individual distributions are not linear, it is obvious that $f_{p_c} \neq f_c$. Whereas f_{p_c} is a unimodal distribution, f_c is a mixture distribution as the average of many generalized beta and triangular distributions, and may have multiple modes. Moreover, in approach 2, identity (11.2) strictly holds, while in approach 1 it does not. Indeed, in approach 1, for several quarters in our sample period, the aggregate variance was smaller than its disagreement component. For this reason, we adopt approach 2 in computing the aggregate variance in the remaining sections.

3.4 Correcting for Bin Size for Variance Decomposition

Before reporting our variance decomposition results, we have to deal with a complication in the survey design that between 1981 Q3 and 1991 Q4 the predefined intervals had length of 2.0 percentage points rather than 1.0 or 0.5 as in other quarters.

First, we show that bin length affects the computed levels of uncertainty, especially when uncertainty is low. A smaller number of bins and longer bin sizes reduce accuracy and information in the forecasts, and can in principle inflate calculated variances (cf. Shoja and Soofi, 2017). For periods before 1981 and after 1991, bin length was 1.0 for GDP growth and inflation. For inflation, the bin length has changed further to 0.5 after 2014 Q1. For periods when bin size is 1.0, we combine probabilities in adjacent bins from the original survey histograms to create new histograms with 2-point bins, refit continuous distributions to them, and then recalculate all relevant measures (cf. Rich and Tracy. 2010).

Figure 11.1 uses a scatterplot of observations for all quarterly uncertainty measures to show that, for the same density forecast, a 2-point bin produces higher aggregate (upper panel) and average (lower panel) variances compared to 1-point bins. This overestimation may partly be due to the fact that in the process of conversion from 1-point bin to 2-point bin histograms, we will have a larger number of cases with only 1 or 2 bins of positive reported probabilities that would necessitate fitting triangular distributions, which may overstate the

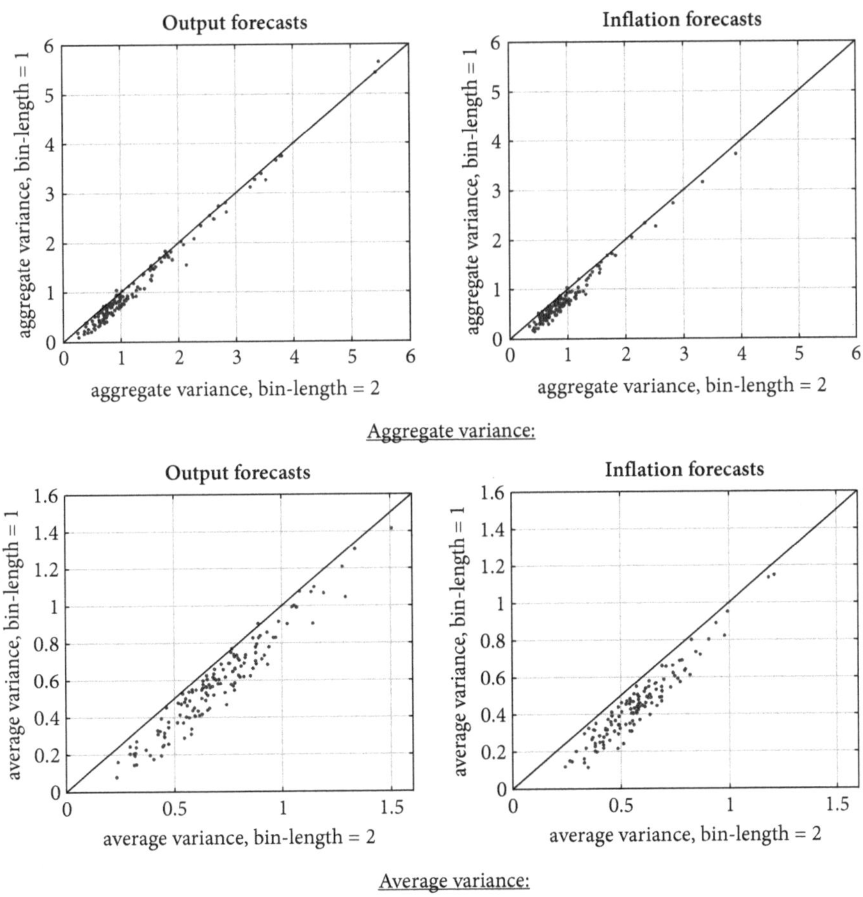

Figure 11.1 Effect of bin size on quarterly moment-based uncertainty.

variances (see Liu and Sheng, 2019). However, by comparing variances based on triangular distributions after converting all histograms to 2-bin histograms, we found that this concern is not an issue in our calculations.

In order to adjust the uncertainty measures during 1981–1991 with a wider bin, we run ordinary least squares (OLS) regressions to approximately convert the uncertainty measures estimated with 2 percent bins to values the same we would have obtained had the bins been 1 percent, like the rest of the sample. We also correct for bin size for 2014 Q1–2017 Q3 inflation forecasts by combining adjacent 0.5 percent bins into 1 percent bins and re-calculate uncertainty measures from the new histograms. Thus, our estimated uncertainty series will have an underlying bin size of 1 percentage point throughout the sample.

We use two regressions to correct average variances for bin-length. In these regressions, dependent variable is the uncertainty measures computed using 1-point bin settings (y), and independent variable is the uncertainty measures computed using 2-point bin settings (x). We regress y on x for the output forecast sample {post-1991Q4} and inflation forecast sample {pre–1981 Q3 and 1992 Q1–2013 Q4}, respectively, and use these coefficients to calculate predicted $\hat{y}$ from x for samples in the period of {1981 Q3 to 1991 Q4}. In this way, estimates from surveys when bin length is 2 points could be adjusted to their 1-point-bin-length equivalents. The regression results are presented in Table 11.1.

With the above regressions, average variances in most quarters during 1981 Q3 and 1991 Q4 are adjusted a little lower by around 0.1 (range—0.09 to 0.13). Since the mapping coefficients were very similar with average variance, we adjusted the aggregate variances by the same magnitude that left the disagreement measures unchanged. During 1981 Q3–1991 Q4, uncertainty measures in most quarters were adjusted a bit lower, but compared to the era after, the values are still high and are not attributed to the larger bin size.

Table 11.1 Regression Results with Variance for Mapping 2-Point Bins to 1-Point Bins: $y = a + bx + u$

	a	b
Average variance, output forecasts (Sample period: 1992Q1–2017Q3)	–0.1468 (0.0147)	1.0375 (0.0167)
Average variance, inflation forecasts (Sample period: pre–1981Q3 plus 1992Q1–2013Q4)	-0.1312 (0.0133)	1.0085 (0.0199)

Note: y is variance when bin length is 1 percent; x is the measured variance after converting bin length to 2 percent. Standard errors are in parentheses.

3.5 Variance Decomposition Results

The time series of aggregate variance, disagreement, and average individual variance for forecast horizons of one, three, six, and eight quarters are presented in Figures 11.2A and 11.2B for output and inflation forecasts, respectively.[8] The eight graphs in Figures 11.2A and 11.2B show how the aggregate variance and its two components evolve over time by forecast horizon.

We summarize the main results as follows:

1. For the period between 1981 Q3 and 1991 Q4, output growth uncertainty as well as inflation forecast uncertainty were exceptionally high. This elevated level of historical uncertainty cannot be explained by the size of the bin being 2.0 rather than 1.0 percentage point.
2. The recent two decades have witnessed persistently low levels of uncertainty and disagreement for both output and inflation forecasts. Forecaster discord tends to soar quickly and can exceed the uncertainty component during recessions and structural breaks. It can be argued that during these periods, forecasters facing conflicting news, interpret them differently and generate higher disagreement. We also note that output uncertainty and disagreement are significantly higher than those of inflation because the former is more difficult to predict (cf. Lahiri and Sheng, 2010b).
3. When we look at the two components of aggregate variance, we find that before 1981 disagreement accounted for most of the aggregate uncertainty, while after 1981, it played a less prominent role. Before 1981, most spikes in aggregate variance were due to sharp increases in disagreement, while between 1981 and 1991, most spikes in aggregate variance were a result of higher average variances.
4. Average variance is much less volatile than disagreement over the full sample, and it picks up only a little during the 1980s. Again, as we showed before, this is not an artifact due to the length of interval being 2.0 percent rather than 1.0 percent during 1981–1991.
5. Uncertainty is higher when forecast horizon is longer. Generally, the average values of all three series tend to decrease as forecast horizon shortens, but over horizons 8 to 6 uncertainty does not decline much. The decline, however, is dramatic as the horizon falls from 5 to 1. We expected this interyear difference because, as more information ("news") comes in the current year, they are directly related to the target year and hence resolve forecasters' uncertainty for current year's outturn more effectively.[9]

[8] These measures for horizons 2, 4, 5, and 7 are not reported here for the sake of brevity, but they are available from the authors upon request.

[9] Note that for some quarters before 1981 Q3 we do not have data for 1-quarter and longer horizon forecasts ($6 \sim 8$ quarters)—as a result we cannot see the potentially high spikes in these volatile quarters, particularly for the longer horizon graphs.

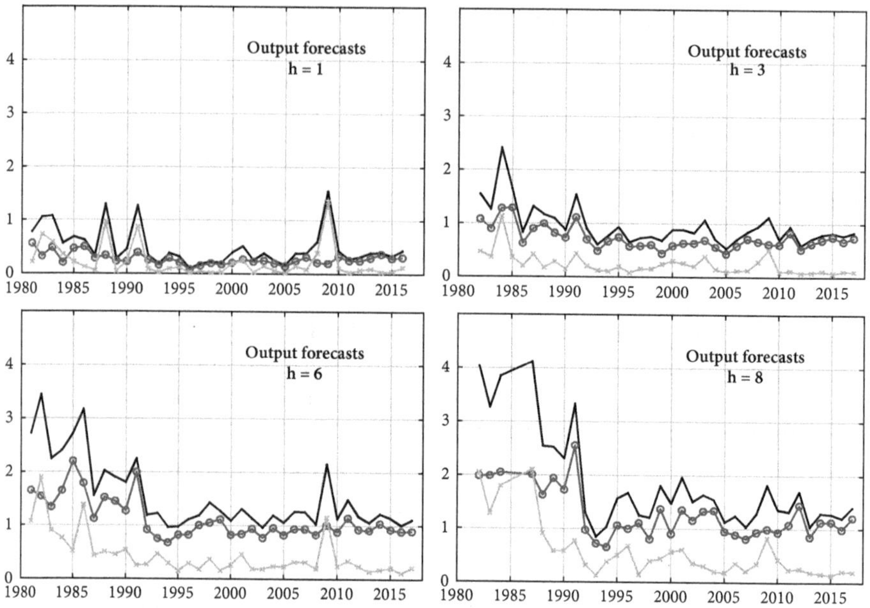

Panel A: Output variance and its components, corrected for bin-length, 1981Q3 to 2017Q3.

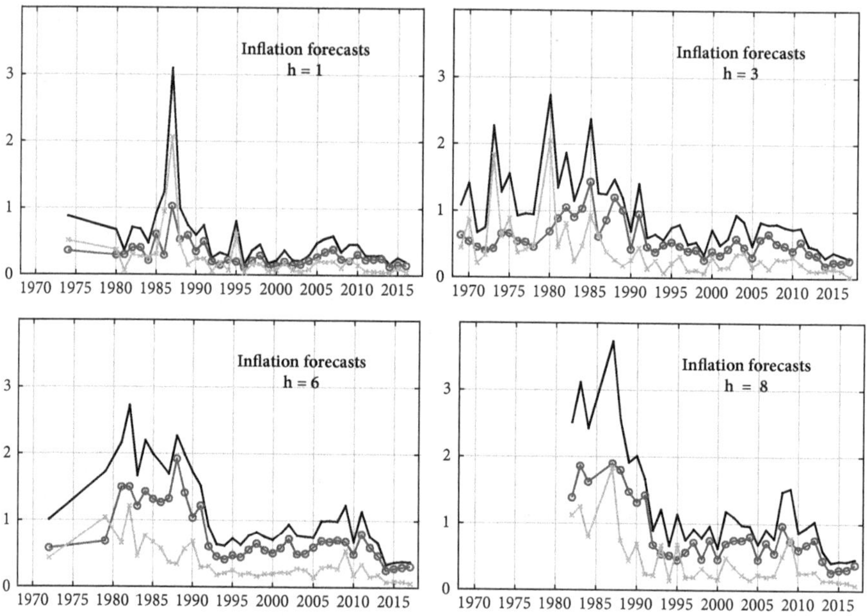

Panel B: Inflation variance and its components, corrected for bin-length, 1968Q4 to 2017Q3.

Figure 11.2 Moment-based uncertainty and its components, corrected for bin length.

Notes: Black line: aggregate variance. Grey line with circles: average variance. Light grey line with crosses: disagreement.

Noting the different dynamics of the three series, we can conclude that disagreement is not a good proxy for uncertainty, especially since the early 1990s. Furthermore, the variance of the point forecasts as a measure of uncertainty has another limitation. Lahiri and Liu (2009), Engelberg et al. (2009), and Clements (2014a) find that point forecasts can deviate from the means/medians of density forecasts and that forecasters tend to provide more favorable point forecasts than their corresponding density central tendencies. Elliott et al. (2005) attribute this to asymmetry in forecaster loss functions. Thus, disagreement in point forecasts may even be a worse proxy than disagreement in density means as a measure of uncertainty.

4. Uncertainty and Information Measures

Recently, info-metrics has been incorporated in forecast analysis.[10] Soofi and Retzer (2002) present a number of different information measures and estimation methodologies. Rich and Tracy (2010) calculate the entropy of forecast densities and show that uncertainty should not be proxied by disagreement. Shoja and Soofi (2017) show that entropy of the consensus distribution can be decomposed into average entropy and disagreement in terms of information divergence, and the decomposition of the aggregate variance could be incorporated in a maximum entropy model based on the first two moments. In this section, we build on the contributions of these researchers and compute entropies and information measures based on fitted generalized beta and triangular distributions that we estimated in the previous section. We decompose the uncertainty in the info-metrics framework, but unlike Shoja and Soofi (2017), we use fitted continuous distributions. In info-metrics, entropy reflects all information contained in a distribution, independent of its shape. The Shannon entropy is defined as

$$H(f) = -\int f(y)\log\left(f(y)\right)dy. \tag{11.4}$$

It measures how close a distribution is to a uniform distribution on the same support. The higher the entropy, the less information the distribution contains, and hence the higher the level of uncertainty. The entropy is akin to the variance of the distribution but embodies more characteristics of the distribution. Different distributions may have the same variance, but their shapes may be different and thus have different entropies. López-Pérez (2015) shows that entropy satisfies several properties as a "coherent risk measure," whereas moment-based

[10] Golan (2018), especially Chapter 4, contains the necessary background literature.

measures such as variance do not. Entropy is superior to variance, especially in cases when the underlying distribution is not unimodal and potentially discontinuous.

We calculate entropy and variance of available density forecasts after fitting generalized beta or triangular distributions as described before, and we show their relationship in Figure 11.3. It suggests a highly significant linear relationship between individual entropies and the logarithm of the variances for majority of the observations. It is not surprising in view of the well-known result that for normally distributed densities with variances σ^2,

$$H = \frac{1}{2}\log\left(2\pi e\sigma^2\right) = \frac{1}{2}\log(2\pi e) + \frac{1}{2}\log\sigma^2 = 1.42 + 0.5\log\sigma^2, \qquad (11.5)$$

Cf. Ebrahimi et al. (1999). However, many of the observations in Figure 11.3 are located below the diagonal line as well. We checked the properties of these distributions and found that they are significantly more skewed and leptokurtic and have longer horizons than those on the diagonal. We also regressed the entropies on the first four moments and their squares from the same individual densities

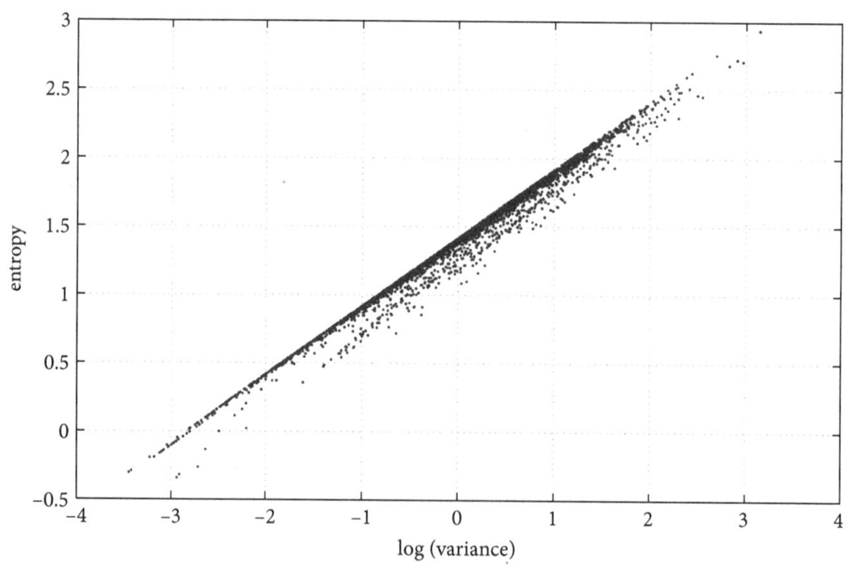

X axis: individual level current year output forecasts log (variance)
Y axis: individual level current year output forecasts entropy

Figure 11.3 Entropy versus log (variance).

Table 11.2 Entropy and Moments of the Density Forecasts, 1992Q1–2017Q3, Output Forecasts

	Horizon 1		Horizon 2		Horizon 3		Horizon 4	
	coefficient	*p*-value	coefficient	*p*-value	coefficient	*p*-value	coefficient	*p*-value
constant	1.4492	0	1.4461	0	1.4489	0	1.4498	0
mean	−0.0019	0.1055	0.0017	0.2467	0.0018	0.1254	0.0022	0.0724
$mean^2$	0.0003	0.2028	−0.0005	0.0878	−0.0004	0.0861	−0.0002	0.4446
log(var)	0.5017	0	0.5020	0	0.5027	0	0.5010	0
$(logvar)^2$	−0.0008	0.2175	−0.0004	0.4074	0.0002	0.7226	−0.0008	0.1317
Skew	0.0013	0.7619	0.0019	0.6370	0.0002	0.9457	−0.0058	0.0420
$Skew^2$	−0.2891	0	−0.2823	0	−0.3031	0	−0.3164	0
excesKurt	0.0837	0	0.0825	0	0.0917	0	0.1026	0
$excesKurt^2$	0.0080	0	0.0068	0	0.0095	0	0.0166	0
observations	461		718		824		849	
adjusted R^2	0.9994		0.9994		0.9991		0.9992	

Note: *p*-values are HAC-corrected. Estimates are based on histograms with more than 2 bins.

for four forecast horizons separately for output forecasts over 1992 Q1 to 2017 Q3. The estimated coefficients, together with their p-values that are robust to heteroscedasticity and autocorrelation (HAC), are reported in Table 11.2. Ebrahimi et al. (1999) used the Legendre series expansion to show that entropy may be related to higher-order moments of a distribution, which, unlike the variance, could offer a much closer characterization of the density distribution. Indeed, the adjusted R-squares in the four horizon-specific regressions, where we also include the squares of the four moments to reflect possible nonlinearity, are very close to one. The pairwise interactions of the moments were not significant at the usual 5 percent level. The significance of the skewness and excess kurtosis terms suggest non-normality of some of the forecast distributions, even though the total marginal contribution of these two factors never exceeded 5 percent for any one of the horizon-specific regressions. The location-invariance property of the entropy is also borne well in that the mean and its square together have statistically insignificant marginal explanatory power for each horizon. Most of the explanatory power comes from the two variance terms—remarkably, the adjusted R^2 falls from 0.999 to 0.237, 0.142, 0.095, and 0.060 in the four horizon-specific regressions, respectively, when log (σ^2) and its square are omitted from the fully specified regression with mean, variance, skewness, and excess kurtosis and their squares. Thus, as horizon increases, the importance of variance terms in explaining entropy increases. At horizons 1 and 2, skewness and kurtosis pick up some of the explanatory power in the absence of log (σ^2) terms. In the above regressions, other parameter estimates are consistent with our expectations under normality. That is, the intercept and the coefficient of log (σ^2) are close to 1.42 and 0.5, respectively. Note that the sample sizes for the four regressions get larger from 461 for horizon 1 to 849 for horizon 4. This is because, due to reduced uncertainty at shorter horizons, the 2-bin histograms are more prevalent for shorter horizons. A dummy to control for the triangular distributions for 2-bin histograms had no effect on our results when the regression reported in Table 11.2 included all histograms.

The divergence between two distributions is related to the entropy. A popular divergence measure is the Kullback–Leibler information measure (KL) or Kullback–Leibler divergence (Kullback and Leibler, 1951). For two distributions whose pdf are $f(x)$ and $g(x)$, the KL divergence is defined as

$$KL\left(f(x),g(x)\right) = \int f(x)\log\frac{f(x)}{g(x)}\,dx. \tag{11.6}$$

The KL information measure is valid only when $f(x)$ is absolutely continuous with respect to $g(x)$. This has some limitations while working with beta-shaped distributions. Besides, the KL information measure as it is used widely is not symmetric to the order of the two-component density distributions, that is, $KL(f(x),g(x)) \neq KL(g(x),f(x))$, if certain conditions are not met.[11]

We adopt another information measure—the Jensen–Shannon divergence (JS). Shoja and Soofi (2017) adopted the JS divergence for mixture models as a measure of disagreement as well as the expected uncertainty reduction in info-metrics. For two distributions $f(x)$ and $g(x)$, the JS divergence is defined as

$$JS\left(f(x),g(x)\right) = \frac{1}{2}\left(\int f(x)\log\frac{f(x)}{m(x)}\,dx + \int g(x)\log\frac{g(x)}{m(x)}\,dx\right), \qquad (11.7)$$

where $m(x) = \frac{1}{2} \times (f(x) + g(x))$. The JS information measure is symmetric to the order of the two-component density distributions and is suitable for cases when the densities $f(x)$ and $g(x)$ are defined over different intervals. The JS divergence has been extensively studied by statisticians since the 1990s (see Lin, 1991 and Minka, 1998).

The information measure can be viewed as a measure of disagreement between forecasters, but it contains more information than "disagreement in means." It captures the differences in all aspects of the distributions rather than the difference in the means only. Following Shoja and Soofi (2017), we develop a parallel analysis between variance decomposition and entropy decomposition.

$$\begin{aligned} \textit{Entropy of aggregate distribution} &= \textit{Average individual entropy} \\ &\quad + \textit{Information measure} \end{aligned} \qquad (11.8)$$

We first obtain an estimate of the entropy of the consensus or aggregate distribution in two steps similar to approach 2 in section 3. We utilize the individual generalized beta and triangular distributions that we have fitted to the individual histograms. We take the average of them to get the consensus distribution $f_c = \frac{1}{N}\sum_{i=1}^{N} f_i$, and we compute the entropy of f_c denoted by $H(f_c)$.

The first term of the right-hand side of Eq. (11.8), "average individual entropy," is obtained by taking the average of entropies of f_i across forecasters. The third term in Eq. (11.8), the "information measure," captures disagreement among f_i in a more holistic manner. With N forecasters, we can compute the JS information measure for $\frac{N(N-1)}{2}$ pairs of distributions, but this approach is computationally inconvenient and unwieldy. Rather, we use the generalized JS information measure, which is defined for more than two distributions as

[11] The original KL paper also introduced a symmetric function.

$$JS(f_1,f_2,...f_N) = H(f_c) - \frac{1}{N}\sum_{i=1}^{N} H(f_i) \tag{11.9}$$

where *H(.)* is the Shannon entropy. This form of JS information measure is closely related to the KL information measure:

$$JS(f_1,f_2,...f_N) = \frac{1}{N}\sum_{i=1}^{N} KL(f_i,f_c) \tag{11.10}$$

We follow Eq. (11.9) and take the difference between $H(f_c)$ and average entropy $\frac{1}{N}\sum_{i=1}^{N} H(f_i)$ to get the information measure.

Similar to section 3 where the variance approach is used, we examine if the heightened entropy during 1981–1991 can be attributable to the longer bin length of 2 percent in this period. To see the effect of this longer bin length on entropy, we recalculated the uncertainty of output and inflation forecasts in periods when bin length is 1 percent, assuming they all had a 2-point bin length as well (in a similar way as with variances in section 3). We combine probabilities in adjacent bins for the original survey histograms to create new histograms with 2-point bins, refit a continuous distribution to each of them, and then recalculate all relevant measures.

In Figure 11.4, we plot the aggregate entropy with 1-point bin length against the same recalculated with 2-point bin length for 2-quarter-ahead forecasts for the sake of illustration. Since we suspect the effects of bin size change are different for different levels of uncertainty, and since uncertainty was high in the 1970s, we include pre-1981Q3 output forecasts as well in this figure, even though the output variable was defined as nominal GNP. Uncertainty, assuming 2-point bin length, is uniformly higher than that using a 1-point after 1991. However, in the 1970s, use of a 2-point bin length does not cause the uncertainty to be much different. These results are very similar for both GDP growth and inflation. It indicates that when uncertainty is low, the structure of bins (bin length) will have a larger effect on the outcome and will cause the entropy measure to be higher. An explanation is that when entropy is high, the densities span over many intervals, and after a change of the bin length from 1 point to 2 points, there are still enough bins with positive probabilities to provide adequate information that will approximate the distribution well. If entropy is low, however, after the change, the number of bins may shrink to 1 or 2, which may alter the choice of the fitted continuous distribution and the estimate of entropy.

Similar to what we did in the last section, we regress average entropy measures calculated from the 1-point bin length (y) on entropy measures recalculated using 2-point bin length (x) for the sample that has a 1-point bin size. We

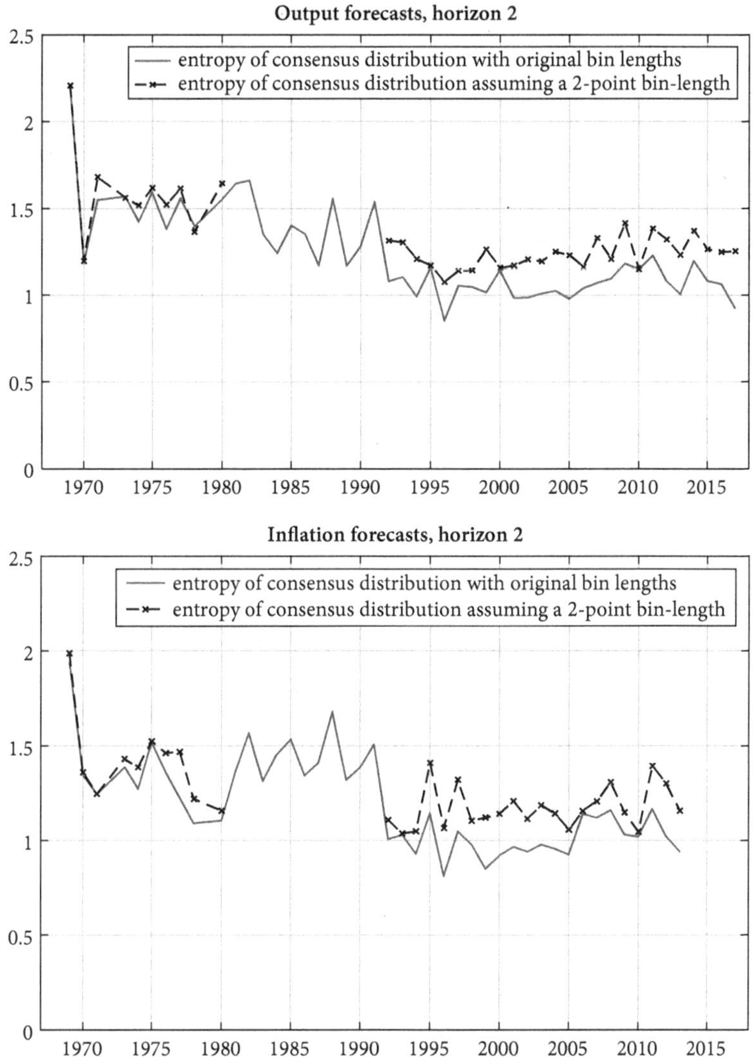

Figure 11.4 Effects of bin length on info-metrics-based uncertainty

then use the coefficients to calculate $\circ y$ using x for a sample in the period 1981Q3–1991Q4, and in this way our estimates from surveys with a 2-point bin length are adjusted to their 1-point bin-length equivalents. We also correct for bin size for 2014Q1–2017Q3 inflation forecasts by combining adjacent 0.5 percent bins into 1 percent bins and recalculate uncertainty measures from the new histograms. We present these regression results in Table 11.3. After the correction for bin length for average entropies, we adjust the aggregate

Table 11.3 Regression Results with Entropy for Mapping 2-Point Bins to 1-Point Bins: $y=a+bx+u$

	a	b
Average entropy, output forecasts (Sample period: 1992Q1–2017Q3)	−0.6987 (0.0415)	1.4615 (0.0362)
Average entropy, inflation forecasts (Sample period: pre-1981Q3 plus 1992Q1–2013Q4)	−0.5859 (0.0476)	1.3657 (0.0457)

Note: y is entropy measured when bin length is 1 percent; x is calculated entropy after converting bin length to 2 percent. Standard errors are in parentheses.

entropy of each quarter by the same values, thus leaving information divergence unchanged.

In Figures 11.5A (output forecasts) and 11.5B (inflation forecasts), the aggregate entropy, average entropy, and information divergence are displayed by forecast horizon. To save space, only horizons 1, 3, 6, and 8 are shown. In these graphs, aggregate entropy is $H(f_c)$, average entropy is $\frac{1}{N}\sum_{i=1}^{N} H(f_i)$, and information divergence is obtained from Eq. (11.8). During 1981—1991, both aggregate entropy and average entropy are now smaller in magnitude than those without any adjustment for bin length, and they are not as high as that in the 1970s.

We can now summarize the main characteristic features of the calculated entropies of the consensus distributions, average entropies, and information measures for real output growth (1981–2017) and inflation (1968–2017) for the different horizons separately:

1. Output and inflation forecasts have similar cyclical variations in uncertainty across all forecast horizons over the sample. The pre-1981 Q3 period for inflation is characterized by the high overall uncertainty (high aggregate entropy), and high disagreement (high information measure). The 1981 Q3–1991 Q4 period is characterized by a big contribution of average entropy to the aggregate entropy. The post-1992 era sees smaller aggregate entropy.
2. As horizon decreases, uncertainty falls, but we found (not reported in the chapter) that the quarterly declines from horizon 8 to horizon 5 is quite small. Only from horizon 4 to 1 does that uncertainty decline sharply. One obvious reason is that, as the quarterly values of the target variable in the current year get announced, part of the target variable becomes known, and hence the uncertainty with the remaining part of the target

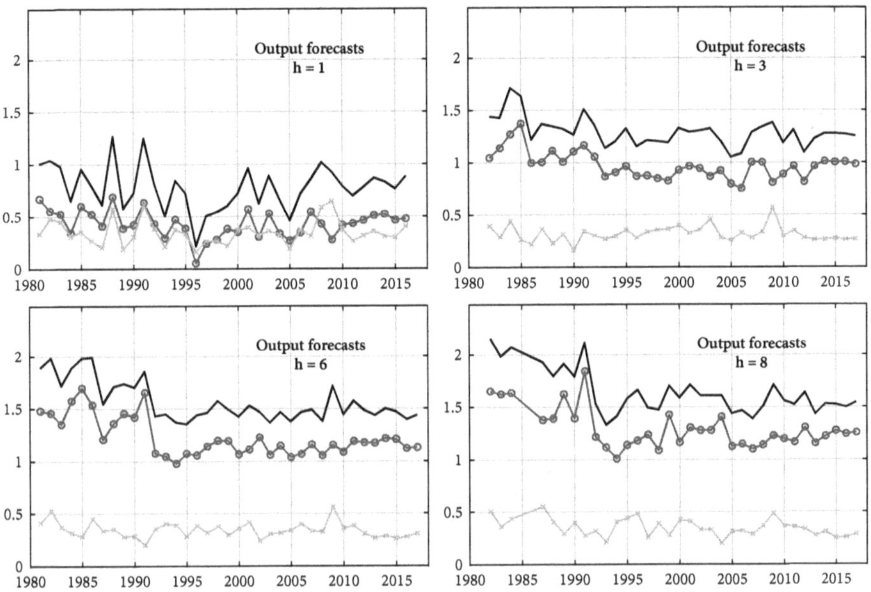

Panel A: Output forecasts, corrected for bin-length, 1981Q3 to 2017Q3.

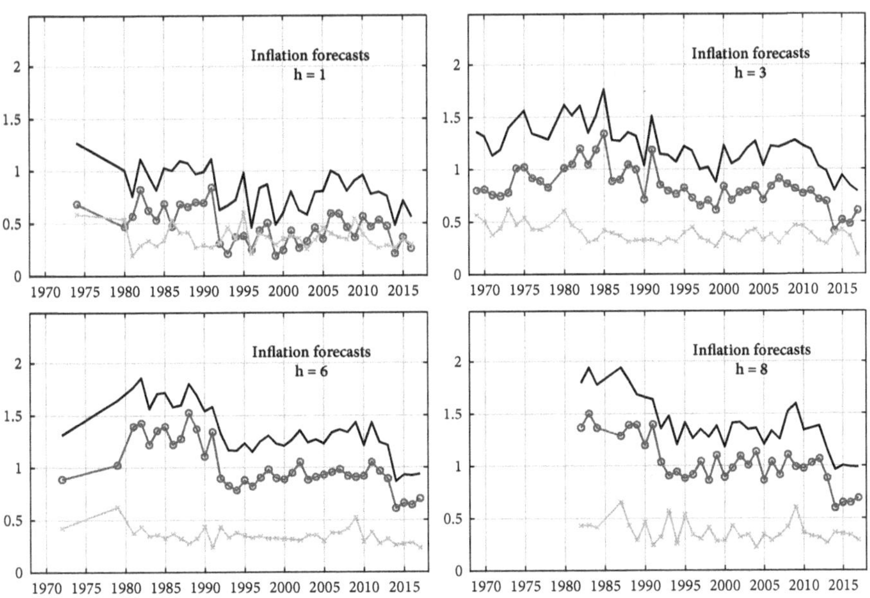

Panel B: Inflation forecasts, corrected for bin-length, 1968Q4 to 2017Q3.

Figure 11.5 Entropy and its components, corrected for bin length.

Black line: entropy of consensus distribution. Grey line with circles: average entropy. Light grey line with crosses: information divergence.

variable decreases. It does not suggest that forecasters are getting better and more efficient over the quarters. Even though we consistently see new information reducing uncertainty in all eight rounds of forecasting, news does not reduce uncertainty very much for next-year forecasts. This is true for both output and inflation.[12]

3. Over all, entropy and information measure in Figures 11.5A and 11.5B are similar to aggregate variance and disagreement in Figures 11.2A and 11.2B. By comparing these figures more closely, we find that from longer to shorter horizons and from the 1970s to the 2000s, uncertainty as measured by variance falls more precipitously than those based on entropy. The highly volatile disagreement in the moments-based measure is much muted in the information measure during the 1970s, primarily because the latter reflects more than just disagreement in means. As a result, the entropy measures are more stable and less volatile than variance measures. The decline in trend uncertainty using entropy is not as significant and sustained as the variance-based measures. We also note that disagreement in means contributed more than 80 percent of the aggregate uncertainty in certain quarters, but disagreement in distributions (information measure) never contributed more than 50 percent of the aggregate entropy. Thus, the relative importance of disagreement toward aggregate uncertainty is more limited with entropy measures compared to variance. These findings are noteworthy regarding the use of info-metrics in measuring uncertainty and forecaster discord.

We conclude section 4 by pointing out that the calculation of entropy using fitted continuous distributions rather than discrete distributions may produce quite different results with or without adjustment for the unequal number of bins (normalized entropy index[13]) as used by Shoja and Soofi (2017) for discrete distributions. For each calculated entropy, the upper panel in Figure 11.6 reports two entropy values for output forecasts using discrete distributions—one without the division by the log of the number of bins (denoted by grey *) and the other with the number of bins adjustment (black •) against the entropy calculated from fitted continuous distributions. The latter are adjusted individually for bin sizes following the logic in the previous section. The lower panel of Figure 11.6 reports the same scattered diagram for inflation forecasts. For a vast majority of distributions in the upper panel, the grey *'s are close to

[12] Clements (2014b) finds that the SPF *ex ante* survey uncertainties overestimate the *ex post* RMSE-based uncertainty for the current year forecasts, but underestimate for next year forecasts.

[13] For a density forecast with k bins, under discrete distribution assumption, the normalized entropy would be the discrete entropy divided by $ln(k)$.

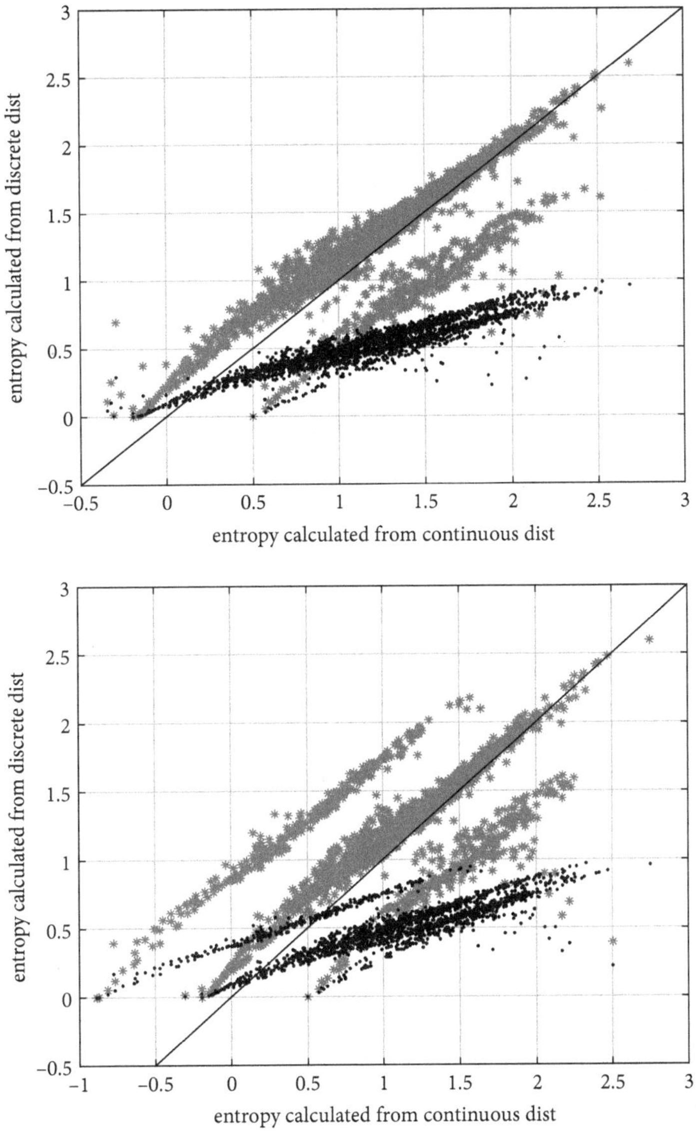

Grey * (stars): X-axis: entropy values calculated from continuous distributions
Y-axis:entropy values calculated from discrete distributions

Black • (dots): X-axis: entropy values calculated from continuous distributions
Y-axis:entropy values calculated from discrete distributions, with adjustment for numberof bins.

Upper panel: current year output forecasts 1968Q4–2017Q3

Lower panel: current year inflation forecasts 1968Q4–2017Q3

Figure 11.6 Scattered plot of entropy from alternative methods (continuous vs. discrete).

the 45° line, indicating that entropies from discrete distributions match well with the estimated entropies based on continuous distributions. However, there is a group of grey *'s that run parallel to but significantly below the diagonal line. We identified the latter group of observations as those coming from histograms between 1981Q3 and 1991Q4 (when survey has 6 bins and bin size 2 percent). Clearly, for these histograms, entropy for discrete distributions is lower than that from a continuous fit. With the correction for the number of bins, the black •'s are no longer separated into two blocks, meaning the correction is successful. The slope of the block of black •'s is smaller than one; this is due mostly to the fact that the correction multiplier is greater than 1.

In the lower panel of Figure 11.6 (inflation), there are three clusters of grey *'s. The group of *'s above the 45° diagonal are from the 2014 Q1–2017 Q3 period (when bin size is 0.5 percent), the group below the 45° diagonal are from the 1981 Q3–1991 Q4 period (when bin size is 2 percent), and the group close to the diagonal are from 1968 Q4–1981 Q2 and 1992 Q1–2013 Q4 (when the bin size is 1 percent). After applying the normalized entropy correction as used by Shoja and Soofi (2017), the group below and on the 45° line merge into the bulk of black •'s; however, the group of *'s for 2014 Q1–2017 Q3 still form a separate group (a thin cluster of black dots whose x-coordinates are between −1 and 1.5). The correction of dividing by *ln(number of bins)* does not match the entropy values from continuous distributions. This is because the calculations based on discrete distributions cannot adjust for the bin size as two discrete distributions that have the same height, but two different bin widths will have exactly the

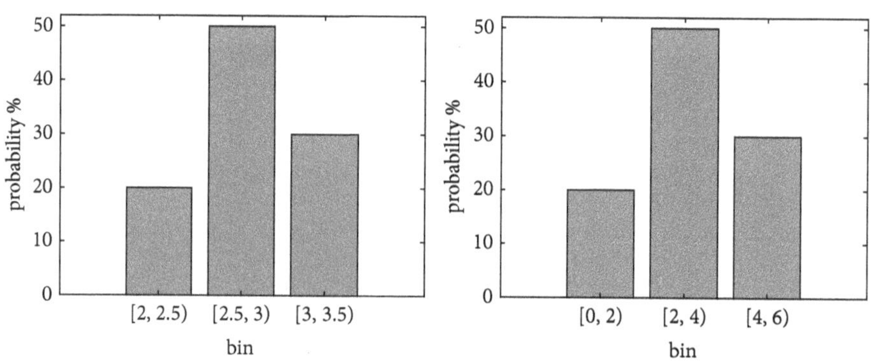

Figure 11.7 Two discrete distributions with the same entropy but different levels of uncertainty.

The above two histograms have the same entropy under discrete distribution, viz.:

$H = -(0.2\ln 0.2 + 0.5\ln 0.5 + 0.3\ln 0.3) = 1.0297$

However, they imply different levels of uncertainty. The forecaster on the left panel is certain that there is a 100 percent probability that the target variable will be between [2 percent and 3.5 percent] while the forecast on the right panel implies only a 50 percent chance that the target variable will fall between [2 percent, 4 percent].

same entropy. We present such an example in Figure 11.7, using two density forecasts. These are from two different surveys and have different predefined bin widths. Both forecasts have the same shape when drawn as histograms. Under the discrete distribution assumption, both forecasts have an entropy of 1.0297; however, they imply different levels of uncertainty. The forecaster for the left panel is certain that there is a 100 percent probability that the target variable will be between [2 percent, 3.5 percent], while the forecast on the right panel implies only a 50 percent chance that the target variable will fall between [2 percent, 4 percent]. Therefore, we can see that it is only coincidental that the normalized entropy successfully corrects the observations during 1981 Q3–1991 Q4 because a larger bin size was accompanied by a smaller number of bins in this period. The number of bins during 2014 Q1–2017 Q3 is 10 and is similar to that in the majority of the histograms, but the bin size in the last sample period is 0.5 percent. As a result, the correction does not take into account this smaller bin size and fails to adjust these entropies at par with others. This leaves a cluster of black dots far from the rest of the sample in the lower panel of Figure 11.6. Thus, even after correcting for the number of bins, discrete entropies are not directly comparable when two histograms have different bin sizes. It is obvious that a larger bin size, *ceteris paribus*, makes the entropy based on discrete distributions smaller than it should be Figure 11.7.

To be consistent with 1981–1991, Rich and Tracy (2010) converted the bin lengths in all other subperiods to 2.0 percent, thereby inflating the entropy. For inflation forecasts, substantial information will be lost if the data since 2014 Q1 are adjusted into 2 percent-bin-size equivalents from 0.5 percent bin size, as it would require combining 4 bins into one. Another disadvantage of the discrete method is that, the estimates are very sensitive to the number of forecasters who attach a probability mass to only one bin. As shown in Figure 11.5 of Shoja and Soofi (2017), such sensitivity creates abrupt spikes in disagreement.

5. Time Series of Uncertainty Measures

After obtaining various uncertainty measures and correcting for bin length for different horizons, we adjust these values for horizons to obtain a single time series of uncertainty. The mean value of uncertainty measures for each horizon is subtracted, and then the average of horizon 4 is added to make all observations horizon 4-equivalent. This way we can have a quarterly time profile of the aggregate macroeconomic uncertainty in the last few decades. We have displayed horizon-corrected uncertainty measures for output and inflation forecasts in Figure 11.8, where we report quarterly average uncertainty as the overall uncertainty measure.

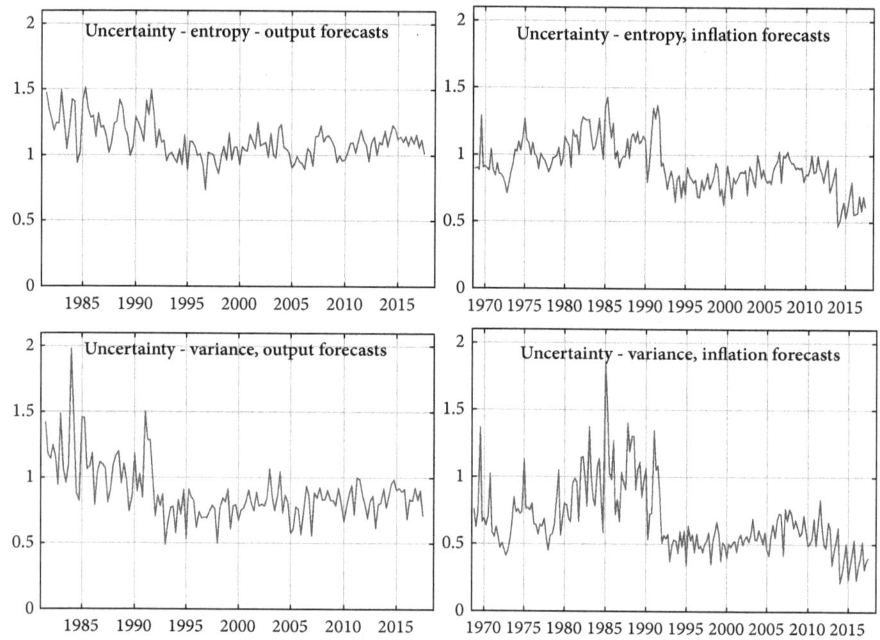

Figure 11.8 Uncertainty (entropy and variance)—corrected for bin length and horizon.

Based on both entropy and variance, output uncertainty fell steadily from the peak in the 1980s to a trough around 1995, but this decline is more dramatic with the variance measure. The fall of uncertainty in the post–1990 era seems to be permanent and is consistent with the macroeconomic uncertainty series generated by a factor stochastic volatility model by Jo and Sekkel (2019).

Uncertainty increased during the 2008 financial crisis and peaked at the end of 2008, which is coincident with the dramatic events of that quarter—Lehman Brother's bankruptcy and the U.S. government's bailout. Note that uncertainty of output forecasts based on entropy remains somewhat elevated during the last two years, suggesting a worry of some risk buildup in the economy.

For inflation forecasts, uncertainty remained high from 1970s until the early 1990s. Then it went into a downward trend. Similar to the output forecast uncertainty, the fall of inflation uncertainty in the 1990s seems to be permanent, and the new peak in the 2008 financial crisis is low compared to 1970/1980 levels. Unlike output forecasts, inflation uncertainty has taken a slow but downward drift beginning in 2010, reflecting a persistently low and stable inflation environment in the U.S. economy.

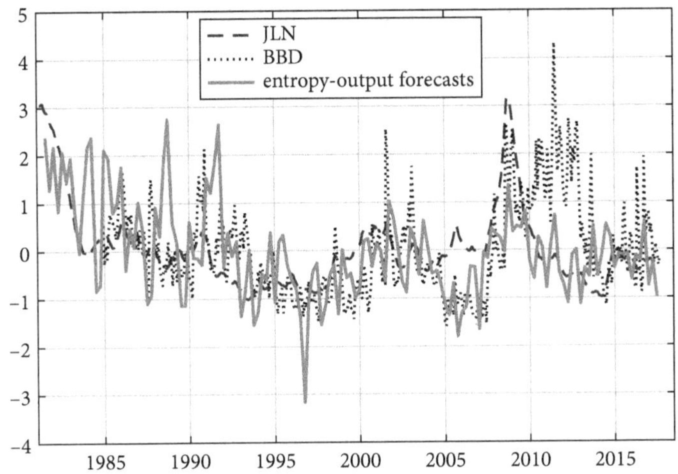

Figure 11.9 Uncertainty compared with other works.

For the sake of comparison, we have plotted our output entropy measure (aggregate entropy) together with two other popular uncertainty indexes from the literature (viz., Jurado, Ludvigson, and Ng, 2015 [JLN] and Baker, Bloom, and Davis, 2016 [BBD]) in Figure 11.9. While doing this, we have normalized each series in terms of their means and variances. JLN depicts general macroeconomic uncertainty, whereas BBD measures economic policy uncertainty. Since these three uncertainty indexes are related to three different underlying target variables, they are not strictly comparable. However, many uncertainty measures have turned out to be correlated, as shown in Kozeniauskas et al. (2018). Figure 11.9 reveals certain common features in these series. The overall cyclical movements of all three series are very similar except that JLN is smoother and BBD exhibits very high and variable values in the post-Great Recession era. Like our entropy measure of uncertainty, both JLN and BBD reached their troughs in the mid-1990s and spiked during 2008 recession. However, the magnitude and duration of the spikes are quite different between the three measures. Our entropy measure did not reach the unprecedented high levels of JLN and BBD during the 2008 crisis. BBD's high policy uncertainty is also persistent in the last few years, even when the economy has recovered. The extraordinary policy environment and subsequent policy changes that existed at the federal level following the financial crisis can explain the unique recent gyrations in BBD.

6. Information Measure and "News"

Economists frequently study how new information or "news" affects the economy. The conventional approach in macroeconomics is to generate "news" from estimated VAR models requiring values of the forecast errors. Revisions in fixed target forecasts as measures of news was first suggested by Davies and Lahiri (1995, 1999) and come closest to the theoretical concept that are obtained in real time using forecasts alone without requiring values of the target variables. The Jensen–Shannon information measure is a more nuanced measure of the same derived from info-metrics. In SPF, forecasters make fixed-target density forecasts, making it possible to check how forecasters update their forecasts quarter to quarter after getting new information. If a forecaster updates forecasts for the same target variable by moving from $f_{i,t,h+1}$ to $f_{i,t,h}$, we can obtain the information divergence of these distributions, based on $JS(f_{i,t,h+1}, f_{i,t,h})$. This measure can be viewed as "news" or "shocks to uncertainty."

"News" of quarter q_t is defined as

$$news = \frac{1}{n_{q_t}} \sum_i JS\left(f_{i,t,h+1}, f_{i,t,h}\right),$$

where n_{q_t} is the number of forecasters who provided fixed-target density forecasts in q_t and the quarter before. The measure is the average across available forecasters of information divergence between the two subsequent fixed-target densities. It is new information that makes the forecasters update their forecasts. Since each JS is computed using the same forecasters in two adjacent quarters, the "news" series will be independent of the varying composition of the SPF panel' cf. Engelberg et al. (2011).

In Figure 11.10, we report news or uncertainty shocks based on the Jensen–Shannon information measure for output forecasts. We find that forecasters were faced with more news in the tumultuous 1970s. Since the mid-1980s, the arrival of news has become less volatile, consistent with the so-called moderation hypothesis. Exceptions are the three recessions of the early 1980s, 2000/2001, and 2008/2009. "News" for real output shows a slightly countercyclical property. It spikes in the recessions when economic data exhibit more negative surprises that lead to large forecast revisions and correspondingly large news values.

The info-metric measure of "news" is quite similar in concept to "revisions in point forecasts." As a local approximation, we regressed "news" on absolute values of first difference of the first four moments at the individual level (first difference between $f_{i,t,h+1}$ and $f_{i,t,h}$) and found that "news" is significantly correlated with the differences (i.e., forecast revisions) in means, variances, skewness. and kurtosis of the density forecasts. The forecast revisions in the means explain most

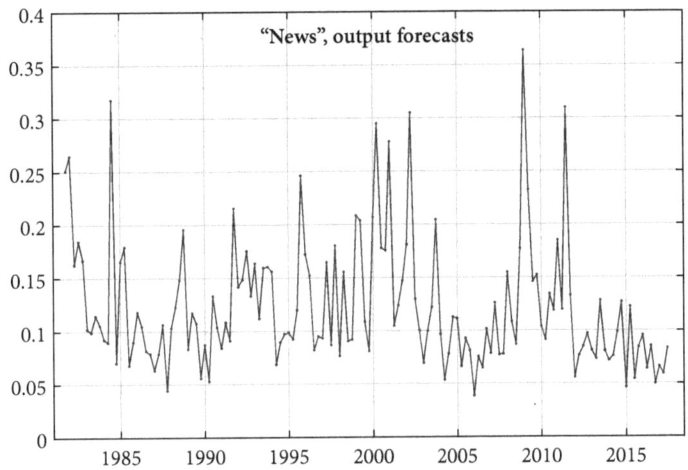

Figure 11.10 Uncertainty shocks or "news" based on JS information.

of the info-metric "news," that is, a shift in the forecast density's mean causes the biggest kick in the information divergence, while changes of variance, skewness, or kurtosis only explain less than 3 percent of the variation, even though their coefficients are statistically significant. This evidence indicates that even though info-metrics-based "news" theoretically reflects changes in higher moments of the forecast densities, in the SPF data it is well approximated by changes in mean forecasts alone.

"News" could affect various aspects of the economy. Engle and Ng (1993) first introduced the "news impact curve." In the last two decades, many authors have studied how "news" impacts on volatility, using mostly high-frequency data from the stock market. Davies and Lahiri (1995) showed a positive relationship between news and inflation volatility and an asymmetric effect of bad news on volatility. With our carefully constructed info-metrics based uncertainty measure, we regressed our uncertainty measure on two types of news to estimate such effects. Our regression model is

$$H_{t,h} = \alpha + \beta_1 \cdot news\_good_{t,h} + \beta_2 \cdot news\_bad_{t,h} + \beta_3 \cdot H_{t,h+1} + \varepsilon_{t,h}, \qquad (11.11)$$

where $H_{t,h}$ is entropy of the consensus forecast density for target year t and horizon h. $News\_good_{t,h}$ is the average of "news" across forecasters for target year t and horizon h, when the mean forecast revision is positive. $News\_bad_{t,h}$ is the average of "news" across forecasters for target year t and horizon h, when the mean forecast revision is negative. $\varepsilon_{t,h}$ is error term with zero mean. The results of the regression are given in Table 11.4.

Table 11.4 Regression of News on Uncertainty
$H_{t,h} = \alpha + \beta_1 \cdot news\_good_{t,h} + \beta_2 \cdot news\_bad_{t,h} + \beta_3 \cdot H_{t,h+1} + \mu_{t,h}$

Output forecasts			Inflation Forecasts		
Parameter	Estimate	*p*-value	Parameter	Estimate	*p*-value
α	0.8563	0.0000	α	0.3827	0.0000
β_1	0.7791	0.0377	β_1	0.8693	0.0342
β_2	0.9997	0.0020	β_2	0.8708	0.0233
β_3	0.2059	0.0055	β_3	0.5735	0.0000

For output forecasts, β_2 is greater than β_1, which is consistent with Davies and Lahiri (1995), indicating greater impact of bad news on output forecast uncertainty. However, the signs of the coefficients β_1 and β_2 are both positive, indicating that both good news and bad news increase uncertainty. Chen and Ghysels (2011) found that in the stock market both very good news (unusual high positive returns) and bad news (negative returns) increase volatility, with the latter having a more severe impact. For inflation forecasts, β_1 is not statistically different from β_2, indicating that bad news does not cause more severe impact on uncertainty than good news of similar size, and both are highly significant. The coefficient of lagged entropy is significant in both regressions, but inflation uncertainty ($\beta_3 = 0.57$) seems to be more persistent than output uncertainty ($\beta_3 = 0.21$). Interestingly, conventional information theory suggests that information is associated with uncertainty reduction. But in our case with fixed target forecasts $A_t = y_{i,t,h} + \varepsilon_{i,t,h}$, where A_t is the actual realized value of the target variable and $\varepsilon_{i,t,h}$ is the forecast error, the variance of A_t is the same over horizons. As h decreases, $y_{i,t,h}$ absorbs more information, and hence the forecast variance will necessarily increase. However, with variance of A_t the same over h, an increase in the entropy of $y_{i,t,h}$ will be compensated by a corresponding reduction in forecast error variance and ex post uncertainty. Under rational forecasts, information is expected to be associated with uncertainty reduction. It is certainly possible that "news" in a particular period may induce forecasters to reevaluate their confidence and increase their perceived *ex ante* uncertainty in the next period.

7. Impact of Real Output Uncertainty on Macroeconomic Variables

In this section, we explore how output forecast uncertainty affects the macro economy. Bloom et al. (2009, 2014), and Bloom et al. (2007, 2014) explore several channels through which uncertainty affects the real economy. Following the literature, particularly Baker et al. (2016) and Jurado et al. (2015), we use a standard VAR model with Cholesky-decomposed shocks to study the effects. We estimate a VAR with the following five quarterly variables: entropy of output forecasts, log of real output (RGDP), log of Non-Farm payroll (NFP), log of private domestic investment (PDI), and Federal Funds rate (FFR). A one standard deviation shock is imposed on the entropy. We include four lags of each variable that minimized the overall AIC. We drew the impulse responses of all five variables to a one standard deviation change in the entropy, keeping the Cholesky ordering the same as above. They are presented in Figure 11.11. We find that an increase of output uncertainty has negative effects on real GDP, payroll employment, and private investment significantly over next 5–10 quarters, after which they tend to revert back to their original values over the next 5 quarters. Thus, uncertainty shocks have a long-lasting effect on the real sector of the economy. Uncertainty itself quickly converges back after the initial shock. The dampening effect of uncertainty shock on the federal funds rate suggests the reaction of the monetary authority to combat the consequent negative effects on the real variables. A change in the ordering of variables does not alter the results. Specifications with fewer variables produce similar results that output uncertainty has negative effects on real GDP growth. The negative relationship between uncertainty and economic activity has been widely reported in recent macroeconomic research; see Bloom (2009, 2014), Bloom et al. (2014), and Jurado et al. (2015). We find that the Jensen–Shannon information measure as news or uncertainty shocks yields qualitatively similar results.

8. Summary and Conclusions

We approximate histogram forecasts reported in the SPF by continuous distributions and decompose moment-based and entropy-based aggregate uncertainty into average uncertainty and disagreement. We adjust for bin sizes and horizon effects, and we report consistent time series of uncertainty in real output and inflation forecasts over the last few decades. We find that adjustments for changes in the SPF survey design are difficult to implement, while working directly with discrete distributions. Even though the broad cyclical movements in uncertainty measures provided by the variance and the entropy of the forecast densities are similar and show significant moderation in the post–1990 era, the former is significantly more variable and erratic than the entropy-based uncertainty. In addition, the variance of mean forecasts as a measure of disagreement is a much larger, more variable component of the variance of the consensus density compared with the Jensen–Shannon measure

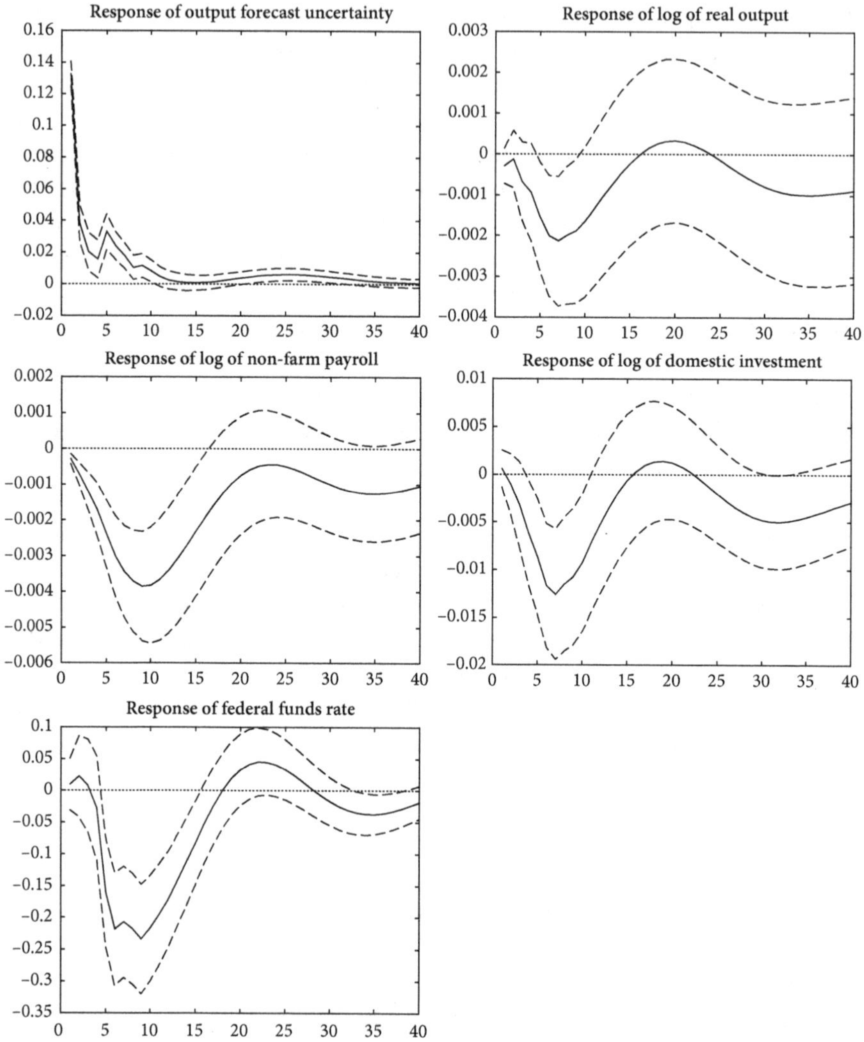

Figure 11.11 Impulse response functions in a five-variable VAR to a one SD shock of output forecast uncertainty (1981 Q3–2017 Q3).

Notes: The fivevariables are entropy of output, log of real output, log of nonfarm payroll, log of private domestic investment, and Federal Funds rate (FFR).

Dashed lines are one standard deviation bands. X-axis is in quarters.

based on aggregate entropy. These results are very similar for both output and inflation.

The cyclical dynamics of our entropy-based output uncertainty is consistent with Jurado et al. (2015), despite the fact that the latter is based on forecast errors of a very large number of macro indicators. Following Baker et al. (2016) and Jurado et al. (2015), we also examine the importance of our output entropy in a five-variable VAR and find that higher uncertainty has significant negative effects on real GDP, employment, investment, and the federal funds rate that lasts anywhere from 5 to 10 quarters. These impulse response results are consistent with recent empirical evidence on the pervasive role uncertainty has in the economy, and they lend credibility to the entropy-based uncertainty measure we presented in this chapter.

The other significant empirical findings are summarized as follows:

> First, since SPF density forecasts provide a sequence of as many as eight quarterly forecasts for a fixed target, the Jensen–Shannon information divergence between two successive forecast densities yields a natural measure of "uncertainty shocks" or "news" at the individual level in a true *ex ante* sense. We find that both bad and good news increase uncertainty and are countercyclical, which is consistent with the previous literature. It is noteworthy that new information in our context increases the horizon-adjusted forecast uncertainty, but correspondingly it should reduce the *ex post* forecast error uncertainty. Somewhat surprisingly, we find that the info-metrics-based "news" is conditioned mostly by the changes in the means of the underlying densities and very little by changes in other higher moments of the density forecasts.

Second, we find that the historical changes in the SPF survey design in the number of bins and bin sizes and how one adjusts for these changes affect uncertainty measures. Bigger bin sizes, *ceteris paribus*, inflate uncertainty estimates; more importantly, the overestimation will be relatively higher during low-uncertainty regimes. We also find that the normalized entropy index for correcting the unequal number of bins by dividing the discrete entropies by the natural log of the number of bins may have certain unexpected distortions. Compared to the entropies based on continuous distributions, the normalized entropies for discrete distributions fail to adjust for bin size, resulting in overestimation during periods when bin size is small.

Finally, even though the Shannon entropy is more inclusive of different facets of a forecast density than just variance, we find that with SPF forecasts it is largely driven by the variance of the densities. An initial indication of this evidence can be found in Shoja and Soofi (2017). However, skewness and kurtosis continue to be statistically significant determinants of entropy, though the total marginal contribution of these two factors is less than 5 percent in a regression of average entropy on the first four moments of the distribution. There are enough non-normal densities in the sample that make these high-order moments nonignorable components of aggregate entropy. All in all, our analysis of individual SPF densities suggests that uncertainty, news shocks, and disagreement measured on the basis of info-metrics are useful for macroeconomic policy analysis because these are, *ex ante*, never revised and reflect information on the whole forecast distributions.

References

Abel, Joshua, Robert Rich, Joseph Song, and Joseph Tracy. (2016). "The Measurement and Behavior of Uncertainty: Evidence from the ECB Survey of Professional Forecasters." *Journal of Applied Econometrics*, 31: 533–550.

Alexander, Carol, Gauss M. Cordeiro, Edwin M. M. Ortega, and José María Sarabia. (2012). "Generalized Beta-Generated Distributions." *Computational Statistics and Data Analysis*, 56(6): 1880–1897.

Baker, Scott R., Nicholas Bloom, and Steven J. Davis. (2016). "Measuring Economic Policy Uncertainty." www.policyuncertainty.com.

Bloom, Nicholas. (2009). "The Impact of Uncertainty Shocks." *Econometrica*, 77(3): 623–685.

Bloom, Nicholas. (2014). "Fluctuations in Uncertainty." *Journal of Economic Perspectives*, 28(2): 153–176.

Bloom, Nicholas, Stephen Bond, and John Van Reenen. (2007). "Uncertainty and Investment Dynamics." *Review of Economic Studies*, 74: 391–415.

Bloom, Nicholas, Max Floetotto, Nir Jaimovich, Itay Saporta-Eksten, and Stephen J. Terry. (2014). "Really Uncertain Business Cycles." *Econometrica*, 86(3): 1031–1065.

Boero, Gianna, Jeremy Smith, and Kenneth F. Wallis. (2008). "Uncertainty and Disagreement in Economic Prediction: The Bank of England Survey of External Forecasters." *Economic Journal*, 118: 1107–1127.

Boero, Gianna, Jeremy Smith, and Kenneth F. Wallis. (2013). "The Measurement and Characteristics of Professional Forecasters' Uncertainty." *Journal of Applied Econometrics*, 30(7): 1029–1046.

Chen, Xilong, and Eric Ghysels. (2011). "News—Good or Bad—and Its Impact on Volatility Predictions over Multiple Horizons." *Review of Financial Studies*, 24(1): 46–81.

Clements, Michael P. (2014a). "US Inflation Expectations and Heterogeneous Loss Functions, 1968–2010." *Journal of Forecasting*, 33(1): 1–14.

Clements, Michael P. (2014b). "Forecast Uncertainty- Ex Ante and Ex Post: U.S. Inflation and Output Growth." *Journal of Business and Economic Statistics*, 32(2): 206–216.

Davies, Anthony, and Kajal Lahiri. (1995). "A New Framework for Analyzing Survey Forecasts Using Three-Dimensional Panel Data." *Journal of Econometrics*, 68(1): 205–227.

Davies, Anthony, and Kajal Lahiri. (1999). "Re-examining the Rational Expectations Hypothesis Using Panel Data on Multiperiod Forecasts." In C. Hsiao, K. Lahiri, L-F Lee, and H. M. Pesaran (eds.), *Analysis of Panels and Limited Dependent Variable Models*. Cambridge: Cambridge University Press, 1999, pp. 226–254.

Draper, David. (1995). "Assessment and Propagation of Model Uncertainty." *Journal of the Royal Statistical Society*, B 57, 45–97.

Ebrahimi, Nader, Esfandiar Maasoumi, and Ehsan S. Soofi. (1999). "Ordering Univariate Distributions by Entropy and Variance." *Journal of Econometrics*, 90: 317–336.

Elliott, Graham, Ivana Komunjer, and Allan Timmermann. (2005). "Estimation and Testing of Forecast Rationality under Flexible Loss." *Review of Economic Studies*, 72(4): 1107–1125.

Engelberg, Joseph, Charles F. Manski, and Jared Williams. (2009). "Comparing the Point Predictions and Subjective Probability Distributions of Professional Forecasters." *Journal of Business and Economic Statistics*, 27: 30–41.

Engelberg, Joseph, Charles F. Manski, and Jared Williams. (2011). "Assessing the Temporal Variation of Macroeconomic Forecasts by a Panel of Changing Composition." *Journal of Applied Econometrics*, 26: 1059–1078.

Engle, Robert F., and Ng, Victor K. (1993) "Measuring and Testing the Impact of News on Volatility." *Journal of Finance*, 48(5): 1749c–1778.

Fresoli, Deigo, Esther Ruiz, and Lorenzo Pascual. (2015). "Bootstrap Multi-step Forecasts of Non-Gaussian VAR Models." *International Journal of Forecasting*, 31(3): 834–848.

Giordani, Paolo, and Paul Söderlind. (2003). "Inflation Forecast Uncertainty." *European Economic Review*, 47(6): 1037–1059.

Golan, Amos. (1991). "The Discrete Continuous Choice of Economic Modeling or Quantum Economic Chaos." *Mathematical Social Sciences*, 21(3): 261–286.

Golan, Amos. (2018). *Foundations of Info-Metrics: Modeling, Inference, and Imperfect Information*. Oxford: Oxford University Press.

Gordy, Michael B. (1998). "A Generalization of Generalized Beta Distributions." Board of Governors of the Federal Reserve System, Washington, DC, Finance and Economics Discussion Series: 18.

Jo, Soojin, and Rodrigo Sekkel. (2019): "Macroeconomic Uncertainty through the Lens of Professional Forecasters." *Journal of Business and Economic Statistics*, 37(3): 436–446.

Jurado, Kyle, Sydney C. Ludvigson, and Serena Ng. (2015). "Measuring Uncertainty." *American Economic Review*, 105(3): 1177–1216.

Kenny, Geoff, Thomas Kostka, and Federico Masera. (2015). "Density Characteristics and Density Forecast Performance: A Panel Analysis." *Empirical Economics*, 48(3): 1203–1231.

Knüppel, Malte. (2014). "Efficient Estimation of Forecast Uncertainty Based on Recent Forecast Errors." *International Journal of Forecasting*, 30(2): 257–267.

Kozeniauskas, Nicholas, Anna Orlik, and Laura Veldkamp. (2018). "What Are Uncertainty Shocks?" *Journal of Monetary Economics*, 100 (2018): 1–15.

Kullback, Solomon, and Richard A. Leibler. (1951). "On Information and Sufficiency." *Annals of Mathematical Statistics*, 22: 79–86.

Lahiri, Kajal, Christie Teigland, and Mark Zaporowski. (1988). "Interest Rates and the Subjective Probability Distribution of Inflation Forecasts." *Journal of Money, Credit, and Banking*, 20(2): 233–248.

Lahiri, Kajal, and Fushang Liu. (2006a). "ARCH Models for Multi-Period Forecast Uncertainty–A Reality Check Using a Panel of Density Forecasts." In D. Terrell, and T. Fomby (eds.), *Econometric Analysis of Financial and Economic Time Series (Advances in Econometrics, Vol. 20 Part 1)*. Bingley: Emerald Group Publishing Limited, 2006a, pp. 321–363.

Lahiri, Kajal, and Fushang Liu. (2006b). "Modeling Multi-Period Inflation Uncertainty Using a Panel of Density Forecasts." *Journal of Applied Econometrics*, 21: 1199–1220.

Lahiri, Kajal, and Fushang Liu. (2009). "On the Use of Density Forecasts to Identify Asymmetry in Forecasters' Loss Functions." *Proceedings of the Joint Statistical Meetings, Business and Economic Statistics*, 2396–2408.

Lahiri, Kajal, and Xuguang Sheng. (2010a), "Measuring Forecast Uncertainty by Disagreement: The Missing Link." *Journal of Applied Econometrics*, 25: 514–538.

Lahiri, Kajal, and Xuguang Sheng. (2010b). "Learning and Heterogeneity in GDP and Inflation Forecasts." *International Journal of Forecasting* (special issue on Bayesian Forecasting in Economics), 26: 265–292.

Lin, Jianhua. (1991). "Divergence Measures Based on the Shannon Entropy." *IEEE Transactions on Information Theory*, 37(1): 145–151.

Liu, Yang, and Xuguang Simon Sheng. (2019). "The Measurement and Transmission of Macroeconomic Uncertainty: Evidence from the U.S. and BRIC Countries." *International Journal of Forecasting*, 35(3): 967–979.

López-Pérez, Víctor. (2015). "Measures of Macroeconomic Uncertainty for the ECB's Survey of Professional Forecasters." In Donduran M, Uzunöz M, Bulut E, Çadirci TO, and Aksoy T (eds.), *Proceedings of the 1st Annual International Conference* on *Social Sciences*, Yildiz Technical University. pp. 600–614.

Mazzeu, João Henrique Gonçalves, Esther Ruiz, and Helena Veiga. (2018). "Uncertainty and Density Forecasts of ARIMA Models: Comparison of Asymptotic, Bayesian, and Bootstrap Procedures." *Journal of Economic Surveys*, 32(2): 388–419.

Minka, Thomas P. (1998). "Bayesian Inference, Entropy, and the Multinomial Distribution." https://tminka.github.io/papers/multinomial.html.

Mitchell, James, and Stephen G. Hall. (2005). "Evaluating, Comparing and Combining Density Forecasts Using the KLIC with an Application to the Bank of England and NIESR 'Fan' Charts of Inflation." *Oxford Bulletin of Economics and Statistics*, 67: 995–1033.

Mitchell, James, and Kenneth F. Wallis. (2011). "Evaluating Density Forecasts: Forecast Combinations, Model Mixtures, Calibration and Sharpness." *Journal of Applied Econometrics*, 26(6): 1023–1040.

Rich, Robert, and Joseph Tracy. (2010). "The Relationships among Expected Inflation, Disagreement and Uncertainty: Evidence from Matched Point and Density forecasts." *Review of Economics and Statistics*, 92: 200–207.

Shoja, Mehdi, and Ehsan S. Soofi. (2017). "Uncertainty, Information, and Disagreement of Economic Forecasters." *Econometric Reviews*, 36: 796–817.

Soofi, Ehsan S., and Joseph J. Retzer. (2002). "Information Indices: Unification and Applications." *Journal of Econometrics*, 107: 17–40.

Zarnowitz, Victor. (2008). "Surviving the Gulag, and Arriving in the Free World: My Life and Times." Westport, CT: Praeger.

Zarnowitz, Victor, and Louis A. Lambros. (1987). "Consensus and Uncertainty in Economic Prediction." *Journal of Political Economy*, 95: 591–621.

12

Reduced Perplexity

A Simplified Perspective on Assessing Probabilistic Forecasts

Kenric P. Nelson

1. Introduction

This chapter introduces a clear and simple approach to assessing the performance of probabilistic forecasts. This approach is important because machine learning and other techniques for decision making are often only evaluated in terms of percentage of correct decisions. Management of uncertainty in these systems requires that accurate probabilities be assigned to decisions. Unfortunately, the existing assessment methods based on "scoring rules" (Good, 1952; Gneiting and Raftery, 2007), which are defined later in the chapter, are poorly understood and often misapplied and/or misinterpreted. The approach here will be to ground the assessment of probability forecasts using information theory (Shannon, 1948; Cover and Thomas, 2006; Golan, 2018), while framing the results from the perspective of the central tendency and fluctuation of the forecasted probabilities. The methods will be shown to reduce both the colloquial perplexity surrounding how to evaluate inferences and the quantitative perplexity that is an information-theoretic measure related to the accuracy of the probabilities.

To achieve this objective, section 2 reviews the relationship between probabilities, perplexity, and entropy. The geometric mean of probabilities is shown to be the central tendency of a set of probabilities. In section 3, the relationship between probabilities and entropy is expanded to include generalized entropy functions (Principe, 2010; Tsallis, 2009). From this, the generalized mean of probabilities is shown to provide insight into the fluctuations and risk-sensitivity of a forecast. From this analysis, a *Risk Profile* (Nelson, Scannell, and Landau, 2011), defined as the spectrum of generalized means of a set of

Kenric P. Nelson, *Reduced Perplexity: A Simplified Perspective on Assessing Probabilistic Forecasts* In: *Advances in Info-Metrics: Information and Information Processing across Disciplines.* Edited by: Min Chen, J. Michael Dunn, Amos Golan, and Aman Ullah, Oxford University Press (2021). © Oxford University Press.
DOI: 10.1093/oso/9780190636685.003.0012

forecasted probabilities, is used in section 4 to evaluate a variety of models for an n-dimensional random variable.

2. Probability, Perplexity, and Entropy

The arithmetic mean and the standard deviation of a distribution are the elementary statistics used to describe the central tendency and uncertainty, respectively, of a random variable. Less widely understood, though studied as early as the 1870s by McAlister (1879), is that a random variable formed by the ratio of two independent random variables has a central tendency determined by the geometric mean rather than the arithmetic mean. Thus, the central tendency of a set of probabilities, each of which is formed from a ratio, is determined by their geometric mean. This property will be derived from information theory and illustrated with the example of the Gaussian distribution.

Instead of using the geometric mean of the probabilities of a distribution to represent average uncertainty, it has been a long-standing tradition within mathematical physics to utilize the entropy function, which is defined as the arithmetic mean of the logarithm of the probabilities. There are at least three important reasons for using entropy to define average uncertainty. Physically, entropy defines the change in heat energy per temperature; mathematically, entropy provides an additive scale for measuring uncertainty; and computationally, entropy has been shown to be a measure of information (Khinchin, 1949, 1957). Unfortunately, using entropy to quantify average uncertainty results in loss of the intuitive relationship between the underlying probabilities of a distribution and a summarizing average probability of the distribution. Perplexity, which determines the average number of uncertain states, provides a bridge between the average probability and the entropy of a distribution. For a random variable with a uniform distribution of N states, the perplexity is N and its inverse, $1/N$, is the average probability. More generally, the average probability P_{avg} and the perplexity PP are related to the entropy $H(\mathbf{p}) = -\sum_{i=1}^{N} p_i \ln p_i$ of a distribution $\mathbf{p} = \left\{p_i : \sum_{i=1}^{N} p_i = 1\right\}$ by

$$P_{avg} \equiv PP^{-1} = \exp(-H(\mathbf{p})) = \exp\left(\sum_{i=1}^{N} p_i \ln p_i\right) = \prod_{i=1}^{N} p_i^{p_i}. \tag{12.1}$$

The expression on the far right is the weighted geometric mean of the probabilities in which the weight appearing in the exponent is also the probability. For a continuous distribution $f(x)$ of a random variable X, these expressions become

$$f_{avg} \equiv PP^{-1} = \exp\left(-H\left(f(x)\right)\right) = \exp\left(\int_{x\in X} f(x)\ln f(x)dx\right), \tag{12.2}$$

where f_{avg} is the average density of the distribution and PP still refers to perplexity.

Figure 12.1 illustrates these relationships for the standard normal distribution. The key point is that by expressing the central tendency of a distribution as a probability (or density for continuous distributions), the context with the original distribution is maintained. For the exponential and Gaussian distributions, translating entropy back to the density domain (Nelson, Umarov, and Kon, 2017) results in the density of the distribution at the location μ plus the scale σ

$$\exp\left(\int_{\mu}^{\infty} \frac{1}{\sigma}e^{-\frac{x-\mu}{\sigma}} \ln\left(\frac{1}{\sigma}e^{-\frac{x-\mu}{\sigma}}\right)dx\right) = \frac{1}{\sigma}e^{-\left(\frac{\mu+\sigma-\mu}{\sigma}\right)} = \frac{1}{\sigma e}, \tag{12.3}$$

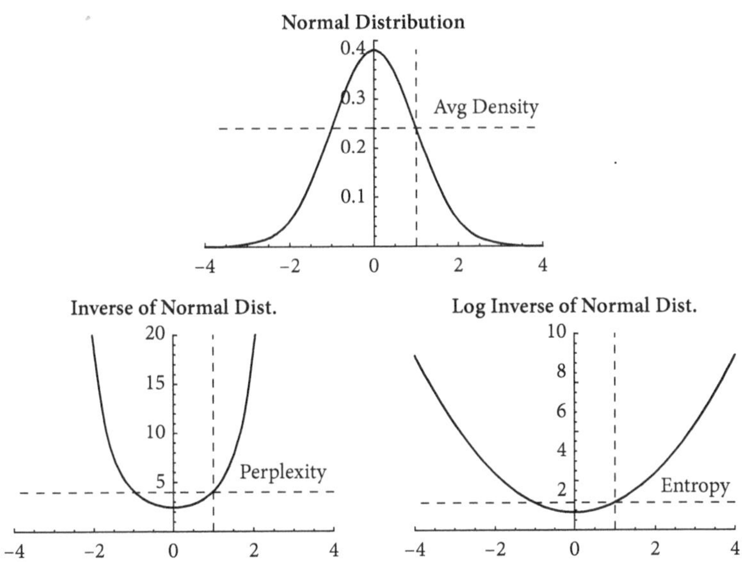

Figure 12.1 Comparison of the average density, perplexity and entropy for the standard normal distribution. Plots of the inverse distribution and the log of the inverse of the distribution provide visualization of the perplexity and entropy. The intersection for each of these quantities with the distribution is at the mean plus the standard deviation.

$$\exp\left(\int_{-\infty}^{\infty}\frac{1}{\sqrt{2\pi}\sigma}e^{-\frac{1}{2}\left(\frac{x-\mu}{\sigma}\right)^2}\ln\left(\frac{1}{\sqrt{2\pi}\sigma}e^{-\frac{1}{2}\left(\frac{x-\mu}{\sigma}\right)^2}\right)dx\right)=\frac{1}{\sqrt{2\pi}\sigma}e^{-\frac{1}{2}\left(\frac{\mu+\sigma-\mu}{\sigma}\right)^2}=\frac{1}{\sqrt{2\pi e}\sigma}. \tag{12.4}$$

Thus, it should be more commonly understood that for these two important members of the exponential family, the average uncertainty is the density at the width of the distribution defined by $f(\mu+\sigma)$. While perplexity and entropy are valuable concepts, it is not common to plot distributions on the inverse scale (perplexity) or the log-inverse scale (entropy). Thus, the intuitive meaning of these quantities is disconnected from the underlying distribution.

Table 12.1 shows the translation of entropy, divergence, and cross-entropy to the perplexity and probability scales. In each case, the translation is via application of the exponential function as in (12.1). The additive combination of logarithmic probabilities translates into a multiplicative combination of the probabilities. The weight on the mean, also a probability, is now a power term. The additive relationship between cross-entropy, entropy, and divergence $H(p,q)=H(p)+D_{KL}\left(p\|q\right)$ is multiplicative in the probability space

$$\begin{aligned}P_{cross\text{-}entropy}&=P_{entropy}P_{divergence}\\&=\left(\prod_{i=1}^{N}p_i^{p_i}\right)\left(\prod_{i=1}^{N}\left(q_i/p_i\right)^{p_i}\right)\\&=\prod_{i=1}^{N}q_i^{p_i}.\end{aligned} \tag{12.5}$$

Jaynes (1957, 2003) established the principle of maximum entropy as a method for selecting a probability distribution such that known constraints were satisfied, but no additional knowledge was represented in the distribution. Two

Table 12.1 Translation of Entropy Functions to Perplexity and Probability Scales

Info-Metric	Entropy Scale	Perplexity Scale	Probability Scale
Entropy	$-\sum_i p_i \ln p_i$	$\prod_i (p_i)^{-p_i}$	$\prod_i (p_i)^{p_i}$
Divergence	$-\sum_i p_i \ln\left(q_i/p_i\right)$	$\prod_i \left(q_i/p_i\right)^{-p_i}$	$\prod_i \left(q_i/p_i\right)^{p_i}$
Cross-Entropy	$-\sum_i p_i \ln q_i$	$\prod_i (q_i)^{-p_i}$	$\prod_i (q_i)^{p_i}$

basic examples are the exponential distribution, which satisfies the constraints that the range is 0 to ∞ and the mean is $\mathbb{E}(X) = \int_0^{\infty} xf(x)dx = \mu$, and the Gaussian distribution, which satisfies a known mean and variance $\mathbb{E}(X^2) - \mathbb{E}(X)^2 = \int_0^{\infty}(x-\mu)^2 f(x)dx = \sigma^2$. Translated to the probability domain, the principle of maximum entropy can thus be framed as a minimization of the weighted geometric mean of the distribution. In section 4, a related principle of minimizing the cross-entropy between a discrimination model and the actual uncertainty of a forecasted random event will be translated to maximizing the geometric mean of the reported probability.

Just as the arithmetic mean of the logarithm of a probability distribution determines the central tendency of the uncertainty or the entropy, the standard deviation of the logarithm of the probabilities, $\sigma_{\ln p}$, is needed to quantify variations in the uncertainty,

$$\sigma_{\ln p} \equiv \left[\sum_{i=1}^{N} p_i(-\ln p_i)^2 - \left(-\sum_{i=1}^{N} p_i \ln p_i\right)^2\right]^{1/2}. \tag{12.6}$$

Unfortunately, the translation to the probability domain ($e^{-\sigma_{\ln p}}$) does not result in a simple function with a clear interpretation. Furthermore, because the domain of entropy is one-sided, just determining the standard deviation does not capture the asymmetry in the distribution of the logarithm of the probabilities. In the next section, the generalized mean of the probabilities is shown to be a better alternative for measuring fluctuations.

3. Relationship between the Generalized Entropy and the Generalized Mean

In this section, the effect of sensitivity to risk (r) will be used to generalize the assessment of probabilistic forecasts. The approach is based on a generalization of the entropy function, particularly the Rényi and Tsallis entropies (Tsallis, 2009; Rényi, 1961; Amari and Nagaoka, 2000). As with the Boltzmann–Gibbs–Shannon entropy, the generalized entropy can be transformed back to the probability domain. The resulting function, derived in Nelson et al. (2017) and summarized in the Appendix, is the weighted generalized mean or weighted p-norm of the probabilities

$$P_r(\mathbf{w},\mathbf{p}) \equiv \begin{cases} \left(\sum_{i=1}^{N} w_i p_i^r\right)^{\frac{1}{r}} & r \neq 0 \\ \prod_{i=1}^{N} p_i^{w_i} & r = 0, \end{cases} \tag{12.7}$$

where the symbol r is used for the power of the mean to avoid confusion with the probabilities and because the power is related to the sensitivity to risk, as discussed below. The weights $\mathbf{w} = \left\{w_i : \sum_{i=1}^{N} w_i = 1\right\}$ are a modified version of the probabilities discussed in the next paragraph. The symbol P_r is used here rather than the traditional symbols of M_r or $\|x\|_r$ for the generalized mean and p-norm, respectively, to emphasize that the result is a probability that represents a particular aggregation of the vector of probabilities. The geometric mean, which is the metric consistent with the Shannon information theory, is recovered when the risk sensitivity is zero ($r = 0$). Positive values of r reduce the influence of low probabilities in the average and are thus associated with risk-seeking, while negative values of r increase the sensitivity to low probabilities and are thus risk-averse.

Several different generalizations of entropy can be shown to transform into the form of (12.7). The Appendix discusses the origin of these generalizations for modeling the statistical properties of complex systems which are influenced by nonlinear coupling. The Tsallis and normalized Tsallis entropy utilize a modified set of probabilities formed by raising the probabilities of the distribution to a power and renormalizing

$$p_i^{(r)}(\mathbf{p}) \equiv \frac{p_i^{1-r}}{\sum_{j=1}^{N} p_j^{1-r}}. \tag{12.8}$$

This new distribution, referred to either as the coupled probability or an escort probability, is the normalized probability of $1-r$ independent events rather than one event. Substituting (12.8) for the weights in (12.7) and simplifying gives the following expression for the weighted generalized mean of a distribution:

$$P_r\left(\mathbf{p}^{(r)},\mathbf{p}\right) = \left(\sum_{i=1}^{N}\left(\frac{p_i^{1-r}}{\sum_{j=1}^{N} p_j^{1-r}}\right)p_i^{r}\right)^{\frac{1}{r}} = \left(\sum_{i=1}^{N} p_i^{1-r}\right)^{\frac{-1}{r}} = P_{-r}(\mathbf{p},\mathbf{p}). \tag{12.9}$$

The normalized probability of $1-r$ events as a weight has the effect of reversing the sign of power r with the original probabilities now the weights, as shown in the rightmost expression.

Figure 12.2, which shows the weighted geometric mean for three different distributions, is plotted in terms of $-r$ so that the visual orientation of graphs is similar to those appearing later regarding assessment of probabilistic forecasts.

The distributions examined are members of the coupled Gaussians, which are equivalent to the Student's t-distribution and discussed in more detail in the Appendix. For consistency, the coupled Gaussian distributions are expressed here in terms of the risk sensitivity r_D, and the subscript D is used to distinguish the parameter of the distribution from the parameter of the generalized mean. The coupled Gaussian with $\mu = 0$ is

$$f(x) = \frac{1}{Z(r_D,\sigma)}\left(1-\left(\frac{r_D}{2+r_D}\right)\frac{x^2}{\sigma^2}\right)_+^{\frac{1}{r_D}}, \tag{12.10}$$

where $(a)_+ \equiv \max(0,a)$, Z is the normalization of the distribution, and σ is the scale parameter of the distribution. For $-2 < r_D < 0$, the distribution is heavy tail, $r_D = 0$ is the Gaussian, and for $r_D > 0$ the distribution is compact-support. Applying the continuous form of the generalized mean (2.8) with the matching power of $r = r_D$ gives the following result:

$$\begin{aligned}
f_{r_D}(f(x,r_D,\sigma)) &= \left(\int_{x\in X} f(x,r_D,\sigma)^{1-r_D}dx\right)^{-\frac{1}{r_D}} \\
&= Z(r_D,\sigma)^{\frac{1-r_D}{r_D}}\left(\int_{x\in X}\left(1-\left(\frac{r_D}{2+r_D}\right)\frac{x^2}{\sigma^2}\right)_+^{\frac{1-r_D}{r_D}}dx\right)^{-\frac{1}{r_D}} \\
&= Z(r_D,\sigma)^{\frac{1-r_D}{r_D}}\left(Z(r_D,\sigma)\left(1-\frac{r_D}{2+r_D}\right)^{-1}\right)^{-\frac{1}{r_D}} \\
&= \frac{1}{Z(r_D,\sigma)}\left(1-\frac{r_D}{2+r_D}\right)^{\frac{1}{r_D}} = f(x=\sigma,r_D,\sigma).
\end{aligned} \tag{12.11}$$

That is, the generalized mean of the coupled Gaussian with a matching risk sensitivity is equal to the density at the mean plus the scale. While not derived here, the equivalence between the generalized maximum entropy principle using the Tsallis entropy and the minimization of the weighted generalized mean is such that the distribution $f(x,r_D,\sigma)$ is the minimization of f_{r_D} given the constraint that the scale is σ.

In Figure 12.2, the weighted generalized mean (WGM) is shown for the Gaussian distribution $r_D = 0$ and two examples of the coupled Gaussian with $r_D = -2/3,\ 1$. As derived in the Appendix, these values of risk sensitivity are conjugate values in the heavy-tail and compact-support domain, respectively. For each of the distributions, the scale is $\sigma = 1$. In order to illustrate the intersection between the distribution and its matching value of the WGM, the mean of each

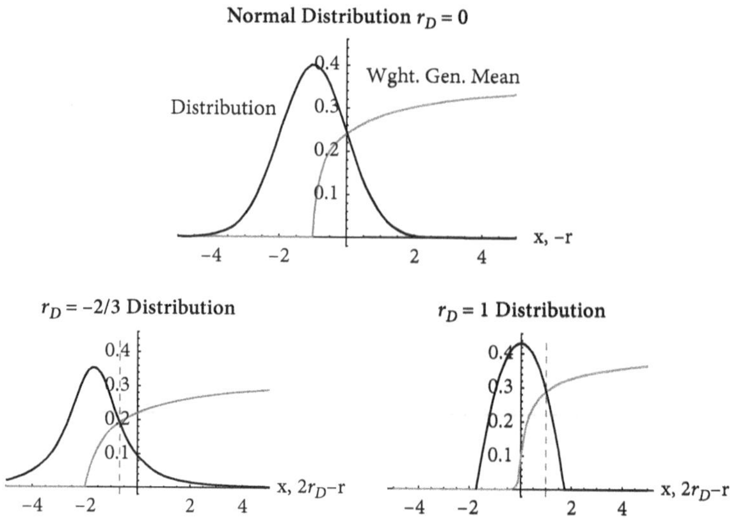

Figure 12.2 Plots of the weighted generalized mean (WGM) overlayed with the distribution which minimizes the WGM at the value $r = r_D$. The mean of each distribution is adjusted to show the WGM intersecting the density at the mean plus width parameter of the distribution. a) Normal distribution N(−1,1) with its WGM. The normal distribution is a coupled Gaussian with $r_D = 0$ and minimizes the WGM at $r = 0$. The WGM at $r = 0$ is equal to the density of the normal at the mean plus standard deviation. b) The coupled Gaussian with $r_D = -2/3$, $\mu = -5/3$, $\sigma = 1$ minimizes the WGM at $r = -2/3$. The orientation of the WGM plot for b and c is inverted and shifted by $2r_D - r$. c) The coupled Gaussian with $r_D = 1$, $\mu = 0$, $\sigma = 1$ minimizes the WGM at $r = 1$. For both b and c the WGM at r_D is equal to the density at the mean plus the generalized standard deviation.

distribution is shifted by $\mu = r_D - \sigma$. The WGM is plotted as a function of $2r_D - r$ rather than r so that the increase in WGM is from left to right, as it will be when evaluating probabilistic forecasts. The coupled exponential distribution and the coupled Gaussian distribution have the following relationship with respect to the generalized average uncertainty

$$\exp_r\left(\int_{\mu}^{\infty} \frac{1}{\sigma}\exp_r\left(-\frac{x-\mu}{\sigma}\right)\ln_r\left(\frac{1}{\sigma}\exp_r\left(-\frac{x-\mu}{\sigma}\right)\right)dx\right) = \frac{e_r^{-\left(\frac{\mu+\sigma-\mu}{\sigma}\right)}}{\sigma} = \frac{1}{\sigma e_r} \tag{12.12}$$

$$\exp_r\left(\int_{-\infty}^{\infty} \frac{1}{Z_r} e_r^{-\frac{(x-\mu)^2}{2\sigma^2}} \ln\left(\frac{1}{Z_r} e_r^{-\frac{(x-\mu)^2}{2\sigma^2}}\right)dx\right) = \frac{1}{Z_r} e_r^{-\frac{(\mu+\sigma-\mu)^2}{2\sigma^2}} = \frac{e_r^{-1/2}}{Z_r}, \tag{12.13}$$

where the subscript D was dropped for readability. These relationships provide evidence of the importance of the generalized mean as an expression of the average uncertainty for nonexponential distributions. The next section demonstrates the use of the generalized mean as a metric to evaluate probabilistic inference.

4. Assessing Probabilistic Forecasts Using a Risk Profile

The goal of an effective probabilistic forecast is to "reduce perplexity"—that is, to enhance decision making by providing accurate information about the underlying uncertainties. Just as the maximum entropy approach is important in selecting a model that properly expresses the uncertainty, minimization of the cross-entropy between a model and a source of data is essential to accurate forecasting. In section 3, the relationship between the weighted generalized mean of a distribution and the generalized entropy functions was established; likewise, the generalized cross-entropy can be translated into the weighted generalized mean in probability space. The result is a spectrum of metrics that modifies the sensitivity to surprising or low-probability events; as such, it is referred to as a *Risk Profile*.

The most basic definition of risk R is the expected cost of a loss L where $p(L)$ is the probability of the loss

$$R = E(L) = \sum_{i=1}^{N} L_i p(L_i). \tag{12.14}$$

The risk can also be defined as the degree of variance or standard deviation for a process, such as an asset price, which has a monetary or more general value. An individual or agent can have different perceptions of risk, expressed as the utility of a loss (or gain). Thus, a risk-averse person would seek to lower exposure to high variances given the same expected loss. With regard to a probabilistic forecast, the cost is being surprised by an event that was forecasted to have a low probability. While a particular application may also assign a valuation to events, with regard to evaluating the quality of the forecast itself, the "surprisal (S)" will be the only cost. A neutral perspective on the risk of being surprised is the information-theoretic measure, the logarithm of the probabilities (Shannon, 1948; Gneiting and Katzfuss, 2014). The expected surprisal cost is the arithmetic average of the logarithmic distance between the forecasted probabilities and a perfect forecast of $p = 1$

$$S = E[S_i] = -\frac{1}{N}\sum_{i=1}^{N}(\ln p_i - \ln 1) = -\frac{1}{N}\sum_{i=1}^{N}\ln p_i. \tag{12.15}$$

The average surprisal is also known as the logarithmic scoring rule or the negative log-likelihood of the forecasts and has the property of being the only scoring rule that is both proper and local. A proper scoring rule is one in which optimization of the rule leads to unbiased forecasts relative to what is known by the forecaster. A local scoring rule is one in which only the probabilities of events that occurred are used in the evaluation. The average surprisal (12.15) can be viewed as the cross-entropy between a model (the reported probabilities) and data (the distribution of the test set). From (12.5) the uncertainty in a forecast is due to underlying uncertainty in the test set (entropy) and errors in the model (divergence). This relationship is used in Nelson (2017) to visualize the quantitative performance of forecasts alongside the calibration curve comparing the forecasted and actual distribution.

The influence of risk seeking and risk aversion in forecasting can be evaluated using a generalized surprisal function, which is defined as

$$\begin{aligned} S_r \equiv E\left[S_{r,i}\right] &= -\tfrac{1}{N}\sum_{i=1}^{N}\left(\ln_r p_i - \ln_r 1\right) = -\sum_{i=1}^{N}\ln_r p_i \\ \ln_r x &\equiv \tfrac{1+r}{r}\left(x^r - 1\right). \end{aligned} \tag{12.16}$$

This generalized logarithmic function is fundamental to the generalization of thermodynamics introduced by Tsallis (2009). Its role in defining a generalized information theory is explained further in the Appendix. The generalized surprisal function is still a local scoring rule, but it is no longer proper. The properties of this function have been studied in economics due to its preservation of a constant coefficient of relative risk. In economics, the variable x of (12.16) is the valuation, and the relative risk aversion ("Risk Aversion," 2017; Simon and Blume, 2006) is defined in terms of $1-r$ since $r=1$ is a linear function and thus is considered to be risk neutral. Here the bias is with respect to the neutral measure of information, namely, $\ln p$ when $r=0$. Thus, for purposes of this discussion, the relative risk sensitivity is defined as

$$r \equiv 1 + p\frac{d^2\left(\ln_r p\right)/dp^2}{d\left(\ln_r p\right)/dp}. \tag{12.17}$$

For negative values of r, the generalized surprisal is risk-averse, since the cost of being surprised goes to infinity faster. This is referred to as the domain of robust metrics, since it encourages algorithms to be conservative or robust in probabilistic estimation. For positive values of r, the measure is risk-seeking and is referred to as a decisive metric since it is more like the cost of making a decision over a finite set of choices, as opposed to the cost of properly forecasting the probability of the decision.

For evaluating a probabilistic forecast, use of the logarithmic or generalized logarithmic scale is needed to ensure that the analysis properly measures the cost of a surprising forecast. Nevertheless, it leaves obscure what is truly desired in an evaluation: knowledge of the central tendency and fluctuation of the probability forecasts. Following the procedures introduced in sections 2 and 3, the generalized scoring rule can be translated to a probability by taking the inverse of the generalized logarithm, which is the generalized exponential

$$\exp_r(x) \equiv \begin{cases} \left(1+\frac{r}{1+r}x\right)_+^{\frac{1}{r}} & r \neq 0 \\ \exp(x) & r = 0, \end{cases} \tag{12.18}$$

where $(a)_+ \equiv \max(0, a)$. Applying (12.18) to (12.16) shows that the generalized mean of the probabilities is the translation of the generalized logarithmic scoring rule to the probability domain

$$P_{r-avg}(\mathbf{p}) \equiv \exp_r(-S_r(\mathbf{p})) = \left(1+\frac{r}{1+r}\left(\frac{1}{N}\sum_{i=1}^{N}\frac{1+r}{r}(p_i^r-1)\right)\right)^{\frac{1}{r}}$$

$$= \begin{cases} \left(\frac{1}{N}\sum_{i=1}^{N} p_i^r\right)^{\frac{1}{r}} & r \neq 0 \\ \prod_{i=1}^{N} p_i^{\frac{1}{N}} & r = 0. \end{cases} \tag{12.19}$$

Thus, the generalized mean of the forecasted probabilities forms a spectrum of metrics that profile the performance of the forecast relative to the degree of relative risk sensitivity. This spectrum is referred to as the *Risk Profile* of the probabilistic forecast. A condensed summary of an algorithms performance is achieved using three points on the spectrum: the geometric mean ($r = 0$) measures the risk-neutral *accuracy*, and the degree of fluctuation is measured by an upper metric called *decisiveness* using the arithmetic mean ($r = 1$) and a lower metric called *robustness* using the $-2/3^{rds}$ mean ($r = -2/3$).

Prior to demonstrating the Risk Profile, a word of caution regarding the use of proper scoring rules, such as the mean-square average, is provided. Starting with Brier (1950), a tradition has grown around the use of noninformation-theoretic measures of accuracy for probabilistic forecasts. Brier himself cannot be faulted, as Shannon's efforts to formulate information theory (Shannon, 1948) were nearly concurrent with Brier's efforts to evaluate weather forecasts. However, the subsequent development of proper scoring rules (Gneiting and Reftery, 2007; Lindley, 1982; Dawid, 2007), which removes the bias in the expectation of a forecast optimized using any convex positive-valued utility function, has led to the impression that an unbiased expectation is the only criterion for evaluating forecasts. While some applications may in fact require a utility function different from information theory, in practice use of the mean-square average because it is "proper" has inappropriately justified avoidance of the rigorous information-theoretic penalties for overconfident forecasts. One way to view this is that, while the first-order expectation is unbiased for a proper scoring rule, all the other moments of the forecasts may still be biased. A rigorous proof of this deficiency would be a valuable contribution as suggested by Jewson (2004). Here, the emphasis is on using the alternative cost functions to complement the negative log-likelihood and thereby to provide insight into how an algorithm responds to risk sensitivity. As derived by Dawid and Musio Dawid (2014), the generalized surprisal with $r = 1$, used here for a measure of decisiveness, becomes the mean-square average scoring rule if the distance between the nonevent forecasts and a probability of zero is included to make a proper score.

To illustrate the Risk Profile in evaluating statistical models, the contrast between robust and decisive models of a multivariate Gaussian random variable is demonstrated. The Student's t-distribution originated from William Gosset's insight (1900) that a limited number of samples from a source known to have a Gaussian distribution requires a model that modifies the Gaussian distribution to have a slower than exponential decay. Again, using the equivalent coupled-Gaussian distribution (2.8), but now for a multidimensional variable the distribution is

$$G_{r_D}(\mathbf{x};\boldsymbol{\mu},\Sigma) \equiv \frac{1}{Z_{r_D}(\Sigma)}\left(1 - \frac{r_D}{2+r_D}(\mathbf{x}-\boldsymbol{\mu})^{\mathsf{T}} \cdot \Sigma^{-1} \cdot (\mathbf{x}-\boldsymbol{\mu})\right)_+^{\frac{1}{r_D}}, \qquad (12.20)$$

where the vectors $\mathbf{x}$ and $\boldsymbol{\mu}$ are the random variable and mean vectors, $\sum$ is the correlation matrix,[1] and Z_r is the normalization. The parameter r_D has a dependence on the dimension, which is explained in the Appendix.

The problems with trying to model a Gaussian random variable using a multivariate Gaussian as the model is shown in Figure 12.3. In this example, ten independent features, which are generated from Gaussian distributions, are modeled as an multivariate Gaussian with a varying number of dimensions based on estimates of the mean and standard deviation from 25 samples. Although reasonable classification performance is achieved (84 percent), the

[1] While $\sum$ is the covariance matrix for a Gaussian distribution, for the coupled Gaussian this matrix is a generalization of the covariance and like the Student's t-distribution is known as the correlation matrix.

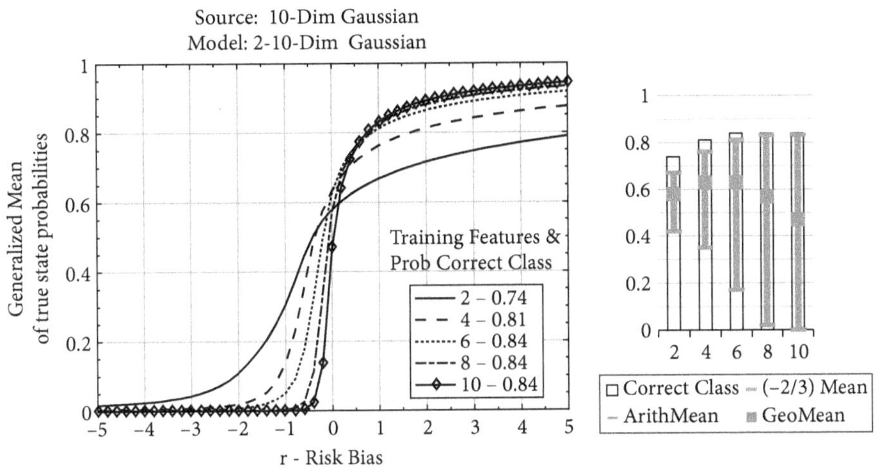

Figure 12.3 A source of 10 independent Gaussian random variables is over-fit given 25 samples to learn the mean and variance of each dimension and a model which is also a multivariate Gaussian. a) The *risk profile* shows that as the number of dimensions increases the model becomes more decisive. b) At 6 dimensions, the classification performance saturates to 84% at and the accuracy of the probabilities reaches is maximum of 63%.

accuracy of the modeled probabilities is reduced beyond six dimensions. Furthermore, the robustness as measured by the $-2/3^{\text{rds}}$ generalized mean drops to zero when all ten dimensions are modeled.

Even without seeking to optimize the coupling value, improvement in the accuracy and robustness of the multivariate model can be achieved using heavy-tail decay. Figure 12.4 shows an example with $r_D = -0.15$ in which the accuracy is improved to 0.69 and is stable for dimensions 6–10. The robustness continues to decrease as the number of dimensions increases but is improved significantly over the multivariate Gaussian model. The classification improves modestly to 86 percent but is not the principal reason for using the heavy-tail model.

The problems with overconfidence in the tails of a model are very visible when a compact-support distribution is used to model a source of data that is Gaussian. In this case, the reporting of $p = 0$ for states that do occur results in the accuracy being zero. An example of this situation is shown in Figure 12.5 in which the distribution power is $r_D = 0.6$. Although the model is neither accurate nor robust in the reporting of the probability of events, it is still capable of a modest classification performance (75 percent for four dimensions and reduced to 67 percent for ten dimensions). Nevertheless, characterization of only the

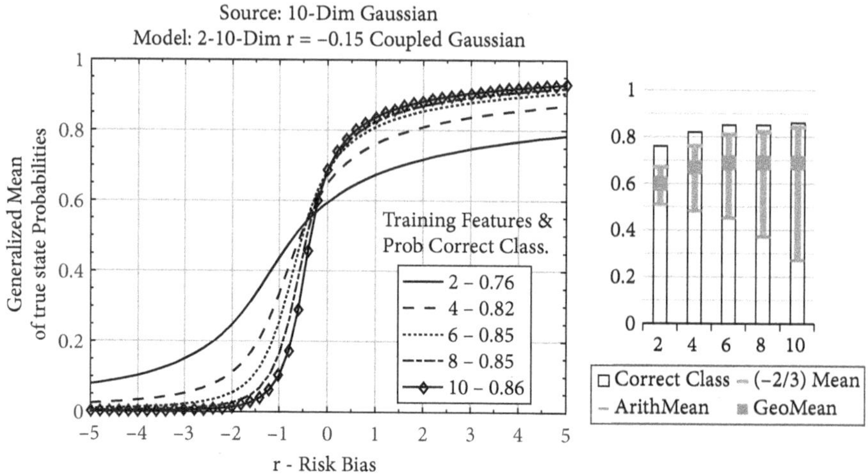

Figure 12.4 The risk of overfitting is reduced by using a heavy-tail coupled-Gaussian. Shown is an example with $r = -0.15$. a) The *RISK PROFILE* shows that the accuracy of 0.69 continues to hold as the dimensions modeled is increased from 6 to 8. b) The percent correct classification (red bar) improves to 86% with 10 dimensions modeled. The robustness does go down as the number of dimensions is increased, but could be improved by optimizing the coupling value used.

classification performance would not show the severity of the problem with inappropriate use of a compact-support model.

5. Discussion and Conclusion

This chapter has shown that translating the results of information theory from the entropy domain to the probability domain can simplify and clarify interpretation of important information metrics. In particular, the basic fact that the entropy of a Gaussian distribution when translated to a density as shown in Eq. (12.4) is equal to the density of the Gaussian at the mean plus the standard deviation should be a widely understood representation of the relationship between the standard deviation and entropy in measuring the central tendency of uncertainty. Unfortunately, while entropy provides the convenience of an additive information measure, the connection to the underlying probabilities of a distribution is often lost.

This disconnect between theory and practical intuition is evident in the confusion associated with evaluating probabilistic forecasts. For most random variables, the "average" is simply the arithmetic mean. Unfortunately, this does

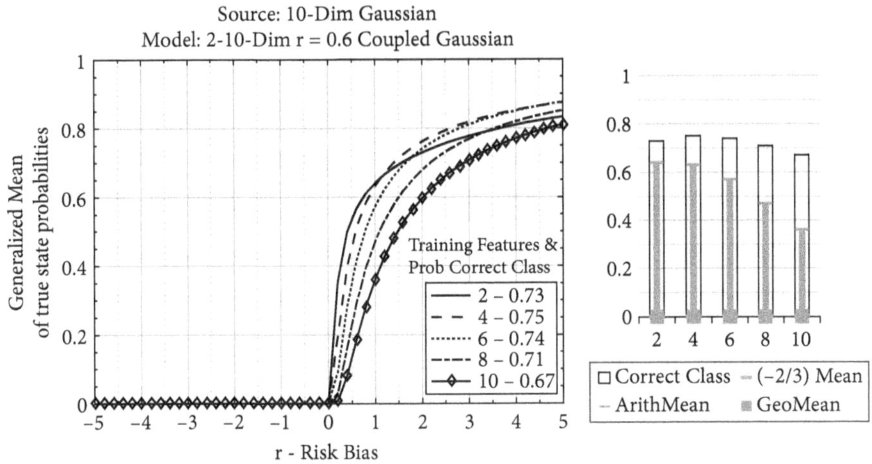

Figure 12.5 Using a compact-support distribution to model a source of data which is Gaussian results in the probability accuracy being zero. a) The risk profile shows that the model using $r = 0.6$ is neither accurate nor robust. b) The classification performance (red bar) is only 75% for the model with four dimensions, but characterization of only the classification performance would not show the severity of the problem with this model.

not hold for random variables that are formed by ratios, of which probabilities are a particularly important example. An elementary principle of probability theory is that the total probability of a set of independent probabilities is their product. So why isn't the nth root of the total probability or the geometric mean of the independent set of probabilities also recognized and taught to be the average? Likewise for a distribution, why isn't the weighted geometric mean in which the weight is also the probability recognized as the central tendency of the distribution? The answer seems to be both the misconception that the arithmetic mean is always the central tendency and the role that entropy serves in translating the geometric mean of probabilities to a domain in which the arithmetic mean is the central tendency.

For the evaluation of probabilistic forecasts, this has created a serious problem in which a variety of different "proper scoring rules" are treated as having equal merit in assessing the central tendency of a forecaster's performance. Only the logarithmic score, which is both proper (unbiased expectation) and local (based on actual events), is sensitive to the accuracy of the full distribution of forecasts. In particular, the Brier or mean-square average, which is a popular alternative, discounts the distance between small probabilities approaching zero. Thus, although the average forecast may be unbiased when optimized using the mean-square average, the distribution of forecasts tends to be overconfident.

A clear example is the allowance of a forecast of impossibility, that is, a reported probability of zero, for events that actually do occur.

The perspective emphasized here is to use the logarithmic score or, equivalently, the geometric mean of the reported probabilities to measure the accuracy of forecasts. The biased scores or, equivalently, the generalized mean of the probabilities is used to measure the fluctuation of the forecasts. The generalized mean of the probabilities is derived from a generalized information theory, which for decision-making models the sensitivity to risk. Rather than making these biased scores proper, their local property is maintained, and they provide a *Risk Profile* that is sensitive to whether the forecasts tend to be under- or overconfident. The arithmetic mean is biased toward decisive forecasting and approximates the classification performance over a small number of decisions. In contrast, means with a negative power are sensitive to the accuracy of rare events and thus provide a measure of the robustness of algorithms. The $-2/3^{\text{rds}}$ mean is conjugate to the arithmetic mean and is thus recommended as a robustness metric. The separation between the arithmetic mean and the $-2/3^{\text{rds}}$ mean of a set of probabilistic forecasts gives an indication of the degree of fluctuations about the central tendency, measured by the geometric or 0th mean.

Identification of a method for assessing probabilistic forecasts on the probability scale opens up other possibilities for integrating analysis with visual representations of performance. Recently, Nelson (2017) showed that a calibration curve comparing reported probabilities and the measured distribution of the test samples can be overlaid with metrics using the generalized mean of the reported and measured probabilities. This approach uses the relationship that the probability associated with cross-entropy is the product of the probabilities associated with entropy and divergence (12.5) to distinguish between sources of uncertainty due to insufficient features versus insufficient models, respectively. As the utility of measuring the generalized mean of a set of probabilities is explored, further innovations can be developed for robust, accurate probabilistic forecasting. These are particularly important for the development of machine learning and artificial intelligence applications, which need to carefully manage the uncertainty in making risky decisions.

Appendix: Modeling Risk as a Coupling of Statistical States

This chapter shows how risk sensitivity r can be used to evaluate the performance of probabilistic forecasts. In describing this assessment method, an effort was made to keep the explanation of the model as simple as possible. This appendix expands on the theoretical origins of the model. The model derives from the Tsallis generalization of statistical mechanics for complex systems (Tsallis, 2009; Wang, 2008; Martinez,

González, and Espíndola, 2009) and utilizes a perspective based on the degree of nonlinear coupling between the statistical states of a system (Nelson et al., 2017; Nelson and Umarov, 2010).

The nonlinearity κ of a complex system increases the uncertainty about the long-range dynamics of the system. In Nelson et al. (2017) the effect of nonlinearity, such as multiplicative noise or variation in the variance, was shown to result in a modification from the exponential family $f(x) \propto e^{-x^{\alpha}}$ to the power-law domain with $\lim_{x\to\infty} f(x) \propto x^{-\frac{\alpha}{r}}$, where the risk sensitivity can be decomposed into the nonlinear coupling and the power and dimension of the variable, $r(\kappa,\alpha,d) = \frac{-\alpha\kappa}{1+d\kappa}$. As the source of coupling κ increases from zero to infinity, the increased nonlinearity results in increasingly slow decay of the tails of the resulting distributions. Negative coupling can also be modeled, resulting in compact-support domain distributions with less variation than the exponential family. The negative domain, which models compact-support distributions, is $-\frac{1}{d} < \kappa < 0$.

The relationship defining the domains of r is also known within the field of nonextensive statistical mechanics as a dual transformation between the heavy-tail and compact-support domains. With the alpha term dropping out, the dual has the following relationship: $\hat{\kappa} \Longleftrightarrow \frac{-\kappa}{1+d\kappa}$. The dual is used to determine the conjugate to the decisive risk bias of 1. Taking $\alpha = 2$ and $d = 1$, the coupling for a risk bias of one is $1 = \frac{-2\kappa}{1+\kappa} \Rightarrow \kappa = -\frac{1}{3}$ and the conjugate values are $\hat{\kappa} = \frac{\frac{1}{3}}{1-\frac{1}{3}} = \frac{1}{2}$ and $\hat{r} = \frac{-2\left(\frac{1}{2}\right)}{1+\frac{1}{2}} = \frac{-2}{3}$.

The risk bias is closely related to the Tsallis entropy parameter $q = 1 - r$ (Nelson and Umarov, 2010; Wang, 2008; Martinez et al., 2009). One of the motivating principles of the Tsallis entropy methods was to examine how power-law systems could be modeled using probabilities raised to the power p_i^q (Tsallis, 1988). As such, q can be thought of as the number of random variables needed to properly formulate the statistics of a complex system, while r represents the deviation from a linear system governed by exponential statistics. When the deformed probabilities are renormalized the resulting distribution can be shown to also represent the probability of a "coupled state" of the system:

$$P_i^r = \frac{p_i^{1-r}}{\sum\limits_{j=1}^{n} p_j^{1-r}} = \frac{p_i \prod\limits_{\substack{k=1\\k\neq i}}^{n} p_k^r}{\sum\limits_{j=1}^{n}\left(p_j \prod\limits_{\substack{k=1\\k\neq j}}^{n} p_k^r\right)}, \tag{A.1}$$

hence use of the phrase "nonlinear statistical coupling."

Just as the probabilities are deformed via a multiplicative coupling, the entropy function is deformed via an additive coupling term. The nonadditivity of the generalized entropy H_κ provides a definition for the degree of nonlinear coupling. The joint coupled-entropy of two independent systems A and B includes a nonlinear term

$$H_\kappa(A,B) = H_\kappa(A) + H_\kappa(B) + \kappa H_\kappa(A) H_\kappa(B). \tag{A.2}$$

For $\kappa = 0$ the additive property of entropy is satisfied by the logarithm of the probabilities. The function that satisfies the nonlinear properties of the generalized entropy is a generalization of the logarithm function referred to as the coupled logarithm

$$\ln_\kappa x \equiv \frac{1}{\kappa}\left(x^{\frac{\kappa}{1+\kappa}} - 1\right), x > 0. \tag{A.3}$$

In the limit when κ goes to zero, the function converges to the natural logarithm. This definition of the generalized logarithm has the property that $\int_0^1 \ln_\kappa p^{-1} dp = 1$; thus, the deformation modifies the relative information of a particular probability while preserving the "total" information across the domain of probabilities.

The inverse of this function is the coupled exponential

$$\exp_\kappa x \equiv (1 + \kappa x)^{\frac{1+\kappa}{\kappa}}. \tag{A.4}$$

A distribution of the exponential family will typically include an argument of the form $-\frac{x^\alpha}{\alpha}$ which is generalized by the relationship $\left(\exp_\kappa x^\alpha\right)^{\frac{-1}{\alpha}} = \exp_\kappa^{\frac{-1}{\alpha}} x^\alpha = (1 + \kappa x^\alpha)^{\frac{1+\kappa}{-\alpha\kappa}}$. The rate of decay for a d-dimensional distribution is accounted for by $\exp_{\kappa,d}^{\frac{-1}{\alpha}} x^\alpha = (1 + \kappa x^\alpha)^{\frac{1+d\kappa}{-\alpha\kappa}}$, neglecting the specifics of the matrix argument. This is the form of the multivariate Student's t-distribution, with κ equal to the inverse of the degree of freedom. When the generalized logarithm needs to include the role of the power and dimension, this is expressed as $\ln_{\kappa,d} x^{-\alpha} \equiv \frac{1}{\kappa}\left(x^{\frac{-\alpha\kappa}{1+d\kappa}} - 1\right)$ or, alternatively, as $\left(\ln_{\kappa,d} x^{-\alpha}\right)^{\frac{1}{\alpha}} = \left(\frac{1}{\kappa}\left(x^{\frac{-\alpha\kappa}{1+d\kappa}} - 1\right)\right)^{\frac{1}{\alpha}}$. The first expression is used here, though research regarding the latter expression has been explored.

A variety of expressions for a generalized entropy function, which, when translated back to the probability domain, lead to the generalized mean of a probability distribution. Generalization of the entropy function can be viewed broadly as a modification of the logarithm function and the weight of the arithmetic mean. The translation back to the probability domain makes use of the inverse of the generalized logarithm, namely, the generalized exponential. The generalized expression for aggregating probabilities is then

$$\begin{aligned} P_\kappa(\mathrm{w}, \mathrm{p}; \alpha, d) &= \exp_{\kappa,d}^{\frac{-1}{\alpha}}\left(H_\kappa(\mathrm{p}; \mathrm{w}, \alpha, d)\right) \\ &= \exp_{\kappa,d}^{\frac{-1}{\alpha}}\left(\sum_{i=1}^{N} w_i \ln_{\kappa,d} p_i^{-\alpha}\right) \\ &= \left(1 + \kappa\left(\sum_{i=1}^{N} \frac{w_i}{\kappa}\left(p_i^{\frac{-\alpha\kappa}{1+d\kappa}} - 1\right)\right)\right)^{\frac{1+d\kappa}{-\alpha\kappa}} \\ &= \left(\sum_{i=1}^{N} w_i p_i^{\frac{-\alpha\kappa}{1+\kappa}}\right)^{\frac{1+d\kappa}{-\alpha\kappa}}, \end{aligned} \tag{A.5}$$

where the weights w_i are assumed to sum to one. In the main text, the focus is placed on the risk bias $r = \frac{-\alpha\kappa}{1+d\kappa}$, which forms the power of the generalized mean. The coupled entropy function is defined using the coupled probability (A.1) for the weights. Other generalized entropy functions use different definitions for the weights and generalized logarithm, but as proven in Nelson et al. (2017) for at least the normalized Tsallis entropy, Tsallis entropy, and Rényi entropy, they all converge to the weighted generalized mean of the distribution

$$\begin{aligned} P_\kappa(\mathbf{p};\alpha,d) &= \left(\sum_{i=1}^{N} \left(\frac{p_i^{1+\frac{\alpha\kappa}{1+d\kappa}}}{\sum_{j=1}^{N} p_j^{1+\frac{\alpha\kappa}{1+d\kappa}}} \right) p_i^{\frac{-\alpha\kappa}{1+d\kappa}} \right)^{\frac{1+d\kappa}{-\alpha\kappa}} \\ &= \left(\frac{\sum_{i=1}^{N} p_i}{\sum_{j=1}^{N} p_j^{1+\frac{\alpha\kappa}{1+d\kappa}}} \right)^{\frac{1+d\kappa}{-\alpha\kappa}} \\ &= \left(\sum_{i=1}^{N} p_i^{1+\frac{\alpha\kappa}{1+d\kappa}} \right)^{\frac{1+d\kappa}{\alpha\kappa}}. \end{aligned} \tag{A.6}$$

The assessment of a probabilistic forecast treats each test sample as an independent, equally likely event. The weights, even using the coupled probability, simplify to one over the number of test samples

$$w_i = \frac{\left({}^1\!/_N\right)^{1+\frac{\alpha\kappa}{1+d\kappa}}}{\sum_{i=1}^{N} \left({}^1\!/_N\right)^{1+\frac{\alpha\kappa}{1+d\kappa}}} = \frac{\left({}^1\!/_N\right)^{1+\frac{\alpha\kappa}{1+d\kappa}}}{N\left(\left({}^1\!/_N\right)^{1+\frac{\alpha\kappa}{1+d\kappa}}\right)} = \frac{1}{N}. \tag{A.7}$$

Thus, the generalized mean used for the Risk Profile has a power with the opposite sign of that used for the average probability of distribution

$$P_\kappa(\mathbf{p};\alpha,d) = \left(\frac{1}{N} \sum_{i=1}^{N} p_i^{\frac{-\alpha\kappa}{1+d\kappa}} \right)^{\frac{1+d\kappa}{-\alpha\kappa}}, \tag{A.8}$$

where the probabilities in this express are samples from a set of forecasted of events.

Acknowledgments

The initial research developing the *Risk Profile* was sponsored by Raytheon IRAD IDS202 2010-2012. Conversations with Ethan Phelps and Herb Landau highlighted the need for the surprisal metric in the assessment of discrimination algorithms. The figures in section 4 were created using Matlab software developed by Brian Scannell.

References

Amari, S. I., and H. Nagaoka. (2000). *Methods of Information Geometry*, vol. 191. Oxford: Oxford University Press.

Brier, G. W. (1950). "Verification of Forecasts Expressed in Terms of Probability." *Monthly Weather Review*, 78(1): 1–3.

Cover, T. M., and J. A. Thomas. (2006). *Elements of Information Theory*, 2nd ed. New York: John Wiley.

Dawid, A. P. (2007). "The Geometry of Proper Scoring Rules." *Annals of the Institute of Statistical Mathematics*, 59(1): 77–93.

Dawid, A., and M. Musio. (2014) "Theory and Applications of Proper Scoring Rules." *Metron*, 72(2): 169–183.

Gneiting, T., and A. E. Raftery. (2007). "Strictly Proper Scoring Rules, Prediction, and Estimation." *Journal of the American Statistical Association*, 102(477): 359–378.

Gneiting, T., and M. Katzfuss. (2014). "Probabilistic Forecasting." *Annual Review of Statistics and Its Applications*, 1(1): 125–151.

Golan, A. (2018). *Foundations of Info-Metrics: Modeling, Inference, and Imperfect information*. New York: Oxford University Press.

Good, I. J. (1952). "Rational Decisions." *Journal of the Royal Statistical Society Series B*, 14: 107–114.

Gossett, W. (1904). "The Application of the 'Law of Error' to the Work of the Brewery.", *Guinness Laboratory Report*, 8, Dublin.

Jaynes, E. T. (1957). "Information Theory and Statistical Mechanics." *Physical Review*, 106(4): 620–630.

Jaynes, E. T. (2003). *Probability Theory: The Logic of Science*. Cambridge: Cambridge University Press.

Jewson, S. (2004). "The Problem with the Brier Score." *Arxiv Prepr. 0401046*, 2004.

Khinchin, A. I. (1949). *Mathematical Foundations of Statistical Mechanics*. New York: Dover Publications.

Khinchin, A. I. (1957). *Mathematical Foundations of Information Theory*. New York: Dover Publications.

Lindley, D. V. (1982). "Scoring Rules and the Inevitability of Probability." *International Statistical Review/Revue Internationale De Statistique*, 50(1): 1–11.

Martinez, A. Souto, R. Silva González, and A. Lauri Espíndola. (2009). "Generalized Exponential Function and Discrete Growth Models." *Physica A: Statistical Mechanics and Its Applications*, 388(14): 2922–2930.

McAlister, D. (1879). "The Law of the Geometric Mean." *Proceedings of the Royal Society London*, 29(196–199): 367–376.

Nelson, K. P. (2017). "Assessing Probabilistic Inference by Comparing the Generalized Mean of the Model and Source Probabilities," *Entropy*, 19(6): 286.

Nelson, K. P., and S. Umarov. (2010). "Nonlinear Statistical Coupling." *Physica A: Statistical Mechanics and Its Applications*, 389(11): 2157–2163.

Nelson, K. P., B. J. Scannell, and H. Landau. (2011). "A Risk Profile for Information Fusion Algorithms." *Entropy*, 13(8): 1518–1532.

Nelson, K. P., S. R. Umarov, and M. A. Kon. (2017). "On the Average Uncertainty for Systems with Nonlinear Coupling." *Physica A: Statistical Mechanics and Its Applications*, 468(15): 30–43.

Principe, J. C. (2010). *Information Theoretic Learning: Rényi's Entropy and Kernel Perspectives*, no. XIV. New York: Springer.

Rényi, A. (1961). "On Measures of Entropy and Information." *Fourth Berkeley Symposium on Mathematical Statistics and Probability*, 1: 547–561.

"Risk Aversion." *Wikipedia*. [Online]. Accessed: September 17, 2017, at: https://en.wikipedia.org/wiki/Risk_aversion.

Shannon, C. E. (1948). "A Mathematical Theory of Communication." *Bell Systems Technological Journal*, 27: 379–423.

Simon, C. P., and L. Blume. (2006). *Mathematics for Economists*. Vinod Vasishtha for Viva Books, 2006.

Tsallis, C. (1988). "Possible Generalization of Boltzmann-Gibbs Statistics." *Journal of Statistical Physics*, 52(1): 79–487.

Tsallis, C. (2009). *Introduction to Nonextensive Statistical Mechanics: Approaching a Complex World*. New York: Springer Science & Business Media.

Wang, Q. A. (2008). "Probability Distribution and Entropy as a Measure of Uncertainty." *Journal of Physics A: Mathematical and Theoretical*, 41: 65004.

PART V

INFO-METRICS IN ACTION II: STATISTICAL AND ECONOMETRICS INFERENCE

This is the second of two parts on info-metrics in action. It deals with statistical and econometrics inference and connects some of the recent developments in info-metrics with cutting-edge classical inferential methods. In line with the information-theoretic methods of inference, which are subsumed within info-metrics, the models developed in this part do not use Shannon entropy as the decision function. Instead, each one uses a different version of a generalized entropy decision function. In Chapter 13, Andrews, Hall, Khatoon, and Lincoln develop an info-metrics method for inference of problems with group-specific moment conditions. This is a common problem in economic modeling where the information at the population level holds only at some group levels; only partial aggregate information is observed. The info-metrics framework provides a way to incorporate this information within the constraints. The decision function used, within the constrained optimization setup, is a generalized entropy one. This inferential model is also known as the information-theoretic, generalized empirical likelihood estimator. Like all info-metrics problems, the primal, constrained optimization model can be converted into its dual, unconstrained model, known also as the concentrated model; it concentrates on the minimal set of the needed parameters. This is also shown in this chapter.

In Chapter 14, Wen and Wu develop a generalized empirical likelihood approach for estimating spatially similar densities, each with a small number of observations. To overcome some of the difficulties that arise with small data and to improve flexibility and efficiency, a kernel-based estimator, which is refined by generalized empirical likelihood probability weights associated with spatial moment conditions, is used.

In Chapter 15, Geweke and Durham deal with the common, yet tough, problem of measuring the information flow in the context of Bayesian updating. That flow can be naturally measured using the Rényi entropy (or divergence) measure, which is a generalized entropy measure. This chapter develops a Monte Carlo integration procedure to measure the Rényi divergence under limited information. This, in turn, provides an estimate of the information flow.

Overall, the chapters in this part, though quite technical, demonstrate the benefits of using the info-metrics framework for solving some of the most common problems in econometrics and statistics. Unlike most of the other parts in this volume, in all of these chapters, a special case of the generalized entropy (or information) measure is used.

13
Info-metric Methods for the Estimation of Models with Group-Specific Moment Conditions

Martyn Andrews, Alastair R. Hall, Rabeya Khatoon, and James Lincoln

1. Introduction

Microeconometrics involves the use of statistical methods to analyze microeconomic issues.[1] In this context, the prefix "micro" implies that these economic issues relate to the behavior of individuals, households, or firms. Examples include the following: how do households choose the amount of their income to spend on consumer goods and the amount to save? How do firms choose the level of output to produce and the number of workers to employ?

The answers to these questions start with the development of an economic theory that postulates an explanation for the phenomenon of interest. This theory is most often expressed via an economic model, which is a set of mathematical equations involving economic variables and certain constants, known as parameters, that reflect aspects of the economic environment such as taste preferences of consumers or available technology for firms. While the interpretation of these parameters is known, their specific value is not. Therefore, in order to assess whether the postulated model provides useful insights, it is necessary to estimate appropriate values for these parameters based on observed economic data.

For microeconometric analyses, three main kinds of data are available: cross-sectional, panel (or longitudinal), and repeated cross-sectional. Cross-sectional data consists of a sample of information on individuals, say, taken at a moment in time. Panel data consists of a sample of individuals who are then observed at regular intervals over time. Repeated cross-section data consists of samples

[1] We dedicate this paper to one of its co-authors, James Lincoln, who tragically died in an accident in August 2019. He was involved in this project as a PhD student, and subsequently joined the department at Manchester as a Lecturer in Econometrics.

Martyn Andrews, Alastair R. Hall, Rabeya Khatoon, and James Lincoln, *Info-metric Methods for the Estimation of Models with Group-Specific Moment Conditions* In: *Advances in Info-Metrics: Information and Information Processing across Disciplines.* Edited by: Min Chen, J. Michael Dunn, Amos Golan, and Aman Ullah, Oxford University Press (2021).
DOI: 10.1093/oso/9780190636685.003.0013

from a population of individuals taken at regular intervals over time. Unlike in the case of panel data, repeated cross-sectional data involves a fresh sample of individuals taken each time period, and so the same individuals are not followed over time.

A number of statistical methods are available for estimation of econometric models. In choosing between them, an important consideration is that the implementation of the estimation method should not require the imposition of restrictions on the statistical behavior of the economics variables beyond those implied by the economic model. If these additional statistical restrictions turn out to be inappropriate, then this may undermine subsequent inferences about the economic question of interest. For example, the estimation method known as Maximum Likelihood (ML) requires the specification of the complete probability distribution of the data, but typically this information is not part of an economic model. As a result, ML is not an attractive choice in this context. While economic models usually do not imply the complete probability distribution, they do imply restrictions on functions of both the economic variables and unknown parameters. These restrictions, known as population moment conditions, can provide the basis for estimation of the parameters.

Lars Hansen was the first person to provide a general framework for population moment-based estimation in econometrics. In his seminal article in *Econometrica* in 1982, Hansen introduced the generalized method of moments (GMM) estimation method.[2] GMM has been widely applied in economics, but with this familiarity has come an understanding that GMM-based inferences may be unreliable in certain situations of interest, such as in the estimation of: models for the returns to education (Bound, Jaeger, and Baker, 1995); stochastic volatility models for asset prices (Andersen and Sørensen, 1996); and covariance models of earnings and hours worked (Altonji and Segal, 1996). This has stimulated the development of alternative methods for estimation based on population moment conditions. Leading examples are the continuous updating GMM (CUGMM; Hansen, Heaton, and Yaron, 1996), empirical likelihood (EL; Owen, 1988; Qin and Lawless, 1994), and exponential tilting (ET; Kitamura and Stutzer, 1997). While all three can be justified in their own right, they can also be regarded as special cases of more general estimation principles: info-metric (IM; Kitamura, 2007; Golan, 2008) or generalized empirical likelihood (GEL; Smith, 1997).

Both GMM and IM/GEL can be straightforwardly applied in the case where the data are a random sample from a homogeneous population, as is typically

[2] Hansen was co-winner of the 2013 Nobel Prize for Economics for his work on empirical analysis of asset pricing models, especially the development of GMM, which has been widely applied in empirical finance.

assumed for cross-section and panel data. In this case, the comparative properties of GMM and IM/GEL are well understood: under certain key assumptions about the information content of the population moment conditions, both estimators have the same large sample (first-order asymptotic) properties.[3] However, IM/GEL estimators exhibit fewer sources of finite sample bias, and IM/GEL-based inference procedures are more robust to circumstances in which the information content of the population moment conditions is low.

In contrast, the case of repeated cross-section data has received far less attention in the literature on moment-based estimation, even though such data is prevalent in the social sciences.[4] While certain GMM approaches have been proposed, to our knowledge IM methods have not been developed.[5] Part of the reason may be due to the fact that the original IM/GEL framework applies to samples from homogeneous populations, but this does not match the assumptions typically applied in econometric analysis of repeated cross-section data. For example, one popular method—Deaton's (1985) "pseudo-panel" approach—requires that the population consist of a number of different homogeneous subpopulations. In a recent paper, Andrews, Hall, Khatoon, and Lincoln (2020) (hereafter AHKL) propose an extension of the GEL framework to allow for estimation and inference based on population moment conditions that hold within the subpopulations. Since the subpopulations are often associated with groups of individuals or firms, the estimator is referred to as group-GEL, GGEL for short. AHKL establish the consistency, first-order asymptotic normality and second-order bias properties of the GGEL estimator, as well as the large sample properties of a number of model diagnostic tests.

We make two contributions in this chapter. The first is to provide an IM counterpart to the GGEL estimator, referred to as GIM, and to compare the computational requirements of the two approaches. Our second contribution is to describe the GGEL-based inference framework in the leading case in which a linear regression model with potentially group-specific parameters is estimated using the information that the expectation of the regression error is zero in each group. This allows us to make a direct comparison with a pseudo-panel estimator, which is a popular approach to linear model estimation based on

[3] Hansen (1982) presents the limiting distribution of the GMM estimator. The limiting distribution of the estimators obtained via the other methods are presented in the following papers: CUGMM, Hansen, Heaton, and Yaron (1996); EL, Qin and Lawless (1994); ET, Kitamura and Stutzer (1997); IM, Kitamura (2007); GEL, Smith (1997).

[4] For example, the UK Government's Data Service identifies key data sets for analysis of various issues relevant to public policy: of the twenty-two data sets identified for their relevance to environmental and energy issues, six consist of repeated cross-sections; of the thirty-four data sets identified for the relevance to health and health-related behavior, thirteen consist of repeated cross-sections; see http://ukdataservice.ac.uk/get-data/themes.aspx.

[5] See *inter alia* Bekker and van der Ploeg (2005), Collado (1997), and Inoue (2008).

repeated cross-section data. Using both theoretical analysis and evidence from a simulation study, it is shown that the GIM estimator yields more reliable inference than the pseudo-panel estimator considered here.

An outline of the chapter is as follows. Section 2 provides a brief review of GMM and IM/GEL where the data are a random sample from a homogeneous population. Section 3 describes Deaton's (1985) pseudo-panel approach, demonstrating how it depends crucially on the population being nonhomogeneous. Section 4 provides the IM version of AHKL's GGEL framework. Section 5 describes the GGEL framework in the context of the linear regression model, with estimation based on the moment condition that the error has mean zero in each group, compares GGEL to the pseudo-panel approach, both analytically and via simulations, and illustrates the methods through an empirical example. Section 6 concludes the chapter.

2. GMM and IM/GEL

In this section, we briefly review the GMM and IM/GEL estimation principles for data obtained as a random sample from a homogeneous population.

Our econometric model is indexed by a vector of parameters that can take values in Θ, a compact subset of $\mathbb{R}^p$. We wish to estimate the true value of these parameters, denoted θ_0. The economic variables are contained in the random vector v with sample space $\mathcal{V}$ and probability measure μ. It is assumed that we have access to a random sample from this population, denoted $\{v_i; i = 1, 2, \ldots, n\}$.

We consider the case where estimation of θ_0 is based on the population moment condition (PMC),

$$E\left[f(\mathrm{v}, \theta_0)\right] = 0, \tag{13.1}$$

where $f: \mathcal{V} \times \Theta \rightarrow \mathbb{R}^q$.

The PMC states that $E\left[f(\mathrm{v}, \theta)\right]$ equals zero when evaluated at θ_0. For the GMM or IM/GEL estimation to have the statistical properties described below, this must be a unique property of θ_0; that is, $E\left[f(\mathrm{v}, \theta)\right]$ is not equal to zero when evaluated at any other value of θ. If that holds, then θ_0 is said to be *globally identified* by $E\left[f(\mathrm{v}, \theta_0)\right] = 0$. A first-order condition for identification is that $\mathrm{rank}\{F(\theta_0)\} = p$, where $F(\theta_0) = E\left[\partial f(\mathrm{v}, \theta) / \partial \theta' \big|_{\theta = \theta_0}\right]$. This condition plays a crucial role in standard asymptotic distribution theory for these estimators. $F(\theta_0)$ is commonly known as the "Jacobian." By definition, the moment condition involves q pieces of information about p unknowns; therefore, identification can only hold if $q \geq p$. For reasons that emerge below, it is convenient to split this scenario into two parts: $q = p$, in which case θ_0 is said to be *just identified*, and $q > p$, in which case θ_0 is said to be *overidentified*.

We illustrate this condition with two popular examples in microeconometrics.

Example 1. Instrumental variable estimation based on cross-section data

Suppose it is desired to estimate θ_0 in the model

$$\mathrm{y} = \mathrm{x}'\theta_0 + \mathrm{u}$$

where y is the dependent variable, **x** is a vector of explanatory variables, and **u** represents an unobserved error term. If $E[\mathbf{u}|\mathbf{x}] = 0$, then θ_0 can be estimated consistently via ordinary least squares (OLS). However, in many cases in econometrics, this moment condition will not hold, with common reasons for its violation being simultaneity, measurement error, or an omitted variable.[6] These problems are commonly circumvented by seeking a vector of variables **z**—each known as an instrument—that satisfies the population moment condition $E[\mathbf{zu}] = 0$ and the identification condition $\mathrm{rank}\{E[\mathbf{zx}']\} = p$.[7] In this case, $\mathrm{v} = (\mathrm{y}, \mathrm{x}', \mathrm{z}')'$ and $f(\mathrm{v}, \theta) = \mathrm{z}(\mathrm{y} - \mathrm{x}'\theta)$.◊

Our second example involves panel data in which individuals are observed for a number of time periods.

Example 2. Dynamic linear panel data models

In a very influential paper, Arellano and Bond (1991) consider estimation of linear dynamic panel data models based on GMM. For panel data, it is convenient to index the random variables by the double index (i, t) where i denotes the cross-section index and t is the time. In such contexts, the relative magnitudes of the two dimensions is important for the statistical analysis. Here it is assumed that it is the cross-section dimension that is (asymptotically) large and the time dimension is fixed. To illustrate Arellano and Bond's (1991) suggested choice of population moment conditions, we consider a simplified version of their model,

$$\mathrm{y}_{i,t} = \theta_0 \mathrm{y}_{i,t-1} + \mathrm{u}_{i,t}, \quad i = 1, 2, \ldots n; t = 2, \ldots T, \tag{13.2}$$

[6] These occur, respectively, if: y and x are simultaneously determined; the true model is $y = x_*'\theta_0 + e$ and $x = x_* + w$; explanatory variables have been omitted from the right-hand side of the regression equation.

[7] As the PMC is linear in θ_0 global identification and first-order identification are identical.

where $\mathbf{y}_{i,t}$ is the scalar dependent variable, θ_0 is a scalar parameter, and $\mathbf{u}_{i,t} = \mathbf{a}_i + \mathbf{w}_{i,t}$. In this case, the unobserved error takes a composite form consisting of an individual effect, $\mathbf{a}_i$ and an idiosyncratic component, $\mathbf{w}_{i,t}$. For each individual, the idiosyncratic error is mean zero, serially uncorrelated, and uncorrelated with the individual effect. The individual effect has mean zero but is correlated with the explanatory variable $\mathbf{y}_{i,t-1}$ by construction here and so is known as a "fixed effect." As a result, $\mathbf{y}_{i,t-1}$ is correlated with the error $\mathbf{u}_{i,t}$, and so OLS estimation of θ_0 based on (13.2) would yield inconsistent estimators. However, it can be shown that the following moment conditions hold: $E\left[\mathbf{y}_{i,t-k}\Delta\mathbf{u}_{i,t}(\theta_0)\right] = 0$ for $k = 2,3,\ldots,t-1$ where $\Delta\mathbf{u}_{i,t}(\theta) = \Delta\mathbf{y}_{i,t} - \theta\Delta\mathbf{y}_{i,t-1}$ and Δ is the first (time) difference operator. The intuition behind the form of the moment conditions is as follows. Since $\Delta\mathbf{u}_{i,t}(\theta_0) = \Delta\mathbf{w}_{i,t}$, first differencing of the error eliminates the fixed effect. While $\Delta\mathbf{w}_{i,t}$ is correlated with $\mathbf{y}_{i,t-1}$ (as both depend on $\mathbf{w}_{i,t-1}$), $\Delta\mathbf{w}_{i,t}$ is uncorrelated with $\mathbf{y}_{i,t-k}$, for $k \geq 2$, as the latter depends on $\mathbf{w}_{i,t-s}$ for $s = k, k+1, \ldots$ and $\mathbf{w}_{i,t}$ is a serially uncorrelated process. In this case, $\mathbf{v}_i = \left(\mathbf{y}_{i,T}, \mathbf{y}_{i,T-1}, \ldots, \mathbf{y}_{i,1}\right)'$ and

$$f(\mathbf{v}_i,\theta) = \left[\Delta\mathbf{u}_{i,3}(\theta)\mathbf{z}'_{i,3},\ \Delta\mathbf{u}_{i,4}(\theta)\mathbf{z}'_{i,4}, \ldots, \Delta\mathbf{u}_{i,T}(\theta)z'_{i,T}\right]'$$

where $\mathbf{z}_{i,t} = \left(\mathbf{y}_{i,t-2}, \mathbf{y}_{i,t-3}, \ldots, \mathbf{y}_{i,1}\right)'$ and so $q = (T-1)(T-2)/2$. Identification holds provided

$$E\left[\Delta\mathbf{y}_{i,2}(\theta)\mathbf{z}'_{i,3},\ \Delta\mathbf{y}_{i,3}(\theta)\mathbf{z}'_{i,4}, \ldots, \Delta\mathbf{y}_{i,T-1}(\theta)\mathbf{z}'_{i,T}\right] \neq 0, \qquad (13.3)$$

which is the case provided $|\theta| < 1$; see Blundell and Bond (1998).◊

The GMM estimator based on (13.1) is defined as

$$\hat{\theta}_{GMM} = \arg\min_{\theta\in\Theta} g_n(\theta)' W_n g_n(\theta),$$

where $g_n(\theta) = n^{-1}\sum_{i=1}^{n} f(v_i,\theta)$ is the sample moment and W_n is known as the *weighting matrix* and is restricted to be a positive semi-definite matrix that converges in probability to W, some positive definite matrix of constants. The GMM estimator is thus the value of θ that is closest to setting the sample moment to zero. The measure of distance for $g_n(\theta)$ from zero depends on the choice of W_n, and we return to this feature below.

The term *info-metric*, which comes from a combination of information and econometric theory, captures the idea that this approach synthesizes work from these two fields. Our implementation of this approach here follows the approach taken by Kitamura (2007). While we use the epithet info-metric, it should be noted that Kitamura (2007) refers to this approach to estimation as generalized minimum contrast (GMC), and it can be viewed as an example of "minimum discrepancy" estimation (Corcoran, 1998).[8] Whichever way we refer to it, the key to this approach is that the PMC is viewed as a constraint on the true probability distribution of data. If **M** is a set of all probability measures, then the subset that satisfies PMC for a given θ is

$$\mathbf{P}(\theta) = \left\{P \in \mathbf{M} : \int f(v,\theta)\, dP = 0\right\},$$

and the set that satisfies the PMC for all possible values of θ is

$$\mathrm{P} = \cup_{\theta \in \Theta} \mathbf{P}(\theta).$$

We proceed here under the assumption that (13.1) holds for some θ_0 and so **P** is not an empty set; however, we return to the nature of this assumption briefly at the end of the section. Estimation is based on the principle of finding the value of θ that makes $\mathbf{P}(\theta)$ as close as possible to the true distribution of data. To operationalize this idea, we work with discrete distributions. Let $\pi_i = P(\mathbf{v} = v_i)$ and $P = [\pi_1, \pi_2, \ldots, \pi_n]$. Assuming no two sample outcomes for $\mathbf{v}$ are the same, the empirical distribution of the data attaches the probability of $1/n$ to each outcome. It is convenient to collect these empirical probabilities into a $1 \times n$ vector $\hat{\mu}$ whose elements are all $1/n$ and whose i^{th} element can thus be interpreted as the empirical probability of the ith outcome. The IM estimator is then defined to be:

$$\hat{\theta}_{IM} = \arg \inf_{\theta \in \Theta} \mathcal{D}_n(\theta, \hat{\mu}),$$

where

$$\mathcal{D}_n(\theta, \hat{\mu}) = \inf_{\hat{P}} D(\hat{P} \| \hat{\mu}),$$

$$\hat{\mathbf{P}}(\theta) = \left\{\hat{P} : \pi_i > 0, \sum_{i=1}^{n} \pi_i = 1, \sum_{i=1}^{n} \pi_i f(v_i,\theta) = 0\right\},$$

and $D(\cdot\|\cdot)$ is a measure of distance. An interpretation of the estimator can be built up as follows. $\hat{\mathbf{P}}(\theta)$ is the set of all discrete distributions that satisfy the PMC for a given value of θ. $\mathcal{D}_n(\theta, \hat{\mu})$ represents the shortest distance between any member of $\hat{\mathbf{P}}(\theta)$, and the empirical distribution for a particular value of θ. $\hat{\theta}_{IM}$ is the parameter value that makes this distance as small as possible over θ. To implement the estimator, it is necessary to specify a distance measure. Following Kitamura (2007), this distance is defined as $n^{-1}\sum_{i=1}^{n} \phi(n\hat{\pi}_i)$ where $\phi(\cdot)$ is a

[8] See Parente and Smith (2014) for further discussion.

convex function.[9] As noted in the introduction, the IM framework contains a number of other estimators as a special case; for EL $\phi(\cdot) = -log(\cdot)$, for ET $\phi(\cdot) = (\cdot)log(\cdot)$, for CUGMM $\phi(\cdot) = 0.5[(\cdot) - 1]^2$. This IM approach emphasizes the idea of economic models placing restrictions on the probability distribution of the data.

Smith (1997) defines the GEL estimator of θ_0 to be

$$\hat{\theta}_{GEL} = \arg \min_{\theta\in\Theta} \sup_{\lambda\in\Lambda_n} C_n(\theta,\lambda),$$

where

$$C_n(\theta,\lambda) = \frac{1}{n}\sum_{i=1}^{n}\left[\rho(\lambda' f_i(\theta)) - \rho_0\right],$$

$\rho(a)$ is a continuous, thrice differentiable and concave function on its domain $\mathcal{A}$, an open interval containing 0, $\rho_0 = \rho(0)$, and λ is an auxiliary parameter vector restricted so that $\lambda' f_i(\theta) \in \mathcal{A}$ (with probability approaching one), for all $(\theta',\lambda')' \in \Theta\times\Lambda_n$ and $i = 1, \ldots, n$.[10] Once again, particular choices of $\rho(\cdot)$ yield the CUGMM, EL and ET estimators: $\rho(a) = log(1-a)$ for EL; $\rho(a) = -e^a$ for ET; $\rho(a)$ quadratic for CUGMM. Within the GEL, the probabilities are defined implicitly. Smith (1997) and Newey and Smith (2004) show that the GEL estimator of π_i is given by

$$\hat{\pi}_{i,GEL} = \frac{\rho_1\left(\hat{\lambda}' f\left(v_i,\hat{\theta}_{GEL}\right)\right)}{\sum_{i=1}^{n}\rho_1\left(\hat{\lambda}' f\left(v_i,\hat{\theta}_{GEL}\right)\right)},$$

where $\rho_1(\overline{\kappa}) = \partial\rho(\kappa)/\partial\kappa|_{\kappa=\overline{\kappa}}$. Newey and Smith (2004) show that $\hat{\pi}_{i,GEL}$ is guaranteed to be positive provided $\hat{\lambda}' f\left(v_i,\hat{\theta}_{GEL}\right)$ is uniformly (in i) small.

While the IM and GEL approaches to estimation are formulated in different ways, there is a connection between them for certain choices of $\phi(\cdot)$ and $\rho(\cdot)$. Specifically, for some constant b, if $\phi(a) = \{a^{b+1} - 1\}/\{b(b+1)\}$ and $\rho(a) = -(1+ba)^{(1+b)/b}/(1+b)$, then GEL is the dual of the IM approach,[11] and so the two estimators coincide.[12] These choices of ϕ(a) and $\rho(a)$ are associated with the Cressie-Read (CR) class of estimators (Cressie and Read, 1984). In these cases, the auxiliary parameter in GEL estimation, λ, is proportional to the Lagrange Multiplier on the constraint that the moment condition holds in the

[9] This is known as the f-divergence between the two discrete distributions, in this case, $\{\hat{p}_i\}$ and $\{\hat{\mu}_i\}$.

[10] Specifically, Λ_n imposes bounds on λ that "shrink" with n, but at a slower rate than $n^{-1/2}$, which is the convergence rate of the GEL estimator for λ.

[11] See Smith (1997) and Newey and Smith (2004).

[12] See Newey and Smith (2004).

IM formulation. In the rest of the chapter, unless otherwise stated, we confine attention to the CR class so that the IM and GEL yield the same estimators; note that this class includes EL, ET, and CUGMM.

A crucial difference between GMM and IM/GEL optimizations is that the latter not only estimates θ_0 but also provides estimated probabilities for the outcomes in the data $\{\hat{\pi}_i\}$ that are constructed to ensure that the sample analog to (13.1) is satisfied at $\hat{\theta}$. As a result, the estimated sample moment is set equal to zero in IM/GEL but it is not in GMM. This difference turns out to be key for understanding key differences in the statistical properties of the estimators, as we now discuss.

If θ_0 is just identified by the PMC, then GMM and IM/GEL estimators are identical, being equal to the method of moments estimator based on(13.1). If θ_0 is overidentified then subject to certain regularity conditions—including global identification and first-order identification—it can be shown that (i) the GMM and IM/GEL are consistent for θ_0; (ii) the so-called two-step GMM estimator and IM/GEL have the large sample distributions given by: $n^{1/2}\left(\hat{\theta}-\theta_0\right) \xrightarrow{d} N(0, V_\theta)$ where $V_\theta = \left\{F(\theta_0)'S(\theta_0)^{-1}F(\theta_0)\right\}^{-1}$ and $S(\theta_0) = Var[f(\mathbf{v}, \theta_0)]$;[13] and (iii) the two-step GMM and IM/GEL estimators achieve the semiparametric asymptotic efficiency bound for estimation of θ_0 based on (13.1), see Chamberlain (1987).

While the first-order asymptotic properties of the estimators are the same, Newey and Smith (2004) show that their second-order properties are different. Specifically, they show that IM/GEL estimators have fewer sources of second-order bias than GMM and that within the IM/GEL class, EL has the fewest sources of bias. This suggests that EL should exhibit the smallest second-order bias.[14] These differences can be traced to the form of the first-order conditions associated with GMM and IM/GEL. Newey and Smith (2004) show that in each case the first-order conditions take the form

$$(\text{Jacobian})' \times (\text{variance of sample moment})^{-1} \times \text{sample moment} = 0,$$

[13] The "two-step" GMM estimator is calculated using a choice of W_n that converges in probability to $S(\theta_0)$. Hansen (1982) shows that this choice leads to a GMM estimator based on (1) with the smallest variance in large samples. Its name comes from the fact that to implement this GMM estimator requires $W_n = \hat{S}^{-1}$ where $\hat{S} \xrightarrow{p} S$, and so a consistent estimator of θ_0 to form $\hat{S}$. As Hansen notes, this can be achieved by the following two-step estimation procedure: re-estimate θ_0 by GMM with a suboptimal choice of W_n and use this to construct $\hat{S}$, then reestimate θ_0 by GMM with $W_n = \hat{S}^{-1}$.

[14] See Andrews, Elamin, Hall, Kyriakoulis, and Sutton (2017) for further discussion of this issue in the context of the model in Example 1 and an empirical illustration of where these differences are important for estimation of a policy parameter in health economics.

with the differences in the estimators arising from how the Jacobian and variance terms are estimated. EL uses the probabilities $\{\hat{\pi}_i\}$ to construct efficient estimators for both; the other members of the GEL class use their associated probabilities to construct an efficient estimator for the Jacobian but use an inefficient estimator for the variance term; and GMM uses an inefficient estimator for both.

Throughout this discussion, it has been assumed that **P** is not an empty set; that is, there exists a value θ_0 at which the PMC is satisfied. While this is guaranteed if $q = p$, it is not if $q > p$, see Hall and Inoue (2003). Within the moment-based estimation literature, the model is referred to as "correctly specified" if $P(\theta)$ is nonempty and is often said to be "misspecified" if $P(\theta)$ is empty. Hall and Inoue (2003) and Schennach (2007) consider the large sample properties of, respectively, GMM and EL estimators in misspecified models.

3. The Pseudo-Panel Data Approach to Estimation Based on Repeated Cross-Sectional Data

In this section, we discuss further the pseudo-panel approach to the estimation of linear regression models based on repeated cross-section data. For ease of exposition, we present this discussion in the context of a specific example involving the relationship between an individual's level of education and subsequent earnings. More specifically, we focus on the estimation of the "returns to education," that is, the impact of an additional year of education on wages.

To this end, suppose we have a repeated cross-section data set containing the values of the log of hourly wages, **y**, and the number of years of education, **ed**, for cross-sections of individuals sampled from a population in each of T consecutive years. We thus index observations by the pair $(j(t),t)$ where t denotes the year and $j(t)$ denotes the jth individual sampled in year t.

Suppose further that wages and education are related by the following model:

$$\mathbf{y}_{j(t),t} = \alpha + \beta \mathbf{ed}_{j(t),t} + \mathbf{u}_{j(t),t}, \tag{13.4}$$

where α, β are unknown parameters, and $\mathbf{u}_{j(t),t}$ is an unobservable error. Within this example, the key parameter of interest is β: 100β equals the implied percentage response in wages to one more year of education. As in our panel data example above, the error is assumed to have a composite form,

$$\mathbf{u}_{j(t),t} = \mathbf{a}_{j(t)} + \mathbf{w}_{j(t),t}. \tag{13.5}$$

The component $\mathbf{a}_{j(t)}$ is an individual-specific effect that captures unobserved characteristics about individual $j(t)$ that may affect the wage earned, and $\mathbf{w}_{j(t),t}$ is the idiosyncratic error. The unobserved characteristic, $\mathbf{a}_{j(t)}$, captures such factors as the innate ability of the individual and government education policy at the time the individual was at school. Both are correlated with education, and so $\mathbf{a}_{j(t)}$ is a fixed effect. We assume that for any individual j in the population the fixed effect is generated via

$$\mathbf{a}_j = \alpha_{c(j)} + \mathbf{a}_j^* \tag{13.6}$$

where $\alpha_{c(j)}$ is an unknown constant that depends on $c(j)$, the birth cohort of individual j, and $\mathbf{a}_j^*$ is a mean-zero random variable that accounts for variation in the fixed effect across individuals from the same birth cohort.

This specification can be justified in our example as follows. There is no reason to suppose that the distribution of innate ability in the population has changed over time, and so the effect of this component on wages is captured by the constant α in (13.4), and the remaining variation contributes to $\mathbf{a}_j^*$. However, government education policy has changed over time, and the systematic component of this change is captured by $\alpha_{c(j)}$ with variation about this level also contributing to $\mathbf{a}_j^*$. Note that this effect is indexed by the cohort of birth because this indicates the calendar years when the individual attended primary and secondary education.

Notice that the individual effect cannot be removed by first differencing as in Example 2 because we do not observe the same individuals in each year. Instead, Deaton (1985) proposes creating a "pseudo-panel" data set from the original repeated cross-section data by constructing (birth) cohort–time averages leading to the estimation of the regression model

$$\bar{y}_{c,t} = \text{“cohort specific intercept”} + \beta\overline{ed}_{c,t} + \text{“error”}, \tag{13.7}$$

where $\bar{(\cdot)}_{c,t}$ represents the sample mean value of $(\cdot)$ in period t for individuals born in cohort c.[15] This is an example of a grouped-data estimation in which the number of groups is $G = CT$, where C is the total number of birth cohorts and T is the total number of time periods.[16] As noted by Angrist (1991), Durbin (1954) shows that OLS regression with group means is equivalent to two-stage least squares (2SLS) estimation using individual-level data, with regressors in

[15] Thus the sample mean is calculated over all $j(t)$ for which $c(j(t)) = c$, where as above $c(j(t))$ denotes the birth cohort of individual $j(t)$.

[16] For ease of exposition, we assume that each time period contains observations from each cohort.

the first stage being a complete set of dummy variables for group membership. Since 2SLS is a GMM estimator,[17] OLS estimation based on group-mean data can be viewed as an estimation method based on population moment conditions.[18]

To present the moment conditions in question, it is convenient to introduce a group index, g. In our example, the "group" is defined by the interaction of the year, t, in which the wage is observed and the birth cohort of the individual concerned, c. Accordingly, we set

$$g = (c-1)T + t, \quad c = 1,2,\dots C; \quad t = 1,2,\dots T,$$

so that, for example, $g = 1$ corresponds to individuals from cohort $c = 1$ in year $t = 1$; $g = T + 1$ corresponds to individuals from birth cohort $c = 2$ in period $t = 1$; and $g = CT$ corresponds to individuals in cohort C in period T. Now define the group-level model

$$\mathbf{y}_g = \mathbf{x}_g'\theta_0 + \mathbf{u}_g \tag{13.8}$$

where $\mathbf{y}_g$ is a random variable modeling the log wage for members of group g, $\mathbf{x}_g = \left[\mathcal{I}_g^{(1)}, \mathcal{I}_g^{(2)}, \dots, \mathcal{I}_g^{(C)}, \mathbf{ed}_g\right]'$, $\mathbf{ed}_g$ denotes the number of years for education for a member of group g, $\mathcal{I}_g^{(c)}$ is an indicator variable that takes the value 1 if group g involves individuals born in cohort c, θ_0 is the parameter vector whose first $p - 1$ elements are the cohort-specific intercepts and whose last element is β (here $p = C + 1$). Following the reasoning in the previous paragraph, OLS estimation based on group-mean data is equivalent to estimation of the group-level model in (13.8) based on the information that

$$E\left[\mathbf{u}_g(\theta_0)\right] = 0, \quad g = 1,2,\dots G, \tag{13.9}$$

where $\mathbf{u}_g(\theta) = \mathbf{y}_g - \mathbf{x}_g'\theta$. In this case, identification requires not only that $p = dim(\theta) < G$ but also that the data follow a different distribution in each group. To demonstrate why this is the case, it is sufficient to consider the case of $C = T = 2$, so that $G = 4$. Expressing (13.9) as a single 4×1 PMC with g^{th} element $E[\mathbf{u}_g(\theta_0)] = 0$, the Jacobian is

[17] For example, see Hall (2005, Chapter 2).

[18] We note that Deaton (1985) does not propose inference based on the OLS estimator but instead a modified version referred to as the Errors in Variables (EVE) estimator. This is because Deaton (1985) considers asymptotics in which the number of groups gets large but the number of observations in each group is finite. Within that framework, Deaton (1985) shows that the OLS estimator is inconsistent. However, if the number of groups is fixed and the number of observations in each group becomes large—the framework we adopt below—then OLS is consistent; see Angrist (1991). See Devereux (2007) for analysis of the second-order properties of the EVE.

$$\begin{bmatrix} 1 & 0 & E[\mathbf{ed}_1] \\ 1 & 0 & E[\mathbf{ed}_2] \\ 0 & 1 & E[\mathbf{ed}_3] \\ 0 & 1 & E[\mathbf{ed}_4] \end{bmatrix}$$

and this matrix must have rank equal to three for θ_0 to be identified.[19] For this rank condition to hold, it must be that $E[\mathbf{ed}_1] \neq E[\mathbf{ed}_2]$ and/or $E[\mathbf{ed}_3] \neq E[\mathbf{ed}_4]$.

As the preceding discussion shows, the pseudo-panel approach to estimation with repeated cross-sectional data involves dividing the population into groups and basing estimation on group-specific moment conditions. Thus, to develop IM estimators for this kind of estimation scenario, we need to extend the IM framework of Section 2 to cover the case in which the population is heterogeneous and consists of homogeneous subpopulations. This is the topic of the next Section.

4. IM Estimation with Group-Specific Moment Conditions

In this section, we extend the IM framework to the case in which the population is heterogeneous, consisting of a finite number of homogeneous subpopulations. In view of the discussion in the previous section, we refer to these subpopulations as "groups."

We first describe the group data structure. It is assumed that (i) there are G groups; (ii) the model for each group g involves a random vector $\mathbf{v}_g$ with probability measure μ_g and sample space $\mathcal{V}_g$; (iii) there is a random sample of n_g draws of $\mathbf{v}_g$ the outcome of which is modeled by the set of random vectors $\{\mathbf{v}_{g,i};\ i = 1, 2, \ldots n_g\}$. We impose the following condition on the groups and how these samples are drawn.

Assumption 1 *(i) $\mathbf{v}_g$ is independent of $\mathbf{v}_h$ for all $g,h = 1,2, \ldots, G$ and $g \neq h$;*
(ii) $\{\mathbf{v}_{g,i};\ i = 1,2, \ldots n_g\}$ is a set of independently and identically distributed random vectors.

Note that Assumption 1 states that observations are independent both across and within groups. This rules out many forms of clustering, as, for example, in worker-firm data. It also rules out serial correlation if g is partly defined on time.

We further assume that the model implies that each group satisfies a set of population moment conditions involving $f_g : \mathcal{V}_g \times \Theta \to \mathbb{R}^{q_g}$.

Assumption 2 *$E[f_g(\mathbf{v}_g, \theta_0)] = 0$ where $\theta_0 \in \Theta \subset \mathbb{R}^p$, for $g = 1, 2, \ldots, G$.*

[19] As the model is linear in θ_0, identification and first-order identification are equivalent.

Note that the moment conditions are allowed to vary by g. This may happen because the functional form of the moment conditions is the same across g, but they are evaluated at group-specific parameters; that is,

$$f_g\left(\mathbf{v}_g,\theta_0\right)=f\left(\mathbf{v}_g,\gamma_g(\psi),\beta\right)$$

where $\theta=(\psi',\beta')'$ and $\gamma_g(\psi)$ is a continuous function of ψ. Our example in Section 3 fits this structure with $\psi=(\psi_1,\psi_2,\ldots,\psi_C)$ and ψ_c denoting the intercept for cohort c so that

$$f_g\left(\mathbf{v}_g,\theta_0\right)=\mathbf{y}_g-\psi_c-\beta\mathbf{ed}_g=\mathbf{y}_g-\gamma_g(\psi)-\beta\mathbf{ed}_g=f\left(\mathbf{v}_g,\gamma_g(\psi),\beta\right),$$

where g involves individuals from cohort c and $\gamma_g(\psi)=\psi_c$. However, the key element is that certain parameters appear in the population moment conditions associated with more than one group.

To present the IM estimator, we need the following additional notation. Let $\mathbf{v}=vec(\mathbf{v}_1,\mathbf{v}_2,\ldots,\mathbf{v}_G)$ and $\mu=\mu_1\times\mu_2\times\ldots\mu_G$; note that Assumption 1 implies μ is the probability measure of $\mathbf{v}$. Now define the following sets of measures: $\mathbf{M}$, the set of all possible probability measures for $\mathbf{v}$;

$$\mathbf{P}(\theta)=\left\{P=P_1\times P_2\times\cdots\times P_G\in\mathbf{M}:\ \int f_g\left(v_g,\theta\right)dP_g=0,g=1,2,\ldots G\right\},$$

the set of all measures for $\mathbf{v}$ which satisfy the population moment conditions in each group for a given value of θ;

$$\mathbf{P}=\cup_{\theta\in\Theta}\mathbf{P}(\theta),$$

the set of all measures for $\mathbf{v}$ that satisfy the PMC in each group for some $\theta\in\Theta$. As for the homogeneous population case in Section 2, estimation is based on the principle of finding the value of θ that makes $\mathbf{P}(\theta)$ as close as possible to true distribution of data and that this approach is operationalized using discrete distributions. To this end, suppose the sample outcome of $\mathbf{v}_{g,i}$ is $v_{g,i}$ for $i=1,2,\ldots n_g$. The total sample size is then $N=\sum_{g=1}^{G}n_g$. Define $\pi_{g,i}=P(\mathbf{v}_{g,i}=v_{g,i})$, and $\hat{P}_g=\left[\pi_{g,1},\pi_{g,2},\ldots,\pi_{g,n_g}\right]$, $\hat{P}=\left[\hat{P}_1,\hat{P}_2,\ldots,\hat{P}_G\right]$. Assuming no two sample outcomes are the same, the empirical distribution of the data is: $\hat{\mu}_{g,i}=n_g^{-1}$; let $\hat{\mu}_g=n_g^{-1}\iota'_{n_g}$ where ι_{n_g} is a $n_g\times 1$ vector of ones, $\hat{\mu}=[\hat{\mu}_1,\ldots,\hat{\mu}_G]$.

The info-metric estimator based on group-specific moment conditions (group IM [GIM]) is then defined to be:

$$\hat{\theta}_{GIM}=\arg\inf_{\theta\in\Theta}\mathcal{D}_N(\theta,\hat{\mu}) \tag{13.10}$$

where

$$\begin{aligned} \mathcal{D}_N(\theta,\hat{\mu}) &= \inf_{\hat{P}(\theta)\in\hat{\mathrm{P}}(\theta)} D\left(\hat{P}(\theta)\middle\|\hat{\mu}\right), \\ \hat{\mathrm{P}}(\theta) &= \left[\hat{\mathrm{P}}_1(\theta),\hat{\mathrm{P}}_2(\theta),\ldots,\hat{\mathrm{P}}_G(\theta)\right], \\ \hat{\mathrm{P}}_g(\theta) &= \left\{\hat{P}_g : \pi_{g,i} > 0, \sum_{i=1}^{n_g}\pi_{g,i} = 1, \sum_{i=1}^{n_g}\pi_{g,i} f\left(v_{g,i},\theta\right) = 0\right\}, \end{aligned}$$

and the distance measure is:

$$D\left(\hat{P}(\theta)\middle\|\hat{\mu}\right) = N^{-1}\sum_{g=1}^{G}\sum_{i=1}^{n_g}\phi\left(n_g\hat{\pi}_{g,i}\right).$$

The natural choices for $\phi(\cdot)$ are the same as for the IM estimator, and the specific choices listed in Section 2 would give grouped-data versions of EL, ET, and CUGMM.

AHKL define the GGEL estimator of θ_0 as

$$\hat{\theta}_{GGEL} = \underset{\theta\in\Theta}{argmin}\ \sup_{\lambda\in\Lambda_N}\sum_{g=1}^{G} w_g\sum_{i=1}^{n_g}\left[\rho\left(\lambda_g' f_{g,i}(\theta)\right) - \rho(0)\right]/N, \qquad (13.11)$$

where $f_{g,i}(\theta) = f_g(v_{g,i}, \theta)$, $\rho(.)$ is a concave function on its domain $\mathcal{A}$, an open interval containing 0, $\rho_0 = \rho(0)$, w_g is a weight and $\lambda = (\lambda_1, \lambda_2, \ldots \lambda_G)'$ is a vector of auxiliary parameters, a vector restricted to lie in the set Λ_N denoting all λ such that with probability approaching 1, $\lambda_g' f_g\left(v_{g,i},\theta\right) \in \mathcal{A}$ for all $(\theta',\lambda')' \in \Theta\times\Lambda_N$ and $g = 1, 2 \ldots G$.[20] Restricting attention to the CR class, the auxiliary parameter λ_g is proportional to the Lagrange Multiplier on the constraint that the group g moment condition holds in the IM formulation (13.10).

We find the info-metric (primal) approach more intuitively appealing because it is formulated explicitly in terms of the PMC placing restrictions on the distribution of the data. However, the GEL approach is often more appealing for the purposes of developing the asymptotic analysis of the estimator.

While either method can be used to calculate the estimator, the primal (GIM) and dual (GGEL) approaches have differing computational requirements. In the primal approach, the optimization is over both the probabilities $\left\{\hat{\pi}_{g,i}; i = 1,2,\ldots n_g; g = 1,2,\ldots G\right\}$ and θ. In the dual approach, the optimization is over λ and θ.[21] While the latter involves fewer parameters, the associated

[20] Setting $\rho(\cdot)$ equal to $log[1 - (\cdot)]$, $-exp(\cdot)$ and $\rho_0 - (\cdot) - 0.5(\cdot)^2$ yields grouped-data versions of EL, ET, and CUGMM, respectively.

[21] The probabilities can be estimated from estimators of λ_g, θ and the data. For this discussion we set $w_g = 1$.

optimizations can be problematic as we now discuss. The computation is performed by iterating between the so-called *inner* and *outer* loops. The inner loop involves optimization over λ for given θ; that is,

$$\hat{\lambda}(\theta) = \arg\sup_{\lambda \in \Lambda_N} \sum_{g=1}^{G} \sum_{i=1}^{n_g} \left[\rho\left(\lambda_g' f_{g,i}(\theta)\right) - \rho(0)\right]/N,$$

and the outer loop involves optimization over θ given λ; that is,

$$\hat{\theta} = \arg\min_{\theta \in \Theta} \sum_{g=1}^{G} \sum_{i=1}^{n_g} \left[\rho\left(\lambda_g'(\theta)' f_{g,i}(\theta)\right) - \rho(0)\right]/N.$$

While the inner loop is well suited to gradient methods because $\rho(\cdot)$ is strictly concave, the outer loop can be more problematic.[22] In terms of calculating the estimators in our context using numerical routines in MATLAB, we have found optimization associated with the primal approach to be far more reliable because, due to the convexity of the primal approach, estimation is robust to the initial parameter starting values provided to the optimizer.[23] In contrast, the solution to the min-max problem in the GEL approach is extremely sensitive to starting values; in cases where the information content of the PMC is low, the optimizing routine would often fail to move away from the initial values.

5. Statistical Properties and Inference

In this section, we describe the asymptotic properties of the GGEL estimator and its associated inference framework.[24] To simplify the exposition, we consider the case in Section 3 where the group-level model is:

$$\mathbf{y}_g = \mathbf{x}_g'\theta_0 + \mathbf{u}_g, \tag{13.12}$$

where $\mathbf{x}_g = \left[1, \mathbf{r}_g'\right]'$, $\mathbf{r}_g$ is a $(p-1) \times 1$ vector of observable random variables, and $\mathbf{y}_g$ and $\mathbf{u}_g$ are scalar random variables. Recall that $\mathbf{u}_g(\theta) = u_g\left(\mathbf{v}_g, \theta\right) = \mathbf{y}_g - \mathbf{x}_g'\theta$ where $u_g : \mathcal{V}_g \times \Theta \to \mathbb{R}$ where $\mathcal{V}_g$ is the sample space of $\mathbf{v}_g = \left(\mathbf{y}_g, \mathbf{x}_g'\right)'$ and $\Theta \subset \mathbb{R}^p$.

[22] For example, see Guggenberger (2008).

[23] Specifically, in the simulations reported in Section 5, the procedure *fmincon* was used to enforce the constraints in Eq. (13.10). Convergence time is greatly improved by providing analytical form for the Jacobian and Hessian of the IM objective function.

[24] To our knowledge, the results presented in this section are new as this chapter and AHKL are the first to consider the GGEL estimator. However, the results can all be viewed as natural extensions of the analogous procedures based on the GEL estimator presented and analyzed in Smith (1997, 2011). For ease of presentation, we set $w_g = 1$.

Estimation is based on the information that

$$E\left[\mathbf{u}_g(\theta_0)\right] = 0, \quad g = 1, 2, \ldots G. \tag{13.13}$$

This model fits into the framework of Section 4 with $\mathbf{v}_g = \left(\mathbf{y}_g, \mathbf{r}_g'\right)'$ and $f_g\left(\mathbf{v}_g, \theta\right) = \mathbf{y}_g - \mathbf{x}_g'\theta$. Equation (13.13) can be written more compactly as

$$E[\mathbf{u}(\theta_0)] = 0, \tag{13.14}$$

where $\mathbf{u}(\theta_0)$ is the $G \times 1$ vector with g^{th} element $\mathbf{u}_g(\theta_0)$. Notice that the Jacobian associated with this population moment condition is

$$F = [B_1, B_2, \ldots, B_G]'$$

where $B_g = E[\mathbf{x}_g]$ and so in this simple model does not depend on θ.

In addition to Assumption 1, the data must satisfy certain regularity conditions.

Assumption 3 *(i)* $E[\boldsymbol{u}_g(\theta_0)] = 0$, $g = 1, 2, \ldots G < \infty$; *(ii)* $E\left[sup_{\theta \in \Theta}\left|\mathbf{u}_g(\theta_0)\right|^{\zeta}\right] < \infty$ *for some* $\zeta > 2$ *and all* $g = 1, 2, \ldots, G$; *(iii)* $Var\left[\mathbf{u}_g\right] = \sigma_g^2 > 0$ *for all* $g = 1, 2, \ldots, G$.

Assumption 4 *(i)* $E[||\boldsymbol{x}_g||] < \infty$ *for* $g = 1, 2, \ldots, G$; *(ii)* $rank\{F\} = p$.

Assumption 3 (i) restates Assumption 2 in the context of our model here (for completeness) and ensures that the PMC is valid. Assumption 4 (ii) implies that θ_0 is both globally and first-order locally identified—and so $p < G$.[25]

The sample sizes from each group are assumed to satisfy the following condition.

Assumption 5 n_g *is a deterministic sequence such that* $n_g/N \to v_g \in (0,1)$ *as* $N \to \infty$.

Notice this assumption implies that the sample size for each group increases with N and so becomes asymptotically large. One consequence of this assumption is that the Weak Law of Large Numbers (WLLN) and Central Limit Theorem (CLT) can be used to deduce the behavior of the group

[25] Assumption 4(ii) is the condition for first-order local identification which is also the condition for global identification as the population moment condition is linear in θ.

averages. Let $\left\{\mathbf{x}_{g,i}\right\}_{i=1}^{n_g}$ and $\left\{\mathbf{u}_{g,i}\right\}_{i=1}^{n_g}$ be random draws from the distributions of $\mathbf{x}_g$ and $\mathbf{u}_g$, respectively, $\bar{(\cdot)}_g = n_g^{-1}\sum_{i=1}^{n_g}(\cdot)_{g,i}, \bar{(\cdot)} = vec\left[\bar{(\cdot)}_1, \bar{(\cdot)}_2, \ldots, \bar{(\cdot)}_G\right]$ and $\hat{\nu}_N = \text{diag}(n_1/N, n_2/N, \ldots, n_G/N)$. Then under Assumptions 1 and 3–5, we can invoke the WLLN and the CLT, respectively, to deduce:

$$\left(n_g/N\right)\bar{x}_g \overset{p}{\to} \nu_g B_g, g = 1, 2, \ldots G \tag{13.15}$$

$$N^{1/2}\hat{\nu}_N \bar{u} \overset{d}{\to} N(0_G, \Psi_u) \tag{13.16}$$

where 0_G denotes the $G \times 1$ null vector, $\Psi_u = \text{diag}\left(\nu_1\sigma_1^2, \nu_2\sigma_2^2, \ldots, \nu_G\sigma_G^2\right)$.

Within our framework, the asymptotic theory is derived under the assumption that the number of groups, G, the number of moment conditions in each group, q_g, and the number of parameters, p, remain constant as the sample size increases. Our simulation results below provide preliminary evidence on the types of settings in which this theory provides a reasonable approximation to finite sample behavior. However, we recognize that in some settings more accurate approximations may be provided by a theory in which G and/or q_g and/or p increase with the sample size. These extensions are currently under investigation.

Finally, for ease of presentation, we confine our attention here to the leading cases of EL, ET and CUGMM estimation.

Assumption 6 *$\rho(\cdot)$ is either log[1 - $(\cdot)$] or - exp$(\cdot)$ or $\{\rho_0 - (\cdot) - 0.5(\cdot)^2\}$.*

It can be shown, using similar arguments to Newey and Smith (2004), that $\hat{\theta}$ is consistent for θ_0.[26] Using this consistency property, it is possible to take a mean value expansion of the first-order conditions of the optimization from which the following result can be deduced.

Proposition 1 *Under Assumptions 1 and 3–6, we have:*

$$\begin{bmatrix} \sqrt{N}\left(\hat{\theta} - \theta_0\right) \\ \sqrt{N}\hat{\lambda} \end{bmatrix} \overset{d}{\to} N\left(\begin{bmatrix} 0_p \\ 0_G \end{bmatrix}, \begin{bmatrix} V_\theta & 0 \\ 0 & V_\lambda \end{bmatrix}\right),$$

where $V_\theta = \left(B'\Psi_u^{-1}B\right)^{-1}, V_\lambda = \Psi_u^{-1} - \Psi_u^{-1}BV_\theta B'\Psi_u^{-1}$, *and* $B = [\nu_1 B_1, \nu_2 B_2, \ldots, \nu_G B_G]'$.

Proposition 1 implies that $\sqrt{N}\left(\hat{\theta} - \theta_0\right)$ and $\sqrt{N}\hat{\lambda}$ converge to normal distributions and are asymptotically independent.

[26] Formal proofs of all results are omitted for brevity but are presented in Khatoon (2014) for the model in this section and in AHKL for the general model described in section 4.

Within this framework, two types of inference are naturally of interest: inference about θ_0 and tests of the validity of the PMC upon which the estimation rests. We now discuss these in turn.

An approximate 100a percent confidence interval for $\theta_{0,k}$, the kth element of θ_0, is given by

$$\left(\hat{\theta}_k \pm z_{1-a/2} se\left(\hat{\theta}_k\right)\right) \tag{13.17}$$

where $\hat{\theta}_k$ is the kth element of $\hat{\theta}$, $z_{1-a/2}$ is the $100(1 - a/2)^{th}$ percentile of the standard normal distribution, $se(\hat{\theta}_k)$ is the kth main diagonal element of $\hat{V}_\theta = \left(\hat{B}'\hat{\Psi}_u^{-1}\hat{B}\right)^{-1}$, $\hat{B} = \hat{\nu}_N\left[\overline{x}_1, \overline{x}_2, \ldots, \overline{x}_G\right]'$, $\hat{\Psi}_u = \hat{\nu}_N \operatorname{diag}\left(\hat{\sigma}_1^2, \hat{\sigma}_2^2, \ldots, \hat{\sigma}_G^2\right)$, $\hat{\sigma}_g^2 = \sum_{i=1}^{n_g} \hat{\pi}_{g,i}\left(y_{g,i} - x_{g,i}'\hat{\theta}\right)^2$, and $\left\{y_{g,i}, x_{g,i}\right\}_{i=1}^{n_g}$ are the sample realizations in group g.

As is apparent from the discussion in Section 3, the estimation exploits the information that the moment condition holds in the population. Typically, this moment condition is derived from some underlying economic and/or statistical model, and so it is desirable to test whether the data is consistent with this moment condition. Within the info-metric framework, this can be done in three ways. First, using Proposition 1, we can base inference on the Lagrange Multipliers $\hat{\lambda}$. Inspection reveals that V_λ is singular, being of rank G—p. One option is to use a generalized inverse of V_λ to construct the test, but following Imbens, Spady, and Johnson (1998) we propose using the asymptotically equivalent and computationally more convenient version of the LM statistic[27]

$$LM = N\hat{\lambda}'\,\hat{\Psi}_u\hat{\lambda}. \tag{13.18}$$

Second, inference can be based directly on the estimated sample moments $\overline{u}\left(\hat{\theta}\right)$. While the optimization forces the weighted sample moments, $\sum_{i=1}^{n_g} \hat{\pi}_{g,i} u_{g,i}\left(\hat{\theta}\right)$, to zero, the estimated sample moments $n_g^{-1}\sum_{i=1}^{n_g} u_{g,i}\left(\hat{\theta}\right)$ are not so constrained but should be approximately zero if the moment condition is valid. This leads to the Wald statistic

$$Wald = N\left\{\hat{\nu}_N\overline{u}\left(\hat{\theta}\right)\right\}'\hat{\Psi}_u^{-1}\hat{\nu}_N\overline{u}\left(\hat{\theta}\right). \tag{13.19}$$

Finally, inference can be based on the optimand. Within the GEL approach, this approach leads to the statistic:

[27] Imbens, Spady, and Johnson (1998) report that this version has better finite sample properties in their simulation study than the version based on the generalized inverse of V_λ. See also Imbens (2002).

$$LR_{GGEL} = -2\sum_{g=1}^{G}\sum_{i=1}^{n_g}\left[\rho\left(\hat{\lambda}_g u_{g,i}\left(\hat{\theta}\right)\right) - \rho(0)\right], \tag{13.20}$$

in which the unconstrained version does not impose the population moment condition and so amounts to $\lambda_g = 0$—and hence $\lambda_g u_{g,i} = 0$—in the GGEL framework.

The Wald and LM statistics are all easily calculated with both the GIM and GGEL approach to estimation, but while the LR_{GGEL} is a natural side product of the dual approach, it is not so with the primal approach. For the latter, more convenient test statistics based on the primal optimand are as follows. If the estimation is performed using EL then, following Qin and Lawless (1994), a suitable statistic is based on the difference between the log-likelihood evaluated at the constrained probabilities and the unconstrained probabilities. Adapting their approach to our setting yields the test statistic:

$$LR_{EL} = -2\sum_{i=1}^{G}\sum_{i=1}^{n_g}\ln\left(n_g\hat{\pi}_{g,i}\right). \tag{13.21}$$

If the estimation is based on ET then, following Imbens, Spady, and Johnson (1998), a suitable statistic can be based on the Kullback–Liebler distance between constrained and unconstrained probabilities. In our setting, this approach yields the test statistic

$$KLIC - R_{ET} = -2\sum_{i=1}^{G}\sum_{i=1}^{n_g} n_g\hat{\pi}_{g,i}\left\{\ln\left(n_g\hat{\pi}_{g,i}\right)\right\}. \tag{13.22}$$

The following proposition gives the limiting distributions of the above tests under the null hypothesis that the moments are valid.[28]

Proposition 2 *If Assumptions 1 and 3–6 hold then: (i) LM, Wald, and LR_{GGEL} are asymptotically equivalent, and they all converge in distribution to χ^2_{G-p} as $N \to \infty$; (ii) LR_{EL} and KLIC—R_{ET} converge in distribution to χ^2_{G-p} as $N \to \infty$.*

The first-order asymptotic properties above are the same for the three variants of GGEL considered here. However, the second-order properties are different. Newey and Smith (2004) derive the second-order bias of GEL estimators for

[28] Part (i) is proved in Khatoon (2014), and part (ii) can be proved by adapting the arguments in Qin and Lawless (1994) (for EL) and Imbens, Spady, and Johnson (1998) (for ET) to our grouped-data context.

the case described in Section 2. Their approach is based on taking a third-order expansion of the first-order conditions of GEL estimation and can be adapted to derive analogous results for the GGEL estimator.[29] To this end, we introduce the following additional restriction.[30]

Assumption 7 *For each* $g = 1, 2, \ldots, G$, *there exists* $b_g(v_g)$ *such that* $E[b_g(v_{g,i})^6] < \infty$, $\left\|\mathbf{x}_g\right\| < b_g\left(\mathbf{v}_g\right)$ *and* $\left\|\mathbf{u}_g\left(\theta_0\right)\right\| < b_g\left(\mathbf{v}_g\right)$.

The second-order bias of the GGEL estimators and model considered in this section is given in the following proposition. For ease of presentation below, we use the following convention regarding the operator $diag(\cdot)$: if A is a $k \times k$ matrix, then $diag(A)$ denotes the $k \times 1$ vector containing the main diagonal elements of A; if a is a $1 \times k$ vector, then $diag(a)$ denotes the $k \times k$ diagonal matrix with main diagonal given by a.

Proposition 3 *Under Assumptions 1 and 3–7,*

$$Bias\left(\hat{\theta}_{GGEL}\right) = -\Xi\left[\mathcal{B}_1 + \left(1 + \frac{\rho_3(0)}{2}\right)\mathcal{B}_2\right]/N, \qquad (13.23)$$

where $\mathcal{B}_1 = \text{diag}[\Xi' M_{xu}]$, $\mathcal{B}_2 = M_u^{(3)}\text{diag}(V_\lambda)$, $\Xi = V_\theta B'\Psi_u^{-1}$, $V_\theta = \left(B'\Psi_u^{-1}B\right)^{-1}$, M_{xu} *is a* $p \times G$ *matrix with* g^{th} *column* $\nu_g E\left[\mathbf{u}_g\mathbf{x}_g\right]$, $M_u^{(3)} = \text{diag}\left(\nu_1\mu_1^{(3)}, \nu_2\mu_2^{(3)}, \ldots, \nu_G\mu_G^{(3)}\right)$, $\mu_g^{(3)} = E\left[\mathbf{u}_g^3\right]$, *and* $\rho_3(0) = \partial^3\rho(\kappa)/\partial\kappa^3\big|_{\kappa=0}$.

As can be seen, the second bias depends on $\rho(\cdot)$ and so is potentially different for different estimators. In fact, the biases of EL, ET, and CUGMM are given, respectively by, $-\Xi\mathcal{B}_1/N$, $-\Xi(\mathcal{B}_1 + 0.5\mathcal{B}_2)/N$ and $-\Xi(\mathcal{B}_1 + \mathcal{B}_2)/N$.[31] So it can be seen that in general, as in Newey and Smith's (2004) analysis, EL has fewer sources of bias than the other two. The formula reveals that in our model the sources of the bias are correlation between $\mathbf{u}_g$ and $\mathbf{x}_g$ and asymmetry of the distributions of $\left\{\mathbf{u}_g\right\}_{i=1}^{n_g}$. Note that within our repeated cross-section model of Section 3, $\mathbf{u}_g$ and $\mathbf{x}_g$ are correlated through the stochastic part of the fixed effect. If $E\left[\mathbf{u}_g^3\right] = 0$ for all g, then the bias is the same for all three estimators.

It is natural to consider whether the GEL approach has similar advantages over GMM in the case of grouped-specific moment conditions as those

[29] See Khatoon (2014).

[30] This is based on Newey and Smith (2004) Assumption 3 but uses the form of u_g here and the fact that we limit attention to EL, ET, and CUGMM.

[31] $\rho_3(0)$ equals –2, –1 and 0, respectively, for EL, ET, and CUGMM.

described for the homogeneous population case in Section 2. For the model in this section, the GMM estimator is

$$\tilde{\theta} = argmin_{\theta \in \Theta} \overline{u}(\theta)' W_N \overline{u}(\theta), \tag{13.24}$$

Following Newey and Smith (2004), the weighting matrix W_N is assumed to satisfy the following conditions.

Assumption 8 *$W_N = diag(W_{N,1}, W_{N,2}, ..., W_{N,G})$ where for $g = 1,2, ..., G$:*

(i) $W_{N,g} = W_g + \zeta_{N,g} + O_p(N^{-1})$; (ii) W_g is positive definite; (iii) $E[\zeta_{N,g}] = 0$; (iv) $E\left[\left\|\zeta_{N,g}\right\|^6\right] < \infty$.

Under our assumptions, the optimal choice of weighting matrix, W_N, is one that converges in probability to Ψ_u^{-1}, and with this choice it is straightforward to show that $\sqrt{N}\left(\tilde{\theta} - \theta_0\right)$ has the same limiting distribution as $\sqrt{N}\left(\hat{\theta} - \theta_0\right)$ (given in Proposition 1). Once again, the second-order properties of the GMM and GEL estimators differ.

Proposition 4 *If Assumptions 1, 3–5 and 8 hold, then*

$$Bias\left(\hat{\beta}_{GMM}\right) = \left[V_\theta \mathcal{A} - \Xi\left(\mathcal{B}_1 + \mathcal{B}_2 + \mathcal{B}_3\right)\right]/N,$$

where $\mathcal{A} = M_{xu}\,\mathrm{diag}(V_\lambda)$, $\mathcal{B}_3$ is a term that depends on M_{xu} and the difference between the first and second step weighting matrices.[32]

A comparison of Propositions 3 and 4 indicates that the grouped-data GMM estimator has more sources of bias than the corresponding GGEL estimator. One of these additional sources is attributable to the two-step nature of GMM estimation and arises if a suboptimal weighting matrix is used on the first step. A similar finding is reported in Newey and Smith (2004), and they argue that these extra sources of bias are likely to translate to an estimator that exhibits more bias in finite samples.

These results can be used to compare GGEL to the pseudo-panel data approach of regression based on group averages.[33] As noted in Section 3, this pseudo-panel approach amounts to estimation of the individual level model via 2SLS using group dummies as instruments and is therefore a GMM estimator

[32] See Khatoon (2014) for details of this term, the details of which are omitted as they are not relevant to the exposition.

[33] The discussion in the paragraph does not cover the EVE estimator; see footnote 17.

based on the moments in (13.13). 2SLS is only equivalent to two-step GMM if the variance of $\mathbf{u}_g$ is the same for all g, and so is only as efficient asymptotically as GGEL under that condition. Even then, the arguments above indicate that finite sample inferences based on the pseudo-panel approach are likely less reliable than those based on GGEL. In the next subsection, we explore whether it is the case in our setting via a small simulation study.

5.1 Simulation Study

Artificial data are generated for groups $g = 1, 2, \ldots, G$ via:

$$\mathbf{y}_g = \delta + \beta \mathbf{r}_g + \mathbf{u}_g \tag{13.25}$$

$$\mathbf{r}_g = \delta_1 + \sum_{j=2}^{G} \mathcal{I}_g^{(j)} \delta_j + \mathbf{a}_g \tag{13.26}$$

where $E[\mathbf{u}_g] = E[\mathbf{a}_g] = 0$, $Var[\mathbf{u}_g] = \sigma_u^2$, $Var[\mathbf{a}_g] = \sigma_a^2$, $Cov[\mathbf{u}_g, \mathbf{a}_g] = \sigma_{u,a} = \rho_{u,a}\sigma_u\sigma_a$, and $\mathcal{I}_g^{(j)}$ is an indicator variable that takes the value one if $j = g$. For the results reported below, we set the parameter values as follows: $\beta = 0.05$, $\delta = 0, \sigma_u^2 = 0.2, \sigma_a^2 = 3.38$, $\rho_{u,a} = 0.2, 0.5, 0.9$, $\delta_1 = 12$, $\delta_2 = 0.9$, and $\delta_j = \delta_{j-1} + 0.9$ for $j = 3, \ldots G$. Notice that within this design, $\mathbf{r}_g$ is correlated with $\mathbf{u}_g$. We report results for when $(\mathbf{u}_g, \mathbf{a}_g)'$ has a bivariate normal and for when $(\mathbf{u}_g, \mathbf{a}_g)'$ has a bivariate Student-t distribution with 7 degrees of freedom.

We also report results for different numbers of groups and sample sizes. Specifically, we consider the scenarios $G = 3, 4, 6, 8$, $N = 96, 144, 312$ with sample size within each group determined via $n_g = N/G$ for all g. Note that within this scheme, there is an inverse relationship between the number of groups and the number of observations within each group. This enables us to examine whether the accuracy of the asymptotic theory depends on just N per se or on the number of groups these observations are spread across. Ten thousand replications are performed for each parameter configuration.

Estimation is based on the moment conditions,

$$E\left[\mathbf{y}_g - \mathbf{x}_g'\theta_0\right] = 0, \text{for } g = 1, 2, \ldots G,$$

where $\mathbf{x}_g = [1, \mathbf{r}_g]'$, $\theta = (\delta, \beta)'$. We consider two versions of GIM: namely EL (with $\phi(\cdot) = -log(\cdot)$) and ET (with $\phi(\cdot) = (\cdot)log(\cdot)$). We also consider estimation based on 2SLS and two-step GMM. Specifically, we report the following statistics: the mean of the simulated distributions of $\hat{\beta}$; the rejection frequency of the 5 percent approximate significance level test of $H_0 : \beta_0 = 0.05$ (its true value) based on the t-statistic, $\hat{\beta}/s.e.\left(\hat{\beta}\right)$;[34] the rejection frequency for LM and $Wald$ (both for EL and ET), LR_{EL}, $KLIC—R_{ET}$ and the overidentifying restrictions test for GMM.[35]

Before discussing the results, we note that there are reasons to expect the finite sample behavior of GMM may be more sensitive to G than EL/ET. Specializing the result in Proposition 4 to the model in our simulations, then the second-order bias of the GMM estimator is:

$$Bias\left(\hat{\beta}_{GMM}\right) = \frac{(G-3)\,\sigma_{u,a}}{N\,R^2_{r,z}\sigma^2_r}, \tag{13.27}$$

where $R^2_{r,z}$ is the population multiple correlation coefficient from the pooled (over g) regression of $\mathbf{r}_g$ on $z_g = \left[\mathcal{J}_g^{(1)}, \mathcal{J}_g^{(2)}, \ldots, \mathcal{J}_g^{(G)}\right]$ and σ^2_r is the population variance of r. In contrast, second-order bias of the GGEL estimator is:

$$Bias\left(\hat{\beta}_{GGEL}\right) = -\frac{\sigma_{u,a}}{NR^2_{r,z}\sigma^2_r}. \tag{13.28}$$

Clearly, *ceteris paribus*, the second-order bias of GMM increases with G (for $G > 3$) but that of GGEL is invariant to G.

Equations (13.27) and (13.28) provide guidance on the relative biases of GMM and GGEL for a given G. However, within our design, these formulae cannot be used to compare the bias of a particular estimator across G because $R^2_{r,z}$ and σ^2_r change with G. Specifically, the term $R^2_{r,z}\sigma^2_r$ takes the values 0.54, 1.01, 2.36, and 4.25 (to 2 dp) for, respectively, $G = 3, 4, 6, 8$. Thus, Eqs. (13.27) and (13.28) suggest the following. For $G = 3$, GMM is approximately unbiased, but GGEL is downward biased. For all other choices of G, GMM is upward biased and GGEL is downward biased: for $G = 4$, GMM and GGEL exhibit similar absolute bias; for $G = 6$, the absolute bias of GMM exhibits three times the absolute bias of GGEL; and for $G = 8$, the multiple is five times.

Tables 13.1–13.3 report the results for the normal distribution for N equal to 96, 144, and 312, respectively, whereas Tables 13.4–13.6 report the results for the Student's t distribution. First, consider the bias. It can be seen that the pattern in the simulated biases corresponds broadly to the pattern predicted above. For $G = 3$ or $G = 4$, GMM tends to exhibit less bias, but for $G = 6$ or $G = 8$, GIM exhibits less bias, with GMM exhibiting relatively larger bias. Interestingly, the

[34] From Proposition 1 it follows that the t-statistic is distributed approximately as a standard normal random variable in large samples.

[35] The GMM overidentifying restrictions test is calculated as N times the minimand on the right-hand side of (13.24). Like the other model specification tests, the overidentifying restrictions test statistic converges to a χ^2_{G-p} distribution under the null hypothesis that the moments are valid.

Table 13.1 Normal Distribution: Results for β, t-Test Based on Conventional Standard Error, Model Specification Test Rejection Rates

$N = 96$	$\rho_{u,a} = 0.2$				$\rho_{u,a} = 0.5$				$\rho_{u,a} = 0.9$			
	$G = 3$	$G = 4$	$G = 6$	$G = 8$	$G = 3$	$G = 4$	$G = 6$	$G = 8$	$G = 3$	$G = 4$	$G = 6$	$G = 8$
EL β	0.0446	0.0475	0.0488	0.0493	0.0378	0.0445	0.0482	0.0489	0.0321	0.0418	0.0465	0.0481
ET β	0.0451	0.0475	0.0488	0.0493	0.0383	0.0446	0.0483	0.0489	0.0321	0.0419	0.0465	0.0481
2SLS β	0.0500	0.0513	0.0518	0.0517	0.0495	0.0539	0.0555	0.0546	0.0505	0.0581	0.0591	0.0584
GMM β	0.0499	0.0513	0.0518	0.0517	0.0495	0.0538	0.0555	0.0546	0.0507	0.0581	0.0590	0.0583
EL rej	0.0451	0.0639	0.0854	0.1002	0.0592	0.0663	0.0910	0.1001	0.0806	0.0679	0.0782	0.0942
ET rej	0.0449	0.0633	0.0839	0.0961	0.0588	0.0656	0.0882	0.0967	0.0795	0.0680	0.0765	0.0920
GMM rej	0.0357	0.0491	0.0656	0.0788	0.0611	0.0662	0.0881	0.0957	0.1095	0.1080	0.1221	0.1302
GMM J-test	0.0432	0.0471	0.0527	0.0493	0.0564	0.0572	0.0568	0.0554	0.0743	0.0803	0.0662	0.0610
EL Wald	0.0380	0.0405	0.0479	0.0498	0.0439	0.0466	0.0485	0.0529	0.0478	0.0521	0.0465	0.0457
ET Wald	0.0378	0.0400	0.0479	0.0495	0.0436	0.0463	0.0481	0.0521	0.0477	0.0519	0.0463	0.0440
EL LM	0.0410	0.0593	0.1213	0.2212	0.0485	0.0628	0.1217	0.2207	0.0533	0.0713	0.1209	0.2196
ET LM	0.0498	0.0774	0.1482	0.2555	0.0581	0.0781	0.1516	0.2550	0.0620	0.0893	0.1516	0.2525
EL LR	0.0421	0.0522	0.0823	0.1235	0.0485	0.0571	0.0821	0.1266	0.0524	0.0649	0.0822	0.1215
ET LR	0.0439	0.0574	0.0914	0.1352	0.0509	0.0603	0.0924	0.1379	0.0547	0.0694	0.0926	0.1330
F-stat	9.1057	11.189	15.570	20.034	9.0731	11.240	15.613	20.207	9.1272	11.238	15.575	20.134

[a] EL β and ET β denote the simulated means of the GIM estimator using, respectively, $\phi(\cdot) = -\log(\cdot)$ and $\phi(\cdot) = (\cdot)\log(\cdot)$; GMM (2SLS) β are the corresponding figures for the two-step GMM (2SLS) estimator; the true value value is $\beta_0 = 0.05$.

[b] EL rej and ET rej are the empirical rejection rates of the tests of $H_0 : \beta_0 = 0.05$ based on the GIM estimators using $\phi(\cdot) = -\log(\cdot)$ and $\phi(\cdot) = (\cdot)\log(\cdot)$, respectively; GMM rej is the corresponding figure based on the two-step GMM estimator; nominal 5% rejection rates.

[c] GMM J-test denotes the empirical rejection rate for the GMM overidentifying restrictions test–see footnote 34; EL (ET) Wald and EL (ET) LM denote the corresponding figure for the Wald test in (Eq. 13.19 and 13.18), respectively, with $\phi(\cdot) = -\log(\cdot)$ $(\phi(\cdot) = (\cdot)\log(\cdot))$; EL LR denotes the t corresponding figure based on LR_{EL} in (Eq. 13.21); ET LR denotes the corresponding figure based on $KLIC - R_{ET}$ in (13.22); the nominal rejection rate is 5%.

[d] F-stat is the average value of the F-statistic for testing $H_0 : \delta_j = 0$ for $j = 2, 3, \ldots, G$ in (13.26).

Table 13.2 Normal Distribution: Results for β, t-Test Based on Conventional Standard Error, Model Specification Test Rejection Rates

$N = 144$	$\rho_{u,a} = 0.2$				$\rho_{u,a} = 0.5$				$\rho_{u,a} = 0.9$			
	$G = 3$	$G = 4$	$G = 6$	$G = 8$	$G = 3$	$G = 4$	$G = 6$	$G = 8$	$G = 3$	$G = 4$	$G = 6$	$G = 8$
EL β	0.0479	0.0483	0.0498	0.0500	0.0431	0.0474	0.0489	0.0494	0.0398	0.0447	0.0480	0.0485
ET β	0.0479	0.0483	0.0498	0.0500	0.0431	0.0474	0.0489	0.0494	0.0398	0.0447	0.0480	0.0485
2SLS β	0.0505	0.0506	0.0517	0.0514	0.0493	0.0531	0.0538	0.0533	0.0505	0.0549	0.0565	0.0555
GMM β	0.0505	0.0507	0.0517	0.0515	0.0494	0.0531	0.0538	0.0533	0.0504	0.0550	0.0566	0.0555
EL rej	0.0487	0.0562	0.0736	0.0786	0.0510	0.0611	0.0716	0.0806	0.0645	0.0643	0.0675	0.0800
ET rej	0.0487	0.0554	0.0728	0.0777	0.0506	0.0615	0.0711	0.0793	0.0646	0.0641	0.0669	0.0793
GMM rej	0.0404	0.0462	0.0617	0.0668	0.0523	0.0641	0.0761	0.0795	0.0852	0.0904	0.0951	0.1059
GMM J-test	0.0537	0.0547	0.0492	0.0488	0.0546	0.0517	0.0563	0.0492	0.0667	0.0683	0.0631	0.0630
EL Wald	0.0488	0.0516	0.0452	0.0464	0.0481	0.0439	0.0508	0.0459	0.0505	0.0489	0.0466	0.0497
ET Wald	0.0485	0.0514	0.0449	0.0463	0.0480	0.0435	0.0505	0.0457	0.0505	0.0488	0.0463	0.0495
EL LM	0.0500	0.0574	0.0835	0.1362	0.0492	0.0530	0.0884	0.1288	0.0515	0.0574	0.0846	0.1401
ET LM	0.0564	0.0732	0.1117	0.1745	0.0555	0.0677	0.1188	0.1673	0.0596	0.0733	0.1145	0.1786
EL LR	0.0508	0.0568	0.0672	0.0884	0.0499	0.0505	0.0722	0.0861	0.0528	0.0559	0.0695	0.0944
ET LR	0.0529	0.0610	0.0750	0.0997	0.0519	0.0537	0.0810	0.0952	0.0542	0.0599	0.0771	0.1068
F-stat	13.179	16.382	22.319	28.767	13.139	16.483	22.274	28.771	13 204	16.424	22.348	28.625

[a] See the notes to Table 13.1.

Table 13.3 Normal Distribution: Results for β, t-Test Based on Conventional Standard Error, Model Specification Test Rejection Rates

$N = 312$	$\rho_{u,a} = 0.2$				$\rho_{u,a} = 0.5$				$\rho_{u,a} = 0.9$			
	$G = 3$	$G = 4$	$G = 6$	$G = 8$	$G = 3$	$G = 4$	$G = 6$	$G = 8$	$G = 3$	$G = 4$	$G = 6$	$G = 8$
EL β	0.0490	0.0496	0.0498	0.0499	0.0469	0.0490	0.0493	0.0499	0.0452	0.0472	0.0490	0.0493
ET β	0.0490	0.0496	0.0498	0.0499	0.0469	0.0490	0.0493	0.0499	0.0452	0.0472	0.0490	0.0493
2SLS β	0.0501	0.0507	0.0508	0.0507	0.0495	0.0518	0.0516	0.0518	0.0501	0.0521	0.0531	0.0527
GMM β	0.0501	0.0507	0.0508	0.0507	0.0495	0.0518	0.0517	0.0519	0.0501	0.0521	0.0531	0.0527
EL rej	0.0525	0.0559	0.0614	0.0624	0.0499	0.0581	0.0597	0.0622	0.0539	0.0552	0.0595	0.0616
ET rej	0.0524	0.0564	0.0614	0.0619	0.0497	0.0581	0.0595	0.0608	0.0540	0.0555	0.0586	0.0598
GMM rej	0.0484	0.0514	0.0572	0.0578	0.0510	0.0589	0.0611	0.0644	0.0658	0.0703	0.0727	0.0732
GMM J-test	0.0502	0.0491	0.0500	0.0511	0.0509	0.0523	0.0488	0.0483	0.0599	0.0555	0.0559	0.0550
EL Wald	0.0483	0.0461	0.0480	0.0482	0.0477	0.0478	0.0451	0.0443	0.0511	0.0458	0.0475	0.0476
ET Wald	0.0483	0.0461	0.0480	0.0478	0.0477	0.0477	0.0449	0.0444	0.0511	0.0459	0.0473	0.0474
EL LM	0.0484	0.0478	0.0570	0.0723	0.0479	0.0508	0.0557	0.0698	0.0520	0.0475	0.0581	0.0730
ET LM	0.0529	0.0565	0.0753	0.1040	0.0525	0.0596	0.0759	0.1019	0.0560	0.0561	0.0767	0.1048
EL LR	0.0494	0.0488	0.0562	0.0638	0.0489	0.0519	0.0546	0.0628	0.0523	0.0491	0.0551	0.0651
ET LR	0.0510	0.0516	0.0612	0.0738	0.0504	0.0541	0.0590	0.0708	0.0533	0.0511	0.0617	0.0734
F-stat	25.102	31.237	43.123	56.038	25.169	31.196	43.191	56.182	25.170	31.103	43.313	55.995

[a] See the notes to Table 13.1.

Table 13.4 t_7-Distribution: Results for β, t-Test Based on Conventional Standard Error, Model Specification Test Rejection Rates

$N = 96$	$\rho_{u,a} = 0.2$				$\rho_{u,a} = 0.5$				$\rho_{u,a} = 0.9$			
	$G = 3$	$G = 4$	$G = 6$	$G = 8$	$G = 3$	$G = 4$	$G = 6$	$G = 8$	$G = 3$	$G = 4$	$G = 6$	$G = 8$
EL β	0.0441	0.0481	0.0495	0.0494	0.0375	0.0452	0.0480	0.0490	0.0327	0.0405	0.0457	0.0476
ET β	0.0443	0.0483	0.0495	0.0494	0.0382	0.0454	0.0480	0.0490	0.0326	0.0407	0.0458	0.0476
2SLS β	0.0497	0.0522	0.0522	0.0517	0.0500	0.0546	0.0550	0.0547	0.0511	0.0572	0.0583	0.0578
GMM β	0.0495	0.0523	0.0522	0.0516	0.0500	0.0547	0.0549	0.0545	0.0510	0.0570	0.0582	0.0574
EL rej	0.0461	0.0689	0.0899	0.1006	0.0613	0.0708	0.0870	0.1041	0.0777	0.0680	0.0841	0.0966
ET rej	0.0451	0.0660	0.0863	0.0960	0.0593	0.0680	0.0817	0.0967	0.0765	0.0666	0.0787	0.0911
GMM rej	0.0362	0.0502	0.0695	0.0745	0.0587	0.0692	0.0810	0.0904	0.1023	0.1079	0.1168	0.1263
GMM J-test	0.0500	0.0514	0.0515	0.0453	0.0575	0.0602	0.0544	0.0517	0.0765	0.0834	0.0735	0.0632
EL Wald	0.0421	0.0404	0.0448	0.0422	0.0453	0.0460	0.0445	0.0453	0.0487	0.0511	0.0478	0.0457
ET Wald	0.0411	0.0395	0.0435	0.0406	0.0446	0.0450	0.0439	0.0437	0.0485	0.0498	0.0468	0.0444
EL LM	0.0512	0.0714	0.1595	0.2747	0.0540	0.0826	0.1578	0.2818	0.0604	0.0892	0.1642	0.2936
ET LM	0.0594	0.0866	0.1767	0.2859	0.0625	0.0961	0.1757	0.2975	0.0690	0.1022	0.1843	0.3086
EL LR	0.0472	0.0558	0.0944	0.1379	0.0505	0.0648	0.0899	0.1446	0.0556	0.0689	0.0986	0.1456
ET LR	0.0494	0.0602	0.0981	0.1395	0.0523	0.0676	0.0928	0.1452	0.0582	0.0718	0.1018	0.1480
R2	0.1676	0.2705	0.4676	0.6179	0.1647	0.2703	0.4677	0.6193	0.1657	0.2697	0.4670	0.6186
F-stat	9.362	11.371	15.811	20.330	9.166	11.360	15.813	20.452	9.233	11.324	15.768	20.391

[a] See the notes to Table 13.1.

Table 13.5 t_7-Distribution: Results for β, t-Test Based on Conventional Standard Error, Model Specification Test Rejection Rates

$N = 144$	$\rho_{u,a} = 0.2$				$\rho_{u,a} = 0.5$				$\rho_{u,a} = 0.9$			
	$G = 3$	$G = 4$	$G = 6$	$G = 8$	$G = 3$	$G = 4$	$G = 6$	$G = 8$	$G = 3$	$G = 4$	$G = 6$	$G = 8$
EL β	0.0476	0.0485	0.0493	0.0496	0.0430	0.0473	0.0490	0.0493	0.0393	0.0454	0.0474	0.0487
ET β	0.0478	0.0486	0.0493	0.0496	0.0432	0.0474	0.0490	0.0493	0.0395	0.0455	0.0475	0.0487
2SLS β	0.0503	0.0509	0.0511	0.0513	0.0497	0.0532	0.0537	0.0531	0.0503	0.0557	0.0560	0.0555
GMM β	0.0503	0.0508	0.0511	0.0512	0.0497	0.0531	0.0536	0.0530	0.0502	0.0555	0.0559	0.0554
EL rej	0.0459	0.0640	0.0759	0.0856	0.0523	0.0620	0.0770	0.0804	0.0658	0.0654	0.0725	0.0843
ET rej	0.0448	0.0624	0.0725	0.0832	0.0507	0.0607	0.0735	0.0750	0.0651	0.0641	0.0705	0.0806
GMM rej	0.0382	0.0515	0.0608	0.0694	0.0528	0.0596	0.0727	0.0744	0.0852	0.0924	0.0935	0.1088
GMM J-test	0.0467	0.0540	0.0505	0.0488	0.0576	0.0576	0.0553	0.0519	0.0686	0.0695	0.0665	0.0630
EL Wald	0.0415	0.0482	0.0451	0.0456	0.0479	0.0482	0.0450	0.0454	0.0494	0.0470	0.0470	0.0486
ET Wald	0.0414	0.0475	0.0441	0.0446	0.0471	0.0475	0.0445	0.0444	0.0491	0.0468	0.0467	0.0484
EL LM	0.0472	0.0655	0.1088	0.1928	0.0540	0.0651	0.1128	0.1916	0.0566	0.0683	0.1180	0.1874
ET LM	0.0548	0.0785	0.1302	0.2206	0.0609	0.0770	0.1348	0.2164	0.0630	0.0813	0.1411	0.2161
EL LR	0.0454	0.0575	0.0742	0.1077	0.0521	0.0591	0.0756	0.1076	0.0545	0.0593	0.0776	0.1047
ET LR	0.0465	0.0602	0.0784	0.1135	0.0546	0.0617	0.0798	0.1134	0.0552	0.0627	0.0838	0.1130
F-stat	13.208	16.600	22.539	29.066	13.344	16.691	22.554	29.101	13.330	16.664	22.466	29.072

[a] See the notes to Table 13.1.

Table 13.6 t_7-Distribution: Results for β, t-Test Based on Conventional Standard Error, Model Specification Test Rejection Rates

$N = 312$	$\rho_{u,a} = 0.2$				$\rho_{u,a} = 0.5$				$\rho_{u,a} = 0.9$			
	$G = 3$	$G = 4$	$G = 6$	$G = 8$	$G = 3$	$G = 4$	$G = 6$	$G = 8$	$G = 3$	$G = 4$	$G = 6$	$G = 8$
EL β	0.0486	0.0491	0.0497	0.0499	0.0467	0.0482	0.0494	0.0498	0.0448	0.0471	0.0490	0.0494
ET β	0.0486	0.0491	0.0497	0.0499	0.0467	0.0482	0.0494	0.0498	0.0448	0.0471	0.0490	0.0494
2SLS β	0.0497	0.0502	0.0506	0.0506	0.0494	0.0510	0.0518	0.0516	0.0498	0.0523	0.0533	0.0528
GMM β	0.0497	0.0502	0.0506	0.0506	0.0495	0.0510	0.0517	0.0516	0.0498	0.0522	0.0532	0.0527
EL rej	0.0502	0.0534	0.0617	0.0698	0.0541	0.0566	0.0630	0.0684	0.0563	0.0523	0.0625	0.0657
ET rej	0.0500	0.0523	0.0595	0.0674	0.0536	0.0552	0.0607	0.0650	0.0558	0.0515	0.0593	0.0639
GMM rej	0.0467	0.0468	0.0531	0.0603	0.0546	0.0564	0.0599	0.0672	0.0684	0.0664	0.0742	0.0783
GMM J-test	0.0491	0.0493	0.0521	0.0503	0.0521	0.0542	0.0514	0.0496	0.0613	0.0654	0.0583	0.0593
EL Wald	0.0461	0.0451	0.0483	0.0471	0.0486	0.0498	0.0462	0.0445	0.0480	0.0538	0.0488	0.0491
ET Wald	0.0460	0.0449	0.0482	0.0463	0.0484	0.0493	0.0457	0.0440	0.0479	0.0534	0.0481	0.0483
EL LM	0.0489	0.0537	0.0747	0.1058	0.0486	0.0550	0.0719	0.0972	0.0517	0.0615	0.0787	0.1020
ET LM	0.0533	0.0629	0.0935	0.1333	0.0547	0.0657	0.0901	0.1193	0.0563	0.0718	0.0964	0.1273
EL LR	0.0475	0.0519	0.0634	0.0735	0.0500	0.0542	0.0607	0.0687	0.0509	0.0586	0.0638	0.0743
ET LR	0.0489	0.0535	0.0678	0.0791	0.0511	0.0564	0.0642	0.0746	0.0518	0.0608	0.0666	0.0785
F-stat	25.299	31.435	43.455	56.365	25.330	31.274	43.485	56.272	25.247	31.357	43.456	56.295

[a] See the notes to Table 13.1.

bias of each estimator is broadly similar across the two-error distributions. All three t-statistics show size distortion as G increases and/or $\rho_{u,a}$ increases. The GMM t-tests have empirical size closer to the nominal size for low degrees of endogeneity ($\rho_{u,a} = 0.2$), but the GIM t-tests have empirical size closer to the nominal size for high degrees of endogeneity ($\rho_{u,a} = 0.9$). For the largest sample size, all the empirical sizes are close to, albeit systematically above, the nominal level. Now consider the overidentifying restrictions tests. The GIM Wald tests exhibit empirical size very close to the nominal level in all settings, and the GMM test has rejects slightly more than it should, but the other tests reject far too often in the smaller samples ($N = 96{,}144$). As would be expected, the degree of over-rejection is reduced as the sample size increases, but it still persists even in the largest sample size ($N = 312$) in the more overidentified models. For example, for $G = 8$, the LR_{EL} and $KLIC—R_{ET}$ tests over-reject by 1.5 percent and 2.5 percent, respectively, in the normal distribution case and by 2.5 percent and 2.8 percent in the Student's t distribution case; the LM tests over-reject by between 2.3 and 7.7 percent.

While our simulation study is limited, the results provide some interesting insights into the comparative properties of the estimators. If the degree of overidentification is small and the degree of endogeneity is low, then the GMM-based inferences tend to be more reliable, although the GIM procedures also seem reasonably reliable. In contrast, if the degree of overidentification is relatively large (four or six in our design) and/or the degree of endogeneity is high, then GIM tends to yield the most reliable inferences. However, the empirical size of the GIM tests about the coefficient value tends to exceed the nominal level by 1–2 percent even in the largest sample size ($N = 312$) considered in our simulation study. Of the three GIM tests of the model specification, the Wald appears the most reliable, as its empirical size is below but close to the nominal level in all cases considered.

In the next subsection, we illustrate these estimators using the textbook (in econometrics) returns-to-education example.

5.2 Empirical Example

Our textbook empirical example is chosen to illustrate the model given by Eqs. (13.25) and (13.26). We estimate the classic returns-to-education example using the data used throughout Wooldridge's (2020) popular introductory econometrics text (see, in particular, Examples 15.1 and 15.2). Since these data are a single cross-section, we cannot use the methods described in Section 3 above, and so we instead consider groupings based on demographic information.

The equation of interest (Eq. 13.25) is a regression of log(wage) ($\mathbf{y}_g$ above) on years-of-education ($\mathbf{r}_g$ above) using various binary measures of family background as instruments for years-of-education. The sample size throughout is 722 individuals.

There is an established tradition of using family background as instruments in this literature. In what follows we combine three such binary variables to form groups of size G = 3, 4, 6, 8, respectively. Using information on the years-of-education for both mother and father, we define a parent to have "high" education if his/her years-of-education is 13 or higher. If so, the dummy variables *feduc* and *meduc* are unity, referring to an individual's father and mother, respectively. The first group, $G = 3$, is defined as to whether neither, one, or both parents have "high" education. The next-sized group is $G = 4$, which happens when both dummies are interacted together (i.e., we distinguish between whether it is the mother or father who has high education when only one parents does). The third binary variable, *sibs,* is defined as unity when the individual has two or more siblings, and zero otherwise. When interacted with the $G = 3{,}4$ groups just defined, we end up with $G = 6, 8$.

The relationship between these different groupings and years-of-education/log wages of the individual can be seen in Table 13.7, which also records the group size. From the cells, we can compute n_g, $g = 1, \ldots, G$ for $G = 3, 4, 6, 8$. There is considerable variation in n_g. When $G = 8$, the smallest group is 13 and the largest 401; when $G = 3$, the smallest is 56 and the largest 567.

To set the scene, the OLS estimate of β, the returns to education, is 0.059, with a robust standard error of 0.0067. When re-estimated using GMM, with years-of-education of both parents continuously measured as the two instruments, the estimate increases to 0.069 (0.0179). The increase in the estimate is typical in the

Table 13.7 Summary Statistics for Group Sizes, Education and Log Wage[a]

feduc=	all		0/1 siblings		2+ siblings	
	0	1	0	1	0	1
meduc= 0	567	56	166	22	401	34
	13.2	15.3	13.7	15.6	13.0	15.0
	6.8	7.0	6.8	7.0	6.7	6.9
meduc= 1	43	56	13	22	30	34
	14.7	15.7	14.5	15.5	14.8	15.9
	6.9	6.9	6.9	6.9	6.8	6.9

[a] Each cell reports group size, average years of education, and average log wage.

literature and implies that $\sigma_{u,a} < 0$. This does not sit comfortably with the idea that **u** partly captures unobserved "ability" and suggests that some other source of endogeneity is present. This issue is discussed in all modern econometrics texts. The model is overidentified by 1, and the Hansen J test is easily not-rejected with a test statistic of 0.745 (p-value = 0.388). The two instruments are not "weak" in that the increase in standard error is relatively modest (by a factor of 2.7), and the first-stage F-statistic is 66.9. (This F-statistic is defined in note d of Table 13.1.)

The fact that the model is well specified means that it is a useful starting place to see what happens when family background is defined on a number of groups rather than using the continuous measures.

In the first panel of Table 13.8, we report GMM estimates of the four models defined by $G = 3, 4, 6, 8$. Three observations stand out. First, the estimate of when $G = 3$ is 0.071 and, as G increases, so does estimated returns-to-education, drifting up to 0.077 when $G = 8$. Second, the degree of overidentification, G–2 obviously increases from column (1) to (4), but the p-value increases as well, meaning that the restrictions are easily not-rejected in all four cases. Third, the explanatory power of the first stage, as more and more group dummies are added, weakens with the F-statistic declining from 67.9 with $G = 3$ to 23.7 with $G = 8$.

In the second panel of Table 13.8, we report EL estimates of the same four models defined by $G = 3, 4, 6, 8$. In each case, the Wald test statistic (see Eq.[13.19]) is almost identical to the Hansen J statistic for GMM. The important takeaway from the EL estimates are that their upwards drift as G increases is

Table 13.8 GMM and EL Estimate of Returns to Education[a]

	(1)	(2)	(3)	(4)
No. of groups G	3	4	6	8
GMM estimates				
β	0.0708	0.0718	0.0763	0.0765
	(0.0175)	(0.0175)	(0.0168)	(0.0166)
Hansen J stat	0.70	1.22	2.29	2.93
Hansen J p	0.40	0.54	0.68	0.82
Hansen J df	1	2	4	6
F-stat	67.89	46.07	31.43	23.65
EL estimates				
β	0.0704	0.0707	0.0753	0.0735
	(0.0175)	(0.0174)	(0.0167)	(0.0165)
Wald stat	0.70	1.22	2.29	2.97
Wald p	0.40	0.54	0.68	0.81
Wald df	1	2	4	6

[a] Robust standard errors in parentheses. $N = 722$.

less pronounced than GMM. This confirms the earlier prediction that the finite sample behavior of GMM may be more sensitive to G than GGEL (see the paragraph that follows Eq. [13.28]).

The fact that the GMM estimate is higher than the GEL estimate for $G = 4, 6, 8$ is consistent with the conclusion in the same paragraph, namely, that GMM is biased upward, whereas GEL is biased downward. However, the gap between the two is very small for $G = 4, 6$. Nonetheless, there is an appreciable gap between 0.0765 and 0.0735 for $G = 8$, and so it would seem that both the theoretical prediction and the simulation results are borne out in this particular data set when we have eight groups. But it is also worth noting that the sample size here is twice as large as that in the simulations and that the correlation between the two error terms is possibly negative, not positive, as assumed in the simulations.

6. Concluding Remarks

In this chapter, we have introduced an IM estimator for the parameters of statistical models using information in population moment conditions that hold at group level, referred to as the group-IM (GIM) estimator. The GIM estimation can be viewed as the primary approach to a constrained optimization. Under certain circumstances that include the leading cases of interest, the estimators can also be obtained via the dual approach to this optimization, known as group-Generalized Empirical Likelihood (GGEL). In a companion paper (AHKL), we provide a comprehensive framework for inference-based GGEL. In this chapter, we compare the computational requirements of the primary and dual approaches. We also describe an inference framework based on GIM/GGEL estimators for a model that naturally arises in the analysis of repeated cross-section data. Using analytical arguments and a small simulation study, it is shown that the GIM/GGEL approach tends to yield more reliable inferences in finite samples than certain extant methods in models where the degree of overidentification of the parameters exceeds one. Our methods are illustrated through their application to the estimation of the returns to education.

References

Altonji, J. G., and Segal, L. M. (1996). "Small Sample Bias in GMM Estimation of Covariance Structures." *Journal of Business and Economic Statistics*, 14: 353–366.

Andersen, T. G., and Sørensen, B. (1996). "GMM Estimation of a Stochastic Volatility Model: A Monte Carlo Study." *Journal of Business and Economic Statistics*, 14: 328–352.

Andrews, M., Hall, A. R., Khatoon, R., and Lincoln, J. (2020). "Generalized Empirical Estimation Based on Group-Specific Moment Conditions." Discussion paper, University of Manchester.

Andrews, M. J., Elamin, O., Hall, A. R., Kyriakoulis, K., and Sutton, M. (2017). "Inference in the Presence Of Redundant Moment Conditions and the Impact of Government Health Expenditure on Health Outcomes in England." *Econometric Reviews*, 36: 23–41.

Angrist, J. (1991). "Grouped-Data Estimation and Testing in Simple Labor-Supply Models." *Journal of Econometrics*, 47: 243–266.

Arellano, M., and Bond, S. R. (1991). "Some Tests of Specification for Panel Data: Monte Carlo Evidence and an Application to Employment Equations." *Review of Economic Studies*, 58: 277–297.

Bekker, P. A., and van der Ploeg, J. (2005). "Instrumental Variable Estimation Based on Grouped Data." *Statistica Neerlandica*, 59: 239–267.

Blundell, R., and Bond, S. (1998). "Initial Conditions and Moment Restrictions in Dynamic Panel Data Models." *Journal of Econometrics*, 87: 115–143.

Bound, J., Jaeger, D. A., and Baker, R. M. (1995). "Problems with Instrumental Variables Estimation When the Correlation Between the Instruments and the Endogenous Explanatory Variable Is Weak." *Journal of the American Statistical Association*, 90: 443–450.

Chamberlain, G. (1987). "Asymptotic Efficiency in Estimation with Conditional Moment Restrictions." *Journal of Econometrics*, 34: 305–334.

Collado, M. D. (1997). "Estimating Dynamic Models from Time Series of Independent Crosssections." *Journal of Econometrics*, 82: 37–62.

Corcoran, S. (1998). "Bartlett Adjustment of Empirical Discrepancy Statistics." *Biometrika*, 85: 965–972.

Cressie, N., and Read, T. R. C. (1984). "Multinomial Goodness-of-Fit Tests." *Journal of the Royal Statistical Society, Series B*, 46: 440–464.

Deaton, A. (1985). "Panel Data from a Time Series of Cross-Sections." *Journal of Econometrics*, 30: 109–126.

Devereux, P. J. (2007). "Improved Errors-in-Variables Estimators for Grouped Data." *Journal of Business and Economic Statistics*, 25: 278–287.

Durbin, J. (1954). "Errors in Variables." *Review of the International Statistical Institute*, 22: 23–32.

Golan, A. (2008). "Information and Entropy Econometrics—A Review and Synthesis." *Foundations and Trends in Econometrics*, 2: 1–145.

Guggenberger, P. (2008). "Finite Sample Evidence Suggesting a Heavy Tail Problem of the Generalized Empirical Likelihood eEstimator." *Econometric Reviews*, 26: 526–541.

Hall, A. R. (2005). *Generalized Method of Moments*. Oxford: Oxford University Press.

Hall, A. R., and Inoue, A. (2003). "The Large Sample Behaviour of the Generalized Method of Moments Estimator in Misspecified Models." *Journal of Econometrics*, 114: 361–394.

Hansen, L. P. (1982). "Large Sample Properties of Generalized Method of Moments Estimators." *Econometrica*, 50: 1029–1054.

Hansen, L. P., Heaton, J., and Yaron, A. (1996). "Finite Sample Properties of Some Alternative GMM Estimators Obtained from Financial Market Data." *Journal of Business and Economic Statistics*, 14: 262–280.

Imbens, G. (2002). "Generalized Method of Moments and Empirical Likelihood." *Journal of Business and Economic Statistics*, 20: 493–506.

Imbens, G., Spady, R. H., and Johnson, P. (1998). "Information Theoretic Approaches to Inference in Moment Condition Models." *Econometrica*, 66: 333–357.

Inoue, A. (2008). "Efficient Estimation and Inference in Linear Pseudo-Panel Data Models." *Journal of Econometrics*, 142: 449–466.

Khatoon, R. (2014). "Estimation and Inference of Microeconometric Models Based on Moment Condition Models." PhD thesis, University of Manchester, Manchester, UK.

Kitamura, Y. (2007). "Empirical Likelihood Methods in Econometrics: Theory and Practice." In R. Blundell, W. K. Newey, and T. Personn (eds.), *Advances in Economics and Econometrics: Ninth World Congress of the Econometric Society*, (pp. 174–237). Cambridge: Cambridge University Press.

Kitamura, Y., and Stutzer, M. (1997). "An Information-Theoretic Alternative to Generalized Method of Moments Estimation." *Econometrica*, 65: 861–874.

Newey, W. K., and Smith, R. J. (2004). "Higher Order Properties of GMM And Generalized Empirical Likelihood Estimators." *Econometrica*, 72: 219–255.

Owen, A. B. (1988). "Empirical Likelihood Ratio Confidence Intervals for a Single Functional." *Biometrika*, 75: 237–249.

Parente, P. M. D. C., and Smith, R. J. (2014). "Recent Developments in Empirical Likelihood and Related Methods." *Annual Review of Economics*, 6: 77–102.

Qin, J., and Lawless, J. (1994). "Empirical Likelihood and General Estimating Equations." *Annals of Statistics*, 22: 300–325.

Schennach, S. M. (2007). "Estimation with Exponentially Tilted Empirical Likelihood." *Annals of Statistics*, 35: 634–672.

Smith, R. J. (1997). "Alternative Semi-Parametric Likelihood Approaches to Generalized Method of Moments Estimation." *Economics Journal*, 107: 503–519.

Smith, R. J. (2011). "GEL Criteria for Moment Condition Models." *Econometric Theory*, 27: 1192–1235.

Wooldridge, J. M. (2020). *Introductory Econometrics: A Modern Approach*. 7th ed. Mason, OH: Thomson-Southwestern Press.

14

Generalized Empirical Likelihood-Based Kernel Estimation of Spatially Similar Densities

Kuangyu Wen and Ximing Wu

1. Introduction

This chapter concerns the estimation of many distinct, yet similar, densities. The densities resemble one another, wherein the similarity generally decreases with spatial distance. Since each density only allows a small number of observations, separate estimation of individual densities suffers from small sample variations. Prime examples include crop yield distributions of many geographic locations (Goodwin and Ker, 1998; Racine and Ker, 2006; Harri et al., 2011) and housing price index distributions by cites (Iversen Jr, 2001; Banerjee et al., 2004; Majumdar et al., 2006).

We aim to design suitable density estimators that are flexible and more efficient than individual estimates. Flexibility can be achieved by utilizing nonparametric density estimators, which are known to excel when there are ample observations. This approach, however, is hampered by the typically small sample size for each individual unit in the kind of empirical investigations under consideration. We therefore propose an information pooling approach that exploits the spatial similarity among the densities. This is made possible by the method of generalized empirical likelihood (GEL) based kernel density estimator (KDE), which has been studied by, among others, Chen (1997) and Oryshchenko and Smith (2013).

The GEL–KDE approach incorporates out-of-sample information in the form of moment restrictions. To incorporate auxiliary information into KDE, Chen (1997) proposed a weighted KDE (WKDE), which replaces the uniform weights with observation-specific weights. These weights are obtained by maximizing the empirical likelihood (EL) subject to moment conditions implied by out-of-sample information. It is well known that empirical likelihood belongs to the family of generalized empirical likelihood; for example, see Imbens (2002)

Kuangyu Wen and Ximing Wu, *Generalized Empirical Likelihood-Based Kernel Estimation of Spatially Similar Densities*
In: *Advances in Info-Metrics: Information and Information Processing across Disciplines*. Edited by: Min Chen, J. Michael Dunn, Amos Golan, and Aman Ullah, Oxford University Press (2021). © Oxford University Press.
DOI: 10.1093/oso/9780190636685.003.0014

and Newey and Smith (2004). Two notable members of GEL are empirical likelihood and exponential tilting (ET). Oryshchenko and Smith (2013) further proposed the WKDE with generalized empirical likelihood weights to incorporate moment conditions and established their statistical properties. It can be shown that the WKDE with GEL weights reduce the asymptotic variance of the KDE. Moreover, the derivation of the optimal bandwidth of the kernel estimation and the generalized empirical likelihood weights can be conducted separately, adding to the practical appeal of this approach.

In this study, we apply the GEL–WKDE to estimation of spatially similar densities. For each density, we construct a set of auxiliary moment conditions that exploit the spatial similarity among neighboring units. Their sample analogs are obtained via a proposed spatial smoothing procedure. We then maximize the generalized empirical likelihood subject to these spatially smoothed moments and apply the implied observation weights to the WKDE, resulting in the spatially smoothed WKDE with GEL weights. We advocate the use of low-order spline-based moments, which are more robust against potential outliers than high-order polynomial moments and allow flexible configuration to suit the need of estimation and investigation.

One notable advantage of the proposed method is that information pooling is achieved without resorting to estimation of joint densities or simultaneous estimation of many densities. For each density, one first constructs its spatially smoothed moments and calculates the GEL weights. Subsequent adjustment to the KDE is applied to each density via a simple reweighting of the conventional KDE.

We conduct several Monte Carlo simulations and show that the spatially smoothed KDE estimates can substantially outperform separate estimates of individual densities, in terms of both global density estimate and tail estimates. The latter is particularly useful given the importance of tail estimation in investment, insurance, and risk management. Lastly, as an empirical illustration, we apply our method to estimate the corn yield densities of Iowa counties.

The remaining text proceeds as follows. Section 2 introduces the weighted kernel density estimators with generalized empirical likelihood weights. Section 3 describes the construction of spatial moment constraints. Section 4 presents the Monte Carlo simulations, followed by an empirical application in Section 5. The last section concludes the chapter.

2. Weighted Kernel Density Estimation

Consider an independently and identically distributed. sample $X_1, X_2, \ldots, X_n$ from a univariate distribution with unknown density f. The standard KDE is defined as

$$\hat{f}(x) = \frac{1}{n}\sum_{i=1}^{n} K_h(x - X_i), \tag{14.1}$$

where K is the so-called kernel function and $K_h(\cdot) \doteq K(\cdot/h)/h$. The standard Gaussian density is a popular choice of the kernel. The bandwidth h controls the degree of smoothness of the KDE. For general treatments of the KDE, see, for example, Silverman (1986) and Wand and Jones (1995).

In addition to the observed data, sometimes auxiliary information regarding the underlying distribution is available. In this chapter, we consider a general framework of incorporating out-of-sample information in kernel density estimation. In particular, we study the situation when out-of-sample information is available in the form of moment conditions. Let $\mathbf{g} = \left(g_1, \ldots, g_q\right)^T$ be a q-dimensional vector of real valued functions with expectation

$$\mathbb{E}\{\mathbf{g}(X_i)\} = \mathbf{b}. \tag{14.2}$$

To incorporate moment constraints in the KDE, Chen (1997) proposed a WKDE

$$\hat{f}_W(x) = \sum_{i=1}^{n} p_i K_h(x - X_i). \tag{14.3}$$

Note that the KDE (14.1) employs a uniform weight $p_i = 1/n, i = 1, \ldots, n$. In contrast, the WKDE applies to the KDE a general weight vector $\mathbf{p} = \left(p_1, \ldots, p_n\right)^T$, whose elements are non-negative and sum to one. In particular, Chen (1997) obtained the weights $\mathbf{p}$ by maximizing the empirical ikelihood (EL) subject to (14.2). Thus, the WKDE with EL weights, denoted by $\hat{f}_{EL}$, reflect auxiliary distributional knowledge.

The empirical likelihood approach, originally proposed by Owen (1988, 1990), is a nonparametric likelihood-based method of estimation and inference. It chooses the sample probability weight p_i for each data point X_i according to the following maximization problem:

$$\begin{aligned}
&\max_{(p_1, \ldots, p_n)} \prod_{i=1}^{n} np_i \\
\text{s.t.} \quad &\sum_{i=1}^{n} p_i \mathbf{g}(X_i) = \mathbf{b}, \\
&0 \leq p_i \leq 1, \quad i = 1, \ldots, n, \\
&\sum_{i=1}^{n} p_i = 1.
\end{aligned} \tag{14.4}$$

The objective function $\prod_{i=1}^{n} np_i$ is called the empirical likelihood because it can be viewed as a nonparametric likelihood. To ease notation, we rewrite the moments as $\mathbf{J}(X_i) = \mathbf{g}(X_i) - \mathbf{b}$ and denote by $\boldsymbol{\lambda} = (\lambda_1, \ldots, \lambda_q)^T$ the Lagrangian multipliers associated with the constraints $\sum_{i=1}^{n} p_i \mathbf{J}(X_i) = 0$. It can be shown that the solution to (14.4) is given by

$$p_i = n^{-1}\{1 + \boldsymbol{\lambda}^T \mathbf{J}(X_i)\}^{-1}, \tag{14.5}$$

with

$$\sum_{i=1}^{n} \frac{\mathbf{J}(X_i)}{1 + \boldsymbol{\lambda}^T \mathbf{J}(X_i)} = 0. \tag{14.6}$$

Chen's empirical likelihood modification of the KDE (1997) can be further generalized to other members of the generalized empirical likelihood (GEL). GEL is an information-theoretic estimation approach; see, for example, Imbens (2002) and Newey and Smith (2004). Specifically, the GEL weights are obtained by minimizing the Cressie-Read power divergence measure subject to the moment constraints, say (14.4):

$$\begin{aligned}
&\min_{(p_1,\ldots,p_n)} \frac{1}{\nu(\nu+1)} \sum_{i=1}^{n} \left((np_i)^{-\nu} - 1 \right) \\
\text{s.t.} \quad &\sum_{i=1}^{n} p_i \mathbf{g}(X_i) = \mathbf{b}, \\
&0 \le p_i \le 1, \quad i = 1, \ldots, n, \\
&\sum_{i=1}^{n} p_i = 1,
\end{aligned} \tag{14.7}$$

where $\nu \in (-\infty, \infty)$. The empirical likelihood belongs to the GEL family with $\nu = 0$. Another notable member is ET, which corresponds to $\nu = 1$. In particular, the WKDE with ET weights, denoted by $\hat{f}_{ET}$, has an appealing information-theoretic interpretation: asymptotically, it minimizes the cross-entropy (or the Kullback–Leibler Information Criterion) between the KDE $\hat{f}$ as the reference density and the set of densities that satisfy the moment constraints (14.2). This procedure can be formulated as:

$$\begin{aligned}
&\min_{\pi} \int \pi(x) \log \frac{\pi(x)}{\hat{f}(x)} dx \\
\text{s.t.} \quad &\int \pi(x) dx = 1,
\end{aligned}$$

$$\int \mathbf{g}(x)\pi(x)dx = \mathbf{b}, \tag{14.8}$$

and its solution takes the form

$$\tilde{f}_{ET}(x) = \hat{f}(x)\exp\{\tilde{\lambda}_0 + \tilde{\lambda}^T \mathbf{g}(x)\},$$

where $\tilde{\lambda}_0$ is a normalization constant such that $\tilde{f}_{ET}$ integrates to unity. It can be shown that the estimator $\hat{f}_{ET}$ is asymptotically equivalent to the solution $\tilde{f}_{ET}$ in the sense that $\hat{f}_{ET} = \tilde{f}_{ET} + o_p\left(n^{-1}\right)$. Thus, the WKDE with ET weights is seen to refine the KDE with a multiplicative adjustment, which consists of the moment functions $\mathbf{g}$ as a basis expansion in the exponent. In particular, if we set $\mathbf{g}$ to be a series of spline-basis functions (as we shall do in the following section), the resultant density estimator $\hat{f}_{ET}$ can be viewed as a hybrid of the KDE and the logspline estimator (Kooperberg and Stone, 1992; Stone et al., 1997).

The asymptotic properties of the WKDE with GEL weights have been examined in Oryshchenko and Smith (2013). They establish, under mild regularity conditions, that

$$\text{bias}\left\{\hat{f}_W(x)\right\} = \text{bias}\left\{\hat{f}(x)\right\} + O\left(n^{-1}\right), \tag{14.9}$$

and

$$\text{var}\left\{\hat{f}_W(x)\right\} = \text{var}\left\{\hat{f}(x)\right\} - \mathbf{J}(x)^T \Sigma^{-1} \mathbf{J}(x) f^2(x) n^{-1} + o\left(n^{-1}\right), \tag{14.10}$$

where Σ is the $q \times q$ covariance matrix of $\mathbf{J}(X_i)$, that is, $\Sigma = E\left[J(X_i)J(X_i)^T\right]$. These results suggest that the difference in bias is of order $O\left(n^{-1}\right)$ between the WKDE and KDE, which is asymptotically negligible. On the other hand, there is an order $O\left(n^{-1}\right)$ reduction in the variance of the WKDE relative to that of the KDE, as the second term of (14.10) is clearly negative. These results are consistent with the general belief that the generalized empirical likelihood decreases an estimator's variance, although in the current case this reduction only occurs in the small-order term due to nonparametric smoothing. Nonetheless, as Chen (1997) and Oryshchenko and Smith (2013) pointed out, the extent of this reduction can be substantial when the sample size is relatively small. We also note that Chen's (1997) bias result is slightly different from (14.9). This is because, when using the EL weights, the $O\left(n^{-1}\right)$ term in (14.9) happens to vanish and is replaced with a smaller-order term $o\left(n^{-1}\right)$. However, this asymptotically desirable property does not always hold for general GEL weights.

An immediate result from (14.9) and (14.10) is that, when considering the global criterion—mean integrated squared error (MISE), we have

$$\text{MISE}\left\{\hat{f}_W(x)\right\} = \text{MISE}\left\{\hat{f}(x)\right\} - n^{-1}\int \mathrm{J}(x)^T \Sigma^{-1} \mathrm{J}(x) f^2(x) dx + o\left(n^{-1}\right).$$

This result leads to an important practical implication: an optimal bandwidth for the KDE remains so for its GEL-weighted counterpart up to order $O\left(n^{-1}\right)$. Therefore, one can conduct the bandwidth selection and the calculation of GEL weights separately, and then construct the subsequent WKDE estimate according to (14.3).

3. Spatially Smoothed Moment Constraints

We apply the GEL-weighted KDE to the estimation of spatially similar densities. Denote by $\mathcal{L}$ a collection of geographic locations. Suppose that for each location $l \in \mathcal{L}$, we observe an i.i.d. random sample $X_i(l), i = 1, \ldots, n$, generated by an unknown distribution with density f_l, where n is rather small. Our task is complicated by the small sample size for each location: separate estimation of individual densities via nonparametric methods may not be satisfactory. Nonetheless, if the densities in the collection are spatially similar, we might be able to improve the estimates via spatial smoothing of the kernel estimates. In this section, we present a simple approach to exploit spatial similarity in kernel density estimations. Our strategy is to incorporate spatial information as moment constraints to kernel density estimates via the WKDE described earlier. In particular, we first construct spatially smoothed moments and then estimate each density nonparametrically subject to the spatial moments using the GEL-guided WDKE.

The implementation of GEL–WKDE starts with the construction of moment conditions. It is well known that a smooth density can be approximated arbitrarily well by a series-type estimator. Thus, natural candidates for $\mathbf{g}$ include some popular basis functions of series estimators, such as the power series, trigonometric series, and splines. Despite the popularity of the power series in the GEL adjustment of kernel estimations, high-order polynomials can be sensitive to possible outliers, especially when sample sizes are small. Instead we focus on the spline basis functions for their noted flexibility and robustness. For instance, the commonly used s-degree truncated power series takes the form

$$\mathbf{g}(x) = \left(1, x, x^2, \cdots, x^s, (x-\tau_1)_+^s, \cdots, (x-\tau_m)_+^s\right),$$

where $(x)_+ = \max(x, 0)$, and $\tau_1 < \cdots < \tau_m$ are spline knots. Due to its piecewise nature, splines are more flexible than power series. At the same time, typical spline estimations employ a low-order piecewise power series with relatively large number of knots. Therefore, they are less sensitive to possible outliers and do not require the existence of high-order moments as power series do. Furthermore, our numerical experiments indicate that imposing power series moment conditions up to the fourth order sometimes impedes the convergence of the optimization. In contrast, moment conditions given by typical linear, quadratic, or cubic spline-basis functions appear to be immune from this numerical difficulty.

An additional advantage of spline basis is that it can be flexibly configured to suit tinvestigation needs. For instance, risk management and insurance are concerned primarily with the lower part of some distribution. Correspondingly, we can place more knots in the lower part of the distribution support. This customized knot placement allows a more focused investigation in the region of interest.

Our next task is to calculate the sample analog of $\mathbb{E}[\mathbf{g}]$ for individual densities f_l's based on their own observations and spatial neighbors. Let $\mathcal{F}$ be the collection of densities $f_l, l \in \mathcal{L}$. Denote by $d(l, l')$ the distance between locations l and l' in $\mathcal{L}$—for example, the Euclidean metric in terms of longitude and latitude. Suppose that the elements in $\mathcal{F}$ resemble their neighbors and that the degree of resemblance gradually declines with spatial distance. We consider the following estimator of spatially smoothed moments, for each location l,

$$\hat{\mathbf{b}}_l = \sum_{l' \in \mathcal{L}} \omega(l, l') \left(\frac{1}{n} \sum_{i=1}^{n} \mathbf{g}(X_i(l')) \right) \equiv \sum_{l' \in \mathcal{L}} \omega(l, l') \bar{\mathbf{b}}_{l'} \tag{14.11}$$

where $\omega(l, l') \geq 0$ and $\sum_{l' \in \mathcal{L}} \omega(l, l') = 1$. Thus $\hat{\mathbf{b}}_l$ is a weighted average of all own sample averages $\bar{\mathbf{b}}_{l'}$ from each location l' in $\mathcal{L}$. The spatial weight $\omega(l, l')$, as a function of spatial distance $d(l, l')$, is constructed as

$$\omega(l, l') \propto \omega^*(l, l') = \exp\left(-\frac{d(l, l')}{\theta_l} \right) 1(l' \in \mathcal{S}_l), \tag{14.12}$$

where $1(\cdot)$ is the indicator function. This configuration assumes that for each location l, there exists a spatial neighborhood $\mathcal{S}_l = \{l' : d(l, l') \leq \delta_l\}$, such that its members are spatially similar, and the similarity gradually decreases with distance. In particular, the spatial similarity is governed by two smoothing parameters: the threshold δ_l determines the size of the neighborhood, while $\theta_l > 0$ governs the decaying rate of spatial similarity. Normalization of the weights then yields

$$\omega(l,l') = \frac{\omega^*(l,l')}{\sum_{l' \in \mathcal{L}} \omega^*(l,l')}. \tag{14.13}$$

In the calculation of spatial moments of f_l, the observations from neighboring locations are included, whose weights of contribution decrease with their distances to location l and are set to be zero if the distances are greater than a certain threshold.

The parameters θ_l and δ_l closely control the amount of spatial smoothing. In this study, we set

$$\theta_l = \frac{1}{|\mathcal{S}_l|} \sum_{l' \in \mathcal{S}_l} d(l,l'), \tag{14.14}$$

where $|\mathcal{S}_l|$ is the number of elements in $\mathcal{S}_l$. Thus, θ_l is seen to be the average distance between location l and its neighbors. The "perimeter" parameter δ_l can be a single-distance threshold such that all locations share the same size of spatial neighborhood, or a location-specific threshold such that each location has the same number of spatial neighbors in its neighborhood.

Equipped with the selected moments and their spatially smoothed sample analogs, one can then estimate the density of each location using the GEL–WKDE. One noted advantage of this approach is that upon the calculation of spatially smoothed moments and their GEL optimization, weighted kernel densities are calculated for each individual location separately. Information pooling is achieved via the calculation of spatially smoothed moments and subsequently incorporated into individual densities with the GEL weights.

4. Monte Carlo Simulations

To explore the performance of the proposed methods and the possible benefits of spatial smoothing in kernel density estimations, we conduct a number of Monte Carlo simulations. We represent a geographical domain by a unit square and construct a 20×20 evenly spaced grid over the square. From the grid, we randomly draw $L = 100$ points without replacement and use them as the geographical locations throughout the simulation. To make our results reproducible, we use `set.seed(1)` in R for this random draw. The placement of these locations is illustrated in Figure 14.1. Since each location is equipped with a two-dimensional coordinate, their pairwise Euclidean distances can be easily obtained.

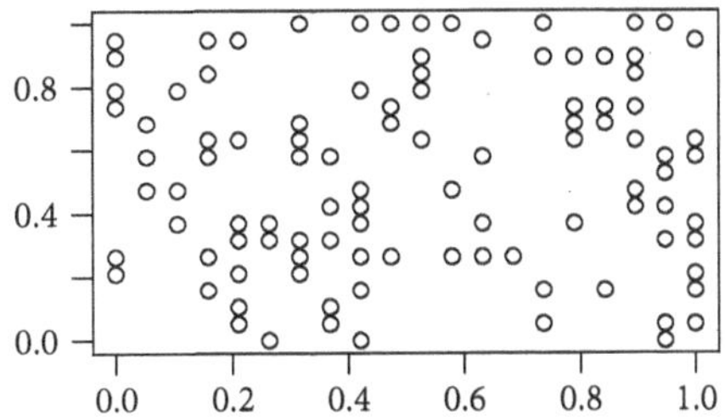

Figure 14.1 Plot of locations.

In the first experiment, we consider densities from the skewed normal distribution family. In particular, for the lth location, we set $f_l \sim \mathcal{SN}\left(\mu_l^{sn}, \sigma_l^{sn}, \alpha_l^{sn}\right)$, where $\left(\mu_l^{sn}, \sigma_l^{sn}, \alpha_l^{sn}\right)$ are the location, scale, and skewness parameters, respectively. To impose spatial structure among f_l's, we generate the distribution parameters according to a spatial error model. In general, an L-dimensional random vector $\mathbf{Y}$ follows the spatial error model $\mathrm{SpE}(\beta, \lambda, \sigma_\varepsilon)$ if

$$\mathbf{Y} = \beta + \eta$$
$$\eta = \lambda \mathbf{W} \eta + \epsilon,$$

where β is seen as the baseline level, λ is the spatial error coefficient, $\mathbf{W}$ is the spatial weight matrix, and the error terms $\epsilon \sim \mathcal{N}\left(0, \sigma_\varepsilon^2 \mathbf{I}_L\right)$ where $\mathbf{I}_L$ is an identity matrix. In this simulation, we follow a common routine in the spatial literature and construct $\mathbf{W}$ through two steps:

1. construct $\mathbf{W}^* = \left[w_{kl}^*\right]$, where $w_{kl}^* = 1$ if $|\, k - l \,| \leq 5$; otherwise $w_{kl}^* = 0$;
2. row standardization, i.e. $\mathbf{W} = [w_{kl}]$, $w_{kl} = w_{kl}^* / \sum_{l=1}^{L} w_{kl}^*$.

Given that ϵ is normally distributed, $\mathbf{Y}$ can be represented as follows:

$$\mathbf{Y} = \beta + (\mathbf{I}_L - \lambda \mathbf{W})^{-1} \epsilon.$$

Specifically, in the simulation, we consider $\boldsymbol{\mu}^{sn} \sim \mathrm{SpE}(-1, 0.8, 0.03)$, and a set of $\boldsymbol{\mu}^{sn}$ values is obtained by generating ϵ with `set.seed(2)` in R. Similarly, $\log(\boldsymbol{\sigma}^{sn}) \sim \mathrm{SpE}(0, 0.8, 0.03)$ and $\boldsymbol{\sigma}^{sn}$ are recovered by generating ϵ with `set.seed(3)`; $\alpha^{sn} \sim \mathrm{SpE}(-0.75, 0.8, 0.03)$ and $\boldsymbol{\sigma}^{sn}$ are obtained by generating ϵ with `set.seed(4)`. We plot the resultant densities in Figure 14.2. It is seen that these densities are similar in shape and at the same time exhibit abundant variations.

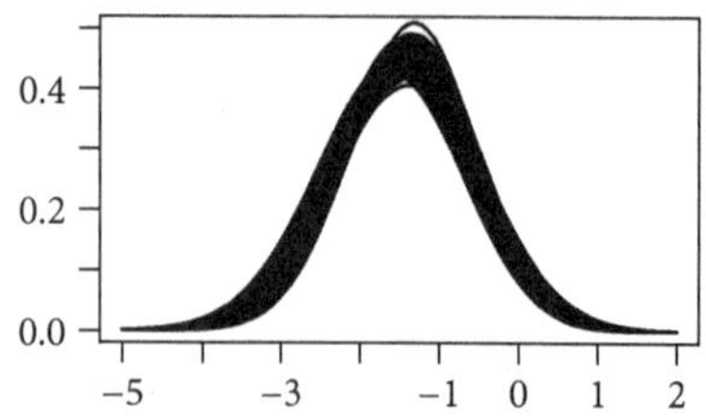

Figure 14.2 Plot of the skewed normal distributions.

For each location l, we generate an i.i.d. sample $\mathbf{X}(l) = \{X_1(l), \cdots, X_N(l)\}$ from the constructed f_l, with sample sizes $N = 30$ and 60 to highlight the small sample context. We then obtain the corresponding KDE and WKDE estimates. For the WKDE, we adopt both EL and ET weights. The moment functions are chosen to be quadratic splines with two knots at the 33rd and 66th percentiles of the pooled sample $\{\mathbf{X}(l) : l = 1, \cdots, L\}$. The threshold δ_l that governs the neighborhood size in spatial smoothing is set to be the distance from location l to its kth nearest neighbor where $k = \sqrt{L} = 10$. For both KDE and WKDE, the bandwidths are selected according to Silverman's rule of thumb. Each experiment is replicated $M = 500$ times.

Let $\tilde{f}_l^{(j)}$ be one of the estimates of f_l from the jth replication, $j = 1, \cdots, M$. We first evaluate the global performance via the average mean integrated squared error across $L = 100$ locations, which is calculated as

$$\text{amise}\left(\tilde{f}\right) = \frac{1}{LM}\sum_{l=1}^{L}\sum_{j=1}^{M}\int\left(\tilde{f}_l^{(j)}(x) - f_l(x)\right)^2 dx,$$

where the integrals are evaluated numerically. We also consider low tail performance, as it is important for risk management and insurance. Let Q_l be the 5 percent quantile of the f_l, $l = 1, \cdots, L$. We calculate the probability $\tilde{\pi}_l^{(j)} = \int_{x \leq Q_l} \tilde{f}_l^{(j)}(x)dx$ for each location l and replication j. The average mean squared error is calculated as

$$\text{amse}(\tilde{\pi}) = \frac{1}{LM}\sum_{l=1}^{L}\sum_{l=1}^{M}\left(\tilde{\pi}_l^{(j)} - 5\%\right)^2.$$

In a second experiment, we construct densities from the family of two-component normal mixtures; that is,

$$f_l = \omega_l^{nm}\phi\left(\mu_{1,l}^{nm}, \sigma_{1,l}^{nm}\right) + \left(1 - \omega_l^{nm}\right)\phi\left(\mu_{2,l}^{nm}, \sigma_{2,l}^{nm}\right),$$

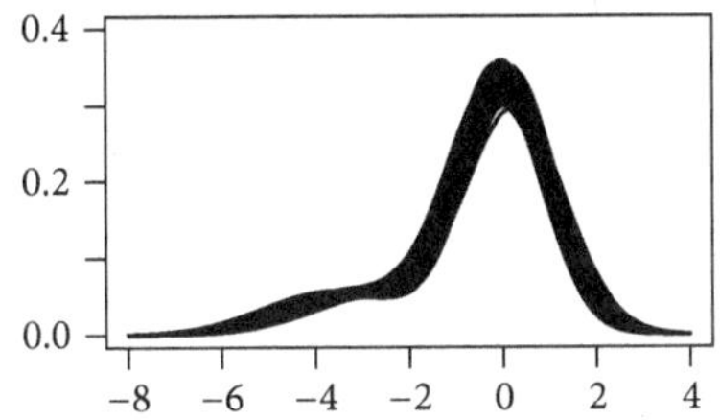

Figure 14.3 Plot of the normal mixture distributions.

where $\phi(\mu,\sigma)$ is the density of a normal distribution with mean μ and standard deviation σ, and $\omega \in (0,1)$ is the mixture probability. As in the first experiment, we construct the parameters according to a spatial error model. Denote $\omega^{nm} = (\omega_1^{nm}, \cdots, \omega_L^{nm})^T$ and analogously for $\boldsymbol{\mu}_1^{nm}$, $\boldsymbol{\mu}_2^{nm}$, $\boldsymbol{\sigma}_1^{nm}$ and $\boldsymbol{\sigma}_2^{nm}$. We set $\Phi^{-1}(\omega^{nm}) \sim \text{SpE}\left(\Phi^{-1}(0.2), 0.6, 0.03\right)$ where Φ^{-1} is the quantile function of the standard normal distribution, and we obtain ω^{nm} with `set.seed(2)` in generating ϵ. Similarly, we set $\boldsymbol{\mu}_1^{nm} \sim \text{SpE}(-3, 0.6, 0.2)$ and obtain $\boldsymbol{\mu}_1^{nm}$ with `set.seed(3)`; set $\boldsymbol{\mu}_2^{nm} \sim \text{SpE}(0, 0.6, 0.1)$ and obtain $\boldsymbol{\mu}_2^{nm}$ with `set.seed(4)`. Let $\log(\boldsymbol{\sigma}_1^{nm}) \sim \text{SpE}\left(\log(1.5), 0.6, 0.03\right)$ and use `set.seed(5)` to obtain $\boldsymbol{\sigma}_1^{nm}$. Finally, let $\log(\boldsymbol{\sigma}_2^{nm}) \sim \text{SpE}(0, 0.6, 0.03)$ and use `set.seed(6)` to obtain $\boldsymbol{\sigma}_2^{nm}$. We plot the second set of densities in Figure 14.3. The subsequent estimation and evaluation steps are the same as those in the first experiment.

In Table 14.1, we present our simulation results, including the amise/amse of KDE and ratios of amise/amse of WKDE relative to those of KDE. It is seen that the WKDE with EL or ET weights have substantially improved upon the KDE, demonstrating the merit of using spatially smoothed moment constraints as auxiliary information. This improvement is of even greater margin regarding tail performance, suggesting its usefulness for risk management and insurance. The performance of the WKDE based on EL weights is essentially identical to, though slightly better than, that based on ET weights. This is not unexpected given the well-known asymptotic equivalence among the members of the general empirical likelihood.

5. Empirical Application

We apply the proposed method to the estimation of crop yield distributions, which is of critical importance to the design of U.S. crop insurance programs. According to the Risk Management Agency of the U.S. Department of Agriculture, the estimated government costs for the federal crop insurance programs were about \$15.8 billion for crop year 2012. Accurate estimation of crop yield

Table 14.1 MISE of Simulation Results

		KDE	WKDE EL	ET
$N = 30$				
Skewed normal	global	0.0176	63.51%	63.89%
	tail	0.0017	56.08%	56.50%
Normal mixture	global	0.0148	77.01%	78.21%
	tail	0.0013	34.66%	34.90%
$N = 60$				
Skewed normal	global	0.0100	62.88%	65.05%
	tail	0.0009	62.95%	63.53%
Normal mixture	global	0.0088	75.49%	76.74%
	tail	0.0006	37.91%	38.11%

[a] "Global" and "tail" refer to average mean integrated square error of density estimation and average mean square error of 5 percent low-tail probability; the results for WKDE are reported relative to those of KDE.

distribution is important to promote the fairness and efficiency of this program; see, for example, Wen et al. (2015) for an in-depth treatment of this topic.

In this empirical illustration, we aim to estimate corn yield densities of ninety-nine Iowa counties. Typically, crop yield data are limited because one single observation is collected each year. Historical corn yield data from 1957 to 2010 are used. Thus, the sample size is $n = 54$ in our case, which is small. Crop yield data usually exhibit an upward trend due to technological advancement. Thus, the raw yield data must be detrended and possibly adjusted for heteroscedasticity to render them approximately i.i.d., see Goodwin and Ker (1998), Harri et al. (2011) and references therein on crop yield regressions. Following a standard procedure used by the USDA Risk Management Agency, we use the Locally Estimated Scattplot Smoothing (LOESS) smoother to estimate the time trend of the yield series of each county and retain the estimated residuals in the subsequent density estimations.

We then apply the WKDE with both EL and ET weights to estimate the yield density of each county. The distance used in spatial smoothing is calculated based on each county's central latitude and longitude. Similarly to the Monte Carlo simulations, we use the quadratic splines with two knots at the 33rd and 66th percentiles of the pooled data from all ninety-nine counties. The threshold δ_l is set to be the distance from county l to its tenth nearest neighbor. We select bandwidths for the KDE according to Silverman's rule of thumb.

For illustration, Figure 14.4 reports the three estimated corn yield densities for Adams County, which is taken as a representative county for Iowa corn

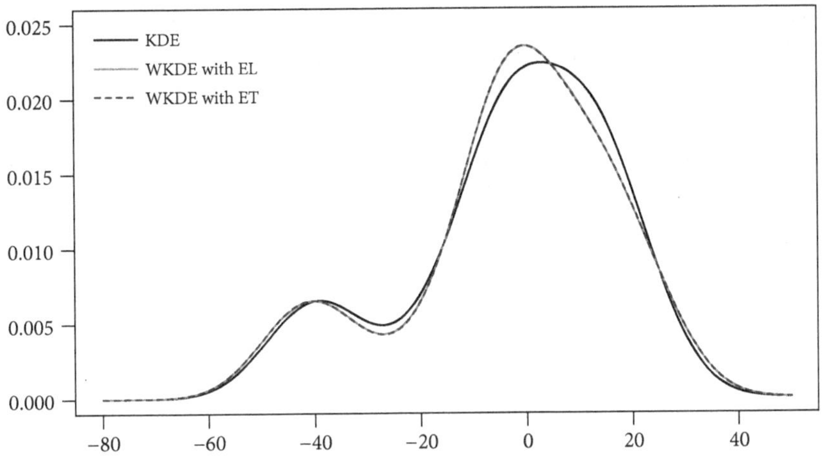

Figure 14.4 Estimates of corn yield density in Adams, Iowa.

production in the literature (Goodwin and Ker, 1998; Ker and Coble, 2003). All three nonparametric estimates exhibit negative skewness, a well-documented fact about yield distributions. They also show evidence of bimodality, which is consistent with an observation by Goodwin and Ker (1998): high yields close to the crop's capacity limit happen frequently, relatively low yields also happen fairly often, while yields between the two extremes are less likely. These rich features, as suggested by the nonparametric estimates, may elude restrictive parametric estimations. A close comparison between the KDE and WKDE results suggest that the two WKDE estimates have shifted the probability mass about the mode leftward compared to the KDE estimate. Also, they slightly increase the probability mass near the tails and decrease the probability mass between the two modes. Finally, we observe that the WKDE with EL weights is almost identical to that with ET weights. This agrees with our observations from the numerical simulations.

6. Concluding Remarks

We have proposed estimators that are suitable to the estimation of distinct, yet similar, densities, each with a small number of observations. Our approach pools information from spatially proximate densities to improve efficiency. In particular, we adjust the standard kernel density estimator with observation-specific weights. These weights are determined via the generalized empirical likelihood method, subject to spatial moment conditions for each density.

Special attention is paid to the empirical likelihood and exponential tilting methods. Our numerical experiments show considerable improvements relative to separate kernel estimates of individual densities and demonstrate the usefulness of the proposed method. We conclude by noting that detailed exploration of the configuration of spline-based moments and optimal selection of spatial tuning parameter(s) can provide useful insight into the proposed approach. Extensions to dependent data, functional data analysis, and applications in spatial social networks may also be topics of interest for future studies.

References

Banerjee, S., Gelfand, A. E., Knight, J. R., and Sirmans, C. F. (2004). "Spatial Modeling of House Prices Using Normalized Distance-Weighted Sums of Stationary Processes." *Journal of Business and Economic Statistics*, 22: 206–213.

Chen, S. X. (1997). "Empirical Likelihood-Based Kernel Density Estimation." *Australian Journal of Statistics*, 39: 47–56.

Goodwin, B. K., and Ker, A. P. (1998). "Nonparametric Estimation of Crop Yield Distributions: Implications for Rating Group-Risk Crop Insurance Contracts." *American Journal of Agricultural Economics*, 80: 139–153.

Harri, A., Coble, K. H., Ker, A. P., and Goodwin, B. J. (2011). "Relaxing Heteroscedasticity Assumptions in Area-Yield Crop Insurance Rating." *American Journal of Agricultural Economics*, 93: 707–717.

Imbens, G. W. (2002). "Generalized Method of Moments and Empirical Likelihood." *Journal of Business & Economic Statistics*, 20: 493–506.

Iversen, Jr. E. S. (2001). "Spatially Disaggregated Real Estate Indices," *Journal of Business and Economic Statistics*, 19: 341–357.

Ker, A. P. and Coble, K. (2003). "Modeling Conditional Yield Densities." *American Journal of Agricultural Economics*, 85: 291–304.

Kooperberg, C., and Stone, C. J. (1992). "Logspline Density Estimation for Censored Data." *Journal of Computational and Graphical Statistics*, 1: 301–328.

Majumdar, A., Munneke, H. J., Gelfand, A. E., Banerjee, S., and Sirmans, C. F. (2006). "Gradients in Spatial Response Surfaces with Application to Urban Land Values." *Journal of Business and Economic Statistics*, 24: 77–90.

Newey, W. K., and Smith, R. J. (2004). "Higher Order Properties of GMM and Generalized Empirical Likelihood Estimators." *Econometrica*, 72: 219–255.

Oryshchenko, V., and Smith, R. J. (2013). "Generalised Empirical Likelihood-Based Kernel Density Estimation." Working Paper.

Owen, A. B. (1988). "Empirical Likelihood Ratio Confidence Intervals for a Single Functional." *Biometrika*, 75: 237–249.

Owen, A. B. (1990), "Empirical Likelihood Ratio Confidence Regions." *Annals of Statistics*, 18: 90–120.

Racine, J., and Ker, A. (2006). "Rating Crop Insurance Policies with Efficient Nonparametric Estimators That Admit Mixed Data Types." *Journal of Agricultural and Resource Economics*, 31: 27–39.

Silverman, B. W. (1986). *Density Estimation for Statistics and Data Analysis*. San Rafael, CA: Chapman and Hall.

Stone, C. J., Hansen, M. H., Kooperberg, C., Truong, Y. K., et al. (1997). "Polynomial Splines and Their Tensor Products in Extended Linear Modeling: 1994 Wald Memorial Lecture." *The Annals of Statistics*, 25: 1371–1470.

Wand, M., and Jones, M. (1995). *Kernel Smoothing*. London: Chapman and Hall.

Wen, K., Wu, X., and Leatham, D. (2015). "Spatially Smoothed Empirical Likelihood Kernel Density Estimation of Crop Yield Distributions." Working Paper.

15

Rényi Divergence and Monte Carlo Integration

John Geweke and Garland Durham

1. Introduction

Information processing is a critical component of effective decision making, where plans are modified as more information becomes available. This includes formal econometrics, where estimates and posterior distributions are updated for the same reason. In these contexts, the economic agent or econometrician is passive, having no control over the magnitude of the information or its implications for the change in the relevant probability distribution. By contrast, in their technical work, economists and econometricians control the rate at which information is introduced and the form of the introduction. This includes both inference (e.g., posterior simulation and the computation of M-estimators) and optimization (e.g., steepest ascent and simulated annealing methods). Much of this technical work involves Monte Carlo integration.

In application, all of these procedures entail a sequence of discrete steps, each characterized by the introduction of incremental information. It is natural to measure the rate of introduction as the divergence of a relevant probability distribution between the start and end of each step. In the passive case, this provides a quantitative characterization of information flow. In the active case of controlled information introduction, alternative measures of divergence provide different bases for control.

This chapter pursues this agenda using the family of divergences introduced by Rényi (1961). Section 2 reviews the definition and those of its well-known properties that are important to the rest of the chapter. It also defines the concept of power concentration, which in turn is central to some important leading approaches to inference and optimization. It develops theory that is essential to the practical implementation of Rényi divergence in algorithms and software.

Section 3 shows that the central theory underlying acceptance and importance sampling, two critical components of Monte Carlo integration, can be stated as finite Rényi divergence of order 0, 2, or infinity between a source

John Geweke and Garland Durham, *Rényi Divergence and Monte Carlo Integration* In: *Advances in Info-Metrics: Information and Information Processing across Disciplines.* Edited by: Min Chen, J. Michael Dunn, Amos Golan, and Aman Ullah, Oxford University Press (2021).
DOI: 10.1093/oso/9780190636685.003.0015

distribution and a distribution of interest. In each case, when the condition is satisfied, the magnitude of the divergence is inversely proportional to the efficiency of the sampling method. Section 3 also shows how either acceptance sampling or importance sampling can be used to measure Rényi divergence between any two distributions, which is of interest for both passive and controlled introduction of information.

It is well recognized that sequential Monte Carlo provides an approach both to accessing posterior distributions and to determining the global mode of a function. Section 4 reviews the theory that underlies this method and shows that convergence conditions amount to finite Rényi divergence of order 2 between single or multiple increments in that algorithm's approach to the introduction of information. Section 4 also shows that the efficient implementation of sequential Monte Carlo that has been developed in the literature amounts to introducing information in a way that targets Rényi divergence of order 2 between the distribution at the start and end of each cycle. In contrast to acceptance and importance sampling, the theory provides no direct link between the magnitude of this measure and the efficiency of the method. This raises the prospect that Rényi divergence of a different order might provide a better foundation for efficiency.

Section 5 illustrates the use of Rényi divergence to monitor passive information introduction, demonstrating its practicability using sequential Monte Carlo in a simple but interesting time series model. It also illustrates controlled information introduction in the context of Bayesian inference, maximum likelihood, and a challenging global optimization problem. For likelihood-based inference performance is unaffected by the choice of the order of Rényi divergence for information introduction. In the global optimization example, in which the objective function is much more irregular, efficiency increases with the chosen order of Rényi divergence. This suggests further investigation in other examples with irregular objective functions and an effort to develop supporting theory if the finding is found to be robust with respect to choice of application.

Proofs of all new results are presented in section 6.

2. Rényi Divergence

2.1 Definition and Properties

Consider probability spaces $(X, \mathcal{F}, P)$ and $(X, \mathcal{F}, Q)$, with P and Q both absolutely continuous with respect to a common measure μ. For any $S \in \mathcal{F}$, $P(S) = \int_S p(x)d\mu(x)$ and $Q(S) = \int_S q(x)d\mu(x)$. Let X_P and X_Q be the supports of P and Q, respectively.

Definition 1. *The Rényi (1961) divergence of order $\alpha \geq 0$ from Q to P is*

$$D_\alpha(P \,\|\, Q) = \frac{1}{\alpha - 1} \log \int_X p(x)^\alpha q(x)^{1-\alpha} d\mu(x).$$

Equivalent expressions are

$$D_\alpha(P \,\|\, Q) = \frac{1}{\alpha - 1} \log \mathrm{E}_Q \left(\frac{p(x)}{q(x)} \right)^\alpha = \frac{1}{\alpha - 1} \log \int_X \left(\frac{p(x)}{q(x)} \right)^\alpha q(x) d\mu(x), \quad (15.1)$$

$$D_\alpha(P \,\|\, Q) = \frac{1}{\alpha - 1} \log \mathrm{E}_P \left(\frac{p(x)}{q(x)} \right)^{\alpha-1} = \frac{1}{\alpha - 1} \log \int_X \left(\frac{p(x)}{q(x)} \right)^{\alpha-1} p(x) d\mu(x). \quad (15.2)$$

There are several notable special cases.

1. $\alpha = 0$. Taking the limit $\alpha \to 0^+$ in (15.1),

$$D_0(P \,\|\, Q) = -\log Q(X_P);$$
$$D_0(P \,\|\, Q) = 0 \iff X_Q \subseteq X_P,$$
$$D_0(P \,\|\, Q) = \infty \iff \mu\left(X_P \cap X_Q\right) = 0.$$

2. $\alpha = 1/2$. The Rényi divergence is symmetric in P and Q,

$$D_{1/2}(P \,\|\, Q) = D_{1/2}(Q \,\|\, P) = -2 \log \int_X [p(x)q(x)]^{1/2} d\mu(x).$$

The Bhattacharyya (1943) coefficient

$$\int_X [p(x)q(x)]^{1/2} d\mu(x),$$

is a monotone decreasing function of $D_{1/2}(P \,\|\, Q)$ and the Hellinger (1909) distance

$$\left\{ \frac{1}{2} \int_X \left[p(x)^{1/2} - q(x)^{1/2} \right]^2 d\mu(x) \right\}^{1/2},$$

is a monotone increasing function of $D_{1/2}(P \,\|\, Q)$.

3. $\alpha = 1$. Taking the limit $\alpha \to 1$ in (15.2)

$$D_1(P \parallel Q) = \int_X \log\left(\frac{p(x)}{q(x)}\right) p(x) d\mu(x), \tag{15.3}$$

the Kullback–Leibler (1951) divergence (or directed distance) from P to Q.

4. $\alpha = 2$. The log of the expected probability density ratio (under P) or expected squared probability density ratio (under Q)

$$\begin{aligned} D_2(P \parallel Q) &= \log \int_X \left[\frac{p(x)}{q(x)}\right] \cdot p(x) d\mu(x) \\ &= \log \int_X \left[\frac{p(x)}{q(x)}\right]^2 \cdot q(x) d\mu(x). \end{aligned} \tag{15.4}$$

Kagan's (1963) divergence between Q and P

$$D_{\chi^2}(P \parallel Q) = \frac{1}{2} \int_X \frac{[q(x) - p(x)]^2}{q(x)} d\mu(x)$$

is a monotone increasing function of $D_2(P \parallel Q)$.

5. $\alpha = \infty$. Taking the limit $\alpha \to \infty$ in (15.1),

$$D_\infty(P \parallel Q) = \log\left[\operatorname*{ess\,sup}_{q(x)} \left(\frac{p(x)}{q(x)}\right)\right];$$

see van Erven and Harremodoës (2014).

Proposition 1. *Some basic properties of Rényi divergence.*

1. *If* $P(X) = Q(X)$ *for all sets* $X \in \mathcal{F}$ *with* $\mu(X) > 0$, *then* $D_\alpha(P \parallel Q) = 0$ *for all* $\alpha \geq 0$ *(by inspection).*
2. *The Rényi divergence* D_α *is weakly monotone increasing as a function of* α *(van Erven and Harremodoës, 2014).*
3. *If the distribution* P^* *has probability density* $p^*(y) = p(Ay)$ *and the distribution* Q^* *has probability density* $q^*(y) = q(Ay)$ *with* A *nonsingular, then* $D_\alpha(P^* \parallel Q^*) = D_\alpha(P \parallel Q)$ *(by inspection).*

2.2 Power Concentration

Sections 4.3 and 5.3 show how power concentration of a probability measure introduces information in an efficient and practical way to solve problems in Bayesian inference and in optimization.

Definition 2. *The power concentration P_r of order $r > 0$ of a probability measure Q with probability density q has probability density $p_r(x) \propto q(x)^{1+r}$.*

The approach in section 4.3 relies on being able to find an order r that achieves a specified Rényi divergence between Q and P^r: that is, it solves the equation

$$D_\alpha(P_r \parallel Q) = D_\alpha^* \tag{15.5}$$

for r. The solution exploits the well-behaved nature of $D_\alpha(P_r \parallel Q)$, established in the following three results. The basic idea is that since $D_\alpha(P_0 \parallel Q) = 0$, if $D_\alpha(P_r \parallel Q)$ is continuous and monotone increasing in r and $D_\alpha^* < \lim_{r\to\infty} D_\alpha(P_r \parallel Q)$ then (15.5) has a unique solution.

Proposition 2. *Monotonicity of $D_\alpha(P_r \parallel Q)$. If Q does not reduce to the uniform distribution, that is,* $\text{var}_Q[q(x)] > 0$, *then the function $D_\alpha(P_r \parallel Q)$ is continuous and monotonically increasing in $r > 0$ for all $\alpha > 0$ (Proof in section 6.1).*

Proposition 3. *Limit of $D_\alpha(P_r \parallel Q)$.* For all $\alpha > 0$, $\lim_{r\to\infty} D_\alpha(P_r \parallel Q) = -\log Q\left(X_Q^*\right)$, *where* $X_Q^* = \left\{x : q(x) = \text{ess sup}_{x\in X_Q} q(x)\right\}$ *(Proof in section 6.2).*

Thus, if Q is a discrete distribution $\lim_{r\to\infty} D_\alpha(P_r \parallel Q) < \infty$. If Q is a continuous distribution, then $\lim_{r\to\infty} D_\alpha(P_r \parallel Q) = \infty$ as long as $Q\left(X_Q^*\right) = 0$. For example, if the distribution Q is $N_k(\mu, \Sigma)$. From Proposition 1, part 3, $D_\alpha(P_r \parallel Q)$ will be the same for all μ and Σ as it is for $\mu = 0$, $\Sigma = I_k$. For the Q distribution $N_k(0, I_k)$, standard manipulations beginning with the definition yield

$$D_\alpha(P_r \parallel Q) = \frac{k}{2(\alpha-1)}\left[\alpha\log(1+r) - \log(1+\alpha r)\right],$$

for $\alpha \in (0,1)\cup(1,\infty)$ and

$$D_1(P_r \parallel Q) = \frac{k}{2}\left[\log(1+r) - r(1+r)^{-1}\right].$$

Further standard if somewhat tedious manipulations confirm $dD_\alpha(P_r \| Q)/dr > 0$, $dD_\alpha(P_r \| Q)/d\alpha > 0$ and $\lim_{r\to\infty} D_\alpha(P_r \| Q) = \infty$ for $\alpha > 0$.

Subsequently, the following generalization of power concentration proves useful.

Definition 3. *Suppose the distribution Q^* has probability density function (p.d.f.) $q^*(x) \propto h(x)q(x)$ with respect to the measure $d\mu(x)$. A partial power concentration P_r^* of order $r > 0$ of Q^* has probability density $p_r^*(x) \propto h(x)q(x)^{1+r}$.*

Proposition 4. *Propositions 2 and 3 remain true when Q^* replaces Q and P_r^* replaces P_r (Proof in section 6.3).*

3. Acceptance and Importance Sampling

The methods of acceptance and importance sampling are widely used in simulation methods of many kinds, either directly or as components of more elaborate procedures. Conditions for convergence, efficiency, and asymptotic normality can be expressed in terms of Rényi divergence, and in turn these methods can be used to provide approximations of $D_\alpha(P \| Q)$ by Monte Carlo integration.

3.1 Acceptance Sampling

Acceptance sampling (also known as rejection sampling) uses a random sample from a source distribution Q, with probability density q, to generate a random sample from a distribution of interest P with probability density p. Thus, given an algorithm for random sampling from Q, it provides an algorithm for random sampling from P. The distributions must have common support X and

$$a = \operatorname*{ess\,sup}_X \left(\frac{p(x)}{q(x)}\right) = D_\infty(P \| Q)$$

must be finite and known.

Algorithm 1. *Acceptance sampling*

1. *Construct the random sample* $x_i \overset{iid}{\sim} Q\ (i = 1, \ldots, n)$.
2. *Construct the random sample* $u_i \overset{iid}{\sim} \text{Uniform}(0,1)\ (i = 1, \ldots, n)$.
3. *Construct the subsample* $\{y_i\} = \{x_i : u_i \le p(x_i)/[a \cdot q(x_i)]\}$ *with* $m \le n$ *elements.*

Then $\{y_i\}$ *is a random sample from* P.

It is straightforward to see why the algorithm works: for any $S \in \mathcal{F}$

$$\frac{\int_S \{p(x)/[a \cdot q(x)]\} q(x) d\mu(x)}{\int_X \{p(x)/[a \cdot q(x)]\} q(x) d\mu(x)} = \frac{a^{-1} \int_S p(x) d\mu(x)}{a^{-1} \int_X p(x) d\mu(x)} = \int_S p(x) d\mu(x). \tag{15.6}$$

The unconditional probability of acceptance in this algorithm is

$$\int_X \frac{p(x)}{a \cdot q(x)} \cdot q(x) d\mu(x) = \frac{1}{a} = D_\infty(P \parallel Q)^{-1},$$

the denominator in (15.6). Since $\mathrm{E}(m|n) = n/a$, it is natural to regard $1/a$ as the efficiency of acceptance sampling.

Both direct and acceptance sampling produce random samples from P. If $\overline{g} = \mathrm{E}_P[g(x)]$ exists, then by the strong law of large numbers

$$\overline{g}_m = m^{-1} \sum_{i=1}^{m} g(y_i) \overset{a.s.}{\to} \overline{g}$$

and if in addition $\sigma_g^2 = \mathrm{var}_P g(x) < \infty$ exists, then

$$m^{1/2} (\overline{g}_m - \overline{g}) \overset{d}{\to} N(0, \sigma_g^2).$$

3.2 Importance Sampling

Hammersly and Handscomb (1964) introduced the importance sampling algorithm in statistics. It is a standard tool in Bayesian econometrics (Kloek and van Dijk, 1978; Geweke, 1989). It also utilizes a source distribution Q with p.d.f. q to target a distribution of interest P with p.d.f. p. It must be possible to produce i.i.d. samples from Q and to evaluate the kernels $k_p \propto p$ and $k_q \propto q$. The constants of proportionality can be different and need not be known, properties that make the algorithm attractive in Bayesian inference.

Algorithm 2. *Importance sampling*

1. *Construct the random sample* $x_i \overset{iid}{\sim} Q \, (i = 1, \ldots, n)$.
2. *Construct the corresponding weights* $w_i = k_p(x_i)/k_q(x_i) \; (i = 1, \ldots, n)$.
3. *For any function* $g(x)$ *for which* $\overline{g} = \mathrm{E}_P[g(x)]$ *exists, the approximation* $\overline{g}^N$ *of* $\overline{g}$ *is*

$$\bar{g}^n = \frac{\sum_{i=1}^n w_i g(x_i)}{\sum_{i=1}^n w_i}.$$

Geweke (1989) showed that if $X_P \subseteq X_Q$, equivalent to $D_0(Q \parallel P) = 0$, then $\bar{g}^n$ is simulation consistent:

$$\bar{g}^n \xrightarrow{a.s.} \bar{g}. \tag{15.7}$$

If, in addition

$$\mathrm{E}_P[w(x)] = \mathrm{E}_Q\left[w(x)^2\right] = \int_X \frac{p(x)^2}{q(x)} d\mu(x) < \infty \tag{15.8}$$

and

$$\mathrm{E}_P\left[g(x)^2 w(x)\right] = \mathrm{E}_Q\left[g(x)^2 w^2(x)\right] = \int_X \frac{g(x)^2 p(x)^2}{q(x)} d\mu(x) < \infty \tag{15.9}$$

then

$$n^{1/2}\left(\bar{g}^n - \bar{g}\right) \xrightarrow{d} N\left(0, \tau_g^2\right).$$

The variance is

$$\tau_g^2 = \mathrm{E}_P\left[\left(g(x) - \bar{g}\right)^2 w^*(x)\right] = \mathrm{E}_Q\left[\left(g(x) - \bar{g}\right)^2 w^{*2}(x)\right]$$

where $w^*(x) = p(x)/q(x)$.

Note that the condition $X_P \subseteq X_Q$ for consistency (15.7) is equivalent to $D_0(Q \parallel P) = 0$. Condition (15.8) for asymptotic normality is equivalent to $D_2(P \parallel Q) < \infty$. Conditions (15.8) and (15.9) can be tedious to verify. In contrast, establishing the stronger sufficient conditions

$$\operatorname*{ess\,sup}_Q w(x) < \infty \iff D_\infty(P \parallel Q) < \infty, \tag{15.10}$$

and

$$\mathrm{var}_P\left[g(x)\right] < \infty \tag{15.11}$$

is often straightforward.

The effective sample size (Liu and Chen, 1998)

$$ESS = \left[\sum_{i=1}^{n} w(x_i)\right]^2 / \sum_{i=1}^{N} w(x_i)^2$$

is widely used as a measure of efficiency. The ratio of ESS to nominal sample size n is relative effective sample size

$$RESS = ESS/n = \frac{\left[n^{-1}\sum_{i=1}^{n} w(x_i)\right]^2}{n^{-1}\sum_{i=1}^{n} w(x_i)^2} \overset{a.s.}{\to} \frac{\left\{\int_X [p(x)/q(x)]\, q(x) d\mu(x)\right\}^2}{\int_X [p(x)/q(x)]^2 q(x) d\mu(x)}$$
$$= \exp\left[-D_2(P \parallel Q)\right]. \qquad (15.12)$$

3.3 Approximation of Rényi Divergence

Expressions (15.1) and (15.2) suggest strategies for approximation of $D_\alpha(P \parallel Q)$, using either acceptance or importance sampling for P or Q. All simulation strategies are predicated on the existence of $D_\alpha(P \parallel Q) < \infty$; yet none of these strategies by themselves can confirm existence (or not). This important analytical requirement has to be addressed on a case-by-case basis.

3.3.1 Direct or Acceptance Sampling

Suppose that a random sample $x_i^q \overset{iid}{\sim} Q\,(i = 1, 2, \ldots)$ can be created, either directly or through acceptance sampling. For $\alpha \in (0,1) \cup (1, \infty)$ with

$$D_\alpha(P \parallel Q) = \frac{1}{\alpha - 1} \log \mathrm{E}_Q \left(\frac{p(x)}{q(x)}\right)^\alpha = \frac{1}{\alpha - 1} \log \int_X \left(\frac{p(x)}{q(x)}\right)^\alpha q(x) d\mu(x),$$

define

$$\widehat{D}_\alpha(P \parallel Q) = \frac{1}{\alpha - 1} \log \left[n^{-1} \sum_{i=1}^{n} \left(\frac{p\left(x_i^q\right)}{q\left(x_i^q\right)}\right)^\alpha \right] \overset{a.s.}{\to} D_\alpha(P \parallel Q).$$

This requires that it be possible to evaluate p and q, but in some applications, including Bayesian inference, only a kernel $k_p(x) \propto p(x)$ and/or $k_q(x) \propto q(x)$ may be known: that is, $k_p(x) = c_p \cdot p(x)$ with c_p unknown, and/or $k_q(x) = c_q \cdot q(x)$ with c_q unknown. Treating the general case in which both are unknown, define $r(x) = k_p(x)/k_q(x)$. If $D_{1+\varepsilon}(P \parallel Q) < \infty$ for some $\varepsilon > 0$, then

$$\hat{r}_n \equiv n^{-1}\sum_{i=1}^{n} r\left(x_i^q\right) \overset{a.s.}{\to} \int_X r(x)q(x)d\mu(x) = \int_X \frac{c_p p(x)}{c_q q(x)} q(x)d\mu(x) = \frac{c_p}{c_q} \equiv r.$$

If, in addition, $D_\alpha(P \parallel Q) < \infty$, then

$$\hat{r}_n^{-\alpha} n^{-1}\sum_{i=1}^{n} r\left(x_i^q\right)^\alpha \overset{a.s.}{\to} \int_X \left[r^{-1}r(x)\right]^\alpha q(x)d\mu(x) = \int_X \left[\frac{p(x)}{q(x)}\right]^\alpha q(x)d\mu(x)$$

and so

$$\widehat{\widehat{D}}_\alpha(P \parallel Q) = \frac{1}{\alpha - 1}\log\left\{\hat{r}^{-\alpha} n^{-1}\sum_{i=1}^{n} r\left(x_i^q\right)^\alpha\right\} \overset{a.s.}{\to} D_\alpha(P \parallel Q).$$

A central limit theorem for the accuracy of the approximation would require the existence of $D_{2\alpha}(P \parallel Q)$. Failure of this condition compromises not only the evaluation of the accuracy of the approximation, but in many applications it also results in an unacceptably slow rate of convergence.

For $\alpha = 1$,

$$\widehat{\widehat{D}}_1(P \parallel Q) = n^{-1}\sum_{i=1}^{n} \log\left[\hat{r}_n^{-1} \cdot r\left(x_i^q\right)\right] \cdot \hat{r}_n^{-1} r\left(x_i^q\right) \overset{a.s.}{\to} \int \log\left(\frac{p(x)}{q(x)}\right) p(x) = D_1(P \parallel Q)$$

as long as $D_1(P \parallel Q) < \infty$. The condition for asymptotic normality (and, typically, for a satisfactory rate of convergence) is

$$\mathrm{E}_Q\left\{\log\left[\frac{p(x)}{q(x)}\right] \cdot \left[\frac{p(x)}{q(x)}\right]\right\}^2 = \mathrm{E}_P\left\langle\left\{\log\left[\frac{p(x)}{q(x)}\right]\right\}^2 \cdot \left(\frac{p(x)}{q(x)}\right)\right\rangle < \infty. \tag{15.13}$$

A procedure beginning with $x_i^p \overset{iid}{\sim} P$ can be constructed in the same way. The convergence condition

$$\mathrm{E}_P\left\{\log\left[\frac{p(x)}{q(x)}\right]\right\}^2 < \infty.$$

then replaces (15.13).

3.3.2 Importance Sampling

Suppose that direct or acceptance sampling is not feasible for either P or Q, but there are importance sampling source distributions Q_p for P with p.d.f. q_p and Q_q for q with p.d.f. q_q. Several conditions, which must be established analytically, are jointly sufficient for a consistent and asymptotically normal approximation of $D_\alpha(P \| Q \|)$.

1. For $\alpha \in (0,1) \cup (1,\infty)$, the existence condition

$$D_{2\alpha}(P \parallel Q) < \infty. \tag{15.14}$$

   (Requirements for $\alpha = 1$ are given subsequently.)
2. Either the condition (15.8) or the condition (15.10) for each pair $\left(P, Q_p\right)$ and $\left(Q, Q_q\right)$;
3. The ability to evaluate $q_p(x)$ and $q_q(x)$, not just kernels of $q_p(x)$ and $q_q(x)$.

Given these three conditions and

$$x_i^u \overset{iid}{\sim} Q_u, \quad w_u(x) \equiv k_u(x)/q_u(x), \quad \hat{c}_u \equiv n^{-1}\sum_{i=1}^{n} w_u\left(x_i^u\right) \ (u = p, q),$$

then

$$\frac{\hat{c}_q}{\hat{c}_p} n^{-1} \sum_{i=1}^{n} \left[\frac{k_p\left(x_i^q\right)}{k_q\left(x_i^q\right)}\right]^{\alpha} w_q\left(x_i^q\right) \text{ and } \frac{\hat{c}_q}{\hat{c}_p} n^{-1} \sum_{i=1}^{n} \left[\frac{k_p\left(x_i^p\right)}{k_q\left(x_i^p\right)}\right]^{\alpha-1} w_p\left(x_i^p\right)$$

are simulation-consistent and asymptotically normal approximations of $D_\alpha(P \parallel Q)$ when $\alpha \in (0,1) \cup (1,\infty)$.

For the case $\alpha = 1$, referring to (15.3)

$$\log\left(\frac{\hat{c}_q}{\hat{c}_p}\right) + n^{-1}\sum_{i=1}^{n} \log\left(\frac{k_p\left(x_i^q\right)}{k_q\left(x_i^q\right)}\right) w_q\left(x_i^q\right) \text{and} \log\left(\frac{\hat{c}_q}{\hat{c}_p}\right) + n^{-1}\sum_{i=1}^{n} \log\left(\frac{k_p\left(x_i^p\right)}{k_q\left(x_i^p\right)}\right)$$

are simulation-consistent and asymptotically normal approximations of $D_1(P \parallel Q)$.

4. Sequential Monte Carlo

Sequentially adaptive Bayesian learning (SABL) is a sequential Monte Carlo algorithm that attacks both Bayesian inference and optimization problems through the controlled introduction of information. SABL builds on theory developed in Douc and Moulines (2008); Geweke and Durham (2019) provide a detailed description of SABL. Section 4.1 introduces only those aspects that are essential in conveying its interaction with Rényi divergence. Section 4.2 shows how SABL provides Rényi divergence measures from a posterior distribution to an updated posterior distribution. Sections 4.3 and 4.4 show how Rényi divergence governs the controlled introduction of information in the algorithm, which is essential to its practical application.

4.1 The Algorithm

Consider the canonical Bayesian inference problem for observable random vectors $y_{1:T} = \{y_1, \ldots, y_T\}$. A model A provides a distribution of observables conditional on an unknown parameter vector,

$$p(y_{1:T}|\theta, A) = \prod_{t=1}^{T} p(y_t|y_{1:t-1}, \theta, A)$$

and a prior distribution Π_0 with p.d.f. $p_0(\theta|A)$ and support Θ. For the observed random vectors $y^o_{1:t}$ the posterior distribution Π_t has p.d.f.

$$\pi_t(\theta) = p(\theta|y^o_{1:t}, A) \propto p_0(\theta|A)\, p(y^o_{1:t}|\theta, A). \tag{15.15}$$

The right side of this expression provides a kernel $k_t(\theta)$ of the posterior p.d.f. $\pi_t(\theta)$ at time t, and the sequence π_t characterizes a sequence of probability distributions $\Pi_t \rightarrow \Pi_T \equiv \Pi$.

For Bayesian updating in real time, this is the natural way to introduce information about θ. But it is not the only way, and in fact the introduction of information is a practical issue in other contexts as well, like optimization. Geweke and Durham (2019) characterize the problem and the algorithm at this greater level of generality. The SABL algorithm introduces information about θ in a series of cycles, and at the end of cycle ℓ information is represented by a set of *n particles*, the triangular array $\left\{\theta^{(\ell)}_{n,i}\right\}$ $(i = 1, \ldots, n)$. Convergence is in the number of particles n, and the theory pertains to the triangular array $\{\theta_{n,i}\}$ $(i = 1, \ldots, n; n = 1, 2, \ldots)$. The algorithm targets a distribution Π for θ; in Bayesian applications, Π is the posterior distribution. Following Douc and Moulines (2008), $\{\theta_{n,i}\}$ is *consistent* for Π and g if $\bar{g}_n \xrightarrow{p} E_{\Pi}(g)$ and $\{\theta_{n,i}\}$ is *asymptotically normal* for Π and g if there exists V_g such that $n^{1/2}(\bar{g}_n - \bar{g}) \xrightarrow{d} N(0, V_g)$. At the end of the last cycle, the particles $\left\{\theta^{(\ell)}_{n,i}\right\}$ $(i = 1, \ldots, n)$ represent the target distribution Π_{ℓ}.

Algorithm 3. *Sequential Monte Carlo (SMC). Given*

1. *the initial distribution Π_0, continuous with respect to μ and with density kernel $k^{(0)}$,*

2. *the intermediate distributions* Π_ℓ, *continuous with respect to* μ *and with density kernels* $k^{(\ell)}$ $(\ell = 1,\dots,L)$,
3. *the target distribution* $\Pi = \Pi_L$, *continuous with respect to* μ *and with density kernel* $k = k^{(L)}$,
4. *Markov kernels* R_ℓ: $\Theta \to \Theta$ *with invariant distribution* Π_ℓ $(\ell = 1,\dots,L)$;

let particles $\theta_{n,i}$ *be drawn as follows:*

- *Initialize: Draw* $\theta_{n,i}^{(0)} \overset{iid}{\sim} \Pi_0$ $(i = 1,\dots,n)$.
- *For cycles* $\ell = 1,\dots,L$
  - *Reweight: Define*

$$w_{n,i}^{(\ell)} = w^{(\ell)}\left(\theta_{n,i}^{(\ell-1)}\right) = k^{(\ell)}\left(\theta_{n,i}^{(\ell-1)}\right)/k^{(\ell-1)}\left(\theta_{n,i}^{(\ell-1)}\right)\ (i = 1,\dots,n).$$

  - *Resample: Draw* $\theta_{n,i}^{(\ell,0)}$ *i.i.d. with*

$$P\left(\theta_{n,i}^{(\ell,0)} = \theta_{n,j}^{(\ell-1)}\right) = w_j^{(\ell)}/\sum_{r=1}^{n} w_r^{(\ell)}.$$

  - *Move: Draw* $\theta_{n,i}^{(\ell)} \sim R_\ell\left(\theta_{n,i}^{(\ell,0)},\cdot\right)$ *independently* $(i = 1,\dots,n)$.

- *Set* $\theta_{n,i} = \theta_{n,i}^{(L)}$ $(i = 1,\dots,n)$.

Douc and Moulines (2008) prove consistency and asymptotic normality, given the following sufficient condition for the kernels $k^{(\ell)}(\theta)$ and a function of interest $g(\theta)$.

Condition 1. *Weak sufficient conditions for the SMC algorithm*

(a) $E_{\Pi_\ell}\left[k^{(m)}(\theta)/k^{(\ell)}(\theta)\right]^2 < \infty (\ell = 0,\dots,m-1; m = 1,\dots,L)$
(b) $E_{\Pi_\ell}\left[g(\theta)k(\theta)/k^{(\ell)}(\theta)\right]^2 < \infty (\ell = 0,\dots,L)$

In practice, it is easier to verify instead:

Condition 2. *Strong sufficient conditions for the SMC algorithm*

(a) *There exists* $\overline{w} < \infty$ *such that*

$$w^{(\ell)}(\theta) = k^{(\ell)}(\theta)/k^{(\ell-1)}(\theta) < \overline{w} (\ell = 1,\dots,L; \theta \in \Theta).$$

(b) $\text{var}_{\Pi_0}\left[g(\theta)\right] < \infty$

It is easy to see that Condition 2 implies Condition 1.

Proposition 5. *(Douc and Moulines, 2008). Given either Condition 1 or Condition 2, the particles* $\{\theta_{n,i}\}$ *in Algorithm 3 are consistent and asymptotically normal for* Π *and* g.

Note that Condition 1 is almost identical to conditions (15.8) and (15.9) for importance sampling, and Condition 2 is the same as conditions (15.10) and (15.11) for importance sampling. That is because the Reweight step in each cycle of Algorithm 3 amounts to importance sampling.

In terms of Rényi divergence, Condition 1(a) is

$$D_2(\Pi_m \parallel \Pi_\ell) < \infty \, (\ell = 0, \ldots, m-1; m = 1, \ldots, L).$$

Algorithm 3 is the starting point for attacking a series of technical issues that must be resolved to produce practical software. These issues are addressed in Durham and Geweke (2014) and Geweke and Durham (2019). Efficiency of the Move step, in particular, requires that the particles $\left\{\theta_{n,i}^{(\ell)}\right\}$ be used to construct the transition density R_ℓ, as anyone who has applied sequential Monte Carlo in practice quickly learns. Durham and Geweke (2014) and Geweke and Durham (2019) resolve these issues and develop the extension of the theory in Douc and Moulines (2008) required to support the procedures. These details are not important for the rest of this chapter, though they are essential in the illustrations in section 5.

4.2 Bayesian Inference with Data Tempering

Real-time Bayesian updating uses (15.15) to augment information about θ. When SABL implements this process, the cycles ℓ correspond one-to-one with observations t: Π_ℓ in Algorithm 3 and Π_t in (15.15) are the same distribution for $\ell = t$. There are $L = T$ cycles, T being sample size, but updating involves only the execution of the last cycle, corresponding to the most recent observation. In the sequential process, this procedure is known as data tempering: it introduces information as it naturally arises. There are other variants on data tempering, described in Durham and Geweke (2014), but the one just identified corresponds most directly to the problem addressed in this section.

Because Algorithm 3 produces the distribution of each Π_t as an intermediate product, it provides the basis for approximating Rényi divergences $D_\alpha(\Pi_t \parallel \Pi_s)(0 \leq s < t \leq T)$. We next develop the details and in the process identify situations in which the approximation will be more accurate or less accurate. Section 5.2 provides an example.

To begin, Durham and Geweke (2014) showed that Algorithm 3 provides good approximations of the marginal likelihood (marginal data density) as a by-product. The argument there, focusing on aspects relevant for the task here, begins by noting that the mean weight in the Reweight step of cycle s is

$$\begin{aligned}\overline{w}_n^{(s)} &\equiv n^{-1}\sum_{i=1}^{n} w_{n,i}^{(s)} = n^{-1}\sum_{i=1}^{n} p\left(y_s^o|y_{1s-1}^o,\theta_i^{(s-1)},A\right)\\ &\overset{a.s.}{\rightarrow}\int_{\theta} p\left(y_s^o|y_{1:s-1}^o,\theta,A\right)p\left(\theta|y_{1:s-1}^o,A\right)d\mu(\theta)\\ &= p\left(y_s^o|y_{1:s-1}^o,A\right).\end{aligned}$$

Hence

$$\prod_{j=s+1}^{t}\overline{w}_n^{(j)} \overset{a.s.}{\rightarrow} \prod_{j=s+1}^{t} p\left(y_j^o|y_{j-1}^o,A\right) = p\left(y_{s+1:t}^o|y_{1:s}^o,A\right).$$

For $s=0$, $t=T$, this is the marginal likelihood $p\left(y_{1:T}^o|A\right)$.

For approximating $D_\alpha\left(\Pi_t \,\|\, \Pi_s\right), \alpha\in(0,1)\cup(1,\infty)$, the integral of interest is

$$\begin{aligned}&\int_\Theta\left[\frac{p\left(\theta|y_{1:t},A\right)}{p\left(\theta|y_{1:s},A\right)}\right]^\alpha p\left(\theta|y_{1:s}^o,A\right)d\mu(\theta)\\ &=\int_\Theta\left[\frac{p(\theta|A)\,p\left(y_{1:t}^o|\theta,A\right)/p\left(y_{1:t}^o|A\right)}{p(\theta|A)\,p\left(y_{1;s}^o|\theta,A\right)/p\left(y_{1:s}^o|A\right)}\right]^\alpha p\left(\theta|y_{1:s}^o,A\right)d\mu(\theta)\\ &=\left[\frac{p\left(y_{1:s}^o|A\right)}{p\left(y_{1:t}^o|A\right)}\right]^\alpha\int_\Theta p\left(y_{s+1:t}^o|y_{1;s}^o,\theta,A\right)^\alpha p\left(\theta|y_{1:s}^o,A\right)d\mu(\theta)\\ &=\left[p\left(y_{s+1:t}^o\right)\right]^{-\alpha}\cdot \mathrm{E}_{\Pi_s}\left[p\left(y_{s+1:t}^o|y_{1;s}^o,\theta,A\right)^\alpha\right]\end{aligned}$$

for which the simulation-consistent approximation is

$$\left[\prod_{j=s+t}^{t}\overline{w}_n^{(j)}\right]^{-\alpha} n^{-1}\sum_{i=1}^{n} p\left(y_{s+1:t}^o|y_{1;s}^o,\theta_{n,i}^{(s)},A\right)^\alpha.$$

For $s=t-1$ this becomes

$$\left[\overline{w}_n^{(t)}\right]^{-\alpha} n^{-1}\sum_{i=1}^{n}\left(w_{n,i}^{(t)}\right)^\alpha.$$

This expression is invariant to scaling the weights $w_i^{(t)}$, and for $w_i^{*(t)} = w_i/\overline{w}_n^{(t)}$ it becomes simply

$$n^{-1}\sum_{i=1}^{n}\left(w_i^{*(t)}\right)^{\alpha}, \tag{15.16}$$

the α'th raw moment of the normalized weights. Unlike the marginal likelihood, Rényi divergence relies only on a kernel of the observables p.d.f. $p(y_t|y_{1:t-1},\theta,A)$.

The quality of the approximation is inversely related to $\mathrm{var}_{\Pi_s}\left[p\left(y^o_{s+1:t}|y^o_{1;s},\theta,A\right)^{\alpha}\right]$, which increases with α and typically increases with the size of $t-s$. It is most likely to be successful for $t-s=1$, and (15.16) indicates that the computations are a trivial by-product of Algorithm 3.

For the case $\alpha=1$, $D_1\left(\Pi_t \parallel \Pi_s\right)$

$$\begin{aligned}
&= \int_{\Theta}\log\left[\frac{\pi_t(\theta)}{\pi_s(\theta)}\right]\pi_t(\theta)\,d\mu(\theta)\\
&= \int_{\Theta}\log\left[\frac{\pi_t(\theta)}{\pi_s(\theta)}\right]\frac{\pi_t(\theta)}{\pi_s(\theta)}\pi_s(\theta)\,d\mu(\theta)\\
&= \int_{\Theta}\log\left[p\left(y^o_{s+1:t}|y^o_{1:s},\theta,A\right)\cdot\frac{p\left(y^o_{1:s}|A\right)}{p\left(y^o_{1:t}|A\right)}\right]\cdot p\left(y^o_{s+1:t}|y^o_{1:s},\theta,A\right)\frac{p\left(y^o_{1:s}|A\right)}{p\left(y^o_{1:t}|A\right)}\pi_s(\theta)\,d\mu(\theta)\\
&= \frac{p\left(y^o_{1:s}|A\right)}{p\left(y^o_{1:t}|A\right)}\int_{\Theta}\log\left[p\left(y^o_{s+1:t}|y^o_{1:s},\theta,A\right)\cdot\frac{p\left(y^o_{1:s}|A\right)}{p\left(y^o_{1:t}|A\right)}\right]p\left(y^o_{s+1:t}|y^o_{1:s},\theta,A\right)\pi_s(\theta)\,d\mu(\theta)\\
&= \left[p\left(y^o_{s+1:t}|y^o_{1:s},A\right)\right]^{-1}\int_{\Theta}\log\left[p\left(y^o_{s+1:t}|y^o_{1:s},\theta,A\right)\right]p\left(y^o_{s+1:t}|y^o_{1:s},\theta,A\right)\pi_s(\theta)\,d\mu(\theta)\\
&\quad -\left[p\left(y^o_{s+1:t}|y^o_{1:s},A\right)\right]^{-1}\log\left[p\left(y^o_{s+1:t}|y^o_{1:s},A\right)\right]\int_{\Theta}p\left(y^o_{s+1:t}|y^o_{1:s},\theta,A\right)\pi_s(\theta)\,d\mu(\theta)\\
&= \left[p\left(y^o_{s+1:t}|y^o_{1:s},A\right)\right]^{-1}\mathrm{E}_{\Pi_s}\left\{\log\left[p\left(y^o_{s+1:t}|y^o_{1:s},\theta,A\right)\right]\cdot p\left(y^o_{s+1:t}|y^o_{1:s},\theta,A\right)\right\}\\
&\quad -\log\left[p\left(y^o_{s+1:t}|y^o_{1:s},A\right)\right].
\end{aligned}$$

For $s=t-1$, this becomes

$$\frac{\mathrm{E}_{\Pi_{t-1}}\left\{\log\left[p\left(y^o_t|y^o_{1:t-1},\theta,A\right)\right]p\left(y^o_t|y^o_{1:t-1},\theta,A\right)\right\}}{p\left(y^o_t|y^o_{t-1},A\right)}-\log\left[p\left(y^o_t|y^o_{1:t-1},A\right)\right].$$

The corresponding simulation-consistent approximation is

$$\frac{n^{-1}\sum_{i=1}^{n}\log\left(w_i^{(t)}\right)w_i^{(t)}}{\overline{w}(t)}-\log\left[\overline{w}(t)\right].$$

For weights normalized to have mean 1, $w_i^{*(t)}$, this expression becomes

$$n^{-1}\sum_{i=1}^{n}\log\left(w_i^{*(t)}\right)w_i^{*(t)}.$$

More generally, the sequential Monte Carlo approximation to a sequence of divergence measures $D_\alpha\left(\Pi_{t_j} \parallel \Pi_{t_{j-1}}\right)(0 = t_1 < \cdots < t_L = T)$ can be constructed in the same way by introducing the information $y^o_{t_{\ell-1}+1:t_\ell}$ in cycle $\ell(\ell = 1, \ldots, L)$. However accuracy decreases roughly exponentially with $t_\ell - t_{\ell-1}$. Because distance measures are not additive, the computations cannot be chained. By contrast, this works for marginal likelihood as described in Durham and Geweke (2014) because log predictive likelihoods sum to log marginal likelihood.

4.3 Bayesian Inference with Power Tempering

This is the specific case of Algorithm 3 in which the kernel

$$k_\ell(\theta) \propto p(\theta|A)p(y_{1:T}|\theta, A)^{r_\ell} \tag{15.17}$$

with $0 < r_1 < \cdots < r_L = 1$. Because

$$k^{(\ell)}(\theta)/k^{(\ell-1)}(\theta) = p(y_{1:T}|\theta, A)^{r_\ell - r_{\ell-1}},$$

Condition 2(a) is satisfied if and only if the likelihood function $p\left(y^o_{1:T}|\theta, A\right)$ is bounded, which is often known to be true or false. If the existence of a finite bound cannot be established then recourse must be had to Condition 1.

In power tempering, the investigator can control the rate of introduction of new information through the design of the sequence $\{r_\ell\}$ in (15.17). Geweke and Durham (2019) chooses r_ℓ so that the relative effective sample size (*RESS*) (15.12) attains a specified value. Target *RESS* values in the interval (0.1, 0.9) are in general equally satisfactory, and the procedure is almost always more robust and efficient than is data tempering. It is more robust because a single outlying observation y^o_t can lead to a very small *RESS* in cycle t, which in turn implies a Move step that is slow. Power tempering, by contrast, controls the rate at which information is introduced in each cycle, producing smooth and reliably efficient execution. In applications (e.g., Duan and Fulop [2015], Geweke [2016], and Geweke and Durham [2019]), that is one reason why power tempering is of interest; section 4.4 provides a second reason.

In the context of section 2.2, this adaptive power tempering is possible because (a) population *RESS* is a monotone transform of $D_2(\Pi_\ell \parallel \Pi_{\ell-1})$, (b) (15.17) is a partial power concentration (Definition 3), and (c) Propositions 2–4 therefore apply. Expression (15.4) for D_2 has a rough correspondence with the term in Condition 1(a) for $\ell = m - 1$, which at least vaguely suggests that targeting *RESS* is a good way to choose r_ℓ.

Propositions 2–4 make it possible to target other $D_\alpha(\Pi_\ell \parallel \Pi_{\ell-1})$ to choose r_ℓ. For cycle ℓ and $\alpha \in (0,1) \cup (1,\infty) r_\ell$ satisfies

$$\int_\Theta \left[\frac{\pi_\ell(\theta)}{\pi_{\ell-1}(\theta)}\right]^\alpha \pi_{\ell-1}(\theta)\, d\mu(\theta) = \exp\left[D_\alpha^* / (\alpha - 1)\right]. \tag{15.18}$$

Note

$$\pi_{\ell-1}(\theta) = \frac{p(\theta|A)\, p(y_{1:T}^o|\theta, A)^{r_{\ell-1}}}{\int_\Theta p(\theta|A)\, p(y_{1:T}^o|\theta, A)^{r_{\ell-1}}\, d\mu(\theta)},$$

$$\pi_\ell(\theta) = \frac{p(\theta|A)\, p(y_{1:T}^o|\theta, A)^{r_{\ell-1}\cdot(1+r)}}{\int_\Theta p(\theta|A)\, p(y_{1:T}^o|\theta, A)^{r_{\ell-1}\cdot(1+r)}\, d\mu(\theta)},$$

and define

$$w_r(\theta) = p(y_{1:T}^o|\theta, A)^{r_{\ell-1}\cdot r}.$$

This is the weight function in the reweight step of cycle ℓ when $r_\ell = r_{\ell-1}(1+r)$. Then

$$\begin{aligned}\frac{\pi_\ell(\theta)}{\pi_{\ell-1}(\theta)} &= w_r(\theta) / \left[\frac{\int_\Theta p(y_{1:T}^o|\theta, A)^{r_{\ell-1}\cdot r} \cdot p(\theta)\, p(y_{1:T}^o|\theta, A)^{r_{\ell-1}}\, d\mu(\theta)}{\int_\Theta p(\theta|A)\, p(y_{1:T}^o|\theta, A)^{r_{\ell-1}}\, d\mu(\theta)}\right] \\ &= w_r(\theta) / \mathrm{E}_{\ell-1}\left[w_r(\theta)\right]\end{aligned}$$

and the left side of (15.18) becomes

$$\int_\Theta \frac{w_r(\theta)^\alpha \pi_{\ell-1}(\theta)\, d\mu(\theta)}{\left[\int_\Theta w_r(\theta)\right]} = \frac{\int_\Theta w_r(\theta)^\alpha \pi_{\ell-1}(\theta)\, d\mu(\theta)}{\left[\int_\Theta w_r(\theta)\, \pi_{\ell-1}(\theta)\right]^\alpha d\mu(\theta)},$$

leading to the equation that is solved by successive bifurcation in cycle ℓ,

$$\frac{n^{-1}\sum_{i=1}^n w_r\left(\theta_{n,i}^{(\ell-1)}\right)^\alpha}{\left[n^{-1}\sum_{i=1}^n w_r\left(\theta_{n,i}^{(\ell-1)}\right)\right]^\alpha} = \exp\left[D_\alpha^* / (\alpha - 1)\right]. \tag{15.19}$$

This equation is invariant to scaling all $w_r\left(\theta_i^{(\ell-1)}\right)$ by a common factor. For weights normalized to have mean 1, the denominator vanishes, leaving the α'th raw moment in the numerator.

For $\alpha = 1$ we target the relation

$$\int_{\Theta} \log\left[\frac{\pi_\ell(\theta)}{\pi_{\ell-1}(\theta)}\right] \pi_\ell(\theta)\, d\mu(\theta)$$
$$= \int_{\Theta} \log\left[\frac{\pi_\ell(\theta)}{\pi_{\ell-1}(\theta)}\right]\left(\frac{\pi_\ell(\theta)}{\pi_{\ell-1}(\theta)}\right)\pi_{\ell-1}(\theta)\, d\mu(\theta) = D_1^*,$$

equivalently

$$\int_{\Theta} \log\left\{\frac{w_r(\theta)}{\mathrm{E}_{\Pi_{\ell-1}}[w_r(\theta)]}\right\} \cdot \frac{w_r(\theta)}{\mathrm{E}_{\Pi_{\ell-1}}[w_r(\theta)]} \cdot \pi_{\ell-1}(\theta)\, d\mu(\theta) = D_1^*.$$

The corresponding simulation-consistent approximation is

$$\frac{\left\{n^{-1}\sum_{i=1}^{n} \log\left[w_r\left(\theta_{n,i}^{(\ell-1)}\right)\right] - \log\left(\overline{w}_r^{(\ell-1)}\right)\right\} w_r\left(\theta_{n,i}^{(\ell-1)}\right)}{n^{-1}\sum_{i=1}^{n} w_r\left(\theta_{n,i}^{(\ell-1)}\right)} = D_1^*,$$

and it can be solved quickly by successive bifurcation. Here, too, the denominator vanishes for normalized weights.

4.4 Optimization with Power Tempering

If the foregoing algorithm is modified by specifying $0 < r_1 < \cdots < r_L = r^* \gg 1$, then it can be interpreted as an algorithm for Bayesian inference that targets a posterior distribution that is much more concentrated than the original, but with the same prior distribution. If the likelihood function is unimodal, then for sufficiently large r^* the particles concentrate near the mode. Indeed, the likelihood function can be replaced by a function $f(\theta)$ to be optimized. In that case, Π_0 is the initial distribution that is used in simulated annealing and other stochastic optimization algorithms. This is another reason why power tempering is of interest, beyond the considerations taken up at the start of section 4.3.

Geweke and Durham (2019) establishes the limiting behavior of the particles for $r^* \to \infty$ and $n \to \infty$, including rates of convergence, and demonstrates superior performance of the algorithm compared to other optimization methods. In that work, the *RESS* metric is used in precisely the same way to determine r_ℓ given $r_{\ell-1}$, and the same question arises: would an alternative metric based on D_α exhibit superior performance? Applications in Section 5.3 provide some evidence on this point.

5. Illustrations

This chapter concludes with applications illustrating the theory and practicability of the methods developed. Section 5.1 sets up the substantive applications. Section 5.2 uses data tempering to measure the divergence in Bayesian updating of a posterior distribution. Section 5.3 investigates the implications for computational efficiency of using alternative Rényi divergence measures for the power-tempering sequence.

5.1 Context

The context for the illustrations consists of two applications from Geweke and Durham (2019). The first is a time series model that is standard but with a nonlinear transformation of the conventional parameters that has a more direct economic interpretation. That model is described in section 5.1.1 and subsequently used in sections 5.2 and 5.3. The second is a member of a suite of problems used in the global optimization literature for comparing different methods. That model is described in section 5.1.2 and subsequently used in section 5.3.

5.1.1 Time Series Model

Sections 5.2 and 5.3 utilize an autoregressive model of order 3

$$y_t = \beta_0 + \beta_1 y_{t-1} + \beta_2 y_{t-2} + \beta_3 y_{t-3} + \varepsilon_t,\ \varepsilon_t \overset{iid}{\sim} N(0,\sigma^2) \tag{15.20}$$

for the logarithm of US real GDP per capita (1970–2014). In (15.20), the parameter space for $\beta = (\beta_1, \beta_2, \beta_3)'$ is restricted so that the generating polynomial $1 - \beta_1 z - \beta_2 z^2 - \beta_3 z^3$ has one real root r_1 and a conjugate pair of complex roots (r_2, r_3). All of the analysis, including the prior distribution, replaces β with the parameters

$$h_s = \log(2)/\log(|r_1|),\quad h_c = \log(2)/\log(|r_2|),\quad p = 2\pi/\tan^{-1}(\mathrm{Im}(r_2)/\mathrm{Re}(r_2)) \tag{15.21}$$

where $\tan^{-1}$ is the principal branch with range $[0, \pi)$. The parameter h_s is the secular half-life of a shock ε_t, the parameter h_c is the cyclical half-life of a shock, and p is the period of the cycle. The reparameterization (15.21) embodies the interpretation motivated by Samuelson (1939) and is further developed in Geweke (1988).

The prior distribution is shown in the following table.

Parameter	Distribution	Centered 90% interval
Intercept $\beta_0(20)$:	$\theta_1 = \beta_0 \sim N(10, 5^2)$	$\beta_0 \in (1.77, 18.22)$
Secular half – life $h_s(21)$:	$\theta_2 = \log(h_s) \sim N(\log(25), 1)$	$h_s \in (5.591, 111.9)$
Cyclical half – life $h_c(21)$:	$\theta_3 = \log(h_c) \sim N(\log(1), 1)$	$h_c \in (0.2237, 22.36)$
Period $p(21)$:	$\theta_4 = \log(p) \sim N(\log(5), 1), p > 2$	$p \in (2.316, 28.46)$
Shock magnitude σ (20) :	$\theta_5 = \log(\sigma) \sim N(\log(0.025), 1)$	$\sigma \in (0.005591, 0.1118)$

Geweke and Durham (2019), section 4.2, provides further discussion and detail.

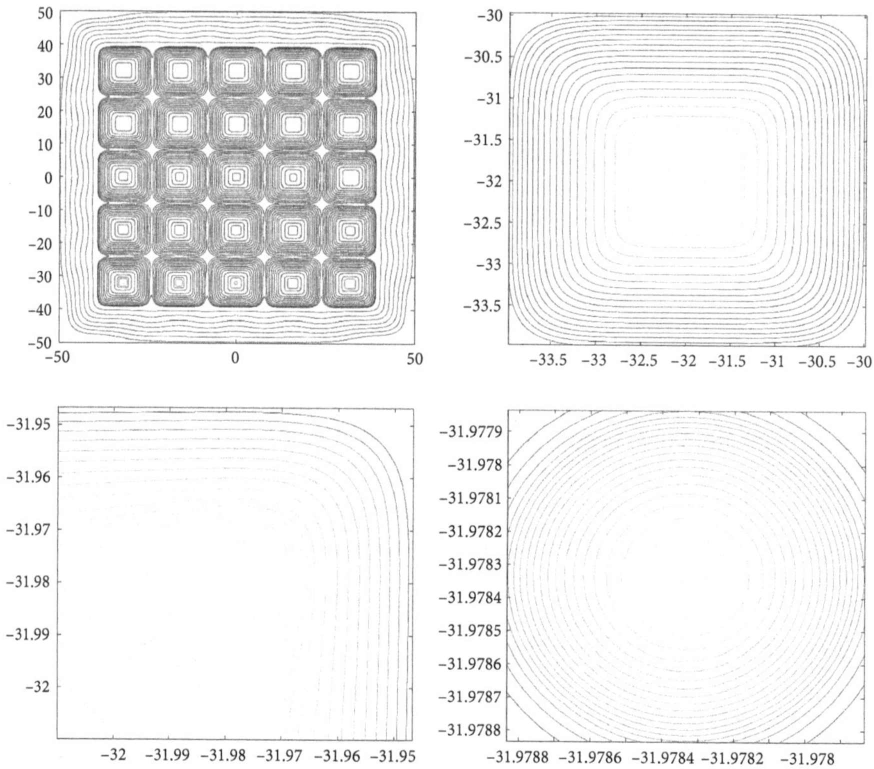

Figure 15.1 DeJong's fifth function.

5.1.2 Optimization Problem

The problem is to optimize De Jong's (1975) fifth function

$$h(x) = -\left[0.002 + \sum_{i=1}^{25} \frac{1}{i + \sum_{j=1}^{2} \left(x_j - a_{ij}\right)^6}\right]$$

where $X = [-50, 50]^2$, $a_{\cdot 1} = e \otimes v$ and $a_{\cdot 2} = v \otimes e$, where $e = (1, 1, 1, 1, 1)$ and $v = (-32, -16, 0, 16, 32)$. Figure 15.1 displays some contours of this function. There are twenty-five local modes. The global mode is in the lower left corner of the northwest panel. Successive panels provide increasing detail in the neighborhood of the global mode.

5.2 Data Tempering

This illustration approximates the Rényi divergence between successive posterior distributions in the time series model described in section 5.1.1, applied to log per capita US real GDP (OECD, 1971–2014). The SABL implementation of sequential Monte Carlo (Algorithm 3) executed the approximations, as described in section 4.2, using $2^{14} = 16,384$ particles, the default number in SABL. Execution time was less than 10 seconds on a laptop and did not exploit graphics processing units.

The top panel in Figure 15.2 shows the approximations; the standard errors of these Monte Carlo approximations, are less than 2 percent of the distance measure itself, on average over all points shown, and the maximum is less than 5 percent. The bottom panel shows first differences in this time series. The first point in both panels corresponds to y_t in 1974 and the rightmost to 2014. Thus, the leftmost point in the top panel shows Rényi divergences between the prior distribution and the posterior distribution with a single observation. Divergencies are shown for $\alpha = 0.5$, 1, 2, 4, 6, and 10, ordered from the bottom plot to the top one by virtue of Proposition 1, part 2.

The behavior of the divergence measures is largely the product of two factors. The first is that divergences are greater early in the sample, when observations have greater potential to modify the posterior distribution than later. The second is that local maxima in divergences correspond to growth rates Δy_t that are either smaller or larger than any previous values of Δy_t in the sample; the 1982 recession and the 2009 financial crisis are especially notable in this regard.

Alternative divergence measures D_α display similar yet distinctive behavior. The local maximum for the 2009 financial crisis exceeds the one for the 1982 recession for $\alpha > 2$, but it is lower for $\alpha \leq 2$. The Rényi divergence between the 1977 and 1978 posterior distributions is lower than that between the 1976

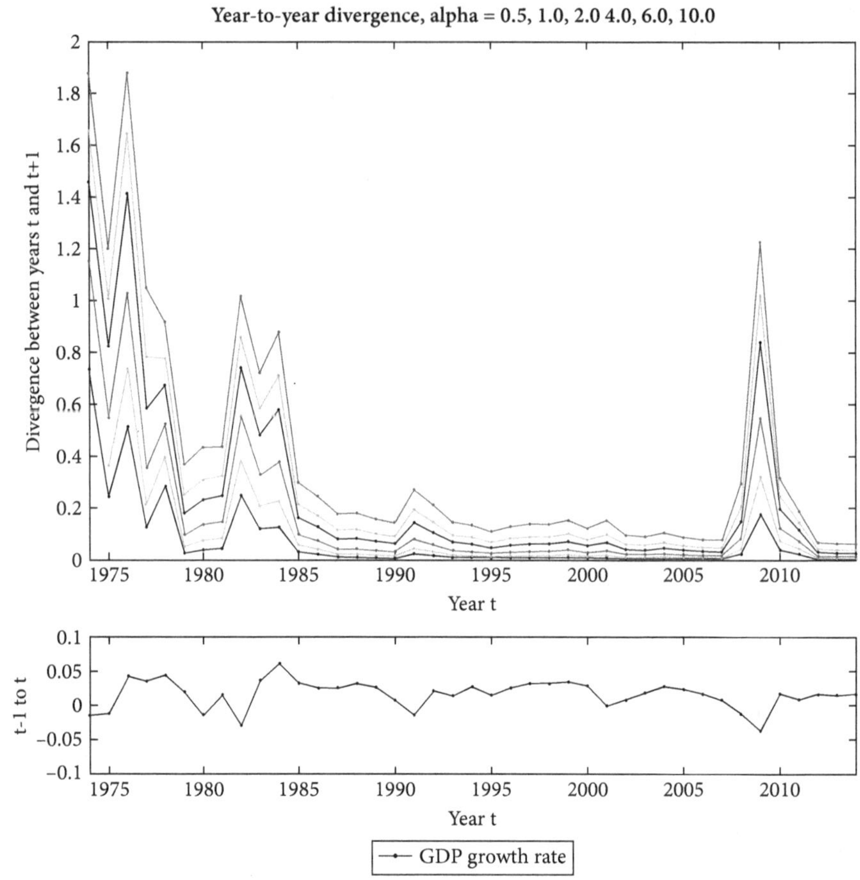

Figure 15.2 Rényi divergences between successive posterior distributions.

and 1977 distributions for $\alpha = 6$ and $\alpha = 10$, but the opposite is true in the other cases.

5.3 Power Tempering

This illustration uses different α in the divergence measure D_α to determine the power increment in the reweight step of Algorithm 3, as described in section 4.3. Different divergence measures imply different targets D_α^* in (15.18) and (15.19). The SABL default *RESS* target is 0.5, based on findings in Durham and Geweke (2014). From (15.18) this is exactly the same as $D_2^* = -\log(0.5) = 0.6931$. Other D_α^* were then computed using the relationships for normal distributions given in the example in section 2.2. The point of the illustration is to gather a little

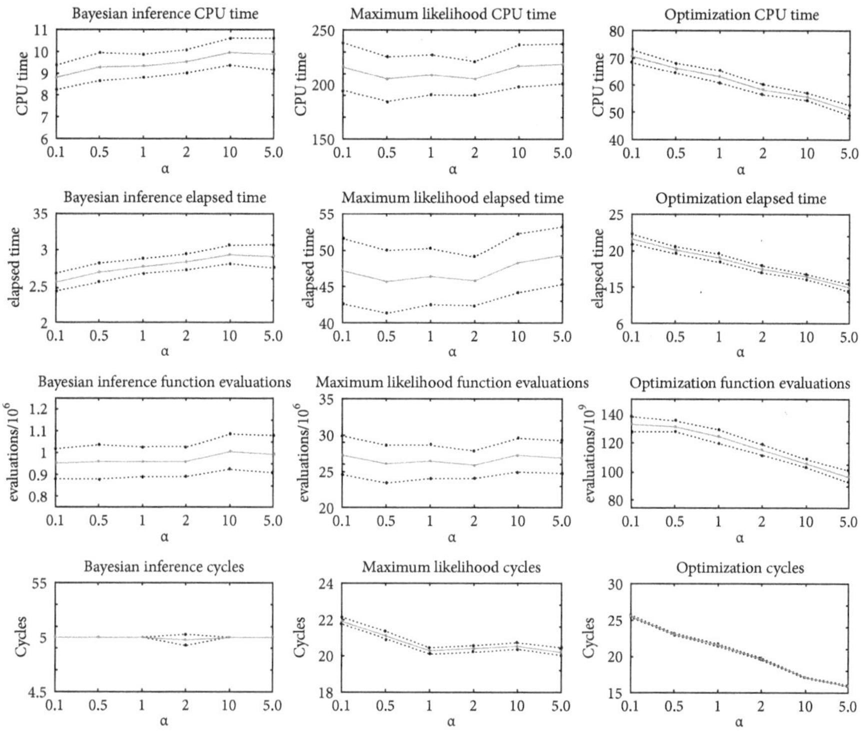

Figure 15.3 Performance measures for alternative D_α in determining the power sequence. Efficiency is inversely related to the measures shown.

evidence on the impact of different α on the efficiency of sequential Monte Carlo for Bayesian inference, maximum likelihood, and optimization in general.

The illustration measures relative efficiency in four different ways: central processing unit (CPU) time, elapsed time, number of function evaluations, and number of cycles in Algorithm 3. These performance measures differ from one execution to another, due to simulation variation in all four cases, as well as to the vagaries of hardware, operating systems, and software for the first two. Therefore, the exercise was repeated forty times in each case, tracking the means and standard deviations over these replications. In Figure 15.3, CPU and elapsed time are recorded in seconds and the number of function evaluations accounts for the separate computations for each of $2^{14} = 16,384$ particles. The solid line is the mean over the forty replications; the dashed lines add and subtract two standard errors for the mean estimate.

For Bayesian inference and maximum likelihood in the model described in section 5.1.1, the evidence is consistent with no more than 10 percent variation in efficiency across the six values of α studied. While $\alpha = 2$ is directly related to the theory for the importance sampling in the reweight step, this conveys no particular efficiency advantage.

For the optimization example (section 4.4), the results are strikingly different: efficiency increases steadily with larger values of α, amounting to about a 35 percent increase for $\alpha = 50$ compared with $\alpha = 0.1$. The increase from $\alpha = 2$ to $\alpha = 50$ is over 20 percent. One possible explanation for this phenomenon resides in the irregularity of the function displayed in Figure 15.1. The arguments of the largest values of the weight function convey revealing information about the distinction between the function values at the different local modes. Indeed, D_∞ depends only on the single largest value of the weight function. In turn, this appears to delay moving from the reweight to resample steps until these nearly unique, most promising arguments are identified, as indicated by the decrease in number of cycles with increase in the value of α in the southeast panel of Figure 15.3.

6. Appendix: Proofs

6.1 Proof of Proposition 2

Let $p_r(x)$ denote the p.d.f. of Π_r. *Proof for the case* $\alpha \in (0,1) \cup (1,\infty)$. Note that

$$p_r(x) = \frac{q(x)^{1+r}}{\int_X q(x)^{1+r} d\mu(x)}$$

and

$$p_r^\alpha(x) q(x)^{1-\alpha} = \frac{q(x)^{1+\alpha r}}{\left[\int_X q(x)^{1+r} d\mu(x)\right]^\alpha}.$$

Referring to the definition of D_α, it suffices to examine the continuous function

$$g(r) = \frac{\int_X q(x)^{1+\alpha r} d\mu(x)}{\left[\int_X q(x)^{1+r} d\mu(x)\right]^\alpha}.$$

Note $g(0) = 1$, corresponding to $D_\alpha(P^0 \parallel Q) = 0$. To show that $D_\alpha(P^r \parallel Q)$ is monotonically increasing in r, it suffices to show that $g(r)$ is monotonically increasing in r when $\alpha > 1$ and monotonically decreasing when $\alpha < 1$.

The function $g(r)$ is differentiable with first derivative

$$g'(r) = N(r)/D(r),$$

where

$$N(r) = \left[\int_X q(x)^{1+r} d\mu(x)\right]^{\alpha} \cdot \alpha \left[\int_X q(x)^{1+\alpha r} \log q(x) d\mu(x)\right]$$
$$-\left[\int_X q(x)^{1+\alpha r} d\mu(x)\right] \cdot \alpha \left[\int_X q(x)^{1+r} d\mu(x)\right]^{\alpha-1} \cdot \left[\int_X q(x)^{1+r} \log q(x) d\mu(x)\right]$$

and

$$D(r) = \left[\int_X q(x)^{1+r} d\mu(x)\right]^{2\alpha} > 0.$$

Collecting terms,

$$N(r) = \alpha \left[\int_X q(x)^{1+r} d\mu(x)\right]^{\alpha} \cdot \left[\int_X q(x)^{1+\alpha r} d\mu(x)\right]$$
$$\cdot \left\{ \frac{\int_X q(x)^{1+\alpha r} \log q(x) d\mu(x)}{\int_X q(x)^{1+\alpha r} d\mu(x)} - \frac{\int_X q(x)^{1+r} \log q(x) d\mu(x)}{\int_X q(x)^{1+r} d\mu(x)} \right\}. \quad (15.22)$$

The expression in parentheses (15.22) is the difference of two weighted averages of $\log q(x)$. The ratio of the weights on the first term to the weights on the second term is $q(x)^{r(\alpha-1)}$. Since $\log q(x)$ is monotone increasing in $q(x)$, Eq. (15.2) is positive if $\alpha > 1$ and negative if $\alpha < 1$.

Proof for the case $\alpha = 1$ (Kullback–Leibler divergence)

$$D_1(P_r \parallel Q) = \int_X \log\left(\frac{p_r(x)}{q(x)}\right) p_r(x) d\mu(x)$$
$$= \int_X \log\left(\frac{q(x)^r}{\int_X q(x)^{1+r} d\mu(x)}\right) \cdot \frac{q(x)^{1+r}}{\int_X q(x)^{1+r} d\mu(x)} d\mu(x)$$
$$= \int_X \log\left[q(x)^r\right] \cdot \frac{q(x)^{1+r}}{\int_X q(x)^{1+r} d\mu(x)} d\mu(x) - \log\left[\int_X q(x)^{1+r} d\mu(x)\right]$$
$$= \frac{\int_X r \log[q(x)]\, q(x)^{1+r} d\mu(x)}{\int_X q(x)^{1+r} d\mu(x)} - \log\left[\int_X q(x)^{1+r} d\mu(x)\right].$$

This function is continuous and differentiable with first derivative

$$dD_1(P_r \parallel Q)/dr = \left[\int_X q(x)^{1+r} d\mu(x)\right]^{-2}$$

$$\cdot\left\{\left[\int_X q(x)^{1+r} d\mu(x)\right]\cdot\int_X\left[q(x)^{1+r}\log q(x)+rq(x)^{1+r}[\log q(x)]^2\right]d\mu(x)\right.$$

$$\left.-\left[r\int_X q(x)^{1+r}\log q(x)d\mu(x)\right]\cdot\left[\int_X q(x)^{1+r}\log q(x)d\mu(x)\right]\right\}$$

$$-\int_X q(x)^{1+r}\log q(x)d\mu(x)/\int_X q(x)^{1+r}d\mu(x)$$

$$=\frac{\int_X q(x)^{1+r}\log q(x)d\mu(x)}{\int_X q(x)^{1+r}d\mu(x)}+\frac{r\int_X q(x)^{1+r}[\log q(x)]^2 d\mu(x)}{\int_X q(x)^{1+r}d\mu(x)}$$

$$-r\cdot\left[\frac{\int_X q(x)^{1+r}\log q(x)d\mu(x)}{\int_X q(x)^{1+r}d\mu(x)}\right]^2-\frac{\int_X q(x)^{1+r}\log q(x)d\mu(x)}{\int_X q(x)^{1+r}d\mu(x)}$$

$$=r\left\{\frac{\int_X q(x)^{1+r}[\log q(x)]^2 d\mu(x)}{\int_X q(x)^{1+r}d\mu(x)}-\left[\frac{\int_X q(x)^{1+r}\log q(x)d\mu(x)}{\int_X q(x)^{1+r}d\mu(x)}\right]^2\right\}$$

$$=r\cdot\left\{\mathrm{E}_{P_r}[\log q(x)]^2-\left[\mathrm{E}_{P_r}\log q(x)\right]^2\right\}=r\cdot\mathrm{var}_{P_r}(\log q(x))>0.$$

Case $\alpha=\infty$. Here

$$D_\infty(P^r\,\|\,Q)=\frac{(Q^*)^r}{\int_X q(x)^{r+1}d\mu(x)}$$

where $Q^*=\text{ess sup}_{q(x)}q(x)$. This is a differentiable function with first derivative

$$dD_\infty(P^r\,\|\,Q)/dr=\frac{\left[\int_X q(x)^{r+1}d\mu(x)\right](Q^*)^r\log Q^*-(Q^*)^r\int_X q(x)^{r+1}\log q(x)d\mu(x)}{\left[\int_X q(x)^{r+1}d\mu(x)\right]^2}$$

whose sign is that of

$$\left[\int_X q(x)^{r+1}\log q(x)d\mu(x)\right]\log Q^*-\int_X q(x)^{r+1}\log q(x)d\mu(x)$$

$$=\int_X\left[\log Q^*-\log q(x)\right]q(x)^{r+1}d\mu(x)>0.$$

6.2 Proof of Proposition 3

If $P_Q\left(X_Q^*\right)>0$, then

$$p_\infty(x)=\lim_{r\to\infty}p_r(x)=\mu\left(X_Q^*\right)^{-1}I_{X_Q^*}(x)\quad\text{and}\quad q(x)=Q\left(X_Q^*\right)/\mu\left(X_Q^*\right)\forall x\in X_Q^*.$$

So

$$p_\infty(x)/q(x) = Q\left(X_Q^*\right)^{-1} I_{X_Q^*}(x).$$

For $\alpha \in (0,1) \cup (1,\infty)$,

$$\begin{aligned}\lim_{r\to\infty} \int_X \left(\frac{p_r(x)}{q(x)}\right)^\alpha q(x)d\mu(x) &= \int_X \left(\frac{p_\infty(x)}{q(x)}\right)^\alpha q(x)d\mu(x)\\ &= \int_{X_Q^*} Q\left(X_Q^*\right)^{-\alpha} q(x)d\mu(x) = Q\left(X_Q^*\right)^{1-\alpha},\end{aligned}$$

and hence

$$\lim_{r\to\infty} D_\alpha\left(P_r \parallel Q\right) = \frac{1}{\alpha-1}\log Q\left(X_Q^*\right)^{1-\alpha} = -\log Q\left(X_Q^*\right).$$

For $\alpha = 1$,

$$\begin{aligned}\lim_{r\to\infty} D_1\left(P_r \parallel Q\right) = D_1\left(P_\infty \parallel Q\right) &= \int_X \log\left(\frac{p_\infty(x)}{q(x)}\right) p_\infty(x)d\mu(x)\\ &= \int_{X_Q^*} \log\left[Q\left(X_Q^*\right)^{-1}\right] d\mu(x) = -\log Q\left(X_Q^*\right).\end{aligned}$$

For $\alpha = \infty$,

$$\lim_{r\to\infty} D_\infty\left(P_r \parallel Q\right) = D_\infty\left(P_\infty \parallel Q\right) = \log \operatorname{ess\,sup}_{x\in Q} \frac{p_\infty(x)}{q(x)} = -\log\left(X_Q^*\right).$$

For $P_Q\left(X_Q^*\right) = 0$ denote $q^* = \text{ess sup}_{x\in Q} q(x)$. Construct the sequence of probability measures Q_ε with p.d.f. $q_\varepsilon(x)$ by first defining

$$q_\varepsilon(x) = c(\varepsilon)\left\{\left[q(x)-\varepsilon\right] I_{X^*_{Q(\varepsilon)}}(x) + q(x) I_{Q-X^*_{Q(\varepsilon)}}\right\}.$$

and then

$$X_Q^*(\varepsilon) = \left\{x : x \in X_Q, q(x) > q^* - \varepsilon\right\},$$

where $c(\varepsilon)$ is the normalizing constant for the probability density q_ε that defines Q_ε; $\lim_{\varepsilon\to 0^-} c(\varepsilon) = 1$. Then

$$\lim_{\varepsilon\to 0} \lim_{r\to\infty} D_\alpha\left(P_r \parallel Q_\varepsilon\right) = 0.$$

6.3 Proof of Proposition 4

Define the measure $d\mu^*(x) = h(x)d^*(x)$. The distribution Q^* has p.d.f. $q^*(x) = q(x)$ with respect to μ^* and the distribution P_r^* has p.d.f. $p_r^*(x) \propto q(x)^{1+r}$ with respect to μ^*. The proofs of Propositions 2 and 3 apply directly with $d\mu^*$ in place of $d\mu$.

References

Bhattacharyya, A. (1943). "On a Measure of Divergence Between Two Statistical Populations Defined by Their Probability Distributions." *Bulletin of the Calcutta Mathematical Society*, 35: 99–109.

DeJong, K. (1975). *An Analysis of the Behaviour of a Class of Genetic Adaptive Systems.* PhD thesis, University of Michigan.

Douc, R., and Moulines, E. (2008). "Limit Theorems for Weighted Samples with Application to Sequential Monte Carlo Methods." *Annals of Statistics*, 36: 2344–2376.

Duan, J., and Fulop, A. (2015). "Density Tempered Marginalized Sequential Monte Carlo Samplers." *Journal of Business and Economic Statistics*, 33: 192–202.

Durham, G., and Geweke, J. (2014). "Adaptive Sequential Posterior Simulators for Massively Parallel Computing Environments." In I. Jeliazkov and D. J. Poirier (eds.), "Bayesian Model Comparison." *Advances in Econometrics*, Vol. 34 (Chapter 1, pp. 1–44). New York: Emerald Group Publishing.

Geweke, J. (1988). "The Secular and Cyclical Behavior of Real GDP in Nineteen OECD Countries, 1957–1983." *Journal of Business and Economic Statistics*, 6: 479–486.

Geweke, J. (1989). "Bayesian Inference in Econometric Models Using Monte Carlo Integration." *Econometrica*, 57: 1317–1340.

Geweke, J. (2016). "Sequentially Adaptive Bayesian Learning for a Nonlinear Model of the Secular and Cyclical Behavior of US Real GDP." *Econometrics*, 4: 1–23.

Geweke, J., and Durham, G. (2019). "Sequentially Adaptive Bayesian Learning for Inference and Optimization." *Journal of Econometrics*, 210: 4–25.

Hammersly, J. M., and Handscomb, D. C. (1964). *Monte Carlo Methods.* London: Methuen.

Hellinger, E. (1909). "Neue Begründung der Theorie quadratischer Formen von unendichvielen Veränderlichen." *Journal für die reine und angewandte Mathematik*, 136: 210–271.

Kagan, A. M. (1963). "On the Theory of Fisher's Amount of Information." *Soviet Mathematics Doklady*, 4: 991–993.

Kloek, T., and van Dijk, H. K. (1978). "Bayesian Estimates of Equation System Parameters: An Application of Integration by Monte Carlo." *Econometrica*, 46: 1–20.

Kullback, S., and Leibler, R. A. (1951). "On Information and Sufficiency." *Annals of Mathematical Statistics*, 22: 79–88.

Liu, J. S., and Chen, R. (1998). "Sequential Monte Carlo Methods for Dynamic Systems." *Journal of the American Statistical Association*, 93: 1032–1044.

Rényi, A. (1961). "On Measures of Information and Entropy." *Proceedings of the Fourth Berkeley Symposium on Mathematics, Statistics and Probability*, 547–561.
Samuelson, P. A. (1939). "Interactions between the Multiplier Analysis and the Principle of Acceleration." *Review of Economics and Statistics*, 21: 75–78.
van Erven, T., and Harremodoës, P. (2014). "Rényi Divergence and Kullback-Leibler Divergence." *IEEE Transactions on Information Theory*, 60: 3797–3820.

PART VI

INFO-METRICS, DATA INTELLIGENCE, AND VISUAL COMPUTING

This part extends the info-metrics framework for solving problems of data intelligence and visual computing. In Chapter 16, Chen tackles the fundamental question of how to evaluate information processing itself. It is one of the basic open problems in data science. The chapter discusses an innovative information-theoretic metric for analyzing the cost-benefit of data intelligence. It provides a new way to understand the data intelligence processes of transforming data to decisions. While the metric was initially proposed for analyzing data intelligence workflows involving machine-centric processes (e.g., statistics and algorithms) and human-centric processes (e.g., visualization and interaction), this chapter presents a set of extended interpretations of the metric by relating it to different cross-disciplinary problems, such as encryption, compression, cognition, language development, and media communication.

In Chapter 17, Feixas and Sbert provide an information-theoretic way to study visual computing. It demonstrates the role of information channels in visual computing. Roughly speaking, a communication channel, known also as an information channel, establishes the shared information between the source (input information) and the receiver (output information). This chapter shows the application of this concept to a number of technical problems in visual computing, including the selection of the best viewpoints of a 3D graphical object, the participation of an image in quasi-homogeneous regions, and the estimation of the complexity of a graphical scene by measuring the amount of information transferred between different parts of the scene.

Overall, the chapters in Part VI demonstrate new ways to evaluate data, information, and the process (i.e., the processing of that information) itself, in a new, innovative way.

16
Cost-Benefit Analysis of Data Intelligence—Its Broader Interpretations

Min Chen

1. Introduction

Data science (or *information science*) is the scientific discipline that studies human and machine processes for transforming data to decisions and/or knowledge. Its main goal is to understand the inner workings of different *data intelligence* processes and to provide a scientific foundation that underpins the design, engineering, and optimization of data intelligence workflows composed of human and machine processes. *Theoretical data science* focuses on the mathematical theories and conceptual models that underpin all aspects of data science and enable abstract modeling of data intelligence workflows. *Applied data science* focuses on the technologies that can be deployed in data intelligence processes and workflows, and used to support the design, engineering, and optimization of data intelligence processes and workflows.

As used here, *data intelligence* is an encompassing term for processes such as statistical inference, computational analysis, data visualization, human–computer interaction, machine learning, business intelligence, collaborative intelligence, simulation, prediction, and decision making. Some of these processes are machine-centric and others are human-centric, while many are integrated processes that capitalize the relative merits of both. Recently, Chen and Golan (2016) proposed an information-theoretic metric for measuring the cost-benefit of a data intelligence workflow or its individual component processes.

Several attempts have been made to evidence, falsify, and exemplify this theoretic proposition. An empirical study was conducted to detect and measure the three quantities of the metric using a visualization process—a type of human-centric data intelligence process (Kijmongkolchai, Abdul-Rahman, and Chen, 2017). The metric was used as the basis for comparing a fully automated machine-learning workflow and a human-assisted machine-learning workflow, as well as for explaining the better models obtained with the latter

Min Chen, *Cost-Benefit Analysis of Data Intelligence—Its Broader Interpretations* In: *Advances in Info-Metrics: Information and Information Processing across Disciplines.* Edited by: Min Chen, J. Michael Dunn, Amos Golan, and Aman Ullah, Oxford University Press (2021).
DOI: 10.1093/oso/9780190636685.003.0016

approach (Tam, Kothari, and Chen, 2017). It was applied to the field of virtual environments, relating different dimensions of virtual reality and mixed reality to the cost-benefit metric and offering explanations about the merits and demerits evidenced in practical applications (Chen, Gaither, John, and McCann, 2019). The concept was also used in a recently proposed ontological framework for reasoning about the relations among symptoms, causes, remedies, and side-effects in the design, development, evaluation, and improvement of data intelligence workflows (Chen, and Ebert, 2019). In addition, a medium-scale elucidation study was conducted to falsify some theories of visualization using 120 arguments in the literature, and the cost-benefit metric survived the scrutiny of this validation effort (Streeb, El-Assady, Keim, and Chen, 2019).

The results of these exercises indicated the explanatory power of this metric in reasoning about the successful and less successful scenarios in several data intelligence workflows, such as visualization, machine learning, and virtual environments, while informing us about some enriched interpretations of this metric in the context of data intelligence as well as some potential interpretations in a broader scope. In this chapter, we report these new insights as an extended elucidation of the cost-benefit metric by Chen and Golan (2016). We are fully aware that confirming these interpretations will require a tremendous amount of effort across several different disciplines. Thus, our aim is to expound a broader set of interpretations of this particular information-theoretic metric and to stimulate new multidisciplinary efforts in advancing our fundamental understanding about data intelligence.

In the remainder of this chapter, we will first briefly describe the cost-benefit metric and the original interpretations that may be derived from this metric in section 2. We will then report two enriched interpretations in relation to encryption and compression in section 3 and to model development in section 4. We will articulate the potential interpretations in relation to perception and cognition in section 5 and to languages and news media in section 6. We will offer our concluding remarks in section 7.

2. The Cost-Benefit Metric for Data Intelligence

2.1 Processes and Transformations

A data intelligence workflow may consist of one or more processes that transform some data to some decisions. Here the term *decision* is a generic placeholder for different types of outcomes that may result from execution of the workflow, such as identifying an object, an event, or a relation; obtaining a fact, a piece of knowledge, or a collection of views; selecting a category, place, time,

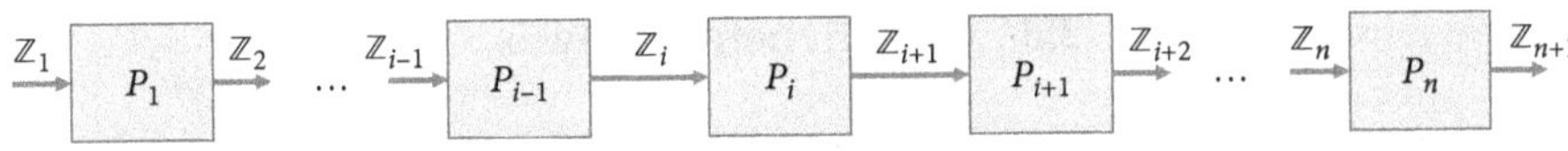

Figure 16.1 A sequentialized representation of a data intelligence workflow.

or an option of any arbitrary type; determining a group, a path, or a course of action; or even unconsciously acquiring a memory, an emotion, or a sense of confidence. The processes in the workflow can be performed by humans, machines, or both jointly. As shown in Figure 16.1, a sequentialized workflow is, in abstraction, a series of processes $P_1, P_2, \ldots, P_i, \ldots, P_n$.

In theory, the steps in Figure 16.1 can be infinitesimally small in time, the resulting changes can be infinitesimally detailed, while the sequence can be innumerably long and the processes can be immeasurably complex. In practice, one can construct a coarse approximation of a workflow for a specific set of tasks. Major iterative steps may be sequentialized and represented by temporally ordered processes for different steps, while minor iterative steps may be combined into a single process. Parallel processes that are difficult to sequentialize, such as voting by a huge number of people, are typically represented by a single macro process.

As shown in Figure 16.1, all of these processes receive input data from a previous state, process the data, and deliver output data to a new state. The input data and output data do not have to have a similar semantic definition or be in the same format. To capture this essence from an information-theoretic perspective, each of these processes in a workflow is referred to as a *transformation*.

2.2 Alphabets and Letters

For each transformation P_i, all possible input data sets to P_i constitute an input *alphabet* $\mathbb{Z}_i$, and an instance of an input is a *letter* of this alphabet. Similarly, all possible output data sets from P_i constitute an output alphabet $\mathbb{Z}_{i+1}$. In the grand scheme of things, we could consider all possible states of the humans and machines involved in a workflow as parts of these input and output alphabets. In practice, we may have to restrict the specification of these alphabets based on those types of data that are explicitly defined for each transformation. Hence, many states, such as human knowledge and operating conditions of machines, are usually treated as external variables that have not been encoded in these alphabets. The presence or absence of these external variables make the cost-benefit analysis interesting as well as necessary.

2.3 Cost-Benefit Analysis

Consider the transformation from alphabet $\mathbb{Z}_i$ to $\mathbb{Z}_{i+1}$ at process P_i in Figure 16.1. Let $\mathcal{H}(\mathbb{Z}_i)$ be the Shannon entropy[1] (Shannon, 1948) of alphabet $\mathbb{Z}_i$ and let $\mathcal{H}(\mathbb{Z}_{i+1})$ be that of $\mathbb{Z}_{i+1}$. The entropic difference between the two alphabets, $\mathcal{H}(\mathbb{Z}_i) - \mathcal{H}(\mathbb{Z}_{i+1})$, is referred to as *alphabet compression*. Chen and Golan observed a general trend of alphabet compression in most (if not all) data intelligence workflows, since the decision alphabet is usually much smaller than the original data alphabet in terms of Shannon entropy.

On the other hand, the reduction of Shannon entropy is usually accompanied by the *potential distortion* that may be caused by the transformation. Instead of measuring the errors of $\mathbb{Z}_{i+1}$ based on a third party and likely-subjective metric, we can consider a reconstruction of $\mathbb{Z}_i$ from $\mathbb{Z}_{i+1}$. If there are external variables (e.g., humans' knowledge) about the data, the context, or the previous transformations, it is possible to have a better reconstruction with the access to such external variables than without. In Chen and Golan (2016), this is presented as one of the main reasons explaining why visualization is useful. The potential distortion is measured by the Kullback–Leibler divergence[2] (Kullback and Leibler, 1951), $\mathcal{D}_{KL}(\mathbb{Z}'_i \| \mathbb{Z}_i)$, where $\mathbb{Z}'_i$ is reconstructed from $\mathbb{Z}_{i+1}$. Furthermore the transformation and reconstruction need to be balanced by the *Cost* involved, which may include the cost of computational and human resources, cognitive load, time required to perform the transformation and reconstruction, adversary cost due to errors, and so on. Together the trade-off of these three measures is expressed in Eq. (16.1):

$$\frac{\text{Benefit}}{\text{Cost}} = \frac{\text{Alphabet Compression} - \text{Potential Distortion}}{\text{Cost}} = \frac{\mathcal{H}(\mathbb{Z}_i) - \mathcal{H}(\mathbb{Z}_{i+1}) - \mathcal{D}_{KL}\left(\mathbb{Z}'_i \middle\| \mathbb{Z}_i\right)}{\text{Cost}} \qquad (16.1)$$

The *Benefit* is measured in the unit of *bit*. While the most generic cost measure is energy, it can be approximated in practice using a monetary measurement or a time measurement. This metric suggests several interpretations about data intelligence workflows, all of which can be supported by practical evidence,

[1] Given an alphabet $\mathbb{Z} = \{z_1, z_2, \ldots, z_n\}$ and a probability mass function $P(z)$ for the letters, with the binary logarithm, the Shannon entropy of the alphabet is defined as $\mathcal{H}(\mathbb{Z}) = -\sum_{i=1}^{n} P(z_i) \log_2 P(z_i)$ (Shannon, 1948).

[2] Given two n-letter alphabets $\mathbb{X}$ and $\mathbb{Y}$, and their respective probability mass functions $P(x)$ and $Q(y)$, the Kullback–Leibler divergence of the two alphabets is defined as $\mathcal{D}_{KL}\left(\mathbb{X} \middle\| \mathbb{Y}\right) = \sum_{i=1}^{n} P(x_i) \log_2 P(x_i)/Q(y_i)$ (Kullback and Leibler, 1951).

though some may not be instinctively obvious. These interpretations include the following:

(a) Losing information (i.e., in terms of Shannon entropy) is a ubiquitous phenomenon in data intelligence processes and has a positive impact on the *Benefit*. For instance, numerical regression can be used to transform an alphabet of n 2D data points in $\mathbb{R}^2$ to an alphabet of k coefficients in $\mathbb{R}$ for a polynomial, and it typically exhibits a positive alphabet compression. A sorting algorithm transforms an alphabet of n data values to an alphabet of n ordered values, and the former has a higher entropy than the latter. A user interaction for selecting a radio button out of k choices transforms an alphabet of $\log_2 k$ bits to an alphabet of 0 bits.

(b) Traditionally, losing information has a negative connotation. This is a one-sided generalization inferred from some observed causal relations that missing useful information leads to difficulties in decision making. This is in fact a paradoxical observation because the assessment of "usefulness" in the cause depends on the assessment of "difficulties" in the effect. The cost-benefit metric in Eq. (16.1) resolves this self-contradiction by treating *alphabet compression* as a positive quality and balancing it with *potential distortion* as a negative quality.

(c) When a transformation P_i is a many-to-one mapping from its input alphabet to its output alphabet, the measure of alphabet compression is expected to be positive. At the same time, the corresponding reconstruction is a one-to-many mapping, with which potential distortion is also expected. Although a typical reconstruction would be based on the maximum entropy principle, this may not be an optimal reconstruction if external variables are present and accessible by the reconstruction (see also (d)). Assume that the letters in the output alphabet $\mathbb{Z}_{i+1}$ contribute equally to the subsequent transformations $P_{i+1}, \dots, P_n$ (see also (f)). With the same amount of alphabet compression, the more faithful the reconstruction, the better the transformation (see also (e)).

(d) Humans' soft knowledge can be used to reduce the potential distortion. As discussed earlier, for any human-centric process, if the humans' soft knowledge is not encoded in the input alphabet of the process, such knowledge will be treated as external variables. For example, consider the process of recognizing a car in an image. When a large portion of the car is occluded by other objects in the scene, most people can perform this task much better than any automated computer vision algorithm that represents the current state of the art. This is attributed to the humans' soft knowledge about various visual features of cars and the context suggested by other objects in the scene. In some decision scenarios, such knowledge may introduce biases into the process. We will discuss this topic further in sections 3–6.

(e) Modifying a transformation P_i may change its *alphabet compression*, *potential distortion*, and *cost*, and may also change the three measures in subsequent transformations $P_{i+1}, \dots, P_n$. Hence, optimizing the cost-benefit of a data intelligence workflow is fundamentally a global optimization. However, while it is necessary to maintain a holistic view about the whole workflow, it is both desirable and practical to improve a workflow through controlled localized optimizations in a manner similar to the optimization of manufacturing and businesses processes.

(f) The cost-benefit of a data intelligence workflow is task-dependent. Such tasks are implicitly encoded in the output alphabets of some transformations, usually toward the end of the sequence $P_1, P_2, \dots, P_n$. For example, if the task of P_n is to select a final decision from the options defined in its output alphabet $\mathbb{Z}_{n+1}$, $\mathbb{Z}_{n+1}$ thus encodes the essence of the task. Some potential distortion at an early transformation P_i $(i < n)$ may affect this selection, but others may not.

3. Relating the Metric to Encryption and Compression

Kijmongkolchai et al. (2017) reported an empirical study to detect and measure the human's soft knowledge used in a visualization process. Their study was designed to evaluate the hypothesis that such knowledge can enhance the cost-benefit ratio of a visualization process by reducing the potential distortion. It focused on the impact of three classes of soft knowledge: (i) knowledge about application contexts, (ii) knowledge about the patterns to be observed, and (iii) knowledge about statistical measures. In each trial, eight time series plots were presented to a participant who was asked to choose a correct time series. Three criteria were used to define the correctness:

1. Matching a predefined application context, which may be the electrocardiogram (ECG), stock price, and weather temperature.
2. Matching a specific visual pattern, which may be a global pattern of the time series (e.g., "slowly trending down," "wandering base line," "anomalous calm," "ventricular tachycardia") or a local pattern within the time series (e.g., "sharp rise," "January effect," "missing a section of data," "winter in Alaska").
3. Matching a statistical measure, which may be a given minimum, mean, maximum, or standard deviation of the time series.

Among the eight time-series plots, the seven distractors (i.e., incorrect or partly correct answers) match zero, one, or two criteria. Before each trial, participants were shown a corresponding newspaper or magazine article where the hints of the application context and specific visual pattern are given. The article was removed from the trial, for which the participants had to recall the hints featured in the article as soft knowledge. The corresponding statistical measure was explicitly displayed during the trial, as it would be unreasonable to demand such memorization. The study showed that the participants made 68.3 percent correct decisions among all responses by successfully utilizing all three classes of knowledge together. In comparison with the 12.5 percent chance, the positive impact of knowledge in reducing potential distortion was evident. The work also proposed a mapping from *Accuracy* and *Response Time* collected in the empirical study to *Benefit* and *Cost* in the cost-benefit metric.

During the design of this study, the authors noticed that the eight time series plots presented in each trial do not exhibit the true probability distribution that would in itself lead participants to a correct decision. Instead, the data alphabet, $\mathbb{Z}_1$, seemed to be intentionally misleading with the maximum entropy of 3 bits (i.e., 12.5 percent chance). Meanwhile, as the corresponding newspaper or magazine article and the statistical measure indicated a correct answer, there was a hidden "truth distribution." This suggests that an extended interpretation is needed to describe the data intelligence workflow in this multiple-choice setup used by numerous empirical studies.

As shown in Figure 16.2a, it is necessary to introduce a *truth alphabet* $\mathbb{Z}_0$ in order to encode the hidden truth distribution. The data sets presented to the participants during the study are from a pretended alphabet $\mathbb{Z}_1$. The desired reconstruction from the decisions made by the participants should be related to $\mathbb{Z}_0$ but not $\mathbb{Z}_1$. In other words, the pretended alphabet $\mathbb{Z}_1$ bears some resemblance to an alphabet of ciphertext in cryptography.

This extended interpretation can also be applied to many real-world data intelligence workflows, as shown in Figure 16.2b. Although alphabet $\mathbb{Z}_1$ in many workflows may not be intentionally misleading, they typically do feature deviation from the truth distribution. At the same time, in some workflows, such as criminal investigations, alphabet $\mathbb{Z}_1$ likely features some intended distortion of the truth distribution in a manner similar to data encryption. Hence, the data intelligence workflow should really be about the real-world alphabet $\mathbb{Z}_0$, which the decision alphabet should ultimately reflect. In addition, the sampled alphabet $\mathbb{Z}_1$ is expected to have much lower entropy than the real-world entropy $\mathbb{Z}_0$. The data capture process thus exhibits some characteristics of both encryption and lossy compression. A data compression method is said to be lossy if the compression process causes information loss and the corresponding decompression process can no longer reconstruct the original data without any distortion.

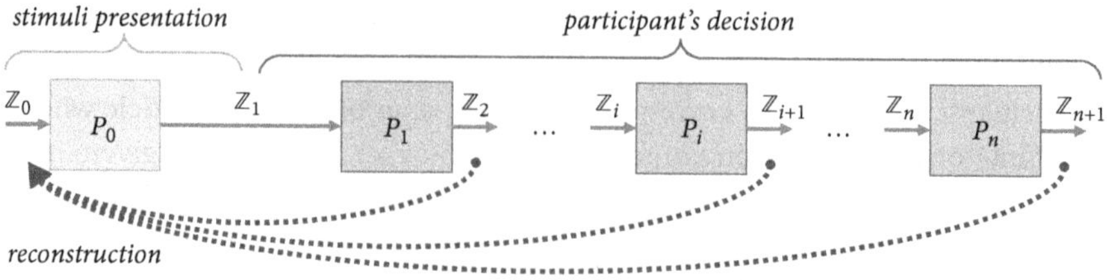

(a) The truth alphabet $\mathbb{Z}_0$ and the pretended alphabet $\mathbb{Z}_1$ in an empirical study.

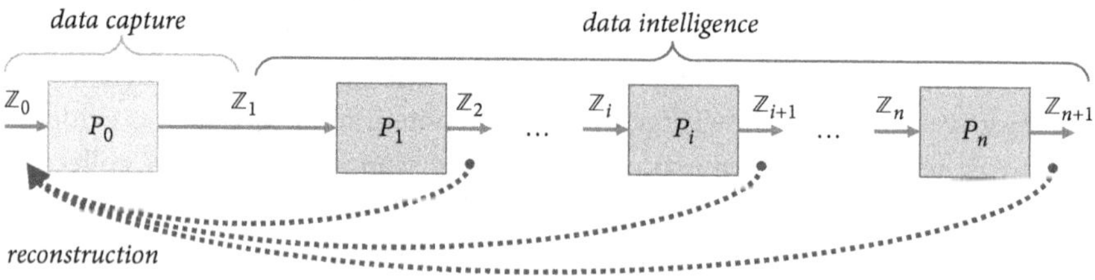

(b) The real-world alphabet $\mathbb{Z}_0$ and the captured data alphabet $\mathbb{Z}_1$ in a data intelligence workflow.

Figure 16.2 In an extended interpretation of the cost-benefit metric, a truth alphabet or real-world alphabet $\mathbb{Z}_0$ is introduced, and the quality of the reconstruction ultimately depends on alphabet $\mathbb{Z}_0$ rather than $\mathbb{Z}_1$.

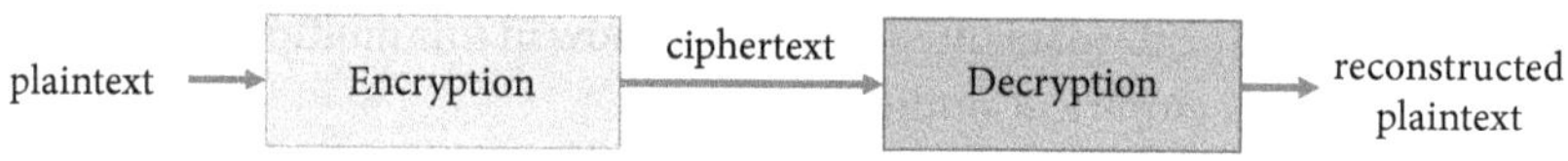

(a) The basic workflow of data encryption and decryption.

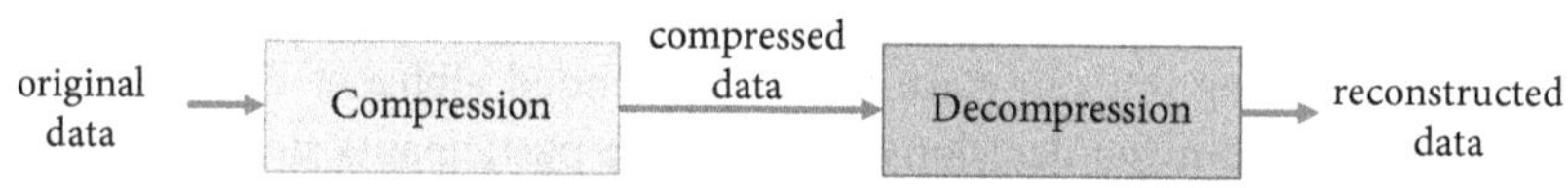

(b) The basic workflow of data compression and decompression.

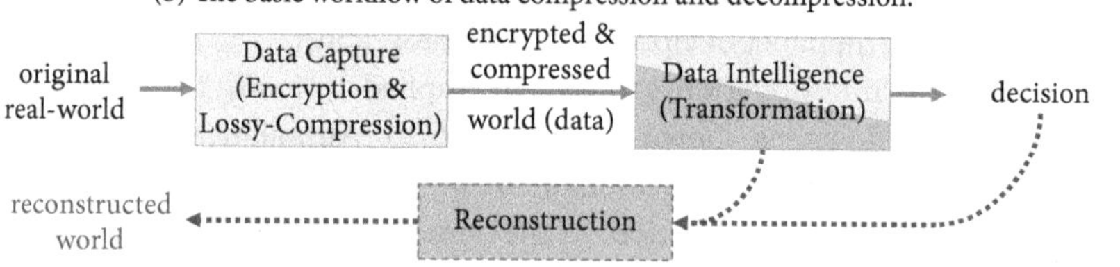

(c) The further abstraction of the data intelligence workflow in Figure 2.

Figure 16.3 Juxtaposing the workflows of data encryption, data compression, and data intelligence.

This naturally suggests that data intelligence is conceptually related to data encryption and data compression, both of which are underpinned by information theory. Figure 16.3 juxtaposes three workflows in their basic forms. The data capture process transforms a real-world alphabet to an encrypted and compressed world that we refer to as data or sampled data. The data intelligence process transforms the data alphabet or sampled data alphabet to a decision alphabet, facilitating further compression in terms of Shannon entropy. Unlike data encryption and data compression, the reconstruction process is not explicitly defined in a data intelligence workflow, possibly because machine-centric processes seldom contain, or are accompanied by, a reverse mapping from an output alphabet to an input alphabet, while humans rarely make a conscious effort to reconstruct an input alphabet. Unconsciously, humans perform this reconstruction all the time, as we will discuss in section 5.

The extended interpretation of the cost-benefit metric instigates that the quality of the decisions made by the data intelligence processing block should be measured by the potential distortion in the reconstruction from the decision alphabet to the original real-world alphabet. Should this reconstruction be defined explicitly, as illustrated in Figure 16.3c, we would be able to draw a parallel among the three processing blocks for *Encryption*, *Compression*, and *Data Capture*; and another parallel among the three blocks for *Decryption*, *Decompression*, and *Reconstruction*. Meanwhile, the characterization of the *Data Intelligence* block is multifaceted. If this is entirely an automated data intelligence workflow, it exhibits the characteristics of *Compression* as the transformation from a data alphabet to a decision alphabet can be seen as a complex form of lossy compression. If this workflow involves humans who likely perform some forms of reconstruction at some stages, the block also exhibits the characteristics of *Decryption* and *Decompression* in addition.

4. Relating the Metric to Model Development

To accompany the proposal of the cost-benefit metric, Chen and Golan (2016) also categorized the tasks of data analysis and data visualization into four levels according to the size of the search space for a decision. The four levels are *Dissemination* of known findings, *Observation* of data, *Analysis* of structures and relations, and *Development* of models. Using the big O notation in computer science, the search spaces of these four levels are characterized by *constant* O(1), *linear* O(n), *polynomial* O(n^k), and *nondeterministic polynomial* (*NP*) (e.g., O(k^n) or O($n!$)), respectively, where n is the number of component alphabets in the largest component alphabet $\mathbb{Z}_1$ and k is the number of letters in the largest component alphabet.

"Model" is an overloaded term. Here we use this term strictly for referring to an executable function in the form of $F : \mathbb{Z}_{in} \rightarrow \mathbb{Z}_{out}$. Hence, a software program is a model, a machine-learned algorithm (e.g., a decision tree and a neural network) is a model, a human's heuristic function is a model, and so on. When a data intelligence workflow in Figure 16.1 is used to obtain a model, the final alphabet $\mathbb{Z}_{n+1}$ consists of all possible models that meet a set of conditions in a specific application context. Although the number of letters in a typical model alphabet seems to be extraordinarily large, many letters do not meet the predefined conditions in the specific application context. Hence, the actual entropy of $\mathbb{Z}_{n+1}$ would be much lower than the maximal amount of entropy derived by enumerating all combinations of the model components. For example, the number of all possible programs with n or fewer lines of code, the number of all possible neural networks with m or fewer neurons, or the number of all possible Bayesian networks with l or fewer nodes and probability values rounded off to k decimal digits would be intractable, but there will only be proportionally a much smaller subset of programs, neural networks, or Bayesian networks that meet the predefined requirements for such a model in a specific application context. However, the main challenge in model development is that the strategy for finding an acceptable model while filtering out the extraordinarily large number of unacceptable models is often not well defined or ineffective. For example, one often refers to programming as "art" and parameter tuning as "black art," metaphorically reflecting the lack of a well-defined or effective strategy in dealing with the NP search space.

Consider an alphabet $\mathbb{M}_{all}$ that contains all possible functions we can create or use. Whether or not we like it, some of these functions are being created using *machine learning* (*ML*). This is not because ML can develop better algorithms or more reliable systems than trained computer scientists and software engineers can. ML is merely a model-developmental tool that helps us write an approximate software function, for which we do not quite know the exact algorithm or would take an unaffordable amount of time to figure it out. In the systems deployed in practical environments, only a small number of components are machine-learned algorithms. Nevertheless, this model-developmental tool is becoming more and more powerful and useful because of the increasing availability of training data and high-performance computing for optimization. With such a tool, we can explore new areas in the *space of functions*, $\mathbb{M}_{ML} \subset \mathbb{M}_{all}$, which is programmable on a stored program computer and where conventional algorithms are not yet found or effective.

In theoretical computer science, a programming language is said to be *Turing complete* if it can describe any single-taped Turing machine and vice versa. We can consider an underlying platform (or framework) used by ML (e.g., neural networks, decision trees, Bayesian networks, support vector machines, genetic algorithms) as a programming language. In such a language, there is a very limited set of constructs (e.g., node, edge, weight). The space of $\mathbb{M}_{ML}$ is a union of all underlying platforms. An ML workflow typically deals with only one such platform or a very small subset of platforms, which determines an alphabet

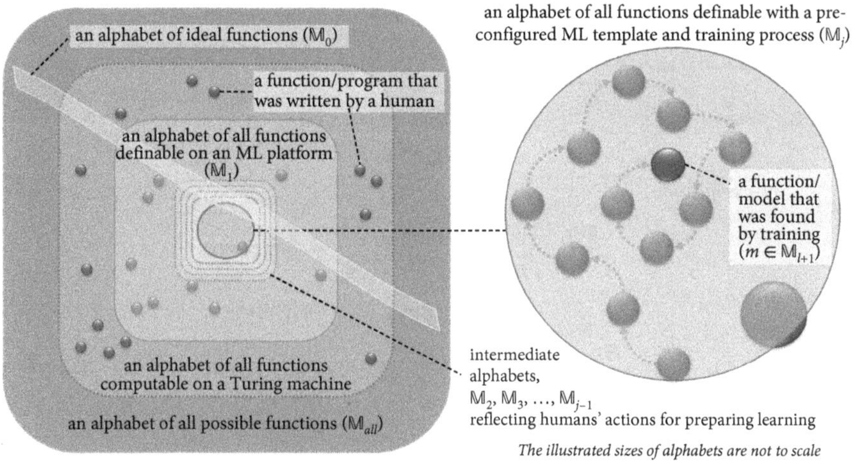

Figure 16.4 A machine learning (ML) workflow searches for a model in the alphabet $\mathbb{M}_1$ determined by an underlying platform. Human- and machine-centric processes work together to enable the search by gradually reducing the size of the search space, which can be measured by the entropy of the corresponding alphabet. The searching is not assured to find an ideal model in $\mathbb{M}_0$, but can examine and test automatically numerous candidature models in a small space. The small search space $\mathbb{M}_j$ is pre-defined by humans, while the search path is determined by the training data and some control parameters.

$\mathbb{M}_1$ (Figure 16.4), that is, the space of all functions definable on the chosen platform(s).

For any practical ML applications, humans usually do the clever part of the programming by defining a *template*, such as determining the candidature variables for a decision tree, the order of nodes in a neural network, or the possible connectivity of a Bayesian network. The ML tool then performs the tedious and repetitive part of the programming for fine-tuning the template using training data to make a function. It is known that most commonly used ML platforms, such as forward neural networks, decision trees, random forests, and static Bayesian networks are not Turing complete. Some others are Turing complete (e.g., recurrent neural networks and dynamic Bayesian networks), but their demand for training data is usually exorbitant. For any ML model, once a template is determined by humans, the *parameter space* that the ML tool can explore is certainly not Turing complete. This implies that the "intelligence" or "creativity" of the ML tool is restricted to the tedious and repetitive part of the programming for fine-tuning the template.

Figure 16.5 illustrates a typical workflow for supervised learning, and its relationship with a conventional data intelligence workflow where a learned model

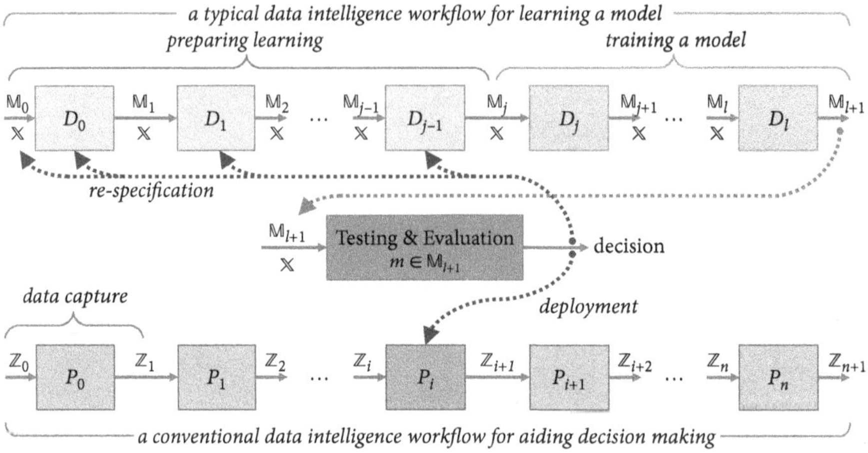

Figure 16.5 The relationship between a typical machine learning workflow (e.g., for learning a decision tree, a neural network, a Markov chain model, etc.) and a conventional data intelligence workflow (e.g., for aiding anomaly detection, market analysis, document analysis, etc.).

will be deployed as an intermediate step P_i. In the conventional data intelligence workflow, each of the other processes, that is, P_w $(w \neq i)$, can be either a machine- or human-centric process. For example, P_0 may be a process for capturing data from a real-world environment (e.g., taking photographs, measuring children's heights, conducting a market survey, and so on). $P_1, \ldots, P_{i-1}$ may be a set of processes for initial observations, data cleaning, statistical analysis, and feature computation. P_i may be an automated process for categorizing the captured data into different classes. $P_{i+1}, \ldots, P_n$ may be a set of processes for combining the classified data with other data (e.g., historical notes, free-text remarks), discussing options of the decision to be made, and finalizing a decision by voting.

Let $\mathbb{X}$ be an alphabet of annotated data for training and testing. It is commonly stored as pairs of data sets and labels corresponding to the input and output alphabets of P_i, respectively. In other words, each $x \in \mathbb{X}$ is defined as a pair (α, β), such that $\alpha \in \mathbb{Z}_i$ and $\beta \in \mathbb{Z}_{i+1}$. It is helpful to note that we do not prescribe that the data sets used for ML must be the raw data captured from the real world, that is, $\mathbb{Z}_1$. This is because (i) a machine-learned model is rarely the sole function in a data intelligence workflow and there may be other transformations before P_i for preprocessing, analyzing, and visualizing data; (ii) captured data sets are often required cleaning before they can be used as part of the data alphabet $\mathbb{X}$ for training and testing and as the input alphabet $\mathbb{Z}_i$ to the machine-learned model P_i; and (iii) it is common to extract features from captured data sets using manually constructed programs and to use such feature data in $\mathbb{X}$ and $\mathbb{Z}_i$ instead of (or in addition to) the captured data sets.

Unlike a conventional data intelligence workflow that transforms the data alphabets $\mathbb{Z}_i$ (as shown in the lower part of Figure 16.5), the ML workflow (shown in the upper part) transforms the model alphabets $\mathbb{M}_j, j = 0, 1, \ldots, l+1$, while making use of different letters in the same data alphabet $\mathbb{X}$. The ML workflow consists of two major stages: *preparing learning* and *model learning.* The stage of preparing learning, which is the clever programming part and is illustrated in Figure 16.5 as $D_0, D_1, \ldots, D_{j-1}$, consists primarily of human-centric processes for selecting an underlying platform, constructing a template, specifying initial conditions of the template, and setting platform-specific control parameters that influence the performance of the processes in the model learning stage.

The stage of model learning is the tedious and repetitive part of the workflow and is illustrated as $D_j, \ldots, D_l$ in Figure16. 5. It consists primarily of machine-centric processes that are preprogrammed to construct a model (or a set of models) step by step with some platforms (e.g., decision trees, random forest, and Markov chain models) or refining model parameters iteratively with other platforms (e.g., regression, neural networks, and Bayesian networks). The parameter refinement approach is illustrated on the right of Figure 16.4, where the model learning stage starts with an initial set of parameters (e.g., the coefficients of a regression model, connectivity of a neural network, probability values in a Bayesian network). The processes, $D_j, D_{j+1}, \ldots, D_l$, at this stage modify these model parameters according to the training data sets in $\mathbb{X}$ encountered as well as the predefined platform-specific control parameters. As a model is defined by a template and a set of model parameters, the progressive changes actualized by these processes steer the search for an optimal model (or a set of optimal models) in the alphabet $\mathbb{M}_j$. This mostly automated stage is often mistaken or misrepresented as the entire data intelligence workflow for machine learning, largely because the more credits are given to the machine-centric processes, the better artificial intelligence appears to be.

After a model (or a set of models) is learned, it is passed onto an independent process for testing and evaluation under the supervision of humans as shown in the center of Figure 16.5, When a learned model is considered to be unsatisfactory, the human effort is redirected to the processes in the stage of preparing learning, where a template may be modified and control parameters may be adjusted. In some cases, the training and testing alphabet $\mathbb{X}$ is considered to be problematic, which may lead to further activities in data capture, data cleaning, feature specification, algorithm designs for feature extraction, and so on. The processes for modifying $\mathbb{X}$ are normally a subset of the processes proceeding P_i, that is, $P_0, P_1, \ldots, P_{i-1}$ in the conventional data intelligence process.

Information-theoretically, we may follow the discussions in section 3 by considering the initial alphabet $\mathbb{M}_0$ as a collection of ideal models that can suitably be deployed as P_i for transforming $\mathbb{Z}_i$ to $\mathbb{Z}_{i+1}$. We can consider the first process D_0 in the data intelligence workflow for learning a model as an encryption by selecting a specific platform; since $\mathbb{M}_1$ consists of all models that can be constructed under this platform, it not only conceals the truth alphabet $\mathbb{M}_0$ but also seldom leads to the discovery of an ideal model in $\mathbb{M}_0$. Meanwhile, selecting a platform facilitates a significant amount of alphabet compression from the space of all possible functions $\mathbb{M}_{all}$. As illustrated in Figure 16.4, when such a compression is lossy, $\mathbb{M}_1$ cannot guarantee that any ideal model in $\mathbb{M}_0$ is recoverable. Further alphabet compression, in a massive amount of entropy, is delivered by the human-centric processes $D_1, \ldots, D_{j-1}$. The candidature models that are contained in $\mathbb{M}_j$ are substantially constrained by the particular template and the control parameters that determine the search strategy and effort.

During the model learning stage, the training data alphabet $\mathbb{X}$ is used to reduce the entropy of D_{j-1} while minimizing the potential distortion. With construction-based platforms, alphabet compression from $\mathbb{M}_j$ to $\mathbb{M}_{l+1}$ is achieved gradually in each construction step (e.g., determining a node in a decision tree or a Markov chain model). With parameter-refinement platforms, the entropy of interim alphabets $\mathbb{M}_{j+1}, \ldots, \mathbb{M}_l$ may reduce gradually if the training manages to converge, but may remain more or less at the same level as $\mathbb{M}_j$ if it fails to converge. Whatever the case, the alphabet $\mathbb{M}_{l+1}$ resulting from the model learning stage typically contains only one or a few models.

In general, the data intelligence workflow for learning a model exhibits the same trend of alphabet compression. The stage of preparing learning consists of knowledge-driven processes for reducing the Shannon entropy of the model alphabet, while human knowledge and heuristics are utilized to minimize the potential distortion. The model learning stage is largely data-driven, where the training algorithm pursues the goal of alphabet compression while various metrics (e.g., the impurity metric in decision tree construction or fitness function in genetic algorithms) are used to minimize the potential distortion. While the automation in the model learning stage is highly desirable for cost reduction, it appears to have difficulties replacing the human-centric processes in the preparing learning stage. Hence, a trade-off has been commonly made between the costs of human resources (e.g., constructing a template in D_1) and the computational cost for searching in a significantly larger alphabet (e.g., $\mathbb{M}_1$). The independent testing and evaluation process, together with the feedback loops to $\mathbb{X}, D_0, D_1, \ldots, D_{j-1}$, provides ways to reduce potential distortion while incurring further cost.

Many in the field of ML also encounter difficulties in obtaining a suitable training and testing alphabet $\mathbb{X}$ that would be a representative sample of the relations between $\mathbb{Z}_i$ and $\mathbb{Z}_{i+1}$. In practice, alphabet $\mathbb{X}$ is often too sparse or has a skewed probability distribution. Some in ML thus introduce human knowledge into the stage of model learning to alleviate the problem. Tam et al. (2017) investigated two practical case studies where human knowledge helped derive better models than fully automated processes at the model learning stage. They made the computational processes as the "observers" and the humans' decisions as the information received by the "observers." By measuring the amount of uncertainty removed due to such information, they estimated the amount of human knowledge available to the stage of model learning.

They considered human knowledges in two categories: *soft alphabets* and *soft models*. Soft alphabets are referred to variables that have not been captured in the data—for example, in their case studies, the knowledge about which facial feature is more indicative about a type of emotion or which feature extraction algorithms are more reliable. Soft models are referred to human heuristic processes for making some decisions, for which the ML workflow does not yet have an effective machine-centric process. These soft models do not have predefined answers but respond dynamically to inputs as a function $F : \mathbb{Z}_{in} \rightarrow \mathbb{Z}_{out}$. For example, in their case studies, (i) given a facial photo (input), a human imagines how the person would smile (output); (ii) given a video featuring a facial expression (input), a human judges if the expression is genuine or unnatural (output); (iii) given a set of points on an axis (input), a human decides how to divide the axis into two or a few sections based on the grouping patterns of the points (output); and (iv) given a section of an axis with data points of different classes, a human predicts if the entanglement can be resolved using another unused axis. They discovered that in all of their case studies, the amount of human knowledge (measured in bits) available to the stage of model learning is much more than the information contained in the training data (also measured in bits). Their investigation confirmed that human knowledge can be used to improve the cost-benefit of the part of the ML workflow that is traditionally very much machine-centric.

Recently, Sacha et al. developed an ontology that mapped out all major processes in ML (Sacha, Kraus, Keim, and Chen, 2019). Built on a detailed review of the relevant literature, this ontology confirms that there are a large number of human-centric processes in ML workflows. While recognizing humans' role in ML does not in any way undermine the necessity for advancing ML as a technology, scientifically such recognition can lead to better understanding about what soft alphabets are not in the training data and what soft models are not available to the automated processes. Practically, it can also stimulate the development of new technical tools for enabling humans to impart their knowledge in ML workflows more effectively and efficiently.

Chen et al. (2019) employed the cost-benefit metric to analyze a wide range of applications of virtual reality and virtual environments, such as theater-based education systems, real-time mixed-reality systems, "big data" visualization systems, and virtual reality training systems. In particular, they examined applications in medicine and sports, where humans' soft models are being learned through virtual reality training systems. In these applications, the primary reason for using virtual reality is the lack of access to the required reality $\mathbb{R}$. For example, it would be inappropriate to train certain medical procedures on real patients. So the training and testing alphabets $\mathbb{X}$ is simulated using virtual reality. While the performance of the ideal models in $\mathbb{M}_0$ can easily be anticipated and described, we have very limited knowledge about the configurations and mechanisms of the models under training (i.e., $m \in \mathbb{M}_1, \mathbb{M}_2, \ldots, \mathbb{M}_j, \ldots, \mathbb{M}_l, \mathbb{M}_{l+1}$). One may prepare a participant in the training with many instructions at the first stage, and the participant may train against $\mathbb{X}$ in numerous iterations in the second stage. However, with limited understanding of the inner workings of the models, the cost-benefit of this form of model development depends hugely on the trade-off of the potential distortion in relating virtual reality $\mathbb{X}$ back to the reality $\mathbb{R}$ and the costs of producing $\mathbb{X}$.

Chen et al. (2019) used information-theoretic analysis to confirm the cost-benefit of this form of human model development. They drew from evidence in cognitive science to articulate that human-centric models for motor coordination are more complex than most, if not all, current machine-centric models. They drew evidence from practical applications to articulate the usefulness and effectiveness of training human models using virtual reality. They pointed out the need to use conventional data intelligence workflow to help understand the inner workings of models under training and to use such understanding to optimize the design of $\mathbb{X}$.

5. Relating the Metric to Perception and Cognition

The human sensory and cognitive system is an intelligent system that processes a huge amount of data every second. For example, it is estimated that our two eyes have some 260 million photoreceptor cells that receive lights as input data. If the light that arrives at each cell is a variable, there are some 260 million input variables. Most televisions and computer displays were designed with an assumption that human eyes can receive up to 50–90 different images per second, that is, about 11–20 milliseconds per image. Some perception experiments showed that human eyes can recognize an image after seeing it for 13 milliseconds. So within a second, the human visual system can potentially process more than 13 billion pieces of data $\left(13 \times 10^9 = 50\ \textit{images per second} \times 260 \times 10^6\ \textit{input variables}\right)$.

Cognitive scientists have already discovered a number of mechanisms that enable the human visual system to perform alphabet compression almost effortlessly. One such mechanism is **selective attention**, which enables humans to distribute limited cognitive resources to different visual signals nonuniformly according to their anticipated importance. In most situations, it works cost-beneficially. For example, when we drive, we do not pay equal attention to all visual signals appearing in various windows and mirrors. We focus mostly on the visual signals in the front windscreen, with some peripheral attention to those signals in the front side windows as well as the rear-view and wing mirrors. For those observed signals, we concentrate on those related to road conditions and potential hazards. At any moment, a huge amount of visual signals available to our eyes do not receive much attention. There is thus a nontrivial amount of potential distortion in reconstructing the actual scene. We are unlikely to notice the color and style of every pedestrian's clothing. Such omission is referred to as *inattentional blindness*. Occasionally, such omission may inattentionally miss some critical signals, causing an accident. However, on balance, selective attention is cost-beneficial as it enables us to keep the cognitive load (i.e., cost) low and to maintain sufficient processing capability for the new visual signals arriving continuously.

Figure 16.6a shows a schematic representation of a family of perceptual and cognitive workflows, where we generalize the terms *encoding* and *decoding* from their narrow interpretation of converting the representations of data from one form to another. Here, "encoding" is a transformation from an input alphabet $\mathbb{Z}_a$ to an output alphabet $\mathbb{Z}_b$, where $\mathbb{Z}_b$ can be an alternative representation of $\mathbb{Z}_a$ as well as a derived alphabet with very different semantics. It includes not only traditional interpretations, such as encoding, compression, and encryption, but also many broad interpretations, such as feature extraction, statistical inference, data visualization, model developments, and some data intelligence processes in the human mind, as we are discussing in this section. On the other hand, "decoding" is an inverse transformation, explicitly or implicitly defined, for an attempt to reconstruct alphabet $\mathbb{Z}_a$ from $\mathbb{Z}_b$. As the prefect reconstruction is not guaranteed, the decoding results in $\mathbb{Z}'_a$ which has the same letters as $\mathbb{Z}_a$ but a different probability distribution.

As illustrated in Figure 16.6b, *selective attention* is an instance of encoding. When a person focus his/her attention on the red object, for example, the process allows more signals (e.g., details of geometric and textual features) to be forwarded to the subsequent cognitive processes. Meanwhile, other objects in the scene receive less attention, and the process forwards less or none of their signals. Selective attention clearly exhibits alphabet compression and inevitably may cause potential distortion in an inverse transformation. For example, should the person close his/her eyes and try to imagine the scene of objects, some objects may be incorrectly reconstructed or not reconstructed at all. Often the person would use his/her knowledge and previous experience about the scene and various objects to fill in some missing signals during the reconstruction.

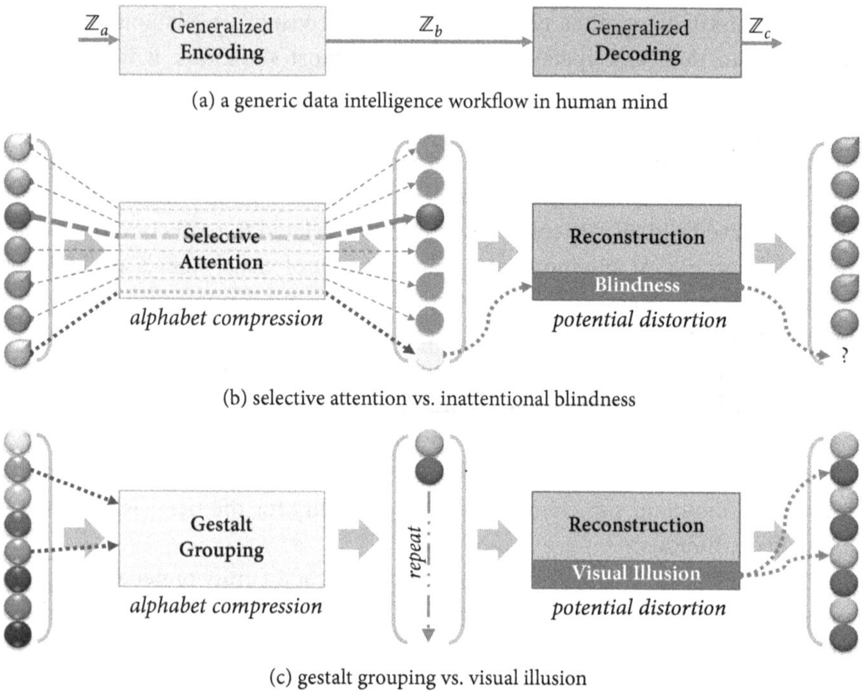

Figure 16.6 *Selective attention* and *gestalt grouping* are two major phenomena in human perception. Information-theoretically, they feature significant alphabet compression and are cost-beneficial, but may occasionally cause potential distortion in the forms of *inattentional blindness* and *visual illusion*.

Inattentional blindness actually refers to a scenario where a specific object in the scene or some features of the object that "should have been" correctly reconstructed happens to be incorrectly reconstructed or not reconstructed at all. As there are numerous objects and features that are not correctly reconstructed, the criteria for "should have been" depend on the objectives of the attentional effort. In many cognitive experiments for evidencing inattentional blindness, participants' attentional effort is often directed to some objectives (e.g., counting the number of ball passing actions) that do not feature the criteria for "should have been" (e.g., spotting a person in an unusual costume). Hence, selective attention is in general cost-beneficial.

Gestalt grouping is another mechanism of alphabet compression. The human visual system intrinsically groups different visual signals into patterns that are usually associated with some concepts, such as shapes, objects, phenomena, and events. This enables humans to remember and think about what is being observed with a smaller alphabet. Although the possible number of letters in the output alphabet after gestalt grouping is usually not a small number, it is a drop in the ocean compared with the number of letters in the input alphabet that represents all possible variations of visual signals. Occasionally, gestalt grouping may lead to visual illusions, where two or more processes for different types of gestalt grouping produce conflicting output alphabets. In general, gestalt grouping functions correctly most of the time, second by second, day by day, and year by year. In comparison, visual illusions are rare events, and pictures and animations of visual illusions were typically handcrafted by experts.

Figure 16c shows an instance of gestalt grouping. Using a one-dimensional version of the chess board illustration as the example input letter, the gestalt grouping process transforms the input to a pattern representation of repeated light-and-dark objects. Such a representation will require less cognitive load to remember and process than the original pattern of eight objects. However, when a person attempts to reconstruct the original pattern, for example, in order to answer the question "Do objects 2 and 5 have the same shade of grey?" a reconstruction error may likely occur. Although visual illustration (potential distortion) is an inevitable consequence of gestalt grouping (alphabet compression), it does not happen frequently to most people. Information-theoretically, this suggests that it must be cost-beneficial for humans to be equipped with the capability of gestalt grouping.

The human visual system consists of many components, each of which is a subsystem. The human visual system is also part of our mind, which is a super-system. In addition to the human visual system, the super-system of the human mind comprises many other systems featuring cost-beneficial data processing mechanisms. Humans' **long-term memory** involves a process of representation for retention and another process of reconstruction for recall. The former facilitates alphabet compression, while the latter introduces potential distortion. It is likely that most humans lack photographic memory because the photographic mechanism would be less cost-beneficial.

Figure 16.7b shows an instance of memorization. A sequence of temporal events may be captured by a person's mind, creating an internal representation. This representation likely records only the main events and the important features of these events. Hence, a significant amount of alphabet compression enables humans to keep the cost of memorization very low. During the memory recall, the sequence of events is reconstructed partly based on the recorded memory representation and partly based on the person's past experience and knowledge of similar sequences of events. Some details may be added in inadvertently and some may be omitted, which leads to memory errors.

Heuristics are powerful functions for humans to make judgments and decisions intuitively and speedily. If these functions are collectively grouped together into a system, each heuristic function can be considered a subsystem. Almost

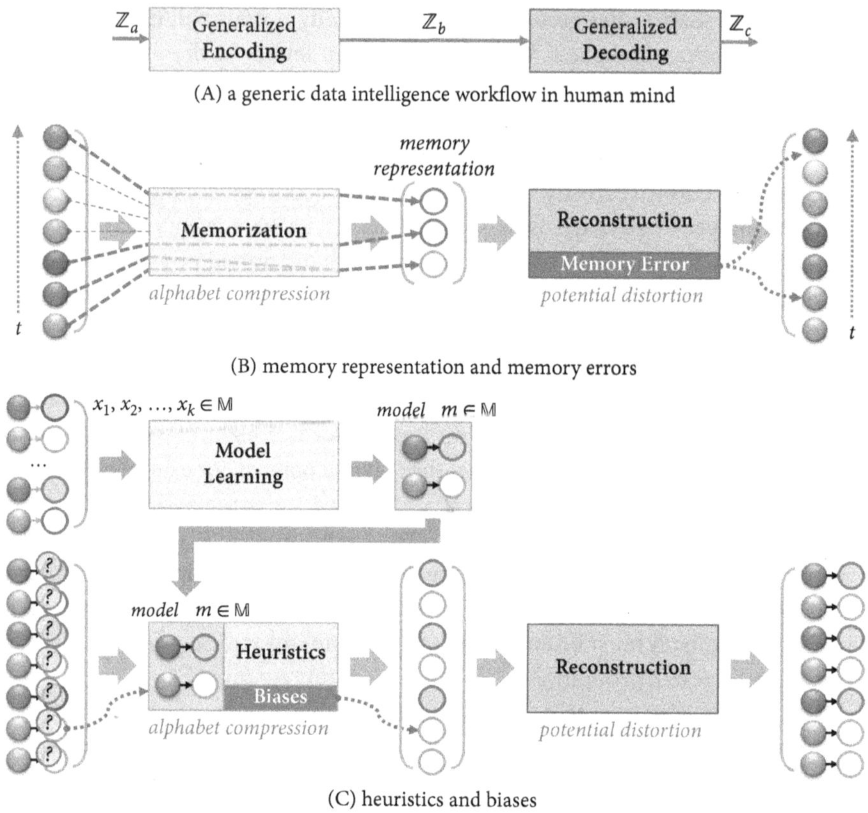

Figure 16.7 *Memorization* and *Heuristics* are two major phenomena in human cognition. Information-theoretically, they feature significant alphabet compression and are cost-beneficial, but may sometimes cause potential distortion in the forms of *memory errors* and *cognitive biases.*

all decision-making processes facilitate alphabet compression since the output alphabet that encodes decision options usually has a smaller amount of uncertainty (i.e., entropy) than the input alphabet that encodes the possible variations of data, for which a decision is to be made. Furthermore, although every situation that requires a judgment or decision mostly differs one from another, somehow humans do not maintain a decision subsystem for every individual situation. This would not be cost-beneficial. Instead, we develop each of our heuristic functions in a way that it can handle decisions in similar situations. Grouping some functions for different situations into a single heuristic function is also a form of alphabet compression. This is the consequence of learning. The potential distortion caused by such functional grouping is referred to as bias. In a biased decision process, a relatively generic heuristic function produces a decision that is significantly different from an ideal function specifically designed for the situation concerned.

Figure 16.7c shows instances of heuristics and biases. Recall the model development workflow in Figure 16.5. It is not difficult to observe the similarity between the two workflows. Figure 16.7c depicts two major workflows, *Model Learning* and *Heuristics/Biases*. A set of letters in $\mathbb{X}$ represent the past experience of some causal relations. A model of the causal relations is learned gradually from the experience. It is then used to make inferences about the effects when some factors of causes are presented. Because the heuristic model is intended to make decisions without incurring too much cognitive cost and may have been learned with sparse experience, the model may feature some overly simplified mappings from causes to effects. Some of this oversimplification may be acceptable, while others may not. The unacceptable oversimplifications are referred to as biases. The criteria for defining the acceptability usually depend on contextual factors such as purposes, tasks, consequences, and social conventions.

Today, many data intelligence systems are computer systems that employ functions, algorithms, and models developed manually or semiautomatically using machine learning. Similar to human heuristics, their design, development, and deployment are intended to make decisions in a cost-beneficial way. Also similar to human heuristics, they facilitate alphabet compression as well as introduce potential distortion inherently. Biases may result from an overgeneralization of some local observations to many situations in a global context, but they can also result from a simplistic application of some global statistics to individual situations in many local contexts (Streeb, Chen, and Keim, 2018). While there are cognitive scientists specialized to study human biases, we also need to pay more attention to computer biases.

Typically, a human's heuristic model is learned over a long period and with controlled or uncontrolled exposure to training data and environments (e.g., learning in schools and judging traffic conditions). Recall the discussions at the end of section 4: with the rapid advances of virtual realty and virtual environment technology, we can design and create virtual training data and environments for developing special-purpose human models.

6. Relating the Metric to Languages and News Media

Consider all the words in a language as an alphabet and each word as a letter. The current version of the Oxford English Dictionary consists of more than 600,000 words. It is a fairly big alphabet. Based on the maximal amount of entropy, they can be encoded using a 20-bit code. The actual entropy is much lower due to the huge variations of word frequencies.

Nevertheless, in comparison with the variations of the objects, events, emotions, attributes, and so on, this alphabet of words represents a massive alphabet compression. For example, using the word "table" is a many-to-one mapping, as the input alphabet contains all kinds of furniture tables, all kinds of row-column tables for organizing numbers, words, figures, actions for having or postponing all kinds of items in a meeting agenda, and so on. In fact, the Oxford English Dictionary has twenty-eight different definitions for the word "table." When including the quotations, the word "table" is described using more than 34,000 words. Interestingly, despite the ubiquity of alphabet compression, in most situations, readers or listeners can perform the reverse mapping from a word to a reality at ease. The potential distortion is not in any way at the level as suggested by the scale of one-to-many mappings. Similarly to what has been discussed in the previous sections, it must be the readers or listeners' knowledge that changes the global probability distribution of the input alphabet to a local one in each situation.

We can easily extrapolate this analysis of alphabet compression and potential distortion to sentences, paragraphs, articles, and books. With a significant amount of alphabet compression and a relatively small chance of potential distortion, written and spoken communications can enjoy relatively low cost by using many-to-one mappings. This is no doubt one of the driving forces in shaping a language.

Not only can written and spoken communications be seen as data intelligence workflows, they also feature processes exhibiting both encryption and compression at the same time, such as the creation and uses of metaphors, idioms, slang, clichés, and puns. To those who have knowledge about these transformed or concealed uses of words, their decoding is usually easy. These forms of figurative speeches often seem to be able to encode more information using shorter descriptions than what would be expressed without the transformation or concealment. In addition, the sense of being able to decode them brings about a feeling of inclusion, enjoyment, or reward. Figure 16.8 shows three examples of "words" commonly encountered in textual communications through the media of mobile phone apps and online discussion forums. They fall into the same generalized encoding and decoding workflow that was also shown in Figures 16.6a and 7a. The cost-benefit analysis can thus be used to reason about their creation and prevalence.

In many ways, these phenomena in languages are rather similar to some phenomena in data visualization. A metro map typically does not correctly convey the geographical locations, routes, and distances. All routes are deformed into straight lines with neatly drawn corners for changing directions. Such transformed and concealed visual encoding of the reality has been shown to be more effective and useful than a metro map with geographically correct

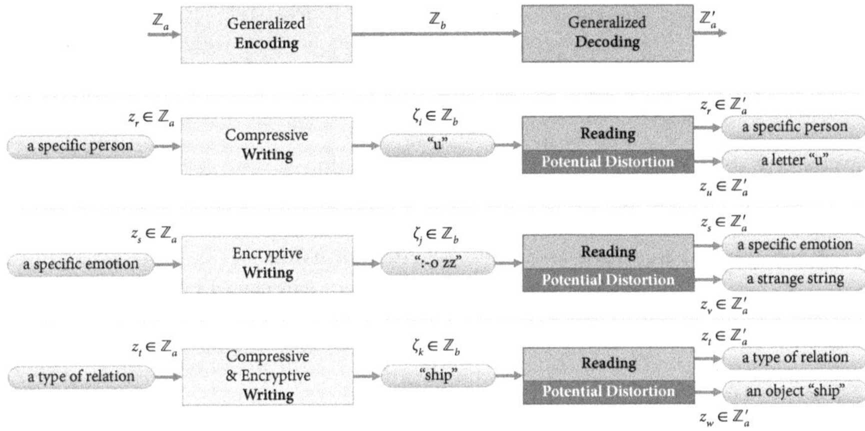

Figure 16.8 Examples of compressive and encryptive encoding in textual communications in some contemporary media such as texting.

representations of locations and routes. From an information-theoretic perspective, this is a form of cost-benefit optimization by delivering more alphabet compression and cost reduction using such encoding and by taking advantage of the fact that human knowledge can reduce potential distortion in decoding.

Languages have indeed evolved with cost-benefit optimization in mind! For example, Chinese writing, which uses a logographic system, has been evolved for more than three millennia (Cao, 2012; Hu, 2014). As illustrated in Figure 16.9a, the early scripts, which exhibit the characteristics of both pictograms and logograms, bear more resemblance to the represented object than to the late scripts. Hence, the early scripts enable relatively easier reconstruction from logograms to the represented objects, incurring less potential distortion in reading for readers of the era without systematic education.

The evolution of the Chinese writing system was no doubt influenced by many time-varying factors, such as the tools and media available for writing, the number of logograms in the alphabet, the typical length of writings, the systematization of education, and standardization. Information-theoretically, the evolution features a trend of decreasing cost for encoding and increasing potential distortion in decoding. However, the improvement of education has alleviated the undesirable problem of increasing potential distortion.

The standardization of the seal scripts in the Qin dynasty (221 BC—206 BC) can be viewed as an effort to reduce the potential distortion and cost of decoding. The regular script first appeared in 151–230 AD and became a de facto standard in the Tang dynasty (618–907 AD). There are currently two Chinese writing systems, namely, *traditional* and *simplified* systems, used by Chinese communities in different countries and regions. The traditional, which was based on

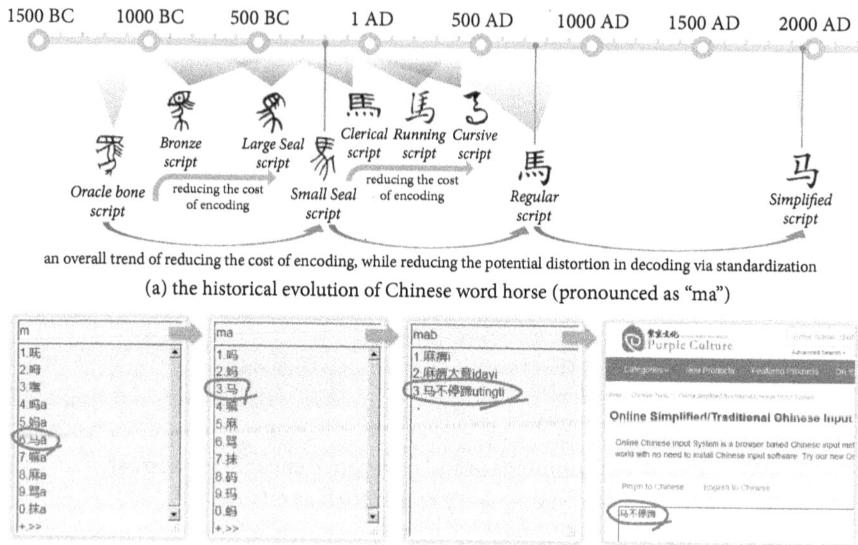

(a) the historical evolution of Chinese word horse (pronounced as "ma")

(b) inputting a Chinese idiom starting with the character horse using a pinyin input tool (Purple Culture).

Figure 16.9 An example showing the reduction of the cost of writing Chinese characters over three millennia. It is common for modern input tools to make recommendations based on a user's partial inputs by using an entropic algorithm.

the regular script, is the official system in Taiwan, Hong Kong, and Macau, and is commonly used in the Chinese Filipino community. The simplified system, which was introduced in the 1950s, is the official system in mainland China and Singapore, and is commonly used in the Chinese Malaysian community.

Information-theoretically, the existence of the two writing systems is obviously not optimal in terms of cost-benefit analysis. However, with the aid of modern computing and Internet technologies, the cost of encoding and the potential distortion in dealing with two writing systems have been reduced significantly from the time when there were no such technologies. Most of modern digital writing systems have an entropic algorithm for character recommendation, and many support both traditional and simplified systems simultaneously. Meanwhile, many websites in Chinese allow readers to switch between the two systems at the click of a button. In many countries and regions where the Chinese language is extensively used, the coexistence of the two systems appears to be gaining ground, perpetuating the dissemination of information more cost-beneficially than during the period when the modern technologies were not yet available and each country or region had to restrict itself to a single system.

As exemplified in Figure16.9b, the large proportion of encoding is now done on computers and mobile phones. The encoding is truly a human–machine cooperation. In Figure 16.9b, a user needs to enter a Chinese idiom with four characters, which individually are translated as horse-not-stop-hoof. It is figurative speech for "continuously making progress" or "continuously working hard." With the full pinyin spelling of each of the four characters, one online pinyin input tool (Chinese-Tools.com) showed 24 optional characters (in the simplified system) for "ma," 25 for"bu," 21 for "ting," and 39 for "ti." The total number of combinations available for the pinyin string "ma bu ting ti" is thus 491,400. Another online tool (Purple Culture), which provides traditional and simplified Chinese characters simultaneously, showed 57 optional characters for "ma," 66 for"bu," 60 for "ting," and 126 for "ti." The total number of combinations available for the string "ma bu ting ti" is thus 28,440,720. However, using an entropic algorithm, we can organize the optional characters and compounding phrases probabilistically based on their frequency of use. If the user needs to enter only the character "horse," it can be done with "m6" or "ma3." By the time the user entered "mab," there were only three options left. The idiom, which is spelled with 10 pinyin letters, requires only 4 key strokes "mab3." In comparison with the handwriting system shown in Figure 16.9a, the cost of encoding is significantly reduced.

In parallel with the phenomenon of cost saving in encoding Chinese characters digitally, we also see the rising popularity of emojis. At the center of data science, the technology of data visualization is pumping out new "words" (e.g., multivariate glyphs), "sentences" (e.g., charts), and new "paragraphs" (e.g., animations) everyday. The modern communication media and computer technologies will continue to reduce the cost of visual encoding. The alphabets of the lexical units, synatactical units, and semantic units of visual representations are rapidly growing in size and complexity. The cost-benefit metric in Eq. (16.1) suggests that efforts such as standardization, systematization, and organized education can alleviate the potential distortion in decoding visual representations of data. A language of data visualization may emerge in the future.

The progressive development of languages brought about the paradigm of organized communication of news (Stephens, 1988). Early news enterprises involved messengers and an organizational infrastructure that supported them. Messengers carried spoken or written words about news, commands, appeals, proclamations, and so on from one place to another. The Olympic event, the *Marathon,* is attributed to the Greek messenger Pheidippides in 490 BC who ran 42 kilometers (about 26 miles) from the battlefield of Marathon to Athens to report the victory. Ancient Persia is widely credited for inventing the postal system for relaying messages. The ancient Mongol army had a communication system called "Yam," which was a huge network of supply points that provided messengers with food, shelter, and spare horses. At that time, the messages transmitted were very compact, and, collectively, in a much smaller number. The cost for delivering the messages was overwhelmingly more than that for consuming the messages.

In contrast, the contemporary news media, which include newspapers, televisions, web-based news services, email newsletters, and social media, generate and transmit astonishingly more messages than the ancient news enterprises. The cost of consuming all received or receivable messages is beyond the capacity of any organization or individual. From an information-theoretic perspective, we can consider the cost-benefit of three typical workflows of news communications as shown in Figures 16.10, 16.11, and 16.12.

Figure 16.10 illustrates a news communication workflow based on an infrastructure for relaying messages. This workflow encapsulates a broad range of mechanisms for delivering news across a few millennia, from ancient messengers on foot or horseback to postal mails, telegraphs, and telephones. After the initial transformation from a real-world alphabet to a description alphabet of, for example, written messages (shown) or spoken messages (not shown), the intermediate transformations focus on transporting messages (i.e., letters in a description alphabet $\mathbb{Z}_{a,i}, i = 1, 2, \ldots, l$). In many situations, the intermediate transformations may involve additional encoding and decoding processes—for instance, digital-analog conversions in telegraphs and telephones, and encryption and decryption for confidential messages. Nevertheless, any semantic transformation is expected to take place only at the writing stage at the beginning

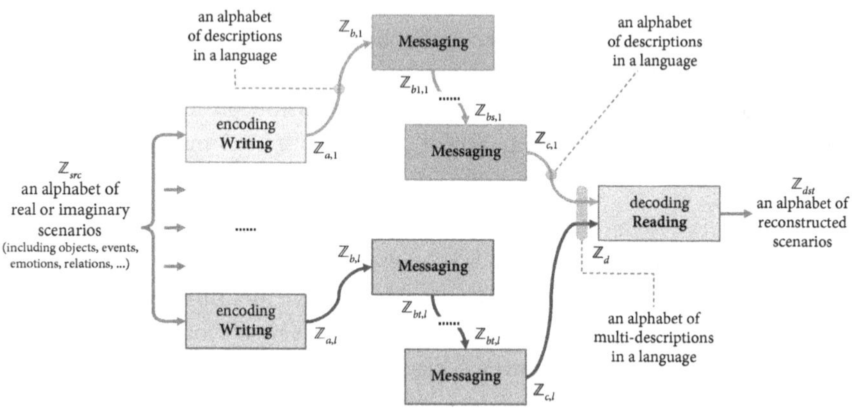

Figure 16.10 A news communication workflow based on an infrastructure for relaying messages. Each message is relayed by a sequence of processes that are not expected to alter the semantics of the message, hence incurring no alphabet compression or potential distortion.

and reading stage at the end of the workflow. In other words, the intermediate transformations from $\mathbb{Z}_{a,i}$ to $\mathbb{Z}_{b[1],i}, \ldots, \mathbb{Z}_{b[\tau],i}$, and then to $\mathbb{Z}_{c,i}$, are designed to have a zero amount of alphabet compression and a zero amount of potential distortion. Here $\tau > 0$ is a path-dependent integer. In terms of the cost-benefit metric for data intelligence, these intermediate relaying transformations bring about zero amount of benefit information-theoretically.

Figure 16.11 illustrates another typical workflow for communicating news to a broader audience (Klapper, 1960). In this workflow, a number of centralized media organizations, such as newspapers, radio stations, and television stations, systematically gather a large collection of descriptions from many sources and produce summary bulletins to be consumed by many readers, listeners, and viewers (only one reader is shown in the figure). The history of such broadcasting workflows can be traced back to ancient Rome in around 131 BC, when *Act Diurna* (Daily Acts) were carved on stone or metal and presented in public places; and to the Han Dynasty of China (206 BC–220 AD), when central and local governments produced *dibao* (official bulletins) and announced them to the public using posters and word of mouth. Information-theoretically, the messages in the alphabets from different sources (e.g., $\mathbb{Z}_{a,i}, i = 1, 2, \ldots, m$) are amalgamated to create more complex alphabets for multi-descriptions, $\mathbb{Z}_{b,j}, j = 1, 2, \ldots, x$. Each centralized organization performs semantically rich transformations that feature a huge amount of alphabet compression in selecting, summarizing, annotating, and enriching the original messages in the input alphabet, resulting in an output alphabet of descriptions $\mathbb{Z}_{c,j}, j = 1, 2, \ldots, x$. To an individual reader (or a listener, or a viewer), there will be potential distortion when the person tries to imagine what was really said in $\mathbb{Z}_{a,i}, i = 1, 2, \ldots, n$.

In comparison with the workflow in Figure 16.10, the workflow in Figure 16.11 can handle more data at lower costs, especially from the perspective of cost-benefit per user. Although the intermediate transformations may introduce potential distortion due to the many-to-one forward mappings, the amount of positive alphabet compression is not only necessary in most cases but also beneficial in general. When the number of media outlets is easily countable, it is not difficult for a reader (or a listener, or a viewer) to become knowledgeable about the editorial coverage, styles, partiality, and so on. Such knowledge may be used to make better reconstruction of the scenarios in $\mathbb{Z}_{src}$ (e.g., by reading between the lines). At the same time, such knowledge may bias decisions as to what to read (or to listen or to view), leading to confirmatory biases.

Figure 16.12 illustrates a contemporary workflow for communicating news through social media. With the rapid development of the Internet and many social media platforms, there are many more intermediate news entities that can perform semantically rich transformations. These entities can collectively

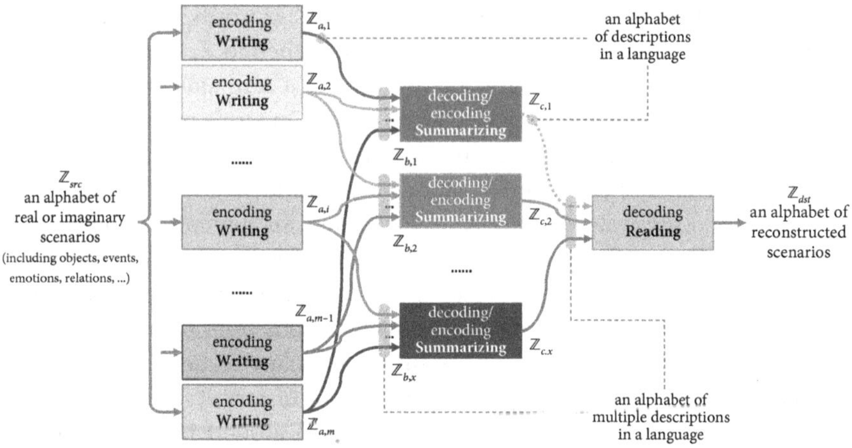

Figure 16.11 A news communication workflow based on centralized media organizations, which deliver semantically-rich transformations from a large number of news reports to a selection of amalgamated and prioritized summary descriptions. The decoding and encoding in the intermediate processes result in both alphabet compression and distortion. While the workflow is commonly used for communicating with a large audience, as an example, only one reader is shown in the figure for the purpose of clarity.

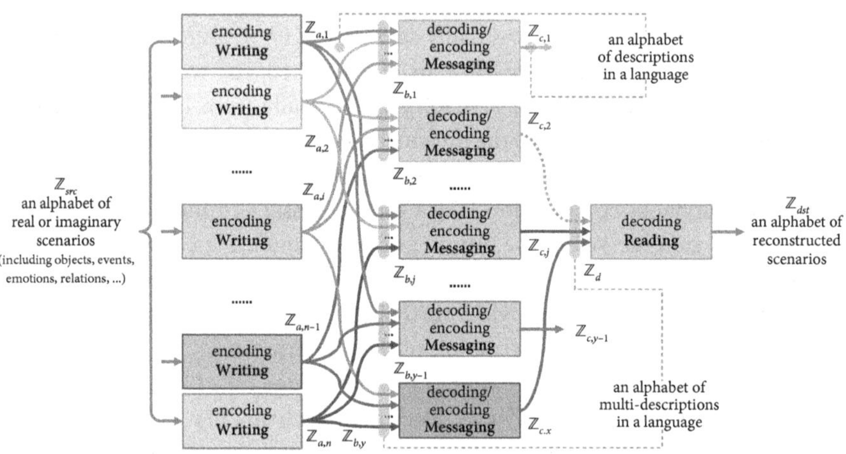

Figure 16.12 A news communication workflow based on social media, where numerous new entities can perform semantically-rich transformations.

access many more descriptions in the description alphabets, $\mathbb{Z}_{a,i}, i = 1, 2, \ldots, n$. Each of these entities can be as simple as a message relay process and as complex as a media organization. They have a nontrivial amount of editorial power in selecting, summarizing, annotating, and enriching the original messages in the input alphabet. They generate overwhelmingly more outputs, $\mathbb{Z}_{c,j}, j = 1, 2, \ldots, y$, than those centralized media organizations in Figure 16.11. For an individual reader (or listener, or viewer), it is no longer feasible to be aware of a reasonable portion of these y outlets, and one has to direct one's selective attention to a tiny portion of them. In comparison with the workflow shown in Figure 16.11, there likely is more selection but less summarization.

In summary, comparing the number of processes that can perform the initial transformations from $\mathbb{Z}_{src}$ in Figures 16.10, 16.11, and 6.12, we have $l \ll m \ll n$. Comparing the number of intermediate media entities that can perform semantically rich transformations in Figures 16.11 and 16.12, we have $x \ll y$. Meanwhile, for each pathway (per user per message) from $\mathbb{Z}_{src}$ to $\mathbb{Z}_{dst}$, the cost of the workflow in Figure 16.10 is significantly higher than that in Figure 16.11, which is further reduced massively by the workflow depicted in Figure 16.12. The main question will no doubt be the assessment of the *Benefit* (i.e., *Alphabet Compressions—Potential Distortion*) in the cost-benefit metric. The intermediate transformations for relaying messages (Figure 16.10) incur little alphabet compression and potential distortion. For both workflows in Figures 16.11 and 16.12, there is a trade-off between the huge amount of alphabet compression and the huge amount of potential distortion. Some social scientists have suggested that the workflow based on social media may incur more potential distortion e.g., uncertainty due to "alternative truth," confirmatory biases, and opinion polarization (Garrett, 2009; Brundidge, 2010; Gentzkow and Shapiro, 2011; Knobloch-Westerwich, 2012; Lee, Choi, Kim, and Kim 2014).

Further qualitative and quantitative analysis will be necessary to estimate the *Benefits* in bits. Nevertheless, as the cost-benefit metric suggests an optimization of the data intelligence workflows in news communications based on the fraction in Eq. (16.1), the cost reduction is certainly a significant factor in setting the trend.

7. Conclusions

In this chapter, we have applied the cost-benefit analysis, which was originally developed as an information-theoretic metric for data analysis and visualization workflows (Chen and Golan, 2016), to a range of data intelligence workflows in machine learning, perception and cognition, and languages and news media. We have postulated and demonstrated the likelihood that the cost-benefit metric in Eq. (16.1) is a fundamental fitness function for optimizing data intelligence workflows that involve machine- and/or human-centric processes. We have also made connections in abstraction between data intelligence workflows and data compression and encryption workflows. Hence, data intelligence can be seen as part of a generalized paradigm of encoding and decoding. We can envisage the prospect that information theory can underpin a broad range of aspects in data science and computer science, and can be used to cogitate a wide range of phenomena in cognitive science, social science, and humanities.

While the broad interpretations that relate the cost-benefit metric in Eq. (16.1) to perception, cognition, languages, and news media will require further scholarly investigations to confirm, the reasonable possibility of its role as a fitness function in these broad interpretations as shown in this chapter provides further evidence that the metric can be used to optimize data intelligence workflows. Of course, it may turn out that there are more general or applicable metrics with even broader explanatory power. As long as this chapter helps stimulate the discovery of such new metrics, these broad interpretations are not just as useless as "speculations" that "make interesting talks at cocktail parties."[3]

References

Brundidge, J. (2010). "Encountering 'Difference' in the Contemporary Public Sphere." *Journal of Communication*, 60(4): 680–700.

Cao, B. H. (2012). *Evolution of Chinese Characters*. [in Chinese].

Chen, M., and D. S. Ebert. (2019). "An Ontological Framework for Supporting the Design and Evaluation of Visual Analytics Systems." *Computer Graphics Forum*, 38(3): 131–144.

Chen, M., K. Gaither, N. W. John, and B. McCann. (2019). "Cost-Benefit Analysis of Visualization in Virtual Environments." *IEEE Transactions on Visualization and Computer Graphics*, 25(1): 32–42.

Chen, M., and A. Golan. (2016). "What May Visualization Processes Optimize?" *IEEE Transactions on Visualization and Computer Graphics*, 22(12):2619–2632.

Chinese-Tools.com. Chinese input method editor, accessed at http://www.chinese-tools.com/tools/ime.html in May 2018.

Garrett, R. K. (2009). "Politically Motivated Reinforcement Seeking." *Journal of Communication*, 59(4): 676–699.

Gentzkow, M., and J. M. Shapiro. (2011). "Ideological Segregation Online and Offline." *Quarterly Journal of Economics*, 126(4): 1799–1839.

Hu, P. A. (2014). *History of Chinese Philology*. Shanghai: Science Literature. [In Chinese].

[3] In his book *A Short History of Nearly Everything* (p. 228, Black Swan, 2004), Bill Bryson told the story about a journal editor who dismissively commented on Canadian geologist Lawrence Morley's proposition about a theory in the context of continental drift.

Kijmongkolchai, N., A. Abdul-Rahman, and M. Chen. (2017). "Empirically Measuring Soft Knowledge in Visualization." *Computer Graphics Forum*, 36(3): 73–85.

Klapper, J. T. (1960). *The Effects of Mass Communication*. New York: Free Press.

Knobloch-Westerwick, S. (2012). "Selective Exposure and Reinforcement of Attitudes and Partisanship before a Presidential Election." *Journal of Communication*, 62(4): 628–642.

Kullback, S., and R. A. Leibler. (1951). "On Information and Sufficiency." *Annals of Mathematical Statistics*, 22(1): 79–86.

Lee, J. K., J. Choi, C. Kim, and Y. Kim. (2014). "Social Media, Network Heterogeneity, and Opinion Polarization." *Journal of Communication*, 64(4): 702–722.

Oxford English Dictionary. (2018). Oxford English Press. Accessed at http://www.oed.com, May 2018.

Purple Culture. (2018). Online Simplifiedtraditional Chinese Input System. Aaccessed in May at https://www.purpleculture.net/online-chinese-input, May 2018.

D. Sacha, D., M. Kraus, D. A. Keim, and M. Chen. (2019). "VIS4ML: An Ontology for Visual Analytics Assisted Machine Learning." *IEEE Transactions on Visualization and Computer Graphics*, 25(1):385–395.

Shannon, C. E. (1948). "A Mathematical Theory of Communication." *Bell System Technical Journal*, 27: 379–423.

Stephens, M. (1988). *A History of News: From the Drum to the Satellite*. New York: Viking.

Streeb, D., M. Chen, and D. A. Keim. (2018). "The Biases of Thinking Fast and Thinking Slow." In G. Ellis (ed.), *Cognitive Biases in Visualizations*, pp. 97–109. New York: Springer.

Streeb, D., M. El-Assady, D. A. Keim, and M. Chen. (In preparation). "Why Visualize?—Untangling a Large Collection of Arguments." *IEEE Transactions on Visualization and Computer Graphics*.

Tam, G. K. L., V. Kothari, and M. Chen. (2017). "An Analysis of Machine- and Human-Analytics in Classification." *IEEE Transactions on Visualization and Computer Graphics*, 23(1): 71–80. VAST2016 Best Paper Award.

17
The Role of the Information Channel in Visual Computing

Miquel Feixas and Mateu Sbert

1. Introduction

In 1948, Claude E. Shannon (1916–2001) published a paper entitled "A Mathematical Theory of Communication" (Shannon, 1948) that marks the beginning of information theory. In his paper, Shannon introduced the concepts of *entropy* and *mutual information*, which since then have been widely used in many fields, such as physics, computer science, neurology, image processing, computer graphics, and visualization. Another fundamental concept of Shannon's work was the *communication* or *information channel*, introduced to model the communication between source and receiver. The communication channel concept was general enough to be applied to any two variables sharing information. In section 2, we show how both source (or input) and receiver (or output) variables are defined by a probability distribution over their possible states and are related by an array of conditional probabilities. These probabilities define the different ways that a state in the output variable can be reached from the states in the input variable. In short, the channel specifies how the two variables share, or communicate, information. The input and output variables can be of any nature, they can be defined or not on the same states, and they can even be the same.

In this chapter, we present some examples of these possibilities. In section 3, we present the application to viewpoint selection. The input and output variables represent all viewpoint positions on a sphere surrounding a virtual 3D object and the polygons of this virtual object, respectively. The conditional probabilities express how much area of each polygon is seen from a given viewpoint. The objective is to obtain the viewpoints that give more information on the object. In section 4, we present the application to image processing. The input and output variables represent the distribution of relative areas of the regions or segments in the image and the color histogram of the image,

Miquel Feixas and Mateu Sbert, *The Role of the Information Channel in Visual Computing* In: *Advances in Info-Metrics: Information and Information Processing across Disciplines.* Edited by: Min Chen, J. Michael Dunn, Amos Golan, and Aman Ullah, Oxford University Press (2021). © Oxford University Press.
DOI: 10.1093/oso/9780190636685.003.0017

respectively. The conditional probabilities express the amount of area of a given image region that corresponds to each histogram bin. In this case, the channel evolves by refining the regions of the image so that with each new partitioning or splitting step the information transfer in the channel increases. In section 5, the application of the information channel we present illustrates the case where the input and output variables are defined on the same states and have the same distribution of probabilities. The application field is global illumination, that is, how a three-dimensional synthetic virtual scene can be realistically illuminated. The states represent the patches or polygons in which the virtual scene is constructed or divided, and the conditional probabilities from a first polygon with respect to a second polygon represent the probability that a ray cast randomly from the first polygon lands in the second polygon. The problem to solve is how to optimally discretize the scene so that the illumination is best represented.

The application of information theory to computer graphics and image processing has been reviewed by Sbert et al. (2009), Escolano et al. (2009), and Feixas et al. (2016). Chen and Jänicke (2010), Wang and Shen (2011), and Chen et al. (2014) have also reviewed the use of information-theoretic measures in the scientific visualization field.

The examples presented in this chapter are of a technological nature. But other examples could be obtained from all the sciences. As an example of the general applicability of the concept, consider the transfer of votes between different parties in successive polls. The input and output variables are the distribution of votes between the different parties, and the conditional probabilities represent the transfer of votes. This information channel could be useful for sociologists to better understand the evolution of voting preferences.

We hope that by presenting all these different examples the readers can realize the power of the information channel concept and might obtain hints on how to apply it to their specific domain.

2. Information Measures and Information Channel

In this section, we briefly describe the most basic information-theoretic measures (Cover and Thomas, 1991; Yeung, 2008), the main elements of an information channel (Cover and Thomas, 1991; Yeung, 2008), and the agglomerative information bottleneck method (Slonim and Tishby, 2000), which is used in some of the applications studied in this chapter.

2.1 Basic Information-Theoretic Measures

Let X be a discrete random variable with alphabet χ and probability distribution $\{p(x)\}$, where $p(x) = \Pr\{X = x\}$ and $x \in \chi$. The distribution $\{p(x)\}$ can also be denoted by $p(X)$. Likewise, let Y be a random variable taking values y in $\mathcal{Y}$.

The *Shannon entropy* $H(X)$ of a random variable X is defined by

$$H(X) = -\sum_{x \in \chi} p(x) \log p(x), \tag{17.1}$$

where all logarithms are base 2 and entropy is expressed in bits. The convention that $0 \log 0 = 0$ is used. $H(X)$, also denoted as $H(p)$, measures the average *uncertainty* or information content of a random variable X.

The *conditional entropy* $H(Y|X)$ is defined by

$$H(Y|X) = \sum_{x \in \chi} p(x) H(Y|x), \tag{17.2}$$

where $p(y|x) = \Pr[Y = y|X = x]$ is the conditional probability and $H(Y|x) = -\sum_{y \in \mathcal{Y}} p(y|x) \log p(y|x)$ is the entropy of Y given x. $H(Y|X)$ measures the average uncertainty associated with Y if we know the outcome of X.

The *relative entropy* or *Kullback–Leibler distance* $D_{KL}(p,q)$ between two probability distributions p and q, that are defined over the alphabet χ, is given by

$$D_{KL}(p,q) = \sum_{x \in \mathcal{X}} p(x) \log \frac{p(x)}{q(x)}. \tag{17.3}$$

The conventions that $0 \log(0/0) = 0$ and $a \log(a/0) = \infty$ if $a > 0$ are adopted.

The *mutual information* $I(X;Y)$ between X and Y is defined by

$$I(X;Y) = H(Y) - H(Y|X) \tag{17.4}$$

$$= \sum_{x \in \mathcal{X}} \sum_{y \in \mathcal{Y}} p(x,y) \log \frac{p(x,y)}{p(x)p(y)}, \tag{17.5}$$

$$= D_{KL}(p(X,Y), p(X)p(Y)) \tag{17.6}$$

where $p(x,y) = \Pr[X = x, Y = y]$ is the joint probability. Mutual information expresses the *shared information* between X and Y. The relations between Shannon's information measures are summarized in the information diagram of Figure 17.1 (Yeung, 2008).

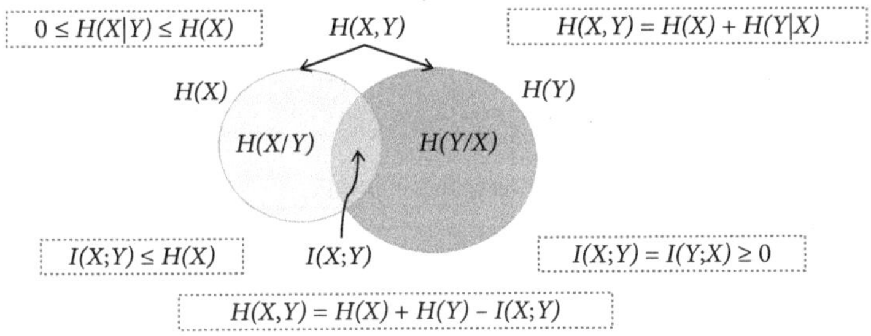

Figure 17.1 The information diagram shows the relationship between Shannon's information measures.

2.2 Information Channel

In this section, we describe the main elements of an information channel. *Conditional entropy* $H(Y|X)$ and *mutual information* $I(X;Y)$ can be thought of in terms of a *communication channel* or *information channel* $X \rightarrow Y$ whose output Y depends probabilistically on its input X (Cover and Thomas, 1991). Conditional entropy and mutual information express, respectively, the uncertainty in the channel output from the sender's point of view and the degree of dependence or information transfer in the channel between variables X and Y, respectively.

Figure 17.2 shows the main elements that configure an information channel. These elements are:

- Input and output variables, X and Y, with their probability distributions $p(X)$ and $p(Y)$, called marginal probabilities, respectively.
- Probability transition matrix $p(Y|X)$ (composed of conditional probabilities $p(y|x)$) which determines the output distribution $p(Y)$ given the input distribution $p(X)$: $p(y) = \sum_{x \in \mathcal{X}} p(x)p(y|x)$. Each row of $p(Y|X)$, denoted by $p(Y|x)$, is a probability distribution.

All these elements are connected by Bayes' rule which relates marginal (input and output), conditional, and joint probabilities: $p(x,y) = p(x)p(y|x) = p(y)p(x|y)$.

In the fields of image processing and neural systems, we find two well-known information channels: the image registration channel and the stimulus-response channel. In the first case, the registration between two images can be modeled by an information channel, where its marginal and joint probability distributions are obtained by simple normalization of the corresponding intensity histograms of the overlap area of both images. This method is based on the conjecture that the optimal registration corresponds to the maximum mutual information

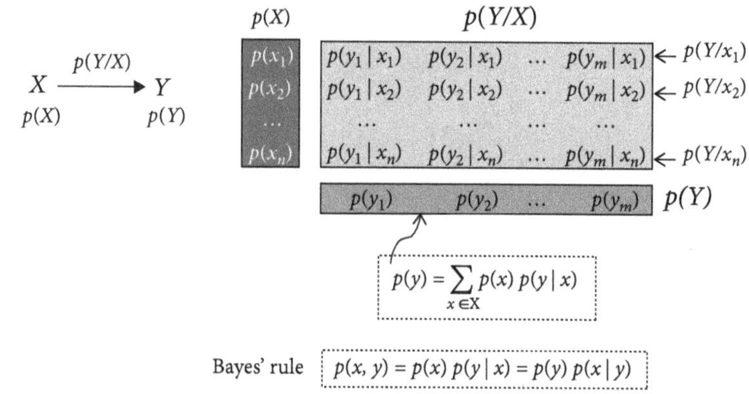

Figure 17.2 Main elements of an information channel. Input and output variables, X and Y, with their probability distributions $p(X)$ and $p(Y)$, and probability transition matrix $p(Y|X)$, composed of conditional probabilities $p\left(y|x\right))$, which determines the output distribution $p(Y)$ given the input distribution $p(X)$. All these elements are connected by Bayes' rule.

between the overlap areas of the two images (Maes et al., 1997; Viola, 1995). In the second case, mutual information between stimulus and response quantifies how much information the neural responses carry about the stimuli, that is, the information shared or transferred between stimuli and responses. The joint probabilities between each possible stimulus and each possible response enable us to compute all the measures associated with the channel (marginal entropies, conditional entropy, and mutual information) and also the specific information associated with each stimulus (or response) (Borst and Theunissen, 1999; Deweese and Meister, 1999).

In sections 3, 4, and 5, we study three different applications that exploit the benefits of modeling a problem as an information channel: viewpoint information channel; image information channel; and scene visibility channel.

2.3 Jensen–Shannon Divergence and Information Bottleneck Method

Two main results derived from the above information measures are the Jensen–Shannon divergence and the agglomerative information bottleneck method. The *Jensen–Shannon divergence* (Burbea and Rao, 1982) measures the dissimilarity between two probability distributions and is defined by

$$JS(\pi_1,\pi_2,\dots,\pi_n;p_1,p_2,\dots,p_n) = H\left(\sum_{i=1}^{n}\pi_i p_i\right) - \sum_{i=1}^{n}\pi_i H(p_i), \tag{17.7}$$

where $p_1, p_2, \dots, p_n$ are a set of probability distributions defined over the same alphabet with prior probabilities or weights $\pi_1, \pi_2, \dots, \pi_n$, fulfilling $\sum_{i=1}^{n}\pi_i = 1$, and $\sum_{i=1}^{n}\pi_i p_i$ is the probability distribution obtained from the weighted sum of the probability distributions $p_1, p_2, \dots, p_n$. From the concavity of entropy (Cover and Thomas, 1991), the following inequality holds

$$JS(\pi_1,\pi_2,\dots,\pi_n;p_1,p_2,\dots,p_n) \geq 0. \tag{17.8}$$

The *information bottleneck method*, introduced by Tishby et al. (1999), is a technique that compresses the variable X into $\hat{X}$ with minimal loss of mutual information with respect to another variable Y. The compressed variable $\hat{X}$ can be understood as a clustering or agglomeration of the states of X, guided by the target of preserving as much information as possible about the control variable Y. This method tries to find the optimal trade-off between accuracy and compression of X when the bins of this variable are clustered.

In the agglomerative information bottleneck method, proposed by Slonim and Tishby (2000), it is assumed that a cluster $\hat{x}$ is defined by $\hat{x} = \{x_1, \dots, x_l\}$, where $x_k \in \mathcal{X}$ for all $k \in \{1, \dots, l\}$, and the probabilities $p(\hat{x})$ and $p(y|\hat{x})$ are defined by

$$p(\hat{x}) = \sum_{k=1}^{l} p(x_k) \tag{17.9}$$

and

$$p(y|\hat{x}) = \frac{1}{p(\hat{x})}\sum_{k=1}^{l} p(x_k)p(y|x_k), \forall y \in \mathcal{Y}. \tag{17.10}$$

From these assumptions, the decrease $\delta I_{\hat{x}}$ in the mutual information due to the merging of $x_1, \dots, x_l$ is given by

$$\delta I_{\hat{x}} = p(\hat{x}) JS(\pi_1,\dots,\pi_l;p_1,\dots,p_l) \geq 0, \tag{17.11}$$

where the weights and probability distributions of the JS-divergence (Eq. [17.7]) are given by $\pi_k = p(x_k)/p(\hat{x})$ and $p_k = p(Y|x_k)$ for all $k \in \{1, \dots, l\}$, respectively. An optimal clustering algorithm should minimize $\delta I_{\hat{x}}$ (Slonim and Tishby, 2000). Figure 17.3 illustrates the main elements of the agglomerative information bottleneck method. Note that Eq. (17.11) expresses the *information loss*

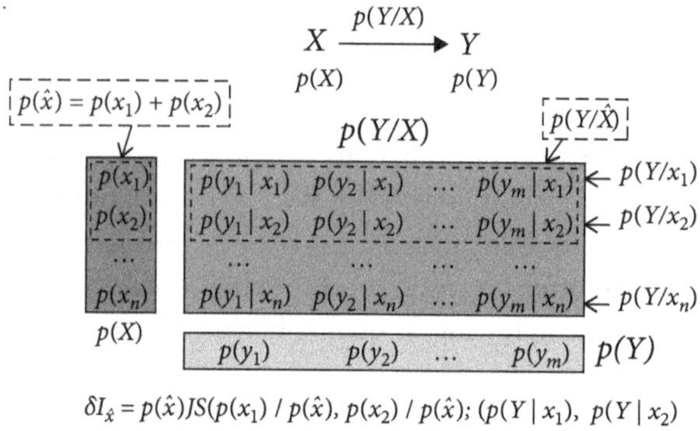

Figure 17.3 Agglomerative information bottleneck method applied to the information channel $X \rightarrow Y$. The information loss $\delta I_{\hat{x}}$ due to the merging of x_1 and x_2 is obtained from the Jensen-Shannon divergence between rows $p(Y|x_1)$ and $p(Y|x_2)$ weighted respectively by $\frac{p(x_1)}{p(\hat{x})}$ and $\frac{p(x_2)}{p(\hat{x})}$. The elements of row $p(Y|\hat{x})$ are given by $p(y|\hat{x})\,2\frac{p(x_1)}{p(\hat{x})}p(y|x_1)+\frac{p(x_2)}{p(\hat{x})}p(y|x_2)\big)$.

when several states are merged in a single state but also the *information gain* when a state $\hat{x}$ is split into several states $x_1, \ldots, x_l$.

In sections 4 and 5, the information bottleneck method is applied to deal with image segmentation and scene patch refinement, respectively. In image segmentation (section 4), the information bottleneck method is applied to calculate the information gain resulting from progressive partitioning. The information bottleneck method is also implicitly used in the accurate computation of scene lighting (section 5) to determine the information gain when a scene patch is subdivided.

3. The Viewpoint Selection Channel

With the objective of obtaining the best views for a virtual 3D object, a *viewpoint information channel* between a set of viewpoints on a sphere surrounding that object (input variable) and its set of polygons (output variable) is introduced to define a set of viewpoint measures. The conditional probabilities of this channel represent the amount of area of each polygon seen from a given viewpoint.

3.1 Viewpoint Entropy and Mutual Information

A viewpoint selection framework is constructed from an information channel $V \rightarrow Z$ between the random variables V (input) and Z (output), which

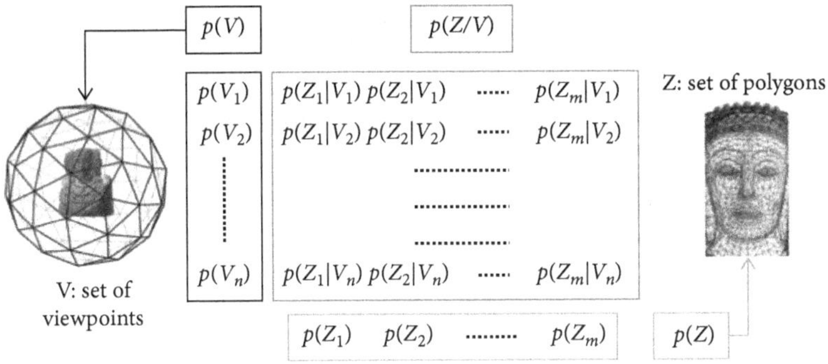

Figure 17.4 Viewpoint information channel $V \rightarrow Z$ between a set of viewpoints ($\mathcal{V}$) and the set of polygons of an object ($\mathcal{Z}$). This channel is constructed from a polygonal model surrounded by a "sphere of viewpoints" and the marginal distribution $p(V)$ is the importance associated with the viewpoints, the conditional probability distribution $p(Z/v_i)$ is given by the normalized area projections of polygons z_j over viewpoint v_i, and the marginal probability distribution $p(Z)$ is obtained using Bayes rule. In this diagram, n and m represent the number of viewpoints and polygons, respectively.

represent, respectively, a set of viewpoints $\mathcal{V}$ and the set of polygons $\mathcal{Z}$ of an object (Feixas, Sbert, and González, 2009). This channel, called the *viewpoint channel*, is defined by a transition probability matrix obtained from the projected areas of polygons at each viewpoint. Viewpoints will be indexed by v and polygons by z. The diagram in Figure 17.4 illustrates the main elements of this channel.

The viewpoint channel can be interpreted as an observation channel where the conditional probabilities represent the probability of "seeing" a determined polygon from a given viewpoint. The three basic elements of this channel are:

- Transition probability matrix $p(Z|V)$, where each element

$$p(z|v) = a_z(v)/A_T(v) \tag{17.12}$$

is obtained from the quotient between $a_z(v)$, the projected area of polygon z at viewpoint v, and $A_T(v)$, the projected area of all polygons over this viewpoint. Conditional probabilities fulfil $\sum_{z\in\mathcal{Z}} p(z|v) = 1$. The background is not taken into account, although it could be considered as another polygon.

- Input distribution $p(V)$, which represents the probability of selecting each viewpoint, is obtained from the normalization of the projected area of the object at each viewpoint:

$$p(v) = A_T(v) / \sum_{v \in \mathcal{V}} A_T(v). \qquad (17.13)$$

It can be interpreted as the probability that a random ray originated at v hits (or "sees") the object or as the importance assigned to each viewpoint v.

- Output distribution $p(Z)$ is given by

$$p(z) = \sum_{v \in \mathcal{V}} p(v) p(z|v), \qquad (17.14)$$

which represents the average projected area of polygon z (i.e., the probability of polygon z to be hit or "seen" by a random ray cast from the viewpoint sphere).

From the previous definitions and Eqs. (17.1, 17.2, 17.3 and 17.4), Shannon's information measures can be defined for the viewpoint channel. We first introduce the viewpoint entropy (Vázquez, Feixas, Sbert, and Heidrich, 2001) and the viewpoint conditional entropy (Feixas et al., 2009). The *viewpoint entropy* (VE) of viewpoint v is defined by

$$H(Z|v) = -\sum_{z \in \mathcal{Z}} p(z|v) \log p(z|v) \qquad (17.15)$$

and measures the degree of uniformity of the projected area distribution at viewpoint v. The maximum viewpoint entropy is obtained when a certain viewpoint can see the maximum number of polygons with the same projected area. The best viewpoint is defined as the one that has maximum entropy (Vázquez et al., 2001). The *conditional entropy* of channel $V \to Z$ is defined by

$$H(Z|V) = -\sum_{v \in \mathcal{V}} p(v) \sum_{z \in \mathcal{Z}} p(z|v) \log p(z|v) = \sum_{v \in \mathcal{V}} p(v) H(Z|v), \qquad (17.16)$$

which is the average of all viewpoint entropies. Both measures $H(Z|v)$ and $H(Z|V)$ tend to infinity when the polygons are infinitely refined. This makes these measures very sensitive to the discretization of the object.

The *mutual information* of channel $V \to Z$ is defined by

$$I(V;Z) = \sum_{v \in \mathcal{V}} p(v) \sum_{z \in \mathcal{Z}} p(z|v) \log \frac{p(z|v)}{p(z)} = \sum_{v \in \mathcal{V}} p(v) I(v;Z) \qquad (17.17)$$

and expresses the degree of dependence or correlation between the set of viewpoints and the object. This equation contains the *viewpoint mutual information* (VMI):

$$I(v;Z) = \sum_{z \in \mathcal{Z}} p(z|v) \log \frac{p(z|v)}{p(z)}, \qquad (17.18)$$

which expresses the degree of dependence between the viewpoint v and the set of polygons. In this context, the best viewpoint is defined as the one that has minimum VMI. This is because the lowest values of VMI correspond to the most representative views, showing the maximum possible number of polygons in a balanced way.

The viewpoint Kullback–Leibler distance can also be applied to viewpoint selection. The *viewpoint Kullback–Leibler distance* (VKL) of viewpoint v is defined by

$$D_{KL}\left(p(Z|v), a(Z)\right) = \sum_{z \in \mathcal{Z}} p(z|v) \log \frac{p(z|v)}{a(z)}, \qquad (17.19)$$

where $a(z)$ is the normalized area of polygon z obtained from the area of polygon z divided by the total area of the object (Sbert, Plemenos, Feixas, and González, 2005). The VKL measure is interpreted as the distance between the normalized distribution of projected areas and the normalized distribution of the areas of polygons. Note that, in this case, the background cannot be taken into account. The minimum value 0 would be obtained when $p(z|v) = a(z)$. Since the goal in this case is to look for the distribution of projected areas as near as possible to the distribution of actual areas, selecting best views means to minimize VKL.

3.2 Results

In this section, we analyze the behavior of VE (Eq.[17.15]), VMI (Eq.[17.18]), and VKL (Eq. [17.19]). To compute these viewpoint quality measures, we need a preprocess step to estimate the projected area of the visible polygons of the object at each viewpoint. Before projection, a different color is assigned to each polygon. The number of pixels with a given color divided by the total number of pixels projected by the object gives us the relative area of the polygon represented by this color (conditional probability $p(z|v)$).

All measures have been computed without taking into account the background and using a projection resolution of 640×480.[1] In all the experiments, the objects are centered in a sphere of 642 viewpoints built from the recursive discretization of an icosahedron, and the camera is looking at the center of

[1] For practical purposes, we use projection on the tangent plane instead of over the sphere of directions.

this sphere. This framework could be extended to any other placement of viewpoints, but the choice of a sphere of viewpoints permits analysis of an object in an isotropic manner. All the measures analyzed here are sensitive to the relative size of the viewpoint sphere with respect to the object.

Three models have been used: a cow (Figure 17.5a), a ship (Figure 17.5b), and the lady of Elche (Figure 17.5c). In Table 17.1, we show the number of polygons of the models used in this section and the cost of the preprocess step (i.e., the cost of computing the probability distributions $p(V)$, $p(Z|V)$, and $p(Z)$). To show the behavior of the measures, the sphere of viewpoints is represented by a thermic scale, where red and blue colors correspond, respectively, to the best and worst views. Note that a high-quality viewpoint corresponds to a high value for VE (Eq. [17.15]) and to a low value for both VKL (Eq. [17.19]) and VMI (Eq. [17.18]).

Figure 17.6 has been organized as follows. Rows (*a*) and (*b*) show, respectively, the behavior of VE and VMI measures. Columns (*i*) and (*ii*) show, respectively, the best and worst views, and columns (*iii*) and (*iv*) show two different projections of the viewpoint spheres. Figure 17.6 illustrates how VMI selects a representative view and how VE chooses to "see" the most highly discretized parts of the cow. While the worst views for VE correspond to the ones that see the less discretized parts, in the VMI case a low-quality view is obtained.

The different behavior between VKL and VMI is shown in Figure 17.7. Remember that the main difference between VMI and VKL is that, while VMI computes the distance between the projected areas of polygons and their average area "seen" by the set of viewpoints, VKL calculates the distance with respect

(a) (b) (c)

Figure 17.5 (a) Cow, (b) ship and (c) lady of Elche wireframe models.

Table 17.1 Number of Triangles of the Models Used and Computational Time (Seconds) of the Preprocess Step for Each Model. (Models are shown in Figure 17.5.)

	Cow	Ship	Lady of Elche
Number of triangles	9,593	47,365	51,978
Computational time	41	62	80

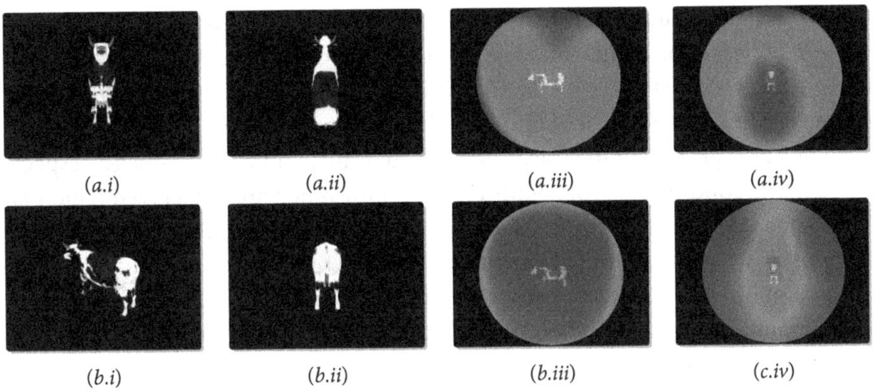

Figure 17.6 (*i*) The most representative (high quality) and (*ii*) the most restricted (low quality) views, and (*iii-iv*) the viewpoint spheres obtained respectively from the (*a*) VE and (*b*) VMI measures. Red colors on the sphere represent the highest quality viewpoints and blue colors represent the lowest quality viewpoints.

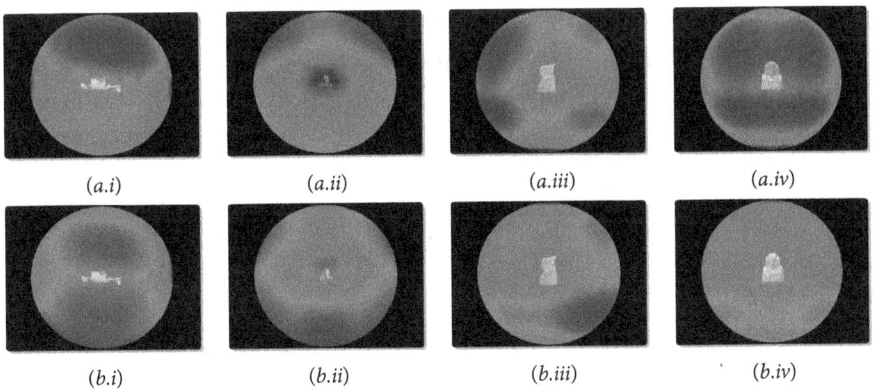

Figure 17.7 Viewpoint spheres obtained respectively from (*a*) VKL and (*b*) VMI measures for the ship and the cow models.

to the actual areas of the polygons. As a result, the reliability of VKL is much affected by the existence of many nonvisible or only partially visible polygons, as in the case of the ship and the lady of Elche models.

4. Image Information Channel

In this section, the information bottleneck method is applied to obtain a greedy algorithm,[2] which splits an image in quasi-homogeneous regions or segments.

[2] A greedy algorithm follows the heuristic of making the locally optimal choice at each step, always searching the most immediate benefit.

This image-splitting algorithm is constructed from an information channel $R \rightarrow B$ between the input variable R and the oputput variable B, which represent, respectively, the set of *regions* $\mathcal{R}$ of an image and the set of intensity *bins* $\mathcal{B}$. The input and output variables represent the distribution of relative areas of the regions or segments in the image and the color histogram of the image, respectively. The conditional probabilities express how much area of a given region is associated with each histogram bin. Figure 17.8 shows the main elements of this channel. At each step, the algorithm chooses the partition that maximizes the information gain and, thus, progressively discovers and reveals the structure of the image. This algorithm was introduced by Rigau et al. (2004) and extended by Bardera et al. (2009).

Given an image with N pixels, N_r regions, and N_b intensity bins, the three basic elements of the channel $R \rightarrow B$ are as follows:

- The conditional probability matrix $p(B|R)$, which represents the transition probabilities from each region of the image to the bins of the histogram, is defined by $p(b|r) = \frac{n(r,b)}{n(r)}$, where $n(r)$ is the number of pixels of region r and $n(r,b)$ is the number of pixels of region r corresponding to bin b. Conditional probabilities fulfil $\sum_{b \in \mathcal{B}} p(b|r) = 1, \forall r \in \mathcal{R}$. This matrix expresses how the pixels corresponding to each region of the image are distributed into the histogram bins.

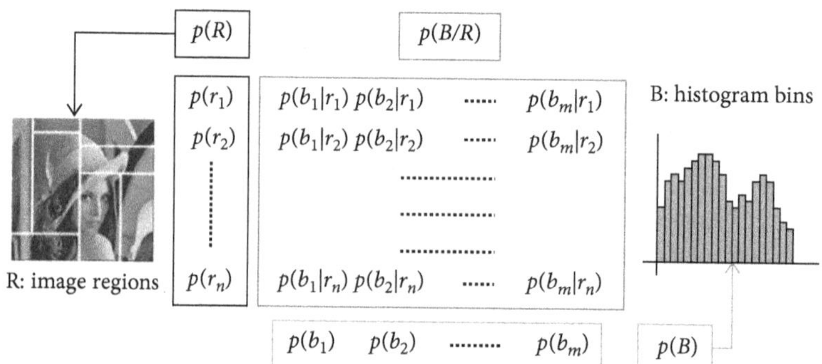

Figure 17.8 Image information channel $R \rightarrow B$ between the regions (or segments) of an image ($\mathcal{R}$) and the image histogram bins ($\mathcal{B}$). This channel is constructed from a digital image partitioned into several regions and the intensity bins of this image. The marginal probability distribution $p(R)$ is given by the relative area of each image region, the conditional probability distribution $p(B/r_i)$ express the amount of area of region i that corresponds to each histogram bin, and the marginal probability distribution $p(B)$ is given by the histogram of color bins. In this diagram, n represent the number of image regions N_r and m the number of intensity bins N_b.

- The input distribution $p(R)$, which represents the probability of selecting each image region, is defined by $p(r) = \frac{n(r)}{N}$ (i.e., the relative area of region r).
- The output distribution $p(B)$, which represents the normalized frequency of each bin b, is given by $p(b) = \sum_{r \in \mathcal{R}} p(r) p(b|r) = \frac{n(b)}{N}$, where $n(b)$ is the number of pixels corresponding to bin b.

From the information bottleneck method (section 2.3), we know that any clustering or quantization over R or B, respectively, represented by $\hat{R}$ and $\hat{B}$, will reduce $I(R;B)$. Thus, $I(R;B) \geq I(R;\hat{B})$ and $I(R;B) \geq I(\hat{R};B)$.

From this channel, we can construct a greedy top-down algorithm that partitions an image in quasi-homogeneous regions (Rigau et al., 2004). We adopt a binary space partitioning (BSP) that takes the full image as the unique initial partition and progressively subdivides it with vertical or horizontal lines chosen according to the maximum MI gain for each partitioning step. Similar algorithms were introduced in the fields of pattern recognition (Sethi and Sarvarayudu, 1982), learning (Kulkarni, 1998), and DNA segmentation (Bernaola et al., 1999). This algorithm has also been applied to analyze the complexity of art paintings (Rigau et al., 2008a) and Van Gogh's work (Rigau et al., 2008b, 2010)

The splitting process is represented over the channel $\hat{R} \to B$, where $\hat{R}$ denotes that R is the variable to be partitioned. Note that this channel varies at each partition step because the number of regions is increased; consequently, the marginal probabilities of $\hat{R}$ and the conditional probabilities of $\hat{R}$ known B also change. For a BSP strategy, the gain of MI due to the partition of a region $\hat{r}$ in two neighbor regions r_1 and r_2, such that

$$p(\hat{r}) = p(r_1) + p(r_2) \tag{17.20}$$

and

$$p(b|\hat{r}) = \frac{p(r_1)p(b|r_1) + p(r_2)p(b|r_2)}{p(\hat{r})}, \tag{17.21}$$

is given by

$$\begin{aligned} \delta I_{\hat{r}} &= I(R;B) - I(\hat{R};B) \\ &= p(\hat{r}) JS\left(\pi_1, \pi_2; p(B|r_1), p\left(B|r_2\right)\right), \end{aligned} \tag{17.22}$$

where $\pi_1 = \frac{p(r_1)}{p(\hat{r})}$ and $\pi_2 = \frac{p(r_2)}{p(\hat{r})}$. JS-divergence $JS\left(\pi_i, \pi_j; p(B|r_1), p\left(B|r_2\right)\right)$ (see Eq. [17.7]) between two regions can be interpreted as a measure of dissimilarity between them with respect to the intensity values. That is, when a region is partitioned, the gain of MI is equal to the degree of dissimilarity between the resulting regions times the size of the region. The optimal partition is determined by the maximum MI gain $\delta I_{\hat{r}}$.

From Eq. (17.4), the partitioning procedure can also be visualized as $H(B) = I(R;B) + H(B|R)$, where $H(B)$ is the histogram entropy and $I(B;R)$ and $H(B|R)$ represent, respectively, the successive values of MI and conditional entropy obtained after the successive partitions. The progressive acquisition of information increases $I(R;B)$ and decreases $H(B|R)$. This reduction of conditional entropy is due to the progressive homogenization of the resulting regions. Observe that the maximum MI that can be achieved is the histogram entropy $H(B)$ that remains constant along the process. The partitioning algorithm can be stopped using a ratio $MIR = \frac{I(R;B)}{H(B)}$ of mutual information gain or a predefined number of regions N_r.

Figure 17.9 shows different decompositions of a Van Gogh painting obtained for several values of the number of regions and only taking into account the luminance. Each region has been painted with the average color corresponding to that region. Observe how, with a relatively small number of regions, the painting composition is already visible (see Figure 17.9c–d), although the details are not sufficiently represented. Figure 17.10 shows two resulting partitions

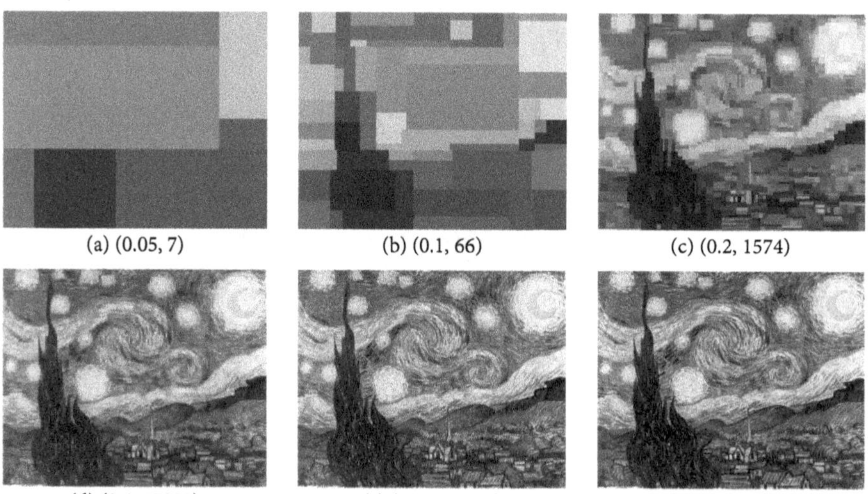

(a) (0.05, 7) (b) (0.1, 66) (c) (0.2, 1574)

(d) (0.4, 17309) (e) (0.8, 246573) (f) (1.0, 789235)

Figure 17.9 Evolution of the decomposition of a Van Gogh painting using the partitioning algorithm. For each figure, the corresponding mutual information ratio (MIR) and the number of regions (N_r) are shown. A total of 789235 regions are needed to achieve the total information.

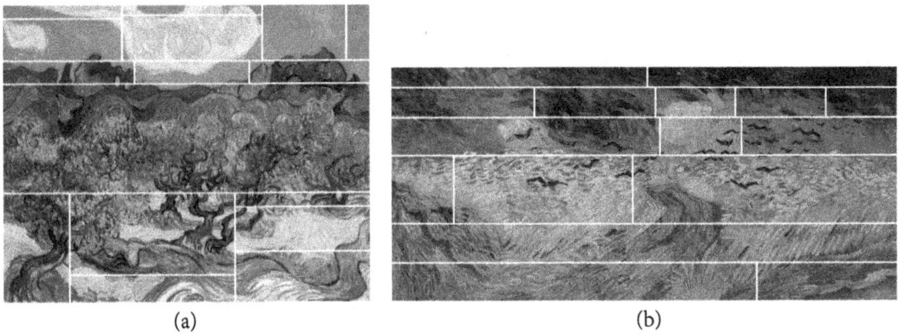

Figure 17.10 Decomposition of two Van Gogh paintings using the splitting algorithm. The number of regions is 16 in both cases.

from two Van Gogh paintings with only sixteen regions. Observe how the successive partitions produce relatively homogeneous regions. This comes from the objective of maximizing the mutual information of the channel, that is, the information gain at each step.

5. Scene Visibility Channel

An important topic in computer graphics is the accurate computation of the *global illumination* in a scene (i.e., the computation of the intensities of light, taking into account all the bounces over the surfaces of a scene). In this section, we deal with global illumination using the *radiosity method*, which only considers diffuse surfaces, where reflected light does not depend on the incoming direction. In the radiosity setting, the scene is discretized into small polygons or patches, and the difficulty in obtaining accurate scene illumination depends mainly on the degree of dependence (or correlation) between all the patches. In this context, entropy and mutual information are used to describe the degree of *randomness* and *dependence* (or *correlation*) in a scene, respectively. We consider that the *scene mutual information*, which measures the amount of information transferred between the different parts of a scene (Feixas et al., 1999; Feixas, 2002), is also an expression of the scene's complexity. At the end of this section, a mutual information-based criterion is integrated into the radiosity algorithm (Hanrahan et al., 1991) to obtain a realistically illuminated scene (Feixas, 2002).

5.1 The Radiosity Method and Form Factors

We will shortly review the radiosity method and the definition of form factor and its properties. The radiosity method, introduced by Goral et al. (1984),

Nishita and Nakamae (1985), and Cohen and Greenberg (1985), solve the problem of illumination in a virtual environment (or scene) of diffuse surfaces, where the radiation emitted by a surface is independent of direction. The radiosity of a patch surface is the light energy leaving this patch per discrete time interval; it can be seen as the combination of emitted and reflected energy for the patch. Thus, the radiosity algorithm computes the amount of light energy transferred among the surfaces of a scene, assuming that the scattering at all surfaces is perfectly diffuse.[3]

To solve the radiosity equation, we can use a finite element approach, discretizing the environment into N_p patches and considering the radiosities, emissivities, and reflectances constant over the patches (Figure 17.11). With these assumptions, the system of radiosity equations is given by (Goral et al., 1984):

$$B_i = E_i + \rho_i \sum_{j=1}^{N_p} F_{ij} B_j, \tag{17.23}$$

where B_i, E_i, and ρ_i are, respectively, the radiosity, emittance (or emissivity), and reflectance of patch i, B_j is the radiosity of patch j, and F_{ij} is the patch-to-patch form factor, dependent only on the geometry of the scene. Form factor F_{ij} between patches i and j expresses the fraction of energy leaving patch i, which goes directly to patch j.[4] In a certain way, the form factor also expresses the degree of visibility between two patches.

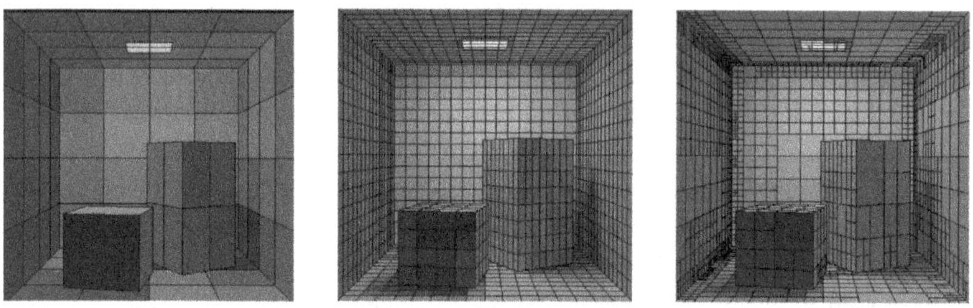

(a) (b) (c)

Figure 17.11 Three different discretisations for the Cornell box scene with 121, 1924, and 1924 patches, respectively.

[3] Contrary to specular reflection, in diffuse reflection an incident ray is reflected at many angles.

[4] Scene meshing not only has to represent illumination variations accurately, but it also has to avoid unnecessary subdivisions of the surfaces that would increase the number of form factors to be computed, and consequently the computational time.

Form factors fulfill the following properties:

1. Reciprocity

$$A_iF_{ij} = A_jF_{ji}, \forall i,j \in \{1,\dots,N_p\}, \tag{17.24}$$

where A_i and A_j are the areas of patches i and j, respectively.

2. Normalized energy conservation

$$\sum_{j=1}^{N_p} F_{ij} = 1, \forall i \in \{1,\dots,N_p\}. \tag{17.25}$$

5.2 Discrete Scene Visibility Channel

In this section, *a scene is modeled from a discrete information channel.* This fact enables us to introduce the notions of entropy and mutual information to study the visibility in a scene (Feixas et al., 2009; Feixas, 2002) and to optimally discretize the scene for a realistic illumination.

From the visibility point of view, a scene can be modeled as a discrete information channel $X \to Y$, where X and Y are discrete random variables with alphabet $\mathcal{X} = \mathcal{Y} = \{1,2,\dots,N_p\}$, corresponding to the set of patches of the scene. Observe that, in this case, the input and output variables represent, respectively, the same scene and have the same probability distributions. The conditional probabilities of this channel represent the amount of rays randomly cast from a polygon lands on another polygon. The diagram in Figure 17.12 shows the main elements of this channel.

This approach is compatible with the study of a scene from an infinite discrete random walk,[5] where at each step an imaginary particle (or ray) makes a transition from its current patch i to a new patch j with *transition probability* F_{ij}, which only depends on the current patch. (For more details, see Feixas, 2002).

The basic elements of the *scene information channel* are the marginal probability distributions of X and Y and the conditional probability matrix $p(Y|X)$:

[5] A Markov discrete random walk is characterized by the transition probabilities between the states. In case the stationary distribution exists, we talk of a stationary Markov chain. Thus, the Markov discrete random walk in a discretized scene is a discrete stationary Markov chain where the states correspond to the patches of a scene, the transition probabilities are the form factors F_{ij}, and the stationary distribution is given by the distribution of relative areas $\{a_i\}$ of patches (Sbert, 1996). Observe that any stationary Markov chain can be interpreted as an information channel.

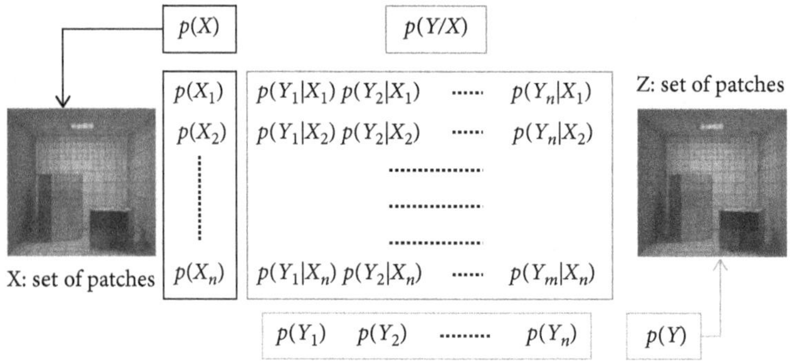

Figure 17.12 Scene information channel $X \to Y$ between the set of scene patches (represented by $\mathcal{X}$ and $\mathcal{Y}$). This channel is constructed from a scene discretized into a set of patches, and the marginal probability distributions $p(X)$ and $p(Y)$ are given by the relative area of patches (i.e., the probability that a ray hits a patch) and the transition probability matrix $p(Y|X)$ is constructed from the the probability that a ray leaving patch i reaches patch j. In this diagram, n represents the number of patches N_p.

- The conditional probability $p(y|x)$ is given by F_{ij}, which represents the probability that a ray leaving patch i reaches patch j. Conditional probabilities fulfil $\sum_{y \in \mathcal{Y}} p(y|x) = 1, \forall x \in \mathcal{X}$, that is, $\sum_{j=1}^{N_p} F_{ij} = 1, \forall i \in \{1, \ldots, N_p\}$.
- The marginal probabilities of input X and output Y, $p(x)$ and $p(y)$, are, respectively, given by the relative areas of patches, a_i and a_j, where $a_i = Pr\{X = i\} = A_i/A_T$ and $a_j = Pr\{Y = j\} = A_j/A_T$ with $i, j \in \mathcal{S}$, A_i and A_j are, respectively, the areas of patches i and j, and A_T is the total area of the scene. Probability $p(x)$ (or $p(y)$) represents the probability that a ray hits a patch i (or j).

These elements can be applied to Eqs. 17.1, 17.2, and 17.4 to obtain Shannon's information measures for a scene. The *discrete scene positional entropy* is defined by

$$H_P = H(X) = H(Y) = -\sum_{i \in \mathcal{S}} a_i \log a_i. \tag{17.26}$$

H_P expresses the average uncertainty on the position (patch) of a ray traveling an infinite random walk. It is the Shannon entropy of the relative area distribution of patches.

The *discrete scene conditional entropy*, called *scene entropy*, is defined by

$$H_S = H(Y|X) = \sum_{i \in \mathcal{S}} a_i H(Y|x) = -\sum_{i \in \mathcal{S}} a_i \sum_{j \in \mathcal{S}} F_{ij} \log F_{ij}. \tag{17.27}$$

The scene entropy can be interpreted as the average uncertainty that remains about the destination patch of a ray when the source patch is known. As the Bayes' theorem is expressed by the reciprocity property of the form factors (Eq. [17.24]), $a_i F_{ij} = a_j F_{ji}$, we obtain that $H_S = H(Y|X) = H(X|Y)$.

The *discrete scene mutual information* is defined by

$$I_S = I(X;Y) = H(Y) - H(Y|X) = \sum_{i \in \mathcal{S}} \sum_{j \in \mathcal{S}} a_i F_{ij} \log \frac{F_{ij}}{a_j} \tag{17.28}$$

and expresses the amount of information that the destination patch conveys about the source patch, and vice versa. It is a measure of the average information transfer or dependence between the different parts of a scene.

It is especially interesting to observe that the scene entropy of the interior of an empty sphere discretized into equal area patches is maximum. In this case, all the form factors are equal $\left(F_{ij} = a_j\right)$ (Feixas et al., 1999; Feixas, 2002), and the uncertainty on the destination patch of a random walk is maximum: $H_S = H_P = \log N_p$ (i.e., no visibility direction is privileged). Note also that the information transfer I_S is zero for any discretization of the sphere since in this case the variables X and Y are independent (i.e., $a_i F_{ij} = a_i a_j$).

The behavior of the entropy and mutual information is illustrated with the scenes of Figure 13. In scenes with the same discretization (Figures 17.13a,b), where we have a cubical enclosure with 512 interior cubes and the same H_P, observe (Table 17.2) that the increase of entropy is compensated by a mutual information decrease, and vice versa. In Figure 17.13b, the increase of mutual information reflects the high correlation created by the small central cubes, while

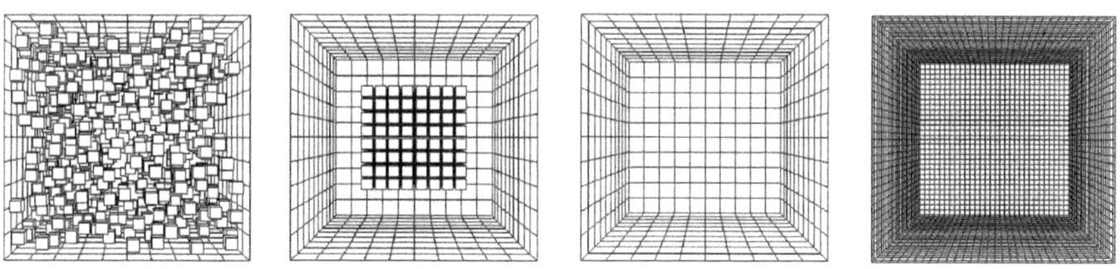

(a) (b) (c) (d)

Figure 17.13 Different scene configurations with entropy and mutual information values in Table 2.

Table 17.2 Entropy and Mutual Information Values for Scenes in Figure 17.13.

Scene	H_S	I_S	H_P
Fig. 13.*a*	6.761	4.779	11.541
Fig. 13.*b*	5.271	6.270	11.541
Fig. 13.*c*	7.838	1.391	9.229
Fig. 13.*d*	10.852	1.547	12.399

in the scene corresponding to Figure 17.13a there is much more uncertainty and less dependence between the small cubes.

Figures 17.13c–d illustrate how entropy and mutual information behave when the number of patches increases. According to Eq.(17.27), the scene entropy goes to infinity when the number of patches also goes to infinity. On the other hand, in this case, the scene mutual information tends to a finite value. This increase of entropy and mutual information is illustrated in Table 17.2 (for Figures 17.13 c–d), where we have a cubical enclosure with two different regular discretizations of their surfaces (600 and 5400 patches, respectively). For these two scenes, $H_P = \log N_p$, as all the patches of each scene have the same area.

5.3 The Refinement Criterion

In this section, we introduce the *scene continuous mutual information* and its relationship with the scene discrete mutual information. The scene continuous mutual information expresses with maximum precision the information transfer (or correlation) in a scene and can be interpreted as the scene complexity, expressing in a certain way the difficulty in achieving a precise discretization. The relationship between the discrete mutual information and the continuous mutual information is used as a refinement criterion for the hierarchical radiosity algorithm (Feixas, 1999, 2002).

Taking into account that a scene is a continuous system, some information is lost; that is, a distortion or error is introduced when a scene is discretized into patches. If we model the scene with a continuous information channel (or with a continuous random walk), the scene continuous mutual information can be obtained from the discrete mutual information (Eq. [17.28]), applying the following substitutions (Feixas, 1999, 2002):

- A_i and A_j are, respectively, substituted by dA_x and dA_y.

- F_{ij} is substituted by $F(x,y)\,dA_y$, where the point-to-point form factor $F(x,y)$ between points x and y is equal to $\cos\theta_x \cos\theta_y / \left(\pi r_{xy}^2\right)$ if x and y are mutually visible and 0 if not; θ_x and θ_y are the angles that the line joining x and y form with the normals at x and y; and r_{xy} is the distance between x and y.

Thus, the *scene visibility continuous mutual information* is given by

$$I_S^c = \int_S \int_S \frac{1}{A_T} F(x,y) \log\left(A_T F(x,y)\right) \mathrm{d}A_x \mathrm{d}A_y, \tag{17.29}$$

where $\mathcal{S}$ represents the set of surfaces of the scene. In the interior of an empty sphere, where $F(x,y) = 1/A_T$, the result obtained is $I_S^c = 0$. Remember that, in a sphere, $I_S = 0$. Thus, $I_S^c = I_S = 0$.

The relationship between the continuous and the discrete scene mutual information can be expressed by these two properties:

- If any patch is divided into two or more patches, the discrete mutual information I_S of the new scene increases or remains the same. This property is based on the information bottleneck method (section 2.3).
- The continuous scene visibility mutual information is the least upper bound to the discrete scene visibility mutual information.

Therefore, $I_S^c - I_S \geq 0$ and I_S converges to I_S^c when the number of patches tends to infinity (and the size of all the patches tends to zero). This difference expresses the loss of information transfer due to the discretization. Thus, between different discretizations of the same scene, we can consider that the most precise will be the one that has a higher discrete mutual information I_S—that is, the one that best captures the information transfer. With this in mind, the *scene discretization error* δ is defined by

$$\delta = I_S^c - I_S. \tag{17.30}$$

To obtain a realistic illumination, the scene discretization has to accurately represent illumination variations, but it has to avoid unnecessary refinements that would increase the computational cost. To achieve this objective, we introduce a mutual-information-based refinement criterion for the hierarchical radiosity algorithm, which is based on the loss of visibility information transfer between two patches due to the discretization.

To obtain a refinement criterion for hierarchical radiosity, we calculate the difference between both continuous and discrete patch-to-patch visibility information transfers. From Eq. (17.28), the term

$$I_{ij} = a_i F_{ij} \log\left(\frac{F_{ij}}{a_j}\right) \tag{17.31}$$

represents the discrete information transfer between patches i and j. It can be seen that $I_{ij} = I_{ji}$. From the continuous visibility mutual information (Eq. [17.29]), the continuous information transfer between patches i and j can be defined by

$$I^c_{ij} = \int_{S_i}\int_{S_j} \frac{1}{A_T} F(x,y) \log\left(A_T F(x,y)\right) \mathrm{d}A_x \mathrm{d}A_y. \tag{17.32}$$

Thus, the visibility discretization error between patches i and j is given by

$$\delta_{ij} = I^c_{ij} - I_{ij} \geq 0. \tag{17.33}$$

This difference gives us the discretization error between two patches and is used as the basis for the mutual-information-based refinement criterion. It can be proved that δ_{ij} is symmetric: $\delta_{ij} = \delta_{ji}$.

As the refinement strategy in hierarchical radiosity deals with one pair of patches at a time, a refinement criterion based on the discretization error between two patches (Eq. [17.33]) is proposed. This error expresses the loss of information transfer or, equivalently, the maximum potential gain of information transfer between two patches. Hence, this difference can be interpreted as the benefit to be gained by refining and can be used as a decision criterion. To create the mutual-information-based refinement criterion (the MI criterion), we multiply $\rho_i B_j$ (from Eq. [17.23]) by the discretization error δ_{ij}. Thus, the MI criterion is given by

$$\rho_i \delta_{ij} B_j < \epsilon. \tag{17.34}$$

The behavior of the MI criterion for a test scene is shown in Figure 17.14. Observe a greater mesh refinement in the corners, edges, and, in general, where a better precision is necessary to obtain accurate shadows, such as the shadows produced by the chairs, tables, and objects on the table.

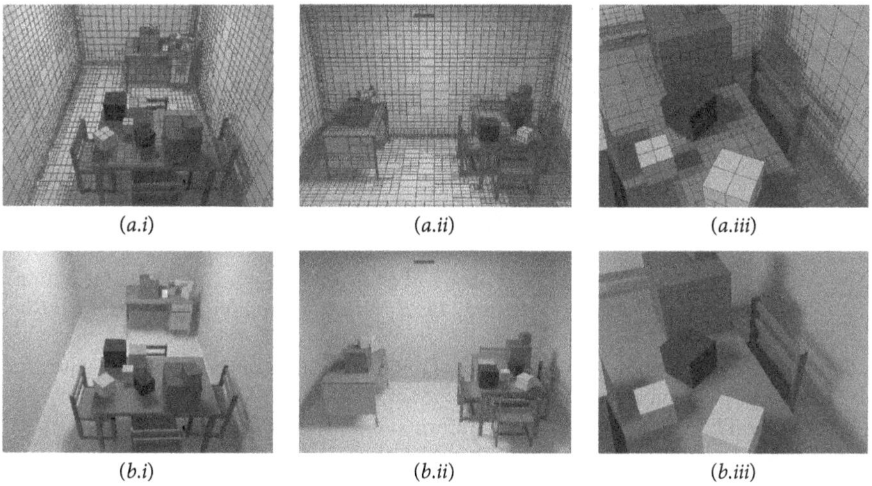

Figure 17.14 Results obtained by the MI criterion with a test scene showing (*a*) the mesh and (*b*) the shaded solution.

6. Conclusions

We have presented the information channel, one of the most powerful tools in information theory, and have shown how to apply it to examples taken from visualization, image processing, and global illumination. To build an information channel between input and output variables, we need to define the input distribution and conditional probabilities. How to obtain them for each problem at hand comes from a mixing of intuition, know-how of the field, and analogy with other applications. Once the information channel is built, we can give a meaningful and useful interpretation to its associated measures, such as mutual information and entropy. The information channel can also be reversed, which allows us to extract yet more useful knowledge, and it can be manipulated to segment input and output variables with a minimum loss of information.

We have illustrated the information channel with three examples: viewpoint selection, image segmentation, and global illumination.

In viewpoint selection, the channel is created projecting the 3D model onto the viewpoints. The viewpoints with minimum associated mutual information, or those with maximum associated conditional entropy, are the ones that give more information about the model.

In image segmentation, a channel is created between the regions of the image, initialized to the whole image as a single region, and the histogram. The bottleneck method is used to segment successively the image while gaining at each step the maximum of mutual information.

In global illumination, the random walk performed by the photons in a scene discretized into patches is interpreted as an information channel between the patches to themselves. The channel allows optimal refinement of discretization and the transport of light.

References

Bardera, A., J. Rigau, I. Boada, M. Feixas, and M. Sbert. (2009). "Image Segmentation Using Information Bottleneck Method." *IEEE Transactions on Image Processing*, 18(7): 1601–1612.

Bernaola, P., J. L. Oliver, and R. Román. (1999). "Decomposition of DNA Sequence Complexity." *Physical Review Letters*, 83(16): 3336–3339.

Borst, A., and F. E. Theunissen. (1999). "Information Theory and Neural Coding." *Nature Neuroscience*, 2(11): 947–957.

Burbea, J., and C. R. Rao. (1982). "On the Convexity of Some Divergence Measures Based on Entropy Functions." *IEEE Transactions on Information Theory*, 28(3): 489–495.

Chen, M., M. Feixas, I. Viola, A. Bardera, H.-W. Shen, and M. Sbert. (2014). *Information Theory Tools for Visualization*. Synthesis Lectures on Computer Graphics and Animation. San Rafael, CA: Morgan & Claypool Publishers.

Chen, M., and H. Jänicke. (2010). "An Information-Theoretic Framework for Visualization." *IEEE Transactions on Visualization and Computer Graphics*, 16: 1206–1215.

Cohen, M. F., and D. P. Greenberg. (1985). "The Hemi-Cube: A Radiosity Solution for Complex Environments." *Computer Graphics (Proceedings of SIGGRAPH '85)*, 19(3): 31–40.

Cover, T. M., and J. A. Thomas. (1991). *Elements of Information Theory*. New York: John Wiley. Wiley Series in Telecommunications.

Deweese, M. R., and Meister, M. (1999). "How to Measure the Information Gained from One Symbol." *Network: Computation in Neural Systems*, 10(4): 325–340.

Escolano, F., P. Suau, and B. Bonev. (2009). *Information Theory in Computer Vision and Pattern Recognition*. 1st ed. New York: Springer.

Feixas, M. (2002). *An Information-Theory Framework for the Study of the Complexity of Visibility and Radiosity in a Scene*. PhD thesis, Universitat Politècnica de Catalunya, Barcelona, Spain.

Feixas, M., A. Bardera, J. Rigau, Q. Xu, and M. Sbert. (2016). *Information Theory Tools for Image Processing*. AK Peters Visualization Series. Boca Raton, FL: A K Peters/CRC Press.

Feixas, M., E. del Acebo, P. Bekaert, and M. Sbert. (1999). "An Information Theory Framework for the Analysis of Scene Complexity." *Computer Graphics Forum*, 18(3): 95–106.

Feixas, M., M. Sbert, and F. González. (2009). "A Unified Information-Theoretic Framework for Viewpoint Selection and Mesh Saliency." *ACM Transactions on Applied Perception*, 6(1): 1–23.

Goral, C. M., K. E. Torrance, D. P. Greenberg, and B. Battaile. (1984). "Modelling the Interaction of Light between Diffuse Surfaces." *Computer Graphics (Proceedings of SIGGRAPH '84)*, 18(3): 213–222.

Hanrahan, D., P., Salzman, and L. Aupperle. (1991). "A Rapid Hierarchical Radiosity Algorithm." *Computer Graphics (Proceedings of SIGGRAPH '91)*, 25(4): 197–206.

Kulkarni, S. R., G. Lugosi, and S. S. Venkatesh. (1998). "Learning Pattern Classification–A Survey." *IEEE Transactions on Information Theory*, 44(6): 2178–2206.

Maes, F., A. Collignon, D. Vandermeulen, G. Marchal, and P. Suetens. (1997). "Multimodality Image Registration by Maximization of Mutual Information." *IEEE Transactions on Medical Imaging*, 16(2): 187–198.

Nishita, T., and E. Nakame. (1985). "Continuous Tone Representation of 3-D Objects Taking Account of Shadows and Interreflection." *Computer Graphics (Proceedings of SIGGRAPH '85)*, 19(3): 23–30.

Rigau, J., M. Feixas, and M. Sbert. (2004). "An Information Theoretic Framework for Image Segmentation." In *IEEE International Conference on Image Processing (ICIP'04)*, 2: 1193–1196. Singapore, Republic of Singapore.

Rigau, J., M. Feixas, and M. Sbert. (2008a). "Informational Aesthetics Measures." *IEEE Computer Graphics and Applications*, 28(2): 24–34.

Rigau, J., M. Feixas, and M. Sbert. (2008b). "Informational Dialogue with Van Gogh's Paintings." In P. Brown, D. W. Cunningham, V. Interrante, and J. McCormack (eds.), *Computational Aesthetics 2008. Eurographics Workshop on Computational Aesthetics in Graphics, Visualization and Imaging*, pp. 115–122, Eurographics Association, Goslar, Germany.

Rigau, J., M. Feixas, M. Sbert, and C. Wallraven. (2010). "Toward Auvers Period: Evolution of van Gogh's Style." In *Proceedings of the Sixth International Conference on Computational Aesthetics in Graphics, Visualization and Imaging*, Computational Aesthetics, 10: 99–106. Eurographics Association, Goslar, Germany, 2010.

Sbert, M. (1996). *The Use of Global Random Directions to Compute Radiosity. Global Monte Carlo Methods*. PhD thesis. Universitat Politècnica de Catalunya, Barcelona, Spain.

Sbert, M., Feixas, J. Rigau, I. Viola, and M. Chover. (2009). *Information Theory Tools for Computer Graphics*. San Rafael, CA: Morgan & Claypool Publishers.

Sbert, M., D. Plemenos, M. Feixas, and F. González. (2005). "Viewpoint Quality: Measures and Applications." In *Computational Aesthetics*, pp. 185–192, eds. László Neumann, Mateu Sbert, Bruce Gooch, Werner Purgathofer, Eurographics Association, Goslar, Germany.

Sethi, I. K., and G. Sarvarayudu. (1982). "Hierarchical Classifier Design Using Mutual Information." *IEEE Transactions on Pattern Analysis and Machine Intelligence*, 4(4): 441–445.

Shannon, C. E. (1948). "A Mathematical Theory of Communication." *Bell System Technical Journal*, 27: 379–423, 623–656.

Slonim, N., and N. Tishby. (2000). "Agglomerative Information Bottleneck." *Proceedings of NIPS-12 (Neural Information Processing Systems)*, pp. 617–623. Cambridge, MA: MIT Press.

Tishby, N., F. C. Pereira, and W. Bialek. (1999). "The Information Bottleneck Method." *Proceedings of the 37th Annual Allerton Conference on Communication, Control and Computing*, pp. 368–377.

Vázquez, P. P., M. Feixas, M. Sbert, and W. Heidrich. (2001). "Viewpoint Selection Using Viewpoint Entropy." *Proceedings of Vision, Modeling, and Visualization 2001*, pp. 273–280, Stuttgart, Germany, Aka GmbH.

Viola, P. A. (1995). *Alignment by Maximization of Mutual Information*. PhD thesis. Cambridge, MA: MIT Artificial Intelligence Laboratory (TR 1548).

Wang, C., and H.-W. Shen. (2011). "Information Theory in Scientific Visualization." *Entropy*, 13(1): 254–273.

Yeung, R. W. (2008). *Information Theory and Network Coding*. New York: Springer.

PART VII

INFO-METRICS AND NONPARAMETRIC INFERENCE

This part uses the tools of info-metrics to develop new approaches for non-parametric inference—inference that is based solely on the observed data. In Chapter 18, Tu develops a new entropy-based approach for nonparametric model averaging. The averaging is based on weights that are determined by maximizing the Shannon entropy's subject to the observed sample moments. A comparison with other averaging methods demonstrates the superior properties of this new info-metrics approach.

In Chapter 19, Mao and Ullah provide a new information-theoretic way for nonparametric estimation of econometric functions. This is a central problem, especially in the social sciences, where we don't know the underlying model, assumptions, or structure of the model. The chapter provides a nice way of tackling that problem, and demonstrates the power of the info-metrics framework in doing so. The chapter also contrasts this new approach with other, more classical, competing approaches.

Overall, this part demonstrates the versatility of the info-metrics framework for solving problems, even problems where the underlying structure, or model, is unknown and therefore one must rely solely on the observed sample information.

18

Entropy-Based Model Averaging Estimation of Nonparametric Models

Yundong Tu

1. Introduction

It has been long recognized that model averaging improves the performance of individual forecast models. The origin of forecast combination dates back to the seminal work of Bates and Granger (1969), and it has spawned a large literature. Some excellent reviews include Granger (1989), Clemen (1989), Diebold and Lopez (1996), Hendry and Clements (2001), and Timmermann (2006). Detailed empirical evidence demonstrating the gains in forecast accuracy through forecast combination can be found in Stock and Watson (1999, 2004, 2005, 2006), Rapach, Strauss, and Zhou (2010) and so on.

About a decade ago, Hansen (2007) proposed an averaging estimation with weights determined by the Mallows criterion. The resulting estimator is easy to implement and enjoys an optimality property, and it was extended to various settings later by Hansen (2008, 2009, 2010) and Hansen and Racine (2012), among others. Other frequentist model averaging methods have been studied by Hjort and Claeskens (2003), Zhang, Wan, and Zou (2013), and Tu and Yi (2017). See Ullah and Wang (2013) for a review. Recently, Zhang, Guo, and Carroll (2015) proposed a model averaging method based on the Kullback–Leibler distance under a homoscedastic normal error term, which is further modified for the heteroscedastic errors. This estimator behaves much like the Mallows model averaging estimator, although it is derived from the Akaike information. However, all these averaging estimators have been applied solely to linear parametric models. Model averaging for nonlinear models is only a recent endeavor; see Hansen (2014), and Henderson and Parmeter (2016). In the parametric (linear or nonlinear) model framework, Tu (2019) proposes an alternative method to construct averaging estimators based on model identification conditions; it is a consequence of optimizing a measure of information uncertainty measured via the Shannon entropy function.

In this chapter, I propose an entropy-based model averaging estimator for nonparametric regression models. In a spirit similar to Tu's model (2019),

Yundong Tu, *Entropy-Based Model Averaging Estimation of Nonparametric Models* In: *Advances in Info-Metrics: Information and Information Processing across Disciplines.* Edited by: Min Chen, J. Michael Dunn, Amos Golan, and Aman Ullah, Oxford University Press (2021).
DOI: 10.1093/oso/9780190636685.003.0018

I consider a model averaging where the averaging process is dictated endogenously by the joint entropy. The entropy functional I use here is the Clausius–Boltzmann–Shannon entropy (Shannon, 1948). The estimation for each model is carried out by local linear least squares (Fan 1992, 1993) using kernel functions. The individual estimates are combined with weights computed by optimizing the joint entropy function that measures the overall uncertainty (weights) of both the model and the observed data. The procedure itself is composed of only one optimization and is practically easy. While the nonparametric model averaging estimator of Henderson and Parmeter (2016) is an average over the choice of kernel, bandwidth selection mechanism and local polynomial order, our model averaging estimator for nonparametric regression is an average across models with different covariates.

Inference based on entropy measures has a long history. (See Golan (2008) for a detailed historical perspective and formulations.) In its current state, all entropic inference models developed after the late 1950s, building on the seminal work of Jaynes (1957), who developed the maximum entropy principle (MEP) as we know it today. For the problem of inferring a probability distribution given a small number of pieces of information (usually in terms of some moments), the MEP determines the most probable probability distribution that is consistent with the information available. This is done by simply maximizing the entropy subject to the observed information and normalization. The MEP forms the basis for the information-theoretic family of estimators, where the traditional Shannon entropy is substituted for one of the generalized entropy functions and the observed information is represented in different forms. See Kitamura and Stutzer (1997) and Golan (2002, 2008) for details and a thorough discussion of the family of information-theoretic methods of inference.

This chapter is organized as follows. Section 2 presents the nonparametric model and provides the details of how to construct a model averaging estimator with the maximum entropy principle. Finite sample experiments carried out in Section 3 compare our averaging estimators with the competitors. Section 4 contains an empirical example illustrating the use of the proposed method. Section 5 concludes the chapter and presents observations regarding future work.

2. Entropy-Based Model Averaging

2.1 Averaging for Linear Models

First, let us consider a linear model

$$y_i = x_{i1}\beta_1 + \cdots + x_{ik}\beta_k + e_i, i = 1, \ldots, T, \tag{18.1}$$

where y_i is the target variable of interest, $x_i = \left(x_{i1}, x_{i2}, \ldots, x_{i(k-1)}, x_{i,k}\right)'$ is the vector of predictors, k is the number of predictors, $\beta = (\beta_1, \beta_2, \ldots, \beta_{k-1}, \beta_k)'$ is the vector of parameters of interest, e_i is the stochastic error term assumed to be uncorrelated with the predictors, and T is the number of observations.

Decompose $x = (x_s', x_{-s}')'$ and $\beta = (\beta_s', \beta_{-s}')'$. In model s, only x_s is used as the predictors, for $s = 1, \ldots, S$. That is, the model s writes

$$y = x_s'\beta_s + e_s, \tag{18.2}$$

which imposes $\beta_{-s} = 0$. This leads to the moment condition

$$Ex_i'\left(y_i - x_{is}'\beta_s\right) = 0. \tag{18.3}$$

Our entropy-based estimator is to solve the following problem, with p_{is} denoting the joint probability of observation i and model s,

$$\text{Max } H(p) = -\sum p_{is} \log p_{is}, \tag{18.4}$$

subject to

$$\sum_i p_{is} x_i' \left(y_i - x_{is}'\beta_s\right) = 0, s = 1, \ldots, S, \tag{18.5}$$

$$\sum_{i,s} p_{is} = 1, \tag{18.6}$$

where $H(p)$ is the entropy functional with $p = \left(p_{is}\right)$.

Based on the estimators $\hat{\beta}_s$ and the estimated optimal probabilities p_{is}^*'s, I propose formulating the entropy-based model averaging (EMA) estimator of β as

$$\hat{\beta}_{EMA} = \sum q_s b_s, \tag{18.7}$$

where $q_s = \sum_i p_{is}^*$ and $b_s = \left(\hat{\beta}_s, \beta_{-s}\right)'$.

2.2 Averaging for Nonparametric Models

The model of interest to a nonparametric practitioner is

$$y_i = m\left(x_{i1}, x_{i2}, \ldots, x_{i(k-1)}, x_{ik}\right) + e_i, \; i = 1, \ldots, T, \tag{18.8}$$

where y_i is the target variable of interest, $x_i = \left(x_{i1}, x_{i2}, \ldots, x_{i(k-1)}, x_{ik}\right)'$ is the vector of predictors, k is the number of predictors, e_i is the stochastic error term assumed to be uncorrelated with the predictors, and $m(\cdot)$ is an unknown smooth function to be estimated. It is well known that the nonparametric estimation of m, by kernel smoothing, series, or spline, suffers from the so-called "curse of dimensionality"; that is, the estimation accuracy deteriorates as the dimension of x_i (i.e., k) increases. Various semiparametric methods are therefore developed as alternative approaches to mitigate such a disadvantage of a pure nonparametric model. Nevertheless, semiparametric models often impose more strict assumptions about the form of m. As an alternative, I consider the model averaging estimation of m based on an information-theoretic point of view.

To introduce our model averaging estimator, some notations are needed. Decompose $x_i = \left(x'_{is}, x'_{i,-s}\right)'$, where the subvector x_{is} collects the predictors (which form a subset of x_i) used in model s, for $s = 1, \ldots, S$. Let $m_s(x_{is})$ be the regression function of y_i on x_{is}, and the associated error term is denoted by e_{is}. That is, model s can be represented by the following regression:

$$y_i = m_s(x_{is}) + e_{is}. \tag{18.9}$$

Note that the above model offers an approximation to the true model in (18.8). Among various situations in which this approximation would produce a reasonable estimate, one special case is when

$$m(x_i) = m_s(x_{is}). \tag{18.10}$$

However, this might be too strong in practice. As a result, I only consider the above model to be an approximation, the idea being to combine various approximations to formulate an average estimate for m.

To be precise, first consider the local linear kernel estimation of $m(x)$ via the following optimization,

$$\left[\hat{m}(x), \hat{m}^{(1)}(x)\right] = \arg\min_{a,b} \sum_{i=1}^{T} \left[y_i - a - b'(x_i - x)\right]^2 k_h(x_i - x),$$

where $k_h(u) = k(u/h)/h$, $k(\cdot)$ is a (product) kernel function and h is a bandwidth parameter that shrinks to 0 as the sample size increases to infinity. $\hat{m}(x)$ is a consistent estimator of $m(x)$, while $\hat{m}^{(1)}(x)$ is consistent for $m^{(1)}(x)$, the first derivative of $m(x)$. Similarly, we can obtain estimates for $m_s(x_s)$ and $m_s^{(1)}(x_s)$. More generally, one can obtain the estimates of $m(x)$ and $m^{(1)}(x)$ by, respectively, $\hat{m}(x;p)$ and $\hat{m}^{(1)}(x;p)$, with a probability p_i assigned to observation i and $p = (p_1, \ldots p_T)$. To be precise, one solves the following probability weighted kernel least squares problem,

$$\left[\hat{m}(x;p), \hat{m}^{(1)}(x;p)\right] = \arg\min_{a,b} \sum_{i=1}^{T} p_i [y_i - a - b'(x_i - x)]^2 k_h(x_i - x).$$

It is known that the first-order conditions of the above optimization are linear in the probabilities, and explicit solutions can be easily found (Li and Racine, 2007).

The model s with regressors x_{is} offers a good estimate when (18.10) holds. This is equivalent to

$$\frac{\partial}{\partial x_{-s}} m(x) = 0. \tag{18.11}$$

This leads to a constraint that can be incorporated into the estimation procedure to be formulated in the maximum entropy paradigm. To do this, let $\delta_s x = x_s$; that is, δ_s is the selection operator that picks out the subset x_s, where s indicates the sth model. With this notation, the constraint (18.11) can be represented by

$$\delta_{-s} m^{(1)}(x) = 0.$$

As a result, I consider the following optimization problem

$$p^* = \arg\max_{p} \sum_{s=1}^{S} \sum_{i=1}^{T} p_{is} \log p_{is},$$

subject to

$$\delta_{-s} \hat{m}^{(1)}(x, p_s) = 0, \text{ for } s = 1, \ldots, S,$$

and

$$\sum_{s=1}^{S} \sum_{i=1}^{T} p_{is} = 1.$$

The EMA estimator for $m(x)$ is defined as

$$\hat{m}_{EMA}(x) = \sum_{s=1}^{S} q_s \hat{m}_s(x_s, p_s^*),$$

where

$$q_s = \sum_{i=1}^{T} p^*_{is}.$$

First, note that the constraints are linear in p's. Thus, the optimization is a convex problem. Lagrangian multipliers can be used to find the unique solution p^*. Second, the key condition for our model averaging to perform well is stated in Eq. (18.11), which guarantees that each model we use for averaging identifies the true function that characterizes the relationship between y and x. Therefore, our averaging estimator averages over models with different covariates all of which can help identify the same function of interest. In contrast, Henderson and Parmeter (2016) only consider the averaging over choice of bandwidth, kernel function, and local polynomial order. Therefore, when the concern is about the uncertainty of which regressors are to be included in the nonparametric models, our averaging method would be preferred. On the other hand, when we are unsure as to which kernel of bandwidth selection mechanism should be used, the averaging method of Henderson and Parmeter (2016) is helpful.

Finally, an argument that follows Tu (2019) in spirit can be used to study the asymptotic properties of $\hat{m}_{EMA}(x)$, although the current setting is more involved and has to deal with local smoothing. This would inevitably require that the underlying functions should be sufficiently smooth, that the observations should be stationary, and that certain conditions on the kernel functions and bandwidths should hold. The detailed treatment of the large sample theory will be investigated separately in a future endeavor. In the following section, I provide some numerical results to illustrate the advantage of the entropy-based model averaging estimator.

3. Simulation

3.1 Linear Models

I first study the performance of the entropy-based model averaging estimators in the class of linear parametric models. I consider the following data-generating process (DGP),

$$\text{DGP1}: \; y_i = x_{i1}\beta_1 + x_{i2}\beta_2 + \cdots + x_{i6}\beta_6 + e_i, i = 1, \dots, n.$$

I set the parameters $\beta_1 = 2$, $\beta_2 = 3$, $\beta_k = 0$, for $k = 3, \dots, 6$. The x_{ik}'s are independently drawn from normal distribution with mean 0 and standard deviation 2, and e_i is independently drawn from normal distribution with mean 0 and standard deviation $\sigma_e = 1, 2, 4$. The sample size n takes values from, 100, 200 and 400. Based on identification of the model parameters, the following models are used to construct the entropy-based model averaging estimator.

$$\begin{aligned} \text{Model 1} &: \; y_i = x_{i1}\beta_1 + x_{i2}\beta_2 + x_{i4}\beta_4 + e_{i1} \\ \text{Model 2} &: \; y_i = x_{i1}\beta_1 + x_{i2}\beta_2 + x_{i5}\beta_5 + e_{i2} \\ \text{Model 3} &: \; y_i = x_{i1}\beta_1 + x_{i2}\beta_2 + x_{i3}\beta_3 + e_{i3} \end{aligned}$$

I compute the mean-squared errors (MSE) of our proposed entropy-based parametric averaging estimator and compare it with that of ordinary least squares (OLS) by the relative MSE (RMSE) as the ratio of the former to the latter. The results for RMSEs evaluated at $x_k = 1$, for $k = 1, \ldots, 6$, over 500 replications, are reported in Table 18.1 (left panel). As the table shows, all the RMSEs are below 1, suggesting that our proposed entropy-based model averaging estimator outperforms the OLS estimator.

I next consider the nonparametric model averaging under DGP 1. I use the models containing the same set of variables as those that have been used in the above experiment (i.e., Model 1, 2, 3), but I leave the functional form (linearity in this case) unspecified. I compare the performance of our entropy-based nonparametric model averaging estimator with the fully nonparametric regression estimator that uses all six predictors. For the computations of this chapter, the second-order Gaussian kernel function is used unless otherwise specified, and the bandwidth for each individual model is selected using cross-validation. Other specifications are kept the same as above. The RMSEs of the averaging estimator compared to the local linear least squares (LLLS) estimator of Fan (1992) are reported in Table 18.1 (right panel). It is observed that the proposed averaging nonparametric estimator outperforms the pure nonparametric estimator for all error variance specifications. The gain of the averaging estimator also seems to increase as the signal-noise ratio decreases (the error variance gets larger).

Our second simulation design is taken from Hansen (2007):

$$DGP2 : y_i = \sum_{j=1}^{\infty} \theta_j x_{ij} + e_i.$$

Table 18.1 Relative Mean-Squared Errors under DGP 1

	Linear models			Nonlinear models		
n	$\sigma_e = 1$	$\sigma_e = 2$	$\sigma_e = 4$	$\sigma_e = 1$	$\sigma_e = 2$	$\sigma_e = 4$
100	0.1992	0.1918	0.2444	0.9939	0.8735	0.7364
200	0.2399	0.2537	0.2131	0.9933	0.9894	0.9581
400	0.2303	0.2273	0.2369	0.9945	0.9926	0.9326

Table 18.2 Relative Mean-Squared Errors under DGP2, as Compared to MMA

	$R^2 = 0.1$				$R^2 = 0.5$			
	$K = 10$		$K = 20$		$K = 10$		$K = 20$	
n	$\alpha = 1.0$	$\alpha = 1.5$	$\alpha = 1.0$	$\alpha = 1.5$	$\alpha = 1.0$	$\alpha = 1.5$	$\alpha = 1.0$	$\alpha = 1.5$
100	0.9101	0.8741	0.9316	0.8155	1.1328	1.2315	1.1333	1.2013
200	0.8140	0.6526	0.8827	0.7179	1.1012	1.1081	1.1099	1.1990
400	0.6323	0.5479	0.7620	0.5648	1.0485	1.0287	1.0819	1.0760

I set $x_{i1} = 1$ to be the intercept; the remaining x_{ij}'s $(j \geq 1)$ are independent and identically distributed (i.i.d.) $N(0,1)$. The error e_i is i.i.d. $N(0,1)$ and independent of x_i. The parameters are determined by the rule $\theta_j = c\sqrt{2\alpha}j^{-\alpha-1/2}$, for $j = 1, \ldots, K$ and $\theta_j = 0$, otherwise. The population $R^2 = c^2/\left(1+c^2\right)$ is controlled by the parameter c. The sample size varies from 100, 200 to 400. The parameter α is chosen from $\{1.0, 1.5\}$. I consider the nested model averaging with our proposed method and compare its performance with that of Hansen's Mallows model averaging (MMA) estimator. The number of models M is determined by the rule (so $M = 13, 17, 22$ for the three sample sizes). The relative mean-squared errors of our estimator compared to Hansen's are reported in Table 18.2 for $R^2 = 0.1$ and 0.5. The superior performance of our estimator is apparently observed as the RMSEs are all smaller than 1 for $R^2 = 0.1$. However, in the experiments with a larger $R^2 = 0.5$, we notice that our proposed method does not perform well in some cases. This is easily explained by our identification condition which states that each submodel should identify the same regression parameters. This requirement is not satisfied in the current setting, with all submodels being approximating instruments that only serve to formulate a reasonable averaged forecast. As R^2 increases, one way to improve the performance of our averaging estimator is to increase the number of regressors in each approximating model so that the identified model becomes closer to the true model. This can be inferred from the fact that as K decreases, our averaging estimator tends to perform better relative to the MMA estimator. In sum, the above reported results show that, as long as the approximations are good enough (low R^2), our proposed method provides a better alternative to the MMA. We note that R^2 is often very low in most cross-sectional models. In practice, a comparison of the pseudo out-of-sample mean-squared forecast errors may be the first step in deciding which averaging method to use.

3.2 Nonlinear Models

I next turn to study of the finite sample properties of our nonparametric averaging estimator in nonlinear models. Consider the following experimental designs,

$$DGP\,3\ :\ y_i = x_{i1}^2\beta_1 + x_{i2}^2\beta_2 + \cdots + x_{i6}^2\beta_6 + e_i;$$
$$DGP\,4\ :\ y_i = \exp\{x_{i1}\beta_1 + x_{i2}\beta_2 + \cdots + x_{i6}\beta_6\} + e_i;$$
$$DGP\,5\ :\ y_i = \exp\{x_{i1}\beta_1\} + \Phi(x_{i2}\beta_2/3) + x_{i3}\beta_3 + \cdots + x_{i6}\beta_6 + e_i.$$

For DGP 3–5, I set the parameters $\beta_1 = 2$, $\beta_2 = 3$, $\beta_k = 0$, for $k = 3, \ldots, 6$. In DGP 5, Φ denotes the standard normal cumulative distribution function. The x_{ik}'s and e_i are independently drawn from standard normal distribution. The sample size n takes values from 100, 200, and 400. Note that DGP 3 is linear in parameters and DGP 4 may be approximated by a model that is linear in parameters after a transformation, which are, however, unknown to the model builder. DGP 5 is nonlinear in both the parameters and variables, and cannot be approximated by any model that is linear in parameters after transformations.

The following models, based on identification of the nonlinear regression function, are used to construct the entropy-based model averaging estimator.

$$\text{Model 1}\ :\ y_i = m_1(x_{i1}, x_{i2}, x_{i4}) + e_{i1}$$
$$\text{Model 2}\ :\ y_i = m_2(x_{i1}, x_{i2}, x_{i5}) + e_{i2}$$
$$\text{Model 3}\ :\ y_i = m_3(x_{i1}, x_{i2}, x_{i3}) + e_{i3}$$

I compare the mean-squared errors of our proposed estimator with that of the local linear least squares (LLLS) estimator of Fan (1992). The comparison for MSE is evaluated at $x_k = 1$, for $k = 1, \ldots, 6$. The results are reported in the left panel of Table 18.3. As the RMSEs are all below 1, the results suggest that our proposed entropy-based model averaging estimator outperforms the LLLS estimator. We further observe that as sample size increases, the averaging estimator generally tends to perform close to the LLLS estimator. This finding is consistent with that of Henderson and Parmeter (2016).

Furthermore, I add the comparison of the least squares cross-validation (LSCV) averaging estimator of Henderson and Parmeter (2016) with the LLLS estimator under DGP 3-5. The construction of the averaging estimator follows the same way as that of Henderson and Parmeter (2016). The relative mean-squared errors of the LSCV averaging estimator compared to those of the LLLS estimator are reported in the middle panel of Table 18.3. It is observed that the LSCV estimator clearly outperforms the LLLS estimator. In addition, it is observed that the LSCV estimator performs slightly worse than our EMA estimator. This finding is as expected because the EMA estimator rules out some irrelevant variables in the construction of models before averaging, while the LSCV averaging includes all the irrelevant regressors.

As a referee points out, the bandwidth selection method of Hall, Li and Racine (2007, hereafter HLR) may also help remove the irrelevant variables. To compare the performance of their estimator with our estimator, I also compute the relative mean-squared errors of the HLR estimator as compared to the LLLS

Table 18.3 Relative Mean-Squared Errors under DGP 3-5, as Compared to LLLS

	EMA			LSCV			HLR		
n	DGP 3	DGP 4	DGP 5	DGP 3	DGP 4	DGP 5	DGP 3	DGP 4	DGP 5
100	0.5871	0.4014	0.5213	0.6325	0.5262	0.5862	0.6413	0.5255	0.5753
200	0.7835	0.8499	0.8304	0.8513	0.8995	0.8841	0.8606	0.8966	0.9021
300	0.9141	0.8801	0.9063	0.9824	0.9521	0.9795	0.9825	0.9443	0.9914

estimator, which are reported in the right panel of Table 18.3. We note that both the HLR estimator and the LLLS estimator use cross-validation to select the bandwidth. However, the difference is that the former estimator allows some of the bandwidths to diverge to infinity and some to diminish to zeros, while the latter only selects the optimal bandwidth from an interval shrinking to zero, that is, choosing the optimal $h = c \times n^{-1/(4+q)}$ by selecting $c \in [-3, 3]$, where q denotes the number of regressors in the regression model. It is observed that the HLR estimator does improve over the LLLS estimator in the current setup where irrelevant regressors appear. However, generally, it performs slightly worse than the LSCV averaging estimator. The averaging seems to be as effective to remove the irrelevant regressors as the bandwidth selection in HLR. On the other hand, the EMA estimator utilizes the averaging and rules out some irrelevant regressors when constructing the individual models for averaging, and outperforms both LSCV and HLM, as expected.

4. An Empirical Example

Next I consider the study of the wage equation (Mincer, 1974) using the nonparametric approach. I used data from the National Longitudinal Survey of Youth (NLSY), a national representative sample of persons between 14 and 22 years of age when the survey began in 1979. The data set contains hourly wages (for 1990), education (years), experience (years), ability as measured by scores on the Armed Forces Qualifying Test (AFQT), family background variables, including reading at home, father's education (years), number of siblings, and school quality. AFQT is the percentile rank at which the respondent performed in the test. School quality is a composite index of the percentile rank of the respondent's high school for the following characteristics: library books per student, teachers per student, counselors per student, fraction of faculty with a master's degree or more, (real) starting salary for teachers with a BA, average daily attendance and the dropout rate for sophomores. Reading at home is a composite index for the family's receipt of magazines, newspapers and the

Table 18.4 Summary Statistics for Variables

	All groups	Male	Female
Hourly wage	11.15 (6.79)	12.01 (7.57)	10.03 (5.40)
Years of education	13.42 (2.24)	13.22 (2.30)	13.67 (2.13)
Years of experience	9.93 (3.29)	10.61 (3.45)	9.05 (2.83)
AFQT score	47.47 (28.22)	46.77 (29.20)	48.37 (26.88)
School quality	0.50 (0.11)	0.50 (0.10)	0.49 (0.11)
Father's education (years)	11.37 (3.81)	11.35 (3.88)	11.38 (3.71)
Number of siblings	3.46 (2.39)	3.50 (2.39)	3.41 (2.38)
Reading at home	2.20 (0.93)	2.20 (0.92)	2.19 (0.94)
Sample size	1933	1088	845

Table 18.5 Out-of-Sample Mean-Squared Forecast Errors

	EMA	LLLS
$n_1 = 100$	0.2133	0.2186
$n_2 = 200$	0.1921	0.1975
$n_3 = 300$	0.2116	0.2223

possession of a library card. Following Griffin and Ganderton (1996), I excluded those respondents who worked less than 20 hours per week, whose hourly wage was below \$2.00, and whose responses remained incomplete in the sample. The total sample used here is 1,933, and the statistics of the data are summarized in Table 18.4. The mean and the standard deviation (in parentheses) are calculated both for the whole sample and for males and females separately.

These covariates were used to explain the return to education. F denotes variables related to family background; A the ability measure (AFQT score), Q school quality, and X all other explanatory variables, including education. The following models used by Griffin and Ganderton (1996), where HW stands for hourly wage, are aggregated to formulate the averaging model:

$$\begin{aligned} \ln HW &= m_1(X,F) + e_1, \\ \ln HW &= m_2(X,A) + e_2, \\ \ln HW &= m_2(X,Q) + e_3. \end{aligned}$$

To evaluate our model performance, I compared the accuracy of (out-of-sample) forecasts of (log hourly) wages produced by our proposed EMA estimator and that of the LLLS estimator for the last n_1 interviewers, for $n_1 = 100,\ 200,\ 300$, respectively. The mean-squared forecast errors are reported in Table 18.5. As can be seen, the averaging estimator produces more accurate forecasts than the local linear estimator, which is robust to the three different sample splits.

5. Conclusion

This chapter proposes an information-theoretic estimator for nonparametric models. The estimator serves as the first nonparametric estimator in this framework and is easy to implement in practice. Finite sample experiments show the superiority of the averaging estimator compared to local linear estimator. The averaging estimator is also shown to well behave in modeling the wage equation using NLSY data.

Several interesting directions can be taken to extend the current work. First, it is theoretically important to establish the asymptotic properties of the proposed averaging nonparametric estimators. Second, imposing certain structures in nonparametric models, such as the single index or the additive structure, would help alleviate the curse of dimensionality in model averaging. In this case, development of a new model averaging strategy is in order. These topics are left for future studies.

Acknowledgments

I thank the editors, two anonymous referees, Professor Amos Golan, Aman Ullah, and participants in Recent Innovations in Info-Metrics: An Interdisciplinary Perspective at American University in 2014 for suggestions to improve the chapter. The data used in the empirical analysis was kindly provided by Shuo Li. This research is supported by the National Natural Science Foundation of China (Grants 71301004, 71472007, 71532001, and 71671002), China's National Key Research Special Program (Grant 2016YFC0207705), the Center for Statistical Science at Peking University, and Key Laboratory of Mathematical Economics and Quantitative Finance (Peking University), Ministry of Education.

References

Bates, J. M., and Granger, C. M. W. (1969). "The Combination of Forecasts." *Operations Research Quarterly*, 20: 451–468.

Clemen, R. T. (1989). "Combining Forecasts: A Review and Annotated Bibliography." *International Journal of Forecasting*, 5: 559–581.

Clements, M. P., and Hendry, D. F. (2001). "Forecasting with Difference-Stationary and Trend-Stationary Models." *Econometrics Journal*, 4: S1–S19.

Diebold, F. X., and Lopez, J. A. (1996). "Forecast Evaluation and Combination." In G.S. Maddala and C.R. Rao (Eds.), *Handbook of Statistics*. North-Holland, Amsterdam: Elsevier.

Fan, J. (1992). "Design-Adaptive Nonparametric Regression." *Journal of the American Statistical Association*, 87: 998–1004.

Fan, J. (1993). "Local Linear Regression Smoothers and Their Minimax Efficiency." *The Annals of Statistics*, 21: 196–216.

Golan, A. (2002). "Information and Entropy Econometrics: Editor's View." *Journal of Econometrics*, 107: 1–15.
Golan, A. (2008). "Information and Entropy Econometrics—A Review and Synthesis." *Foundations and Trends in Econometrics*, 2: 1–145.
Granger, C. W. J. (1989). "Combining Forecasts—Twenty Years Later." *Journal of Forecasting*, 8: 167–173.
Griffin, P., and Ganderton, P. T. (1996). "Evidence on Omitted Variable Bias in Earnings Equations." *Economics of Education Review*, 15: 139–148.
Hall, P., Li, Q., and Racine, J. S. (2007). "Nonparametric Estimation of Regression Functions in the Presence of Irrelevant Regressors." *Review of Economics and Statistics*, 89(4): 784–789.
Hansen, B. E. (2007). "Least Squares Model Averaging." *Econometrica*, 75: 1175–1189.
Hansen, B. E. (2008). "Least Squares Forecast Averaging." *Journal of Econometrics*, 146: 342–350.
Hansen, B. E. (2009). "Averaging Estimators for Regressions with a Possible Structural Break." *Econometric Theory*, 35: 1498–1514.
Hansen, B. E. (2010). "Averaging Estimators for Autoregressions with Near Unit Root." *Journal of Econometrics*, 158: 142–155.
Hansen, B. E. (2014). "Nonparametric Sieve Regression: Least Squares, Averaging Least Squares, and Cross-Validation." In *The Oxford Handbook of Applied Nonparametric and Semiparametric Econometrics and Statistics*, (pp. 215–248). New York: Oxford University Press.
Hansen, B. E., and Racine, J. S. (2012). "Jackknife Model Averaging." *Journal of Econometrics*, 167(1): 38–46
Henderson, D. J., and Parmeter, C. F. (2016). "Model Averaging over Nonparametric Estimators." In G. Gonzalez-Rivera, R. C. Hill, and T. H. Lee (eds.), *Advances in Econometrics*, 36: (pp. 539–560). Bingley, UK: Emerald Group Publishing Limited.
Hjort, N., and Claeskens, G. (2003). "Frequentist Model Average Estimators." *Journal of the American Statistical Association*, 98: 879–899.
Jaynes, E. T. (1957). "Information Theory and Statistical Mechanics." *Physics Review*, 106: 620–630.
Kitamura, Y., and Stutzer, M. (1997). "An Information-Theoretic Alternative to Generalized Method Of Moment Estimation." *Econometrica*, 66: 861–874.
Li, Q., and Racine, J. S. (2007). "Nonparametric Econometrics: Theory and Practice." Princeton, NJ: Princeton University Press.
Mincer, J. (1974). "Schooling, Experience, and Earnings." Human Behavior and Social Institutions, No. 2.
Rapach, D. E., Strauss, J., and Zhou, G. (2010). "Out-of-Sample Equity Premium Prediction: Combination Forecasts and Links to the Real Economy." *Review of Financial Studies*, 23: 821–862.
Shannon, C. E. (1948). "A Mathematical Theory of Communication." *Bell System Technical Journal*, 27: 379–423.
Stock, J. H., and Watson, M. W. (1999). "A Comparison of Linear and Nonlinear Univariate Models for Forecasting Macroeconomic Time Series." In Engle, White, (Eds.), Cointegration, Causality and Forecasting: A Festschrift for Clive W. J. Granger. London: Oxford University Press.
Stock, J. H., and Watson, M. W. (2004). "Combination Forecasts of Output Growth in a Seven Country Data Set." *Journal of Forecasting*, 23: 405–430.

Stock, J. H., and Watson, M. W. (2005). An Empirical Comparison of Methods for Forecasting Using Many Predictors. Working Paper. NBER.

Stock, J. H., and Watson, M. W. (2006). "Forecasting with Many Predictors." In Elliott, Granger, and A. Timmermann (eds.), *Handbook of Economic Forecasting*, (Chapter 10). Elsevier.

Tsallis, C. (1988). "Possible Generalization of Boltzmann-Gibbs Statistics." *Journal of Statistical Physics*, 52: 479–487.

Timmermann, A. (2006). "Forecast Combinations." In Elliott, Granger, and Timmermann (eds.), *Handbook of Economic Forecasting*, (Chapter 4). Elsevier.

Tu, Y. (2019). "Model Averaging Partial Effect (MAPLE) Estimation of Large Dimensional Data." Working Paper. Peking University.

Tu, Y., and Yi, Y. (2017). "Forecasting Cointegrated Nonstationary Time Series with Time-Varying Variance." *Journal of Econometrics*, 196(1): 83–98.

Ullah, A., and Wang, H. (2013). "Parametric and Nonparametric Frequentist Model Selection and Model Averaging." *Econometrics*, 1(2): 157–179.

Zhang, X., Wan, A., and Zou, G. (2013). "Model Averaging by Jackknife Criterion in Models with Dependent Data." *Journal of Econometrics*, 172: 82–94.

Zhang, X., Zou, G., and Carroll, R. J. (2015). "Model Averaging Based on Kullback–Leibler Distance." *Statistica Sinica*, 25: 1583–1598.

19

Information-Theoretic Estimation of Econometric Functions

Millie Yi Mao† and Aman Ullah

1. Introduction

In the literature of estimation, specification, and testing of econometric models, many parametric assumptions have been made. First, parametric functional forms of the relationship between independent and dependent variables are usually assumed to be known. For example, a regression function is often considered to be linear. Second, the variance of the error terms conditional on the independent variables is specified to have a parametric form. Third, the joint distribution of the independent and dependent variables are conventionally assumed to be normal. Last but not least, in many econometric studies, the independent variables are considered to be nonstochastic. However, parametric econometrics has drawbacks since particular specifications may not capture the true data-generating process. In fact, the true functional forms of econometric models are hardly known. Misspecification of parametric econometric models may therefore result in invalid conclusions and implications. Alternatively, data-based econometric methods can be adopted to avoid the disadvantages of parametric econometrics and implemented into practice. One widely used approach is the nonparametric kernel technique; see Ullah (1988), Pagan and Ullah (1999), Li and Racine (2007), and Henderson and Parmeter (2015). However, nonparametric kernel procedures have some deficiencies, such as the "curse of dimensionality" and a lack of efficiency due to a slower rate of convergence of the variance to zero. Accordingly, we propose a new information-theoretic (IT) procedure for econometric model specification by using a classical maximum entropy formulation. This procedure is consistent and efficient, and is based on minimal distributional assumptions.

Shannon (1948) derived the entropy (information) measure, which is similar to that of Boltzmann (1872) and Gibbs (1902). Using Shannon's entropy measure, Jaynes (1957a, 1957b) developed the maximum entropy principle to infer probability distribution. Entropy is a measure of a variable's average information content, and its maximization subject to some moments and normalization

Millie Yi Mao and Aman Ullah, *Information-Theoretic Estimation of Econometric Functions* In: *Advances in Info-Metrics: Information and Information Processing across Disciplines*. Edited by: Min Chen, J. Michael Dunn, Amos Golan, and Aman Ullah, Oxford University Press (2021). © Oxford University Press.
DOI: 10.1093/oso/9780190636685.003.0019

provides a probability distribution of the variable. The resulting distribution is known as the maximum entropy distribution; for more details, see Zellner and Highfield (1988), Ryu (1993), Golan et al. (1996), Harte et al. (2008), Judge and Mittelhammer (2011), and Golan (2018). The joint probability distribution based on the maximum entropy approach is a purely data-driven distribution in which parametric assumptions are avoided. This distribution can be used to determine the regression function (conditional mean) and its response function (derivative function), which are of interest to empirical researchers. This determination is the main goal of this chapter.

We organize this chapter in the following order. In section 2, we present the IT-based regression and response functions using a bivariate maximum entropy distribution. A recursive integration process is developed for their implementation. In section 3, we carry out simulation examples to illustrate the small sample efficiency of our methods, and then we present an empirical example of Canadian high school graduate earnings. In section 4, we present asymptotic theory on our IT-based regression and response function estimators. In section 5, we draw conclusions and provide potential future extensions. The mathematical details of the algorithm used in section 2 and the proofs of asymptotic properties of the IT-based estimators are shown in the Appendix.

2. Estimation of Distribution, Regression, and Response Functions

We consider $\{y_i, x_i\}, i = 1, \dots, n$ independent and identically distributed observations from an absolutely continuous bivariate distribution $f(y, x)$. Suppose that the conditional mean of y given x exists and that it provides a formulation for the regression model as

$$\begin{aligned} y &= E(y|x) + u \\ &= m(x) + u, \end{aligned} \tag{19.1}$$

where the error term u is such that $E(u|x) = 0$, and the regression function (conditional mean) is

$$E(y|x) = m(x) = \int_y y \frac{f(y,x)}{f(x)} dy. \tag{19.2}$$

When the joint distribution of y and x is not known, which is often the case, we propose the IT-based maximum entropy method to estimate the densities of the random variables and introduce a recursive integration method to solve the conditional mean of y given x.

2.1 Maximum Entropy Distribution Estimation: Bivariate and Marginal

Suppose x is a scalar and the marginal density of it is unknown. Our objective is to approximate the marginal density $f(x)$ by maximizing the information measure (Shannon's entropy) subject to some constraints. That is,

$$\underset{f}{\text{Max}}\ H(f) = -\int_x f(x)\log f(x)dx,$$

subject to

$$\int_x \phi_m(x)f(x)dx = \mu_m = E\phi_m(x), m = 0, 1, \ldots, M,$$

where $\phi_m(x)$ are known functions of x. $\phi_0(x) = \mu_0 = 1$. See, for example, Jaynes (1957a, 1957b) and Golan (2018). The total number of constraints is $M + 1$. In particular, $\phi_m(x)$ can be moment functions of x. We construct the Lagrangian

$$\mathcal{L}(\lambda_0, \lambda_1, \ldots, \lambda_M) = -\int_x f(x)\log f(x)dx + \sum\nolimits_{m=0}^{M}\lambda_m\left(\mu_m - \int_x \phi_m(x)f(x)dx\right),$$

where $\lambda_0, \lambda_1, \ldots, \lambda_M$ represent Lagrange multipliers. The solution has the form

$$f(x) = \exp\left[-\sum\nolimits_{m=0}^{M}\lambda_m\phi_m(x)\right] = \frac{\exp\left[-\sum_{m=1}^{M}\lambda_m\phi_m(x)\right]}{\int_x \exp\left[-\sum_{m=1}^{M}\lambda_m\phi_m(x)\right]dx}$$

$$\equiv \frac{\exp\left[-\sum_{m=1}^{M}\lambda_m\phi_m(x)\right]}{\Omega(\lambda_m)},$$

where λ_m is the Lagrange multiplier corresponding to constraint $\int_x \phi_m(x)f(x)dx = \mu_m$, and λ_0 (with $m = 0$) is the multiplier associated with the normalization constraint. With some simple algebra, it can be easily shown that $\lambda_0 = \log\Omega(\lambda_m)$ is a function of other multipliers. Replacing $f(x)$ and λ_0 into $\mathcal{L}(\lambda_0, \lambda_1, \ldots, \lambda_M) = \mathcal{L}(\boldsymbol{\lambda})$, we get

$$\mathcal{L}(\boldsymbol{\lambda}) = \sum\nolimits_{m=1}^{M}\lambda_m E\phi_m(x) + \lambda_0.$$

The Lagrange multipliers are solved by maximizing $\mathcal{L}(\boldsymbol{\lambda})$ with respect to λ_m's. The above inferred density is based on minimal information and assumptions. It is the flattest density according to the constraints. In this case, the Lagrange multipliers not only are the inferred parameters characterizing the density function, but also capture the amount of information conveyed in each one of the constraints relative to the rest of the constraints used. They measure the strength of the constraints.

In particular, when $M = 0$, $f(x)$ is a constant, and hence x follows a uniform distribution. When the first moment of x is known, $f(x)$ has the form of an exponential distribution. When the first two moments of x are known, $f(x)$ has the form of a normal distribution. Furthermore, if more moment information is given, that is, $M \geq 3$, to estimate the Lagrange multipliers, we use the Newton method considered in the literature. See Mead and Papanicolaou (1984) and Wu (2003).

In the bivariate case, the joint density of y and x is obtained from maximizing the information criterion $H(f)$ subject to some constraints. Here, we assume the moment conditions up to fourth order are known. Then

$$\operatorname*{Max}_{f} H(f) = -\int_x \int_y f(y,x) \log f(y,x)\, dydx \tag{19.3}$$

subject to

$$\int_x \int_y y^{m_1} x^{m_2} f(y,x)\, dydx = \mu_{m_1 m_2} = E\left(y^{m_1} x^{m_2}\right), 0 \leq m_1 + m_2 \leq 4. \tag{19.4}$$

We construct the Lagrangian

$$\begin{aligned}\mathcal{L}(\boldsymbol{\lambda}, \lambda_{00}) = &-\int_x \int_y f(y,x) \log f(y,x)\, dydx \\ &+ \sum_{m_1=0}^{4} \sum_{m_2=0}^{4} \lambda_{m_1 m_2} \left(\mu_{m_1 m_2} - \int_x \int_y y^{m_1} x^{m_2} f(y,x)\, dydx \right),\end{aligned} \tag{19.5}$$

where $\boldsymbol{\lambda} = \left(\lambda_{m_1 m_2}\right)_{14\times1}$ for all $1 \leq m_1 + m_2 \leq 4$. The solution of the joint density distribution yields the form

$$\begin{aligned} f(y,x) &= \exp\left[-\sum_{m_1+m_2=0}^{4} \lambda_{m_1 m_2} y^{m_1} x^{m_2}\right] \\ &= \frac{\exp\left[-\sum_{m_1+m_2=1}^{4} \lambda_{m_1 m_2} y^{m_1} x^{m_2}\right]}{\int_x \int_y \exp\left[-\sum_{m_1+m_2=1}^{4} \lambda_{m_1 m_2} y^{m_1} x^{m_2}\right] dydx} \equiv \frac{\exp\left[-\sum_{m_1+m_2=1}^{4} \lambda_{m_1 m_2} y^{m_1} x^{m_2}\right]}{\Omega\left(\lambda_{m_1 m_2}\right)}, \end{aligned} \tag{19.6}$$

where $\lambda_{m_1 m_2}$ is the Lagrange multiplier that corresponds to the constraint $\int_x \int_y y^{m_1} x^{m_2} f(y,x)\,dydx = \mu_{m_1 m_2}$, and $\lambda_{00} = \log \Omega\left(\lambda_{m_1 m_2}\right)$ (with $m_1 + m_2 = 0$) is the multiplier associated with the normalization constraint which is a function of other multipliers. See, for example, Golan (1988, 2018) and Ryu (1993).

For deriving our results in section 2, we rearrange the terms in $f(y,x)$ and write

$$f(y,x) = \exp\left[-\left(\lambda_{04}x^4 + \lambda_{03}x^3 + \lambda_{02}x^2 + \lambda_{01}x + \lambda_{00}\right)\right] \\ \times \exp\left\{-\left[\lambda_{40}y^4 + \lambda_{30}(x)y^3 + \lambda_{20}(x)y^2 + \lambda_{10}(x)\right]\right\}y \tag{19.7}$$

where

$$\lambda_{30}(x) = \lambda_{30} + \lambda_{31}x, \lambda_{20}(x) = \lambda_{20} + \lambda_{21}x + \lambda_{22}x^2, \\ \lambda_{10}(x) = \lambda_{10} + \lambda_{11}x + \lambda_{12}x^2 + \lambda_{13}x^3.$$

Replacing $f(y,x)$ and λ_{00} into $\mathcal{L}(\boldsymbol{\lambda}, \lambda_{00}) = \mathcal{L}(\boldsymbol{\lambda})$, we obtain the Lagrange multipliers by maximizing

$$\mathcal{L}(\boldsymbol{\lambda}) = \sum\nolimits_{m_1+m_2=1}^{4} \lambda_{m_1 m_2}\mu_{m_1 m_2} + \lambda_{00}. \tag{19.8}$$

The marginal density of x is computed by integrating $f(y,x)$ over the support of y,

$$f(x) = \int_y f(y,x)\,dy \\ = \exp\left[-\left(\lambda_{04}x^4 + \lambda_{03}x^3 + \lambda_{02}x^2 + \lambda_{01}x + \lambda_{00}\right)\right] \\ \times \int_y \exp\left\{-\left[\lambda_{40}y^4 + \lambda_{30}(x)y^3 + \lambda_{20}(x)y^2 + \lambda_{10}(x)\right]\right\}y. \tag{19.9}$$

We note that $f(x) = f(x,\boldsymbol{\lambda})$ and $f(y,x) = f(y,x,\boldsymbol{\lambda})$. When the Lagrange multipliers $\boldsymbol{\lambda}$ are estimated as $\hat{\boldsymbol{\lambda}}$ from (19.8). we get $\hat{f}(x) = f(x,\hat{\boldsymbol{\lambda}})$ and $\hat{f}(y,x) = f(y,x,\hat{\boldsymbol{\lambda}})$.

Although the above results are written under fourth-order moment conditions in (19.4), they can be easily written when $0 \leq m_1 + m_2 \leq M$. We have considered fourth-order moment conditions without any loss of generality since they capture data information on skewness and kurtosis.

2.2 Regression and Response Functions

Based on the bivariate maximum entropy joint distribution (19.7) and the marginal density (19.9), the conditional mean (regression function) of y given x is represented as

$$m(x) = E(y|x) = \int_y y\frac{f(y,x)}{f(x)}dy$$
$$= \frac{\int_y y\exp\{-[\lambda_{40}y^4 + \lambda_{30}(x)y^3 + \lambda_{20}(x)y^2 + \lambda_{10}(x)y]\}dy}{\int_y \exp\{-[\lambda_{40}y^4 + \lambda_{30}(x)y^3 + \lambda_{20}(x)y^2 + \lambda_{10}(x)y]\}dy}. \tag{19.10}$$

Given the values of the Lagrange multipliers, we define

$$F_r(x) \equiv \int_y y^r \exp\{-[\lambda_{40}y^4 + \lambda_{30}(x)y^3 + \lambda_{20}(x)y^2 + \lambda_{10}(x)y]\}dy. \tag{19.11}$$

where $r = 0, 1, 2, \ldots$. The regression function $m(x)$ thus takes the form

$$m(x) = m(x, \boldsymbol{\lambda}^*) = \frac{F_1(x)}{F_0(x)} = \frac{F_1(x, \boldsymbol{\lambda}^*)}{F_0(x, \boldsymbol{\lambda}^*)}, \tag{19.12}$$

where $\boldsymbol{\lambda}^* = (\lambda_{m_1 m_2})_{10\times 1}$ for all $1 \le m_1 + m_2 \le 4$ except λ_{0m_2} for $m_2 = 1, \ldots, 4$. When the Lagrange multipliers are estimated from (19.8) by the Newton method,

$$\hat{m}(x) = m(x, \hat{\boldsymbol{\lambda}}^*) = \frac{F_1(x, \hat{\boldsymbol{\lambda}}^*)}{F_0(x, \hat{\boldsymbol{\lambda}}^*)}. \tag{19.13}$$

This is the IT nonparametric regression function estimator. Furthermore, the response function $\beta(x) = \frac{dm(x)}{dx}$ (derivative) can be written as

$$\beta(x) = \beta(x, \boldsymbol{\lambda}^*) = \frac{F_1'(x, \boldsymbol{\lambda}^*)F_0(x, \boldsymbol{\lambda}^*) - F_1(x, \boldsymbol{\lambda}^*)F_0'(x, \boldsymbol{\lambda}^*)}{F_0^2(x, \boldsymbol{\lambda}^*)}, \tag{19.14}$$

and its estimator is given by

$$\hat{\beta}(x) = \beta(x, \hat{\boldsymbol{\lambda}}^*) \tag{19.15}$$

We note that $F_r'(x)$ represents the first derivative of $F_r(x)$ with respect to x, $r = 0, 1, 2, \ldots$.

2.3 Recursive Integration

Solving the exponential polynomial integrals in the numerator and denominator from (19.10) in explicit forms is unlikely. Numerical methods can be used to solve the problem by integrating the exponential polynomial function at each value of x. However, for large sample size, numerical methods are computationally quite expensive and hence are not satisfactory. We have developed a recursive integration method that can not only solve the conditional mean $m(x)$ but also reduce the computational cost significantly.

According to the definition of $F_r(x)$ in (19.11), the changes in F_0, F_1 and F_2 are given by

$$\begin{aligned} F_0' &= -\lambda_{30}'(x)F_3 - \lambda_{20}'(x)F_2 - \lambda_{10}'(x)F_1 \\ F_1' &= -\lambda_{30}'(x)F_4 - \lambda_{20}'(x)F_3 - \lambda_{10}'(x)F_2 \\ F_2' &= -\lambda_{30}'(x)F_5 - \lambda_{20}'(x)F_4 - \lambda_{10}'(x)F_3, \end{aligned} \tag{19.16}$$

where $\lambda'(x)$ denotes the first derivative of $\lambda(x)$ with respect to x. Due to the special properties of (19.11), integrals of higher-order exponential polynomial functions can be represented by those of lower orders. Based on this fact, F_3, F_4, and F_5 in (19.16) are replaced by the linear combinations of F_0, F_1, and F_2, resulting in a system of linear equations:

$$\begin{aligned} F_0'(x) &= \Lambda_{00}(x)F_0(x) + \Lambda_{01}(x)F_1(x) + \Lambda_{02}(x)F_2(x) \\ F_1'(x) &= \Lambda_{10}(x)F_0(x) + \Lambda_{11}(x)F_1(x) + \Lambda_{12}(x)F_2(x) \\ F_2'(x) &= \Lambda_{20}(x)F_0(x) + \Lambda_{21}(x)F_1(x) + \Lambda_{22}(x)F_2(x). \end{aligned} \tag{19.17}$$

The derivations of (19.16 and 19.17) are provided in Appendix A.1. Starting from an initial value x_0, for a very small increment h, we trace out $F_0(x)$, $F_1(x)$ and $F_2(x)$ over the entire range of x

$$\begin{aligned} F_0(x_0+h) &\approx F_0(x_0) + F_0'(x_0)h \\ F_1(x_0+h) &\approx F_1(x_0) + F_1'(x_0)h \\ F_2(x_0+h) &\approx F_2(x_0) + F_2'(x_0)h \end{aligned} \tag{19.18}$$

The IT estimators $\hat{m}(x)$ in (19.13) and $\hat{\beta}(x)$ in (19.15) are thus evaluated using (19.17 and 19.18), with $\boldsymbol{\lambda}^*$ replaced by $\hat{\boldsymbol{\lambda}}^*$. The results for the finite domain integration are similar to the above, which are provided in Appendix A.2.

3. Simulation and Empirical Examples

Here we first consider two data-generating processes (DGP) to evaluate the performance of our proposed IT estimator of response function in sections 3.1

and 3.2. Then we present our illustrative empirical example to study regression and response functions in section 3.3.

3.1 Data-Generating Process 1: Nonlinear Function

The true model considered is a nonlinear function[1]

$$y_i = -\frac{1}{5}\log\left(e^{-2.5} + 2e^{-5x_i}\right) + u_i \tag{19.19}$$

where $i = 1, 2, \ldots, n$, the variables y_i and x_i are in log values, and x_i are independent and identically drawn from uniform distribution with mean 0.5 and variance $\frac{1}{12}$. The error term u_i follows independent and identical normal distribution with mean 0 and variance 0.01.

The goal is to estimate the response coefficient $\beta(x) = \frac{\partial y}{\partial x}$. Two parametric approximations considered are

$$\begin{aligned} Linear &: \ y_i = \beta_0 + \beta_1 x_i + u_i \\ Quadratic &: \ y_i = \beta_0 + \beta_1 x_i + \beta_2 x_i^2 + u_i. \end{aligned}$$

These two parametric models are not correctly specified. Thus, one can expect that the estimation of the response coefficients may be biased. Besides these two parametric models, local constant nonparametric estimation of the response coefficient is also of interest as a comparison with our IT method estimator. The local constant (Nadaraya-Watson) nonparametric kernel estimator is $\tilde{m}(x) = \sum y_i w_i(x)$, where $w_i(x) = \frac{K((x_i - x)/b)}{\sum K((x_i - x)/b)}$ in which $K(\cdot)$ is a kernel function and b is the bandwidth; for example, see Pagan and Ullah (1999). We have used normal kernel and cross-validated bandwidth. The bias and root mean square error (RMSE) results from linear function, quadratic approximation, local constant nonparametric method, and IT method are reported in Table 19.1, averaged over 1,000 replications of sample size 200. The values of the response coefficients shown are evaluated at the population mean of x, which is 0.5. Standard errors are given in parentheses. True value of the response coefficient is $\beta\,(x = 0.5) = 0.6667$.

The biases for nonparametric kernel and IT estimators are smaller than those under linear and quadratic approximations. However, nonparametric estimation yields a larger RMSE compared with the three other methods. Even though both nonparametric and IT estimations have the advantage of avoiding the

[1] This simulation example is similar to Rilstone and Ullah (1989).

Table 19.1 Bias and RMSE Comparison under DGP 1

	Linear	Quadratic	Nonparametric	IT
$\beta(x)=\frac{\partial y}{\partial x}$	0.6288	0.6296	0.6468	0.6550
	(0.0276)	(0.0263)	(0.0904)	(0.0268)
Bias	0.0379	0.0371	0.0199	**0.0117**
RMSE	0.0469	0.0455	0.0926	**0.0292**

Table 19.2 Bias and RMSE Comparison under DGP 2

	Linear	Quadratic	Nonparametric	IT
$\beta(x)=\frac{\partial y}{\partial x}$	1.0009	1.0009	1.0105	1.0014
	(0.0250)	(0.0251)	(0.1133)	(0.0284)
Bias	0.0009	0.0009	0.0105	0.0014
RMSE	0.0250	0.0251	0.1138	0.0284

difficulties associated with the functional forms, results have indicated that the IT method outperforms the nonparametric method. This may be because the rate of convergence for MSE to zero for the IT estimator is n^{-1} whereas that of nonparametric kernel estimator is known to be $(nb)^{-1}$ where b is small (Li and Racine [2007]).

3.2 Data-Generating Process 2: Linear Function

Now the true data-generating process is a linear function:

$$y_i = 2 + x_i + u_i \tag{19.20}$$

where $i = 1, 2, \ldots, n$, x_i and u_i follow the same distributions as in DGP 1. Comparisons are made with linear, quadratic approximations and nonparametric estimation. The results on bias and RMSE shown in Table 19.2 are averaged over 1,000 replications of sample size 200. The values of the response coefficients shown are evaluated at $x = 0.5$.

When the true DGP is linear in x, it is not surprising that linear approximation has the smallest bias and RMSE. The IT method has much smaller bias and RMSE than that under the nonparametric kernel estimation. Even though the true relationship between x and y is linear, considering the IT estimator based on the first four moments is still successful in capturing the linearity. For example, $\beta(x)$ based on IT, compared to nonparametric kernel, is closer to $\beta(x)$ based on the true linear DGP. This is similar to the result in DGP 1. Since true DGP in practice is not known, IT provides a better option to be used.

3.3 Empirical Study: Canadian High School Graduate Earnings

To further illustrate the superiority of the maximum entropy method, we conduct the study of the average logwage conditional on the age using the 1971 data set of 205 Canadian high school graduate earnings. According to Eq. (19.10), the variable y denotes the logwage of high school graduates and x denotes the age. As a comparison, local constant and local linear nonparametric estimations, as well as quadratic and quartic approximations, are considered

$$\begin{aligned} Quadratic &: y=\beta_0+\beta_1 x+\beta_2 x^2+u \\ Quartic &: y=\beta_0+\beta_1 x+\beta_2 x^2+\beta_3 x^3+\beta_4 x^4+u. \end{aligned}$$

The local linear nonparametric kernel estimators $m^*(x)$ and $\beta^*(x)$ are obtained by minimizing the local linear weighted squared losses $\sum(y_i-m(x)-(x_i-x)\beta(x))^2 K((x_i-x)/b)$ with respect to $m(x)$ and $\beta(x)$. We note that minimizing local constant weighted-squared losses $\sum(y_i-m(x))^2 K((x_i-x)/b)$ with respect to $m(x)$ provides the local constant nonparametric kernel estimator $\tilde{m}(x)$ used in sections 3.1 and 3.2. We use our IT method to show the plot of the estimated $\hat{m}(x)$ of logwage in Figure 19.1. An illustration of numerical calculations based on the IT method is given in Appendix A.3.

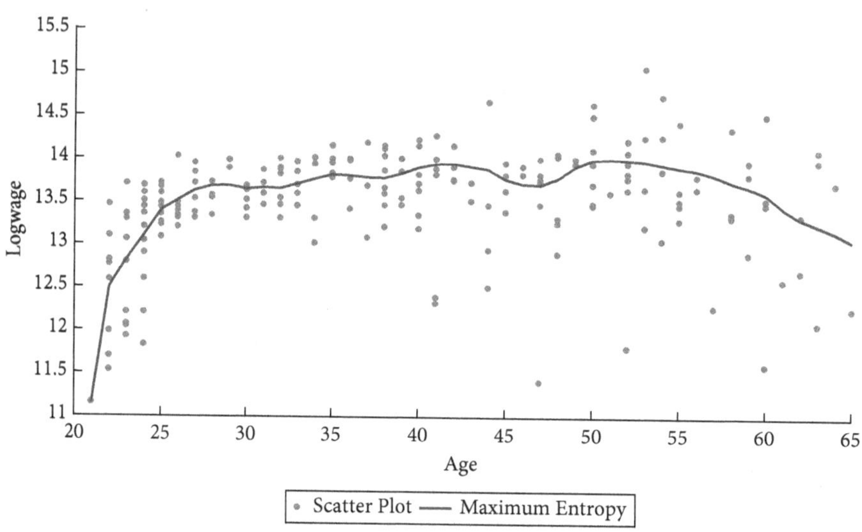

Figure 19.1 Logwage and Age Maximum Entropy Conditional Mean.

From Figure 19.1, log earning grows rapidly from age 21 to age 25. Before around age 45, the growth speed of logwage is slowed down. A depth at age 47 is displayed. From age 47 to age 65, log earning rises and then declines smoothly. It is shown that the IT estimation captures the tail observations very well; see Appendix A.3 for numerical calculations in tails. The average response coefficient $\hat{\beta}$ over the range of age is approximately 0.0434. The average response coefficients under local constant and local linear nonparametric kernel methods are 0.0315 and 0.0421, respectively. Under quadratic approximation, the average response coefficient is 0.0195. We use the average response coefficient under quartic approximation as a benchmark, which is 0.0461. Average $\hat{\beta}$ under the IT method is larger than that under quadratic approximation, local constant, and local linear estimations, which shows that the IT method is advantageous over the rest considered.

4. Asymptotic Properties of IT Estimators and Test for Normality

4.1 Asymptotic Normality

First, we define

$$\begin{aligned}
\underset{14\times 1}{\mathbf{Z}_i} &= \left(y_i, x_i, y_i^2, x_i^2, y_i^3, x_i^3, y_i^4, x_i^4, y_i x_i, y_i x_i^2, y_i^2 x_i, y_i x_i^3, y_i^3 x_i, y_i^2 x_i^2\right)^T, \\
\underset{14\times 1}{\hat{\boldsymbol{\mu}}} &= \left(\hat{\mu}_{10}, \hat{\mu}_{01}, \hat{\mu}_{20}, \hat{\mu}_{02}, \hat{\mu}_{30}, \hat{\mu}_{03}, \hat{\mu}_{40}, \hat{\mu}_{04}, \hat{\mu}_{11}, \hat{\mu}_{12}, \hat{\mu}_{21}, \hat{\mu}_{13}, \hat{\mu}_{31}, \hat{\mu}_{22}\right)^T, \\
\underset{14\times 1}{\boldsymbol{\mu}} &= \left(\mu_{10}, \mu_{01}, \mu_{20}, \mu_{02}, \mu_{30}, \mu_{03}, \mu_{40}, \mu_{04}, \mu_{11}, \mu_{12}, \mu_{21}, \mu_{13}, \mu_{31}, \mu_{22}\right)^T,
\end{aligned} \tag{19.21}$$

where $\hat{\mu}_{m_1 m_2} = \frac{1}{n}\sum_{i=1}^{n} y_i^{m_1} x_i^{m_2}$, $\mu_{m_1 m_2} = E\left(y_i^{m_1} x_i^{m_2}\right)$, $m_1, m_2 = 0,1,2,3,4$ and $1 \le m_1 + m_2 \le 4$, and all the bold letters represent vectors. Suppose the following assumptions hold.

1. $\mathbf{Z}_i, i = 1, \ldots, n$ are independent and identically distributed from $(\boldsymbol{\mu}, \Sigma)$.
2. $\Sigma = \mathrm{COV}(\mathbf{Z}_i)$ is assumed to be positive semi-definite, where the diagonals of Σ are

$$\mathrm{Var}\left(y_i^{m_1} x_i^{m_2}\right) = \mu_{(2m_1)(2m_2)} - \mu_{m_1 m_2}^2,$$

and the off-diagonals of Σ are

$$\mathrm{Cov}\left(y_i^{m_1} x_i^{m_2}, y_i^{m_1^*} x_i^{m_2^*}\right) = \mu_{(m_1+m_1^*)(m_2+m_2^*)} - \mu_{m_1 m_2}\mu_{m_1^* m_2^*}.$$

3. $\mu_{(m_1+m_1^*)(m_2+m_2^*)} < \infty, \forall m_1, m_2, m_1^*, m_2^* = 0,1,2,3,4,\ m_1 + m_2 \le 4,\ m_1^* + m_2^* \le 4.$

Now we present the following proposition.

Proposition 1 *Under assumptions 1 to 3, as n goes to* ∞,

$$\sqrt{n}(\hat{\boldsymbol{\mu}} - \boldsymbol{\mu}) \sim N(\mathbf{0}, \Sigma). \tag{19.22}$$

The proof of this proposition is given in Appendix B.1.

Now, suppose the unique solution for each Lagrange multiplier exists. Then from (19.8) the vector $\boldsymbol{\lambda} = (\lambda_{10}, \lambda_{01}, \lambda_{20}, \lambda_{02}, \lambda_{30}, \lambda_{03}, \lambda_{40}, \lambda_{04}, \lambda_{11}, \lambda_{12}, \lambda_{21}, \lambda_{13}, \lambda_{31}, \lambda_{22})^T$ can be expressed as a function of $\boldsymbol{\mu}$, that is,

$$\boldsymbol{\lambda} = g(\boldsymbol{\mu}) \text{ and } \hat{\boldsymbol{\lambda}} = g(\hat{\boldsymbol{\mu}}). \tag{19.23}$$

Since from Proposition 1, $\sqrt{n}(\hat{\boldsymbol{\mu}} - \boldsymbol{\mu}) \sim N(\mathbf{0}, \Sigma)$ as $n \to \infty$, it follows that

$$\sqrt{n}(\hat{\boldsymbol{\lambda}} - \boldsymbol{\lambda}) \sim N\left(\mathbf{0}, g^{(1)}(\boldsymbol{\mu}) \Sigma g^{(1)}(\boldsymbol{\mu})^T\right) \text{ as } n \to \infty, \tag{19.24}$$

where $g^{(1)}(\boldsymbol{\mu}) = \frac{\partial g(\boldsymbol{\mu})}{\partial \boldsymbol{\mu}^T}$ is the first derivative of $g(\boldsymbol{\mu})$ with respect to $\boldsymbol{\mu}$. See Appendix B.1.

Using the results in Proposition 1 and (19.24), $\sqrt{n}(\hat{\boldsymbol{\lambda}}^* - \boldsymbol{\lambda}^*) \sim N(\mathbf{0}, g^{*(1)}(\boldsymbol{\mu}) \Sigma g^{*(1)}(\boldsymbol{\mu})^T)$ as $n \to \infty$, where $\boldsymbol{\lambda}^* = g^*(\boldsymbol{\mu})$ and $g^{*(1)}(\boldsymbol{\mu}) = \frac{\partial g^*(\boldsymbol{\mu})}{\partial \boldsymbol{\mu}^T}$. We get the following proposition for $\hat{m}(x)$ and $\hat{\beta}(x)$.

Proposition 2 *Under assumptions 1 to 3 and (19.24), the asymptotic distributions of* $\hat{m}(x) = m(x, \hat{\boldsymbol{\lambda}}^*)$ *and* $\hat{\beta}(x) = \beta(x, \hat{\boldsymbol{\lambda}}^*)$ *are given as* $n \to \infty$,

$$\sqrt{n}(m(x, \hat{\boldsymbol{\lambda}}^*) - m(x, \boldsymbol{\lambda}^*)) \sim N\left(\mathbf{0}, m^{(1)}(x, \boldsymbol{\lambda}^*) g^{*(1)}(\boldsymbol{\mu}) \Sigma g^{*(1)}(\boldsymbol{\mu})^T m^{(1)}(x, \boldsymbol{\lambda}^*)^T\right), \tag{19.25}$$

where $m^{(1)}(x, \boldsymbol{\lambda}^*) = \frac{\partial m(x, \boldsymbol{\lambda}^*)}{\partial \boldsymbol{\lambda}^{*T}}$ *is the first derivative of* $m(x, \boldsymbol{\lambda}^*)$ *with respect to* $\boldsymbol{\lambda}^*$. *And*

$$\sqrt{n}(\beta(x, \hat{\boldsymbol{\lambda}}^*) - \beta(x, \boldsymbol{\lambda}^*)) \sim N\left(\mathbf{0}, \beta^{(1)}(x, \boldsymbol{\lambda}^*) g^{*(1)}(\boldsymbol{\mu}) \Sigma g^{*(1)}(\boldsymbol{\mu})^T \beta^{(1)}(x, \boldsymbol{\lambda}^*)^T\right), \tag{19.26}$$

where $\beta^{(1)}(x, \boldsymbol{\lambda}^*) = \frac{\partial \beta(x, \boldsymbol{\lambda}^*)}{\partial \boldsymbol{\lambda}^{*T}}$ *is the first derivative of* $\beta(x, \boldsymbol{\lambda}^*)$ *with respect to* $\boldsymbol{\lambda}^*$.

The proof of Proposition 2 is given in Appendix B.1. Also, we note that the convergence rates of $m(x,\hat{\boldsymbol{\lambda}}^*)$ and $\beta(x,\hat{\boldsymbol{\lambda}}^*)$ are each $\sqrt{n}$.

4.2 Testing for Normality

When the true distribution $f(x,y)$ is normal, the Lagrange multipliers for moments with orders higher than two are equal to zero, that is, $\lambda_{ij}=0, \forall i+j>2$. $\boldsymbol{\lambda}$ contains nine elements with orders higher than two. Testing whether (x,y) are jointly normal is equivalent to testing the null

$$H_0 : R\boldsymbol{\lambda} = \mathbf{0},$$

where R is a 9×14 matrix with elements $R(1,6), R(2,7), \cdots, R(9,14)=1$ and the rest elements $=0$. We develop a Wald test statistic

$$W=(R\hat{\boldsymbol{\lambda}})'\left(V(R\hat{\boldsymbol{\lambda}})\right)^{-1}(R\hat{\boldsymbol{\lambda}}),$$

where $V(R\hat{\boldsymbol{\lambda}})=RV(\hat{\boldsymbol{\lambda}})R'$ in which $V(\hat{\boldsymbol{\lambda}})$ is the asymptotic variance. Since $R\hat{\boldsymbol{\lambda}}$ is asymptotically normal from (19.24), it follows that W is χ^2_9 asymptotically. Conclusion is based on the comparison between the calculated value of W and the tabulated Chi-square distribution critical value. When the true distribution $f(x,y)$ is normal, the relationship between x and y is linear. Therefore, it is also a test for linearity. In our empirical example in section 3.3, we compute the Wald statistic $W\approx 13548$. At 1% significance level, we reject the null hypothesis. Thus, we conclude that the relationship between age and logwage is nonlinear. Alternatively, one can test the null hypothesis using the entropy-ratio test.

5. Conclusions

In this chapter, we have estimated the econometric functions through an IT method, which is nonparametric. Two basic econometric functions, regression and response, have been analyzed. The advantages of using the IT method over parametric specifications and nonparametric kernel approaches have been explained by the simulation and empirical examples. It can be a useful tool for practitioners due to its simplicity and efficiency. Asymptotic properties are established. The IT-based estimators are shown to be $\sqrt{n}$ consistent and normal. Thus, it has a faster rate of convergence compared to the nonparametric kernel procedures.

Based on what has been developed in this chapter, further work can be done in the future; for example, the bivariate regression approach introduced in this chapter can potentially be extended to the multivariate case. Next, we can extend our chapter's IT analysis for conditional variance and conditional covariance functions, among other econometric functions. Furthermore, the IT method may be carried over from Shannon's information theory to Kullback and Leibler (1951) divergence, which has been discussed in the literature by, among other researchers, Golan et al. (1996), Judge and Mittelhammer (2011), Golan (2018), Ullah (1996). Chakrabarty et al. (2015), Maasoumi and Racine (2016), and Racine and Li (2017) have approached quantile estimation problems using nonparametric kernel methods. Similarly, the maximum entropy-based probability distributions derived in our chapter may be adopted for nonparametric quantile estimation problems. In addition, our IT-based estimator of conditional mean can be applied to the nonparametric component in semiparametric models, such as the partial linear model. Along with all these, other future work may be to explore links between IT-based density and the log-spline density considered in Stone (1990). Moreover, it would be useful to establish connections of our asymptotically χ^2 distributed Wald's type normality test in section 4.2 with those of Neyman's smooth test considered in Ledwina (1994) and Inglot and Ledwina (1996) and the entropy-ratio test on the lambdas, which is two times the difference of the objective functions (with/without imposing the null hypothesis) and asymptotically distributed as χ^2; see, e.g., Golan (2018, p. 96). The IT approach for specifying regression and response functions considered here may open a new path to address specification and other related issues in econometrics with many applications.

Appendix A: Calculations

A.1. Recursive Integration

When the range for y is from $-\infty$ to $+\infty$, define the following integrals as functions of x.

$$F_r(x) \equiv F_r \equiv \int_{-\infty}^{+\infty} y^r \exp\{-[\lambda_{40}y^4 + \lambda_{30}(x)y^3 + \lambda_{20}(x)y^2 + \lambda_{10}(x)y]\}dy,$$

where $r = 0, 1, 2, \ldots$. In particular,

$$F_0(x) \equiv F_0 \equiv \int_{-\infty}^{+\infty} \exp\{-[\lambda_{40}y^4 + \lambda_{30}(x)y^3 + \lambda_{20}(x)y^2 + \lambda_{10}(x)y]\}dy$$

$$F_1(x) \equiv F_1 \equiv \int_{-\infty}^{+\infty} y\exp\{-[\lambda_{40}y^4 + \lambda_{30}(x)y^3 + \lambda_{20}(x)y^2 + \lambda_{10}(x)y]\}dy$$

$$F_2(x) \equiv F_2 \equiv \int_{-\infty}^{+\infty} y^2 \exp\{-[\lambda_{40}y^4 + \lambda_{30}(x)y^3 + \lambda_{20}(x)y^2 + \lambda_{10}(x)y]\}dy.$$

Suppose that λ_{40} is positive. First, solve for F_3.

$$\begin{aligned}
0 &= \int_{-\infty}^{+\infty} d\exp\{-[\lambda_{40}y^4 + \lambda_{30}(x)y^3 + \lambda_{20}(x)y^2 + \lambda_{10}(x)y]\} \\
&= \int_{-\infty}^{+\infty} (-4\lambda_{40}y^3 - 3\lambda_{30}(x)y^2 - 2\lambda_{20}(x)y - \lambda_{10}(x)) \\
&\quad e^{-[\lambda_{40}y^4 + \lambda_{30}(x)y^3 + \lambda_{20}(x)y^2 + \lambda_{10}(x)y]}dy \\
&= -4\lambda_{40}F_3 - 3\lambda_{30}(x)F_2 - 2\lambda_{20}(x)F_1 - \lambda_{10}(x)F_0 \\
F_3 &= -\frac{1}{4\lambda_{40}}(3\lambda_{30}(x)F_2 + 2\lambda_{20}(x)F_1 + \lambda_{10}(x)F_0)
\end{aligned}$$

Second, solve for F_4.

$$\begin{aligned}
F_0 &= \int_{-\infty}^{+\infty} \exp\{-[\lambda_{40}y^4 + \lambda_{30}(x)y^3 + \lambda_{20}(x)y^2 + \lambda_{10}(x)y]\}dy \\
&= ye^{-[\lambda_{40}y^4 + \lambda_{30}(x)y^3 + \lambda_{20}(x)y^2 + \lambda_{10}(x)y]}\Big|_{-\infty}^{+\infty} \\
&\quad - \int_{-\infty}^{+\infty} yde^{-[\lambda_{40}y^4 + \lambda_{30}(x)y^3 + \lambda_{20}(x)y^2 + \lambda_{10}(x)y]} \\
&= \int_{-\infty}^{+\infty} (4\lambda_{40}y^4 + 3\lambda_{30}(x)y^3 + 2\lambda_{20}(x)y^2 + \lambda_{10}(x)y) \\
&\quad e^{-[\lambda_{40}y^4 + \lambda_{30}(x)y^3 + \lambda_{20}(x)y^2 + \lambda_{10}(x)y]}dy \\
&= 4\lambda_{40}F_4 + 3\lambda_{30}(x)F_3 + 2\lambda_{20}(x)F_2 + \lambda_{10}(x)F_1 \\
F_4 &= \frac{1}{4\lambda_{40}}(-3\lambda_{30}(x)F_3 - 2\lambda_{20}(x)F_2 - \lambda_{10}(x)F_1 + F_0) \\
&= -\frac{3\lambda_{30}(x)}{4\lambda_{40}}F_3 - \frac{\lambda_{20}(x)}{2\lambda_{40}}F_2 - \frac{\lambda_{10}(x)}{4\lambda_{40}}F_1 + \frac{1}{4\lambda_{40}}F_0.
\end{aligned}$$

Replace F_3 with $-\frac{1}{4\lambda_{40}}(3\lambda_{30}(x)F_2 + 2\lambda_{20}(x)F_1 + \lambda_{10}(x)F_0)$.

$$\begin{aligned}
F_4 = &\left(\frac{9\lambda_{30}^2(x)}{16\lambda_{40}^2} - \frac{\lambda_{20}(x)}{2\lambda_{40}}\right)F_2 \\
&+ \left(\frac{3\lambda_{30}(x)\lambda_{20}(x)}{8\lambda_{40}^2} - \frac{\lambda_{10}(x)}{4\lambda_{40}}\right)F_1 \\
&+ \left(\frac{3\lambda_{30}(x)\lambda_{10}(x)}{16\lambda_{40}^2} + \frac{1}{4\lambda_{40}}\right)F_0
\end{aligned}$$

Third, solve for F_5.

$$\begin{aligned}
F_1 &= \int_{-\infty}^{+\infty} y\exp\{-[\lambda_{40}y^4 + \lambda_{30}(x)y^3 + \lambda_{20}(x)y^2 + \lambda_{10}(x)y]\}dy \\
&= \int_{-\infty}^{+\infty} \exp\{-[\lambda_{40}y^4 + \lambda_{30}(x)y^3 + \lambda_{20}(x)y^2 + \lambda_{10}(x)y]\}d\left(\frac{1}{2}y^2\right) \\
&= \frac{1}{2}y^2 e^{-[\lambda_{40}y^4+\lambda_{30}(x)y^3+\lambda_{20}(x)y^2+\lambda_{10}(x)y]}\Big|_{-\infty}^{+\infty} \\
&\quad - \int_{-\infty}^{+\infty} \frac{1}{2}y^2 de^{-[\lambda_{40}y^4+\lambda_{30}(x)y^3+\lambda_{20}(x)y^2+\lambda_{10}(x)y]} \\
&= \int_{-\infty}^{+\infty} \frac{1}{2}y^2 \left(4\lambda_{40}y^3 + 3\lambda_{30}(x)y^2 + 2\lambda_{20}(x)y + \lambda_{10}(x)\right) \\
&\quad e^{-[\lambda_{40}y^4+\lambda_{30}(x)y^3+\lambda_{20}(x)y^2+\lambda_{10}(x)y]}dy \\
&= 2\lambda_{40}F_5 + \frac{3}{2}\lambda_{30}(x)F_4 + \lambda_{20}(x)F_3 + \frac{1}{2}\lambda_{10}(x)F_2 \\
F_5 &= -\frac{1}{2\lambda_{40}}\left(\frac{3}{2}\lambda_{30}(x)F_4 + \lambda_{20}(x)F_3 + \frac{1}{2}\lambda_{10}(x)F_2 - F_1\right) \\
&= -\frac{3\lambda_{30}(x)}{4\lambda_{40}}F_4 - \frac{\lambda_{20}(x)}{2\lambda_{40}}F_3 - \frac{\lambda_{10}(x)}{4\lambda_{40}}F_2 + \frac{1}{2\lambda_{40}}F_1.
\end{aligned}$$

Replace F_3 and F_4.

$$\begin{aligned}
F_5 = &\left(-\frac{27\lambda_{30}^3(x)}{64\lambda_{40}^3} + \frac{3\lambda_{30}(x)\lambda_{20}(x)}{4\lambda_{40}^2} - \frac{\lambda_{10}(x)}{4\lambda_{40}}\right)F_2 + \\
&\left(-\frac{9\lambda_{30}^2(x)\lambda_{20}(x)}{32\lambda_{40}^3} + \frac{3\lambda_{30}(x)\lambda_{10}(x)}{16\lambda_{40}^2} + \frac{\lambda_{20}^2(x)}{4\lambda_{40}^2} + \frac{1}{2\lambda_{40}}\right)F_1 + \\
&\left(-\frac{9\lambda_{30}^2(x)\lambda_{10}(x)}{64\lambda_{40}^3} - \frac{3\lambda_{30}(x)}{16\lambda_{40}^2} + \frac{\lambda_{20}(x)\lambda_{10}(x)}{8\lambda_{40}^2}\right)F_0
\end{aligned}$$

Define

$$\begin{aligned}
F_0' &\equiv \frac{dF_0(x)}{dx}, F_1' \equiv \frac{dF_1(x)}{dx}, F_2' \equiv \frac{dF_2(x)}{dx} \\
\lambda_{30}'(x) &\equiv \frac{d\lambda_{30}(x)}{dx}, \lambda_{20}'(x) \equiv \frac{d\lambda_{20}(x)}{dx}, \lambda_{10}'(x) \equiv \frac{d\lambda_{10}(x)}{dx}
\end{aligned}$$

First, solve for F_0'.

$$\begin{aligned}
F_0' &\equiv \frac{d}{dx}\int_{-\infty}^{+\infty} \exp\{-[\lambda_{40}y^4 + \lambda_{30}(x)y^3 + \lambda_{20}(x)y^2 + \lambda_{10}(x)y]\}dy \\
&= \int_{-\infty}^{+\infty} \frac{d}{dx}\exp\{-[\lambda_{40}y^4 + \lambda_{30}(x)y^3 + \lambda_{20}(x)y^2 + \lambda_{10}(x)y]\}dy \\
&= \int_{-\infty}^{+\infty} (-\lambda_{30}'(x)y^3 - \lambda_{20}'(x)y^2 - \lambda_{10}'(x)y)\, e^{-[\lambda_{40}y^4+\lambda_{30}(x)y^3+\lambda_{20}(x)y^2+\lambda_{10}(x)y]}dy \\
&= -\lambda_{30}'(x)F_3 - \lambda_{20}'(x)F_2 - \lambda_{10}'(x)F_1
\end{aligned}$$

Replace F_3 with $-\frac{1}{4\lambda_{40}}(3\lambda_{30}(x)F_2 + 2\lambda_{20}(x)F_1 + \lambda_{10}(x)F_0)$.

$$F_0' = \left(\frac{3\lambda_{30}'(x)\lambda_{30}(x)}{4\lambda_{40}} - \lambda_{20}'(x)\right)F_2 + \left(\frac{\lambda_{30}'(x)\lambda_{20}(x)}{2\lambda_{40}} - \lambda_{10}'(x)\right)F_1 + \frac{\lambda_{30}'(x)\lambda_{10}(x)}{4\lambda_{40}}F_0$$

Second, solve for F_1'.

$$\begin{aligned}
F_1' &\equiv \frac{d}{dx}\int_{-\infty}^{+\infty} y\exp\{-[\lambda_{40}y^4 + \lambda_{30}(x)y^3 + \lambda_{20}(x)y^2 + \lambda_{10}(x)y]\}dy \\
&= \int_{-\infty}^{+\infty} \frac{d}{dx}y\exp\{-[\lambda_{40}y^4 + \lambda_{30}(x)y^3 + \lambda_{20}(x)y^2 + \lambda_{10}(x)y]\}dy \\
&= \int_{-\infty}^{+\infty} (-\lambda_{30}'(x)y^4 - \lambda_{20}'(x)y^3 - \lambda_{10}'(x)y^2)\, e^{-[\lambda_{40}y^4+\lambda_{30}(x)y^3+\lambda_{20}(x)y^2+\lambda_{10}(x)y]}dy \\
&= -\lambda_{30}'(x)F_4 - \lambda_{20}'(x)F_3 - \lambda_{10}'(x)F_2
\end{aligned}$$

Replace F_3 and F_4.

$$\begin{aligned}
F_1' = &\left(-\frac{9\lambda_{30}'(x)\lambda_{30}^2(x)}{16\lambda_{40}^2} + \frac{\lambda_{30}'(x)\lambda_{20}(x)}{2\lambda_{40}} + \frac{3\lambda_{20}'(x)\lambda_{30}(x)}{4\lambda_{40}} - \lambda_{10}'(x)\right)F_2 + \\
&\left(-\frac{3\lambda_{30}'(x)\lambda_{30}(x)\lambda_{20}(x)}{8\lambda_{40}^2} + \frac{\lambda_{30}'(x)\lambda_{10}(x)}{4\lambda_{40}} + \frac{\lambda_{20}'(x)\lambda_{20}(x)}{2\lambda_{40}}\right)F_1 + \\
&\left(-\frac{3\lambda_{30}'(x)\lambda_{30}(x)\lambda_{10}(x)}{16\lambda_{40}^2} - \frac{\lambda_{30}'(x)}{4\lambda_{40}} + \frac{\lambda_{20}'(x)\lambda_{10}(x)}{4\lambda_{40}}\right)F_0
\end{aligned}$$

Third, solve for F_2'.

$$
\begin{aligned}
F_2' &\equiv \frac{d}{dx}\int_{-\infty}^{+\infty} y^2 \exp\{-[\lambda_{40}y^4 + \lambda_{30}(x)y^3 + \lambda_{20}(x)y^2 + \lambda_{10}(x)y]\}\,dy \\
&= \int_{-\infty}^{+\infty} \frac{d}{dx} y^2 \exp\{-[\lambda_{40}y^4 + \lambda_{30}(x)y^3 + \lambda_{20}(x)y^2 + \lambda_{10}(x)y]\}\,dy \\
&= \int_{-\infty}^{+\infty} \left(-\lambda_{30}'(x)y^5 - \lambda_{20}'(x)y^4 - \lambda_{10}'(x)y^3\right) e^{-[\lambda_{40}y^4+\lambda_{30}(x)y^3+\lambda_{20}(x)y^2+\lambda_{10}(x)y]}\,dy \\
&= -\lambda_{30}'(x)F_5 - \lambda_{20}'(x)F_4 - \lambda_{10}'(x)F_3
\end{aligned}
$$

Replace F_5, F_4 and F_3.

$$
\begin{aligned}
F_2' = &\left(\begin{gathered}\frac{27\lambda_{30}'(x)\lambda_{30}^3(x)}{64\lambda_{40}^3} - \frac{3\lambda_{30}'(x)\lambda_{30}(x)\lambda_{20}(x)}{4\lambda_{40}^2} + \frac{\lambda_{30}'(x)\lambda_{10}(x)}{4\lambda_{40}} \\ - \frac{9\lambda_{20}'(x)\lambda_{30}^2(x)}{16\lambda_{40}^2} + \frac{\lambda_{20}'(x)\lambda_{20}(x)}{2\lambda_{40}} + \frac{3\lambda_{10}'(x)\lambda_{30}(x)}{4\lambda_{40}}\end{gathered}\right) F_2 + \\
&\left(\begin{gathered}\frac{9\lambda_{30}'(x)\lambda_{30}^2(x)\lambda_{20}(x)}{32\lambda_{40}^3} - \frac{3\lambda_{30}'(x)\lambda_{30}(x)\lambda_{10}(x)}{16\lambda_{40}^2} - \frac{\lambda_{30}'(x)\lambda_{20}^2(x)}{4\lambda_{40}^2} \\ - \frac{\lambda_{30}'(x)}{2\lambda_{40}} - \frac{3\lambda_{20}'(x)\lambda_{30}(x)\lambda_{20}(x)}{8\lambda_{40}^2} + \frac{\lambda_{20}'(x)\lambda_{10}(x)}{4\lambda_{40}} + \frac{\lambda_{10}'(x)\lambda_{20}(x)}{2\lambda_{40}}\end{gathered}\right) F_1 + \\
&\left(\begin{gathered}\frac{9\lambda_{30}'(x)\lambda_{30}^2(x)\lambda_{10}(x)}{64\lambda_{40}^3} + \frac{3\lambda_{30}'(x)\lambda_{30}(x)}{16\lambda_{40}^2} - \frac{\lambda_{30}'(x)\lambda_{20}(x)\lambda_{10}(x)}{8\lambda_{40}^2} \\ - \frac{3\lambda_{20}'(x)\lambda_{30}(x)\lambda_{10}(x)}{16\lambda_{40}^2} - \frac{\lambda_{20}'(x)}{4\lambda_{40}} + \frac{\lambda_{10}'(x)\lambda_{10}(x)}{4\lambda_{40}}\end{gathered}\right) F_0
\end{aligned}
$$

Equations (19.16) and (19.17) are thus obtained.

A.2. Finite Integral Range

When the range for y $[a(x), b(x)]$ is varying based on x, define the following functions.

$$
F_r(x) \equiv F_r \equiv \int_{a(x)}^{b(x)} y^r \exp\{-[\lambda_{40}y^4 + \lambda_{30}(x)y^3 + \lambda_{20}(x)y^2 + \lambda_{10}(x)y]\}\,dy
$$

where $r = 0, 1, 2, \ldots$ Define the following functions of x.

$$A_0(x) \equiv A_0 \equiv \exp\left\{-\left[\lambda_{40}a(x)^4 + \lambda_{30}(x)a(x)^3 + \lambda_{20}(x)a(x)^2 + \lambda_{10}(x)a(x)\right]\right\}$$

$$B_0(x) \equiv B_0 \equiv \exp\left\{-\left[\lambda_{40}b(x)^4 + \lambda_{30}(x)b(x)^3 + \lambda_{20}(x)b(x)^2 + \lambda_{10}(x)b(x)\right]\right\}$$

$$A_1(x) \equiv A_1 \equiv a(x)\exp\left\{-\left[\lambda_{40}a(x)^4 + \lambda_{30}(x)a(x)^3 + \lambda_{20}(x)a(x)^2 + \lambda_{10}(x)a(x)\right]\right\}$$

$$B_1(x) \equiv B_1 \equiv b(x)\exp\left\{-\left[\lambda_{40}b(x)^4 + \lambda_{30}(x)b(x)^3 + \lambda_{20}(x)b(x)^2 + \lambda_{10}(x)b(x)\right]\right\}$$

$$A_2(x) \equiv A_2 \equiv a(x)^2\exp\left\{-\left[\lambda_{40}a(x)^4 + \lambda_{30}(x)a(x)^3 + \lambda_{20}(x)a(x)^2 + \lambda_{10}(x)a(x)\right]\right\}$$

$$B_2(x) \equiv B_2 \equiv b(x)^2\exp\left\{-\left[\lambda_{40}b(x)^4 + \lambda_{30}(x)b(x)^3 + \lambda_{20}(x)b(x)^2 + \lambda_{10}(x)b(x)\right]\right\}$$

$$L_{0,a}(x) \equiv a'(x)\exp\left\{-\left[\lambda_{40}a(x)^4 + \lambda_{30}(x)a(x)^3 + \lambda_{20}(x)a(x)^2 + \lambda_{10}(x)a(x)\right]\right\}$$

$$L_{0,b}(x) \equiv b'(x)\exp\left\{-\left[\lambda_{40}b(x)^4 + \lambda_{30}(x)b(x)^3 + \lambda_{20}(x)b(x)^2 + \lambda_{10}(x)b(x)\right]\right\}$$

$$L_{1,a}(x) \equiv a'(x)a(x)\exp\left\{-\left[\lambda_{40}a(x)^4 + \lambda_{30}(x)a(x)^3 + \lambda_{20}(x)a(x)^2 + \lambda_{10}(x)a(x)\right]\right\}$$

$$L_{1,b}(x) \equiv b'(x)b(x)\exp\left\{-\left[\lambda_{40}b(x)^4 + \lambda_{30}(x)b(x)^3 + \lambda_{20}(x)b(x)^2 + \lambda_{10}(x)b(x)\right]\right\}$$

$$L_{2,a}(x) \equiv a'(x)a(x)^2\exp\left\{-\left[\lambda_{40}a(x)^4 + \lambda_{30}(x)a(x)^3 + \lambda_{20}(x)a(x)^2 + \lambda_{10}(x)a(x)\right]\right\}$$

$$L_{2,b}(x) \equiv b'(x)b(x)^2\exp\left\{-\left[\lambda_{40}b(x)^4 + \lambda_{30}(x)b(x)^3 + \lambda_{20}(x)b(x)^2 + \lambda_{10}(x)b(x)\right]\right\}$$

The expressions for $F_0'(x)$, $F_1'(x)$, and $F_2'(x)$ are modified as

$$\begin{aligned}
F_0'(x) &= \Lambda_{00}(x)F_0(x) + \Lambda_{01}(x)F_1(x) + \Lambda_{02}(x)F_2(x) + C_0(x) \\
F_1'(x) &= \Lambda_{10}(x)F_0(x) + \Lambda_{11}(x)F_1(x) + \Lambda_{12}(x)F_2(x) + C_1(x) \\
F_2'(x) &= \Lambda_{20}(x)F_0(x) + \Lambda_{21}(x)F_1(x) + \Lambda_{22}(x)F_2(x) + C_2(x)
\end{aligned}$$

where Λ's denote the corresponding coefficients. C's are defined as follows, which contain the age x and its logwage range $[a(x), b(x)]$.

$$C_0(x) = -\frac{\lambda'_{30}(x)}{4\lambda_{40}}(A_0 - B_0) + L_{0,b}(x) - L_{0,a}(x)$$
$$C_1(x) = -\frac{\lambda'_{30}(x)}{4\lambda_{40}}(A_1 - B_1) + \left(\frac{3\lambda'_{30}(x)\lambda_{30}(x)}{16\lambda_{40}^2} - \frac{\lambda'_{20}(x)}{4\lambda_{40}}\right)(A_0 - B_0) + L_{1,b}(x) - L_{1,a}(x)$$
$$C_2(x) = -\frac{\lambda'_{30}(x)}{4\lambda_{40}}(A_2 - B_2) + \left(\frac{3\lambda'_{30}(x)\lambda_{30}(x)}{16\lambda_{40}^2} - \frac{\lambda'_{20}(x)}{4\lambda_{40}}\right)(A_1 - B_1)$$
$$+\left(-\frac{9\lambda'_{30}(x)\lambda_{30}^2(x)}{64\lambda_{40}^3} + \frac{\lambda'_{30}(x)\lambda_{20}(x)}{8\lambda_{40}^2} + \frac{3\lambda'_{20}(x)\lambda_{30}(x)}{16\lambda_{40}^2} - \frac{\lambda'_{10}(x)}{4\lambda_{40}}\right)(A_0 - B_0)$$
$$+ L_{2,b}(x) - L_{2,a}(x)$$

Since

$$F_0(x_0 + h) \approx F_0(x_0) + F_0'(x)h$$
$$F_1(x_0 + h) \approx F_1(x_0) + F_1'(x)h$$
$$F_2(x_0 + h) \approx F_2(x_0) + F_2'(x)h$$

for a given initial value x_0 and a small increment h, the functions of x, $F_0(x)$, $F_1(x)$ and $F_2(x)$ can be traced out. At each x value, the logwage(y) limits $a(x)$, $b(x)$ can be obtained through Taylor expansion in the neighborhood of a certain data point x^*,

$$a(x) \approx a(x^*) + a'(x^*)(x - x^*) + \frac{1}{2!}a''(x^*)(x - x^*)^2 + \frac{1}{3!}a'''(x^*)(x - x^*)^3 + \cdots$$
$$b(x) \approx b(x^*) + b'(x^*)(x - x^*) + \frac{1}{2!}b''(x^*)(x - x^*)^2 + \frac{1}{3!}b'''(x^*)(x - x^*)^3 + \cdots.$$

Derivatives of $a(x)$, $b(x)$ can be approximated by

$$a'(x) = \frac{a(x+1) - a(x-1)}{2}$$
$$a''(x) = \frac{a(x+2) - 2a(x) + a(x-2)}{4}$$
$$a'''(x) = \frac{a(x+3) - 3a(x+1) + 3a(x-1) - a(x-3)}{8}.$$

It is similar for the upper limit $b(x)$.

A.3. Empirical Study: An Illustration of Calculations

The logwage range is approximated by Taylor expansion. For example, at age 30, logwage range $[a(30), b(30)]$ is estimated twice in the neighborhood of age 29 and 31,

$$a(30) \approx a(29) + a'(29)(30-29) + \frac{1}{2!}a''(29)(30-29)^2$$

$$a(30) \approx a(31) + a'(31)(30-31) + \frac{1}{2!}a''(31)(30-31)^2$$

$a(30)$ is the average of these two estimates. $b(30)$ is calculated in the same way. For ages x at the two tails, range $[a(x), b(x)]$ is averaged by Taylor expansions in the neighborhood of several data points. For example, the starting age is 21 in the data set. Logwage range $[a(21), b(21)]$ is estimated by averaging Taylor expansions in the neighborhood of age 22, 23, and 24. The initial values $F_0(21), F_1(21)$,and $F_2(21)$ are computed by integration with the approximated range $[a(21), b(21)]$, that is, $x_0 = 21$ in the algorithm above. For small enough value h, sequences of $F_0(x), F_1(x)$, and $F_2(x)$ are obtained by the recursive algorithm. Thus, $m(x)$ is computed by taking ratios of $F_1(x)$ and $F_0(x)$ at every age x. Integration is needed only once, at the initial age, 21.

Appendix B: Asymptotic Properties of IT Estimators

B.1. Proof of Proposition 1 and (Eq. [19.24])

From (Eq. [19.21]),

$$\sqrt{n}(\hat{\mu}-\mu) = \sqrt{n}\left(\frac{1}{n}\sum_{i=1}^{n} \mathbf{Z}_i - \mu\right) = \frac{1}{\sqrt{n}}\left(\sum_{i=1}^{n} \mathbf{Z}_i - \mu\right).$$

The multivariate characteristic function is

$$\begin{aligned}
\varphi_{\sqrt{n}(\hat{\mu}-\mu)}(\mathbf{t}) &= \varphi_{\frac{1}{\sqrt{n}}\left(\sum_{i=1}^{n} \mathbf{Z}_i - \mu\right)}(\mathbf{t}) \\
&= \varphi_{\mathbf{Z}_1-\mu}\left(\frac{\mathbf{t}}{\sqrt{n}}\right)\varphi_{\mathbf{Z}_2-\mu}\left(\frac{\mathbf{t}}{\sqrt{n}}\right)\cdots\varphi_{\mathbf{Z}_n-\mu}\left(\frac{\mathbf{t}}{\sqrt{n}}\right) \\
&= \left[\varphi_{\mathbf{Z}_1-\mu}\left(\frac{\mathbf{t}}{\sqrt{n}}\right)\right]^n \\
&= E\left[e^{i\left(\frac{\mathbf{t}}{\sqrt{n}}\right)'(\mathbf{Z}_1-\mu)}\right],
\end{aligned}$$

where $\mathbf{t}$ is a column vector. By Taylor's Theorem,

$$\varphi_{\mathbf{Z}_1-\mu}\left(\frac{\mathbf{t}}{\sqrt{n}}\right) = 1 - \frac{1}{2n}\mathbf{t}'\Sigma\mathbf{t} + O(\mathbf{t}^3), \mathbf{t} \to 0.$$

Since $e^x = \lim_{n\to\infty}\left(1+\frac{x}{n}\right)^n$,

$$\varphi_{\sqrt{n}(\hat{\mu}-\mu)}(\mathbf{t}) = \left[1-\frac{1}{2n}\mathbf{t}'\Sigma\mathbf{t}+O(\mathbf{t}^3)\right]^n \rightarrow \exp\left(-\frac{1}{2}\mathbf{t}'\Sigma\mathbf{t}\right) \text{as } n\rightarrow\infty.$$

Thus, $\sqrt{n}(\hat{\boldsymbol{\mu}}-\boldsymbol{\mu}) \sim N(\mathbf{0},\Sigma)$ as given in Proposition 1.

Now to obtain the result in (Eq. [19.24]), we write the first-order approximation of $\hat{\boldsymbol{\lambda}}$ as

$$\begin{aligned}
\hat{\boldsymbol{\lambda}} &= g(\hat{\boldsymbol{\mu}}) \\
&\simeq g(\boldsymbol{\mu}) + \left.\frac{\partial g(\hat{\boldsymbol{\mu}})}{\partial \hat{\boldsymbol{\mu}}^T}\right|_{\hat{\boldsymbol{\mu}}=\boldsymbol{\mu}} (\hat{\boldsymbol{\mu}}-\boldsymbol{\mu}) \\
&= g(\boldsymbol{\mu}) + g^{(1)}(\boldsymbol{\mu})(\hat{\boldsymbol{\mu}}-\boldsymbol{\mu}). \\
\sqrt{n}(\hat{\boldsymbol{\lambda}}-\boldsymbol{\lambda}) &= \sqrt{n}((g(\hat{\boldsymbol{\mu}})-g(\boldsymbol{\mu}))) \\
&\simeq \sqrt{n}\left(g^{(1)}(\boldsymbol{\mu})(\hat{\boldsymbol{\mu}}-\boldsymbol{\mu})\right).
\end{aligned}$$

Since $\sqrt{n}(\hat{\boldsymbol{\mu}}-\boldsymbol{\mu}) \sim N(\mathbf{0},\Sigma)$ as $n\rightarrow\infty$,

$$\sqrt{n}(\hat{\boldsymbol{\lambda}}-\boldsymbol{\lambda}) \sim N\left(\mathbf{0}, g^{(1)}(\boldsymbol{\mu})\Sigma g^{(1)}(\boldsymbol{\mu})^T\right).$$

The convergence rate of $\hat{\boldsymbol{\lambda}}$ is $n^{1/2}$. This is the result in(19.24).

B.2. Asymptotic Normality of Maximum Entropy Joint Density, Regression Function, and Response Function

Using first-order approximation of the estimated maximum entropy joint density,

$$\begin{aligned}
f(y,x,\hat{\boldsymbol{\lambda}}) &\simeq f(y,x,\boldsymbol{\lambda}) + \frac{\partial f(y,x,\hat{\boldsymbol{\lambda}})}{\partial \hat{\boldsymbol{\lambda}}^T}\Big|_{\hat{\boldsymbol{\lambda}}=\boldsymbol{\lambda}} (\hat{\boldsymbol{\lambda}}-\boldsymbol{\lambda}) \\
&= f(y,x,\boldsymbol{\lambda}) + f^{(1)}(y,x,\boldsymbol{\lambda})(\hat{\boldsymbol{\lambda}}-\boldsymbol{\lambda}). \\
\sqrt{n}(f(y,x,\hat{\boldsymbol{\lambda}})-f(y,x,\boldsymbol{\lambda})) &= \sqrt{n}\left(f^{(1)}(y,x,\boldsymbol{\lambda})(\hat{\boldsymbol{\lambda}}-\boldsymbol{\lambda})\right) \\
&= \sqrt{n}\left(f^{(1)}(y,x,\boldsymbol{\lambda})g^{(1)}(\boldsymbol{\mu})(\hat{\boldsymbol{\mu}}-\boldsymbol{\mu})\right).
\end{aligned}$$

Since $\sqrt{n}(\hat{\boldsymbol{\mu}}-\boldsymbol{\mu}) \sim N(\mathbf{0},\Sigma)$ as $n\rightarrow\infty$,

$$\sqrt{n}(f(y,x,\hat{\boldsymbol{\lambda}})-f(y,x,\boldsymbol{\lambda})) \sim N\left(\mathbf{0}, f^{(1)}(y,x,\boldsymbol{\lambda})g^{(1)}(\boldsymbol{\mu})\Sigma g^{(1)}(\boldsymbol{\mu})^T f^{(1)}(y,x,\boldsymbol{\lambda})^T\right).$$

The convergence rate of $f(y,x,\hat{\boldsymbol{\lambda}})$ is $n^{1/2}$.

The maximum entropy regression function of x and $\hat{\boldsymbol{\lambda}}^*$ is approximated by

$$\begin{aligned} m(x,\hat{\boldsymbol{\lambda}}^*) &\simeq m(x,\boldsymbol{\lambda}^*) + \frac{\partial m(x,\hat{\boldsymbol{\lambda}}^*)}{\partial \hat{\boldsymbol{\lambda}}^{*T}}\Big|_{\hat{\boldsymbol{\lambda}}^*=\boldsymbol{\lambda}^*} (\hat{\boldsymbol{\lambda}}^* - \boldsymbol{\lambda}^*) \\ &= m(x,\boldsymbol{\lambda}^*) + m^{(1)}(x,\boldsymbol{\lambda}^*)(\hat{\boldsymbol{\lambda}}^* - \boldsymbol{\lambda}^*). \\ \sqrt{n}\left(m(x,\hat{\boldsymbol{\lambda}}^*) - m(x,\boldsymbol{\lambda}^*)\right) &= \sqrt{n}\left(m^{(1)}(x,\boldsymbol{\lambda}^*)(\hat{\boldsymbol{\lambda}}^* - \boldsymbol{\lambda}^*)\right) \\ &= \sqrt{n}\left(m^{(1)}(x,\boldsymbol{\lambda}^*)g^{*(1)}(\boldsymbol{\mu})(\hat{\boldsymbol{\mu}} - \boldsymbol{\mu})\right). \end{aligned}$$

Since $\sqrt{n}(\hat{\boldsymbol{\mu}} - \boldsymbol{\mu}) \sim N(\mathbf{0}, \Sigma)$ as $n \to \infty$,

$$\sqrt{n}\left(m(x,\hat{\boldsymbol{\lambda}}^*) - m(x,\boldsymbol{\lambda}^*)\right) \sim N\left(\mathbf{0}, m^{(1)}(x,\boldsymbol{\lambda}^*)g^{*(1)}(\boldsymbol{\mu})\Sigma g^{*(1)}(\boldsymbol{\mu})^T m^{(1)}(x,\boldsymbol{\lambda}^*)^T\right).$$

The convergence rate of $m(x,\hat{\boldsymbol{\lambda}}^*)$ is $n^{1/2}$.

Similarly, it can be shown that

$$\sqrt{n}\left(\beta(x,\hat{\boldsymbol{\lambda}}^*) - \beta(x,\boldsymbol{\lambda}^*)\right) \sim N\left(\mathbf{0}, \beta^{(1)}(x,\boldsymbol{\lambda}^*)g^{*(1)}(\boldsymbol{\mu})\Sigma g^{*(1)}(\boldsymbol{\mu})^T \beta^{(1)}(x,\boldsymbol{\lambda}^*)^T\right).$$

Acknowledgments

An earlier version of this work was first presented at a conference organized by Info-Metrics Institute, American University in November 2016, and then in its economics department in September 2017. The authors are thankful to Duncan Foley and other participants for their valuable comments. We are grateful to Amos Golan for many constructive and helpful suggestions. The comments from co-editors and a referee were also helpful.

References

Boltzmann, L. (1872). "Weitere Studien über das Wärmegleichgewicht unter Gasmolekülen" [Further Studies on the Thermal Equilibrium of Gas Molecules]. *Sitzungsberichte der Akademie der Wissenschaften, Mathematische-Naturwissenschaftliche Klasse*, 66: 275–370. Wien: k. und k. Hof- und Staatsdruckerei.

Chakrabarty, M., Majumder, A. and Racine, J. S. (2015). "Household Preference Distribution and Welfare Implication: An Application of Multivariate Distributional Statistics." *Journal of Applied Statistics*, 42: 2754–2768.

Gibbs, J. W. (1902). *Elementary Principles in Statistical Mechanics*. New Haven, CT: Yale University Press.

Golan, A. (1988). "A Discrete Stochastic Model of Economic Production and a Model of Fluctuations in Production—Theory and Empirical Evidence." PhD thesis, University of California, Berkeley.

Golan, A. (2018). *Foundations of Info-Metrics: Modeling, Inference, and Imperfect Information*. New York: Oxford University Press.

Golan, A., Judge, G., and Miller, D. (1996). *Maximum Entropy Econometrics: Robust Estimation with Limited Data*. New York: John Wiley.

Harte, J., Zillio, T., Conlisk, E., and Smith, A.B. (2008). "Maximum Entropy and the State-Variable Approach to Macroecology." *Ecology*, 89: 2700–2711.

Henderson, D., and Parmeter, C. (2015). "Applied Nonparametric Econometrics." Cambridge: Cambridge University Press.

Inglot, T., and Ledwina, T. (1996). "Asymptotic Optimality of Data-Driven Neyman's Tests for Uniformity." *The Annals of Statistics*, 24: 1982–2019.

Jaynes, E. T. (1957a). "Information Theory and Statistical Mechanics." *Physical Review*, 106: 620–630.

Jaynes, E. T. (1957b). "Information Theory and Statistical Mechanics II." *Physical Review*, 108: 171–190.

Judge, G., and Mittelhammer, R. (2011). *An Information Theoretic Approach to Econometrics*. Cambridge: Cambridge University Press.

Kullback, S., and Leibler, R. A. (1951). "On Information and Sufficiency." *The Annals of Mathematical Statistics*, 22: 79–86.

Ledwina, T. (1994). "Data-Driven Version of Neyman's Smooth Test of Fit." *Journal of the American Statistical Association*, 89: 1000–1005.

Li, Q., and Racine, J. (2007). *Nonparametric Econometrics: Theory and Practice*. Princeton, NJ: Princeton University Press.

Maasoumi, E., and Racine, J. S. (2016). "A Solution to Aggregation and an Application to Multidimensional 'Well-Being' Frontiers." *Journal of Econometrics*, 191: 374–383.

Mead, L. R., and Papanicolaou, N. (1984). "Maximum Entropy in the Problem of Moments." *Journal of Mathematical Physics*, 25: 2404–2417.

Pagan, A., and Ullah, A. (1999). *Nonparametric Econometrics*. Cambridge: Cambridge University Press.

Racine, J. S., and Li, K. (2017). "Nonparametric Conditional Quantile Estimation: A Locally Weighted Quantile Kernel Approach." *Journal of Econometrics*, 201: 72–94.

Rilstone, P., and Ullah, A. (1989). "Nonparametric Estimation of Response Coefficients." *Communications in Statistics-Theory and Methods*, 18: 2615–2627.

Ryu, H. K. (1993). "Maximum Entropy Estimation of Density and Regression Functions." *Journal of Econometrics*, 56: 397–440.

Shannon, C. E. (1948). "A Mathematical Theory of Communications." *Bell System Technical Journal*, 27: 379–423, 623–656.

Stone, C. (1990). "Large-Sample Inference for Log-Spline Models." *The Annals of Statistics*, 18: 717–741.

Ullah, A. (1988). "Non-Parametric Estimation of Econometric Functionals." *The Canadian Journal of Economics*, 21:625–658.

Ullah, A. (1996). "Entropy, Divergence and Distance Measures with Econometric Applications." *Journal of Statistical Planning and Inference*, 49: 137–162.

Wu, X. (2003). "Calculation of Maximum Entropy Densities with Application to Income Distribution." *Journal of Econometrics*, 115: 347–354.

Zellner, A., and Highfield, R. A. (1988). "Calculation of Maximum Entropy Distributions and Approximation of Marginalposterior Distributions." *Journal of Econometrics*, 37: 195–209.

Index

Note: Page numbers followed by "*f*" or "*t*" refer to figures or tables respectively.